DVD Studio Pro 4

The Complete Guide to DVD Authoring with Macintosh

DVD Studio Pro 4

The Complete Guide to DVD Authoring with Macintosh

Bruce Nazarian, MPSE

McGraw-Hill
New York   Chicago   San Francisco   Lisbon
London   Madrid   Mexico City   Milan   New Delhi
San Juan   Seoul   Singapore   Sydney   Toronto

The **McGraw·Hill** Companies

Copyright © 2006 by The McGraw-Hill Companies, Inc. All rights reserved. Printed in the United States of America. Except as permitted under the United States Copyright Act of 1976, no part of this publication may be reproduced or distributed in any form or by any means, or stored in a database or retrieval system, without the prior written permission of the publisher.

2 3 4 6 7 8 9 0   DOC/DOC   0 1 2 1 0 9 8 7 6

P/N 147789-6
PART OF
ISBN 0-07-147015-8

The sponsoring editor for this book was Stephen S. Chapman and the production supervisor was Richard C. Ruzycka. It was set in Times Roman by Patricia Wallenburg. The art director for the cover was Anthony Landi.

Printed and bound by RR Donnelley.

McGraw-Hill books are available at special quantity discounts to use as premiums and sales promotions, or for use in corporate training programs. For more information, please write to the Director of Special Sales, McGraw-Hill Professional, Two Penn Plaza, New York, NY 10121-2298. Or contact your local bookstore.

This book is printed on acid-free paper.

Information contained in this work has been obtained by The McGraw-Hill Companies, Inc. ("McGraw-Hill") from sources believed to be reliable. However, neither McGraw-Hill nor its authors guarantee the accuracy or completeness of any information published herein, and neither McGraw-Hill nor its authors shall be responsible for any errors, omissions, or damages arising out of use of this information. This work is published with the understanding that McGraw-Hill and its authors are supplying information but are not attempting to render professional services. If such services are required, the assistance of an appropriate professional should be sought.

I'd like to dedicate this book to those of you who have, at one time or another, asked: "Ok, so . . . when are YOU going to write a book about DVD?" . . . to those intrepid souls who looked a new technology squarely in the face and said, without reservation, "I can do that!" . . . to all those who generously gave of their time and knowledge to help me learn the ropes about many things . . . to all of you who have put time into helping your DVD compatriots online at www.recipe4dvd.com . . . to the Cubmaster, because he was too cool for words . . . to those who know I care, even when they're far away . . . to Jessica Amanda, who made me smile . . . and to Nilceia, who always provides inspiration and makes me smile . . .

CONTENTS

Foreword	xxxv
Preface	xxxvii
Acknowledgments	xxxvix
DVD Studio Pro 4	xli
Introduction	xliii

Chapter 1	What Is a DVD?	1
	Goals	2
	What Is a DVD, Anyway?	2
	What Is an HD DVD?—What Is a Blu-ray Disc?	2
	What Makes Up a DVD?	2
	How DVDs Work	3
	One Finite Limit to Your DVD's Size	4
	Making a DVD Play Back	4
	The Remote Control Is the Key to DVD Navigation	4
	Transport Controls	4
	Presentation Options	4
	Menu Navigation Arrows, and Enter Key	5
	Navigating to and from Menus: Using TITLE, MENU; RETURN	5
	Using the Remote Control or Mouse for DVD Navigation	6
	DVD Disc Sizes (How Much Video Will My Disc Hold?)	7
	Single-Sided Discs	7
	Double-Sided Discs	8
	Mini-DVDs	8
	A Quick Note about Recording Your Own DVDs	8
	Making DVD Discs with Data—DVD-ROM, DVD-Hybrids	8
	What DVD Formats Are Available?	9
	DVD's Logical Formats	9
	DVD's Physical Formats	9

Contents

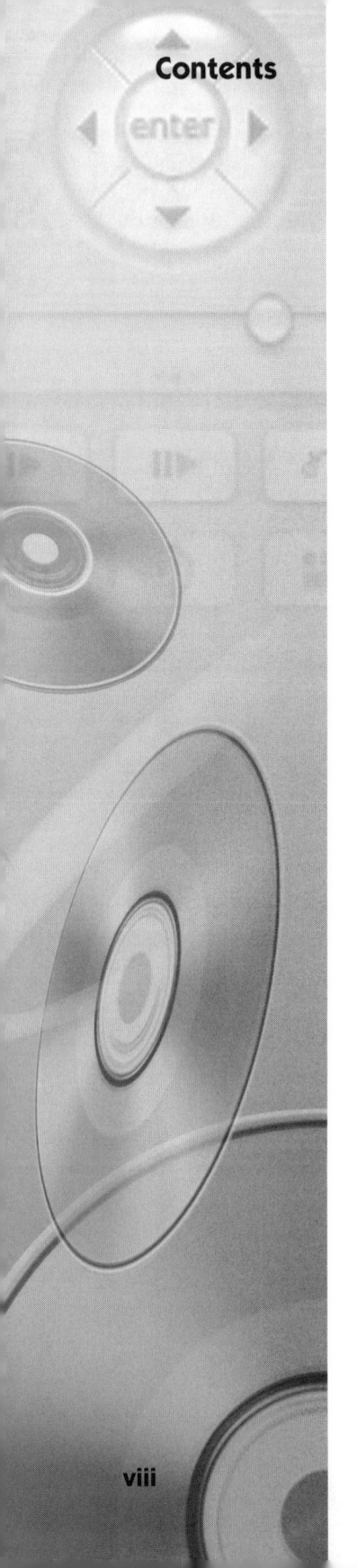

Burning to DVD-R (A)	10
DVD-R (A)	11
DVD-R (G)	11
DVD-RW	11
DVD+R	11
DVD+RW	11
DVD+R DL	11
DVD-RAM	12
Disc Size—Gigabytes (GB) versus Billions (BB)	12
DVD Workflow	13
1—Plan the Project, then Acquire the Assets You Need	14
2—Encode the Video	14
3—Author the Project	15
4—Build the Project	15
5—Deliver the Project—Burn a DVD or Write a DLT	15
Today's Workflow Is Different	16
OK, I'm Going to Replicate—What Now?	16
How Do I Plan My DVD?	17
Helpful Data for Use While Calculating a Bit Budget	17
Significance of Encode Rates	18
What Is HD DVD and Blu-ray Disc?	18
First on the Block for HD DVD Authoring	19
NTSC for HD DVD	19
PAL for HD DVD	19
Some Notes on How This Book Is Arranged	20
Summary	21

Chapter 2 What Is DVD Studio Pro 4?   23

Goals	24
Using Earlier DVD Studio Pro Projects in DVD Studio Pro 4	24
What Is DVD Studio Pro 4?	24
What's New in DVD Studio Pro 4?	25
New!—HD DVD Authoring—First in the World!	25
New!—24P Compatibility	25
New!—VTS Editor in the User Interface	26
New!—Templates for HD	26
New!—Integrated Encoding for HD MPEG2	26

Contents

New!—Integrated Encoding for AC3 in Compressor 2	26
New!—Dual-Layer Burning Support (Sans Hassles)	26
New!—Support for Blue Laser Burners	26
New!—Simulation to Digital Cinema Desktop	26
New!—Simulation to an External Video Monitor	26
New!—Support for the DTS 6.1 Audio Format	27
New!—Simulation to an External Audio Monitor	27
New!—GPRM Partitioning for Scripting	27
New!—Dual-Layer Breakpoints Can Be Set in the DVD-ROM Zone	27
New!—Better Access to Menu Loop Points	27
Getting Started with DVD Studio Pro 4	27
Which Applications Are Installed	28
Documentation Files Installed	28
Extras Installed	28
Templates, Shapes, Patches, Fonts	28
Launching DVD Studio Pro 4 for the First Time	29
Setting the Default Configuration	29
Getting Acquainted with Your Options	29
Establishing Project Basics	30
DVD Studio Pro 4 Preferences	32
Project Preferences	32
General Preferences	33
Menu Preferences	35
Track Preferences	36
Alignment Preferences	38
Text Preferences	39
Colors Preferences	40
Simulator Preferences	41
Destinations Preferences	42
Encoding Preferences	44
Summary	46

Chapter 3   The Workspace   47

Goals	48
The DVD Studio Pro 4 Workspace	49
Using the Interface Configurations	49

ix

Contents

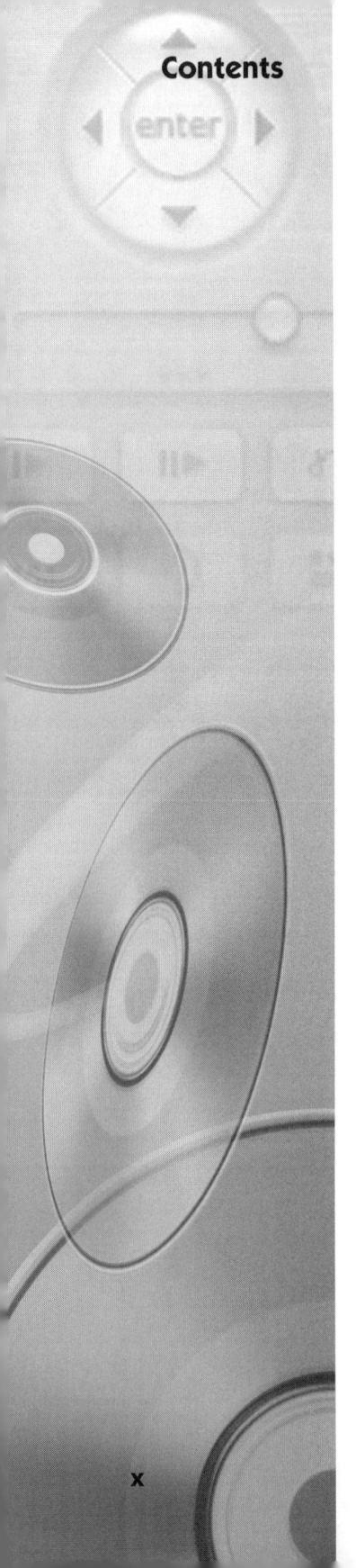

Sizing the Main Window for Best Effectiveness	49
The Basic Configuration	50
The Extended Configuration	50
The Advanced Configuration	51
The Inspector	52
Showing and Hiding the Inspector	52
The Palette	53
Showing and Hiding the Palette	53
Using Windows, Quadrants, and Tabs	54
Reducing or Enlarging the Main Window Size	54
Quadrants Divide the Main Window	54
Resizing Window Halves	54
Resizing All Quadrants Simultaneously	55
Resizing Quadrants Independently of Each Other	55
Unlinking the Quadrants from Each Other	55
Dragging an "Unlinked" Quadrant Boundary	56
Hiding and Revealing Quadrants	56
Dragging to Hide a Visible Quadrant	56
Dragging to Reveal a Hidden Quadrant	57
Unlinking One Boundary Doesn't Unlink the Other!	58
Dragging "Linked" Quadrants	58
Re-Linking the "Unlinked" Boundaries—Returning to Normal	59
Moving Tabs around in the Workspace	59
Moving Tabs between Quadrants by Dragging	59
Moving Tabs between Quadrants Using Control-Click	60
Turning Tabs into Freestanding Windows	60
Tearing Off Tabs to Become Windows	60
Restoring a Torn-Off Tab to a Quadrant	60
Using Keyboard Shortcuts to Select/Display Tabs/Windows	61
Using the Window Buttons—Close, Minimize, Maximize	61
Using Workspace Configurations	62
Managing Configurations	62
Recalling Workspace Configurations Quickly	62
The Palette—A Virtual Media Browser	63
Templates, Styles, and Shapes	63
Video, Audio, and Stills	64

Contents

Palette Default Entries Are Tied to the Current User	65
Adding Your Own Elements in the Palette, or Removing Them	65
Changing the Size of the Thumbnail Images	66
Palette Entries Stay in the Palette	66
The Video Tab	66
Using Movie Files in Your DVD	67
The Audio Tab	67
Using Music and Sound Files in Your DVD	67
Creating Dolby Digital (AC3) Files	67
The Stills Tab	68
What Kind of Graphics Stills Can You Use?	68
Organizing Assets Using the Palette	68
Configurations	68
The Basic Configuration Close-Up	68
The Menu Editor Window	69
The Toolbar	70
More about The Toolbar	71
Customizing the Toolbar	72
Adding Tools to the Toolbar (or Removing Them)	72
Moving Tools around in the Toolbar	73
Saving the Toolbar Configuration	74
Extra Menu Editor Tools	75
Context-Sensitive Drop Palettes	76
Hey! The Graphical View Window Is BACK!	76
Workspace Tabs Close-Up	78
The Outline Tab	79
Adding New Elements to the Outline	79
The Story Tab	80
Adding a New Story	81
The Menu Tab	81
The Connections Tab	81
The Viewer Tab	82
The Assets Tab	83
The Log Tab	84
The Track Tab	84
The Slideshow Tab	84
The Script Tab	85
About the Inspectors	86
To Select a Project Element for Inspection	87
To Select a Project Asset for Inspection	88

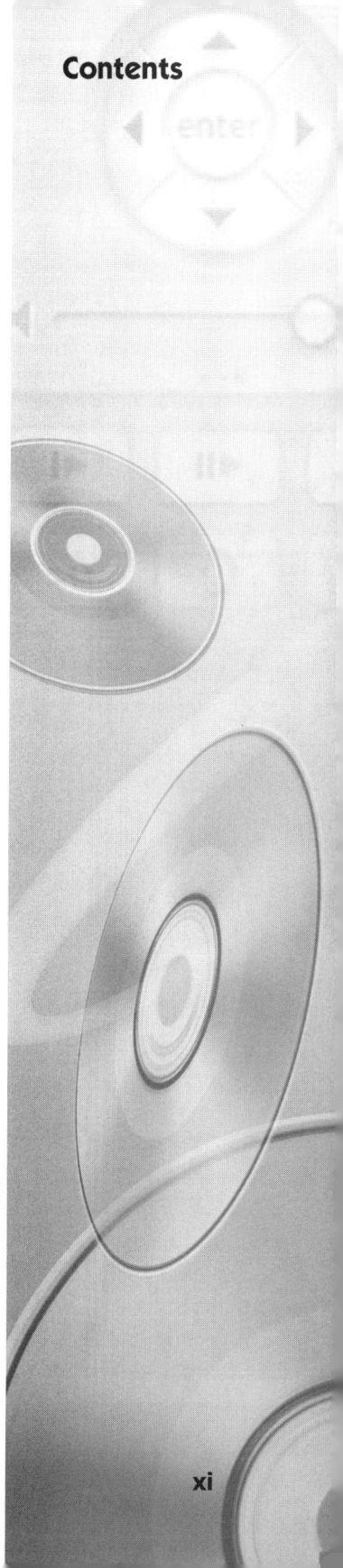

Contents

Summary	89

Chapter 4 Assets — 91

Goals	92
What Are Assets?	92
What Kind of Assets Are There?	92
Video Assets	92
Using Video Files as DVD Assets	93
Audio Assets	94
About Audio Files	94
Still Assets	94
Subtitle Assets	95
Other Assets—Hybrid (DVD-ROM)	
Deliverable Files	95
Organizing Your Assets	96
Organizing Assets in Folders before	
Importing	96
Organizing Assets Using the Palette	96
The Assets Tab	96
Where Do the Assets Go?	97
Creating a New Folder within the Assets Tab	98
Adding Assets to the Assets Tab	98
To Add Assets Using an Import Command	98
Initiating the Asset Import	99
Adding Assets by Drag and Drop	99
Importing Files or Folders by Drag-and-Drop	99
How DVD Studio Pro Organizes Assets	100
Checking Imported Asset Files for Usability	100
What Does Importing Really Do?	100
Locating an Asset File in the Finder	100
Seeing More Details in the Assets Tab	101
Displaying More Data in the Assets Tab	101
Missing Assets...	102
Fixing Missing Assets	102
Relinking One Missing Asset File	103
Relinking an Entire Folder, or More	
than One File	103
Asset File Properties	103
Removing a File or Folder	104
To Remove a Folder or File from	
the Assets Tab	104

Adding Assets by Dragging Directly into Elements		104
Dragging an Asset into the Track Editor		105
Dragging an Asset into the Slideshow Editor		105
Summary		106

Chapter 5   Making Video   107

Goals	108
Before You Author, You Need Video Content	109
Before You Author, Do You Have to Encode?	109
MPEG-1, MPEG-2, MPEG-4—What Video Can I Use?	109
Elementary Streams, My Dear Watson	110
About Encoding	110
Hardware Encoders	110
Which Encoding Approach to Use?	110
Software Encoders	111
Software Encoding Options	111
Compressor 2	112
NTSC-to-PAL and PAL-to-NTSC Conversions Now Possible	113
Qmaster for Distributed Encoding	113
DVD Studio Pro 4 Onboard MPEG-2 Encoder	115
Final Cut Pro—Export to MPEG-2	115
QuickTime Pro Player	115
Hardware Encoder Options	116
About MPEG-2	117
There's a Lot of Data in That Tiny MPEG File	118
Image Enhancement before or during Encoding	119
Some Hints for Successful MPEG Encoding	119
Encoding for DVD in Apple's World	121
Prepping for Export from Final Cut Pro	121
Choice of Video Resolution	122
Rendering	122
Setting In and Out Times	122
Setting DVD Markers in Final Cut Pro	122
Know Your Marker Types	122
Compression Markers	123
Chapter Markers	123
Exporting into MPEG-2 from Final Cut Pro Using Compressor	124

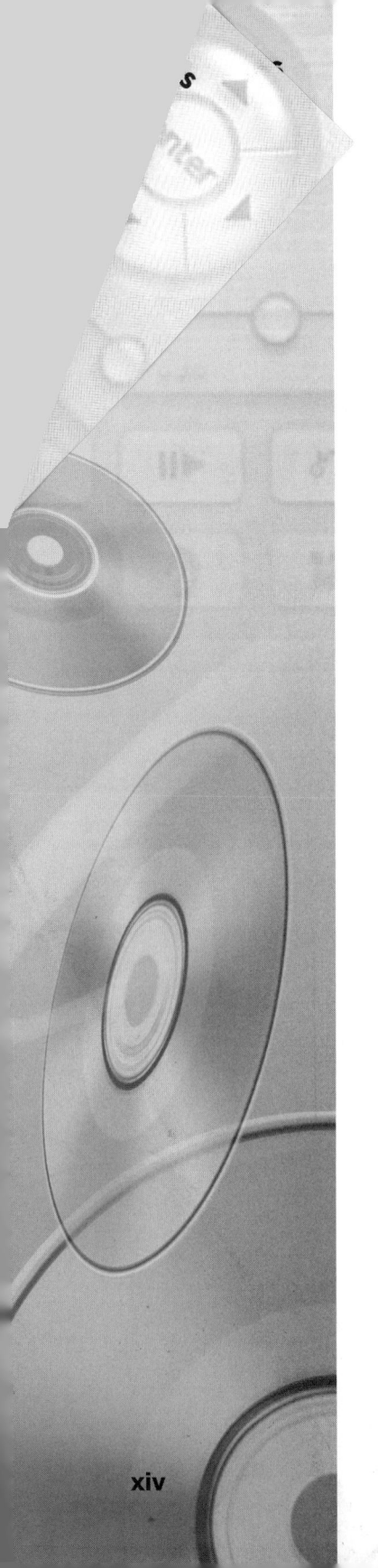

Using Compressor to Encode (Quick Start)	124
Encoding Using DVD Studio Pro's Onboard MPEG Encoder	127
Creating an MPEG-2 File Using DVD Studio Pro's Encoder	127
Encode Status Indicators	128
Exporting from Final Cut Pro Using QuickTime	128
Exporting Requires QuickTime "Export Components"	128
MPEG Encoding in the QuickTime Player	129
Setting up the QuickTime MPEG Encoder "Video" Options	130
What Happens during MPEG Encoding	132
Face to Face with GOPs—the MPEG "Group Of Pictures"	133
GOP?	133
Results of the MPEG Encode	133
QuickTime MPEG-2 Encoder Performance Benchmarks	133
Using QuickTime MPEG-2 Encoding in OS 9	134
Alternative Encoders	135
Using a DVD Recorder as an Encoder	135
Summary	137

Chapter 6	**Preparing Audio Assets**	**139**
	Goals	140
	What Audio Formats Can I Use with DVD Studio Pro?	140
	PCM FILES	140
	AIFF FILES (Audio Interchange File Format)	140
	.WAV FILES	141
	Dolby Digital (AC-3)	141
	Dolby Surround (4-ch L, C, R, S) [AC-3-encoded]	142
	MPEG Audio (".MPA")	143
	DTS (supported since DVD Studio Pro 3.02)	143
	Audio Formats Upcoming for HD DVD and BD	144
	What Audio Format Should I Use for My DVD?	144
	How to Create AIFF Audio Files (in General)	145
	How to Create an AIFF File Using Final Cut Pro	145

Contents

How to Create an AIFF File Using iMovie	146
Encoding AC-3 Audio Streams Using A.Pack	147
About the A.Pack Batch List	148
What Is A.Pack?	148
Why Use Dolby Digital?	148
A.Pack Quick Encode Guide for Stereo Files	148
Preparing to use A.Pack	149
Using the A.Pack Instant Encoder	149
In the Audio Tab	149
In the Bitstream Tab	150
In the Preprocessing Tab	150
The File Grid (Input Channels)	150
Loading the File Grid	150
Settings Preferences in the Audio Tab (in Detail)	150
Setting the Target System	150
Setting the Audio Coding Mode	151
Setting the Data Rate (the All-Important Encode Rate)	151
Setting the Dialog Normalization Setting	152
Setting the Bit Stream Mode Parameter	152
Settings in the Bitstream Tab (in Detail)	153
Settings in the Preprocessing Tab	155
Setting the Compression Factor	155
Setting the RF Overmodulation Protection	155
Setting the Digital Deemphasis Parameter	156
Setting the Low-Pass Filter (for Full Bandwidth Channels)	156
Setting the DC Filter (for Full BW Channels)	157
Setting the Low-Pass Filter (for LFE Channel)	157
Setting the Surround Phase Shift	157
Setting the 3 dB Attenuation in Surround Channels	158
Using the A.Pack Batch List	158
Setting the A.Pack Batch List Encode Job Settings	161
Monitoring Your Dolby Digital Encodes	161
The Log Window	162
Encoding AC-3 Audio Streams Using Compressor 2	163
What Is Compressor 2?	163
Why Use Compressor 2 and Dolby Digital?	163

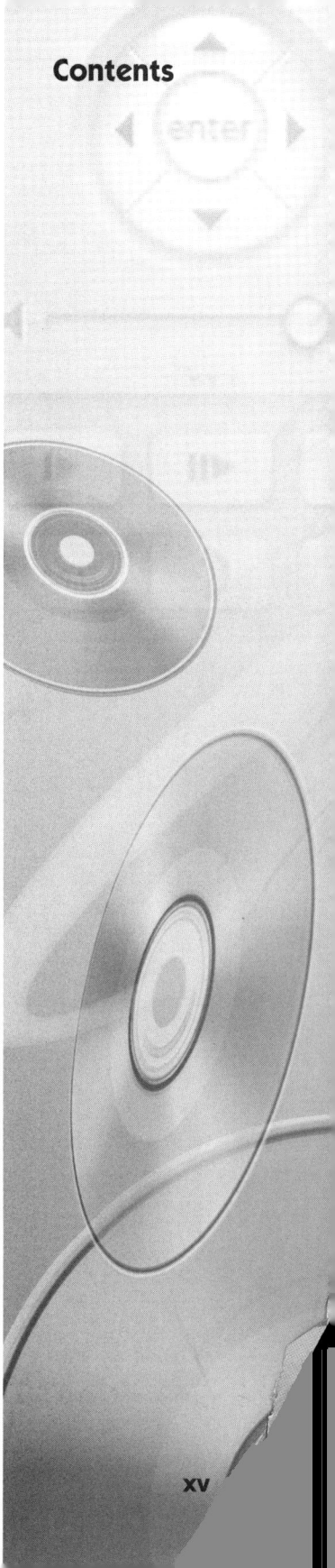

Contents

About the Compressor 2 Batch List	163
Launching Compressor 2	163
Using Compressor 2 for Encoding Stereo Files	164
Loading Stereo Files	164
Selecting an Encode Format Preset	164
Set Destination and Output Filename	165
Using the Inspector for a Closer Look	166
Verifying the Encode Preset Settings	166
Submitting the Job for Encoding	167
Using Compressor 2 for Surround Sound Encoding	167
Loading a Surround Sound Group	168
Loading the File Grid	171
Dolby Encoding Preferences Outlined in Detail	171
Dolby Preferences in the Audio Tab	171
Setting the Target System	171
Setting the Audio Coding Mode	172
Verify the Sample Rate of the Source File(s)	173
Setting the Data Rate (the All-Important Encode Rate)	173
Setting the Dialog Normalization Setting	173
Setting the Bit Stream Mode Parameter	174
Settings in the Bitstream Tab (in Detail)	174
Settings in the Preprocessing Tab (in Detail)	176
Setting the Compression Factor	176
Setting the RF Overmodulation Protection	177
Setting the Digital Deemphasis Parameter	177
Setting the Low-Pass Filter (for LFE Channel)	178
Setting the Low-Pass Filter (for Full Bandwidth Channels)	178
Setting the DC Filter (for Full Bandwidth Channels)	178
Setting the Surround Phase Shift	179
Setting the 3 dB Attenuation in Surround Channels	179
Encoding DTS Streams for DVD	179
A Simple Overview of the DTS Process	180
Preparing Files for Packing	180
Encoding a DTS Packed File into a DTS Bitstream	180
Summary	182

Chapter 7 Authoring Tracks 183

Goals	184
Authoring Tracks	184
What Is a Track?	184
Track Structure and Limits	185
How Many Tracks Do I Need for My DVD?	185
I'm Not Sure How to Structure My DVD (Yet)	186
Playing Video Clips Individually	186
Playing Video Clips Sequentially	186
Playing Video Clips Sequentially and Individually	187
When Can a "Nonseamless" Transition Look "Seamless"?	188
Tracks and Chapters and Stories	188
Working with Tracks	189
The Track Editor	189
The Track Inspector	189
You Need Separate Video and Audio for DVD Authoring	189
Creating a Track—(Track Editor Basics)	189
What Do I Need to Create a Track?	189
Creating a Track without an Asset	189
Adding a Video Asset to an Empty Track	190
Find Matching Audio for a Video Asset Automatically	192
Adding Audio Streams by Hand	193
Adding AC-3 Audio Streams	193
Some Assets to Practice With	193
What about Markers?	193
Can I Still Use the Track Inspector to Set the Video Asset?	194
Creating a Track with an Asset Automatically	194
Adding Additional Clips to a Track	196
Duplicating Clips in a Track	196
You Can Author with Movie Files as Well as MPEG Streams	197
Creating a Track Using the Menu Editor	197
You Don't Need a Menu to Make a DVD...	198
Using the Track Editor	198
About the Track Tab Controls	198
Stream Areas	199
Organizing the Stream Display	200

Contents

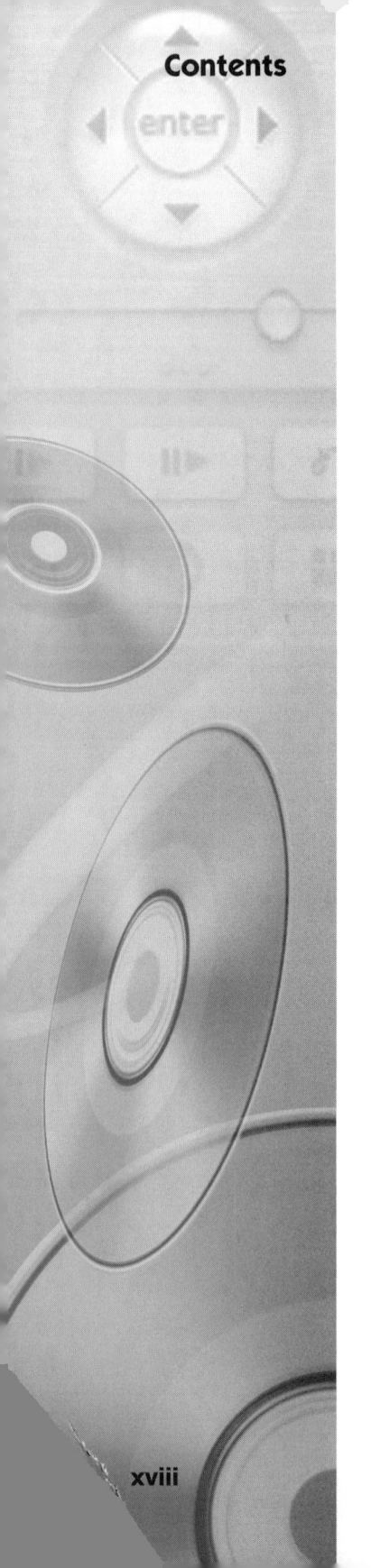

Customizing the Stream Area Display	200
More Track Display Selection Tools	202
How the Timeline Works—Zero-Based versus Asset-Based TC	204
Trimming Assets in the Timeline	204
Why Trim an Asset Stream?	205
Trimming Streams Using the Asset Inspector	205
Track Inspectors	206
The Track Inspector General Tab	207
Playback Options	207
The Track Inspector Other Tab	208
The Track Inspector User Operations Tab	208
The Transition Inspector	209
The Clip Inspector—Viewing Clip Properties	210
Clip Conflicts	211
About DVD Data Rates	212
About Multi-Angle Tracks (and Mixed-Angle Tracks, Too)	213
About Mixed-Angle Tracks	214
Previewing a Track in the Viewer	214
Controlling the Viewer	215
Keyboard Commands for Viewer Controls	215
Configuring Streams for Preview	215
Previewing Angles	216
How Angle Playback Works in Viewer	216
How the Playhead Works	216
The Viewer Is for Quick Checks Only...	217
How to Make Your Tracks Do More	217
Summary	218

Chapter 8 Creating Markers and Stories   219

Goals	220
About Markers	220
What Is a Marker?	220
Markers as Chapters	220
Markers for Buttons over Video	220
Marker for a Dual-Layer Breakpoint	221
Markers for DVD@ccess Functions	221
Markers to Define Story Segments	221
Creating Markers in Tracks	221

About Marker Locations	221
Defining Markers in Tracks with Multiple Clips	223
Importing Markers	224
Importing Markers from a Text File	224
About Marker Types	225
Chapter Markers (Purple)	225
Button Highlight Markers (Orange)	225
Dual-Layer Break Markers (the Black Dot)	225
Cell Markers (Green)	226
About the Marker Inspector	226
Top Area	226
More about Marker Names	226
Marker Inspector General Tab	226
Marker Inspector User Operations Tab	228
Previewing Markers	228
Some Important Marker Do's and Don'ts	228
Interactive Markers (Buttons over Video)	229
About Stories	229
How Do Stories Work?	229
Creating Stories	230
The Story Editor Tab	231
Using the Story Editor to Define the Story Segments	231
To Add a Story Marker	231
To Move an Existing Story Marker	232
To Delete a Story Marker	232
To Reassign the Track Marker Assigned to a Story Marker	232
To Complete the Story Definition	232
Defining the Story by Defining the Marker Order	233
About the Story Inspector	233
Top Area	233
To Rename the New Story	233
To Redefine the Story's End Jump Action	233
General Tab	233
The User Operations Tab	234
Setting Story Marker Properties	234
Summary	236
Markers	236
Stories	236

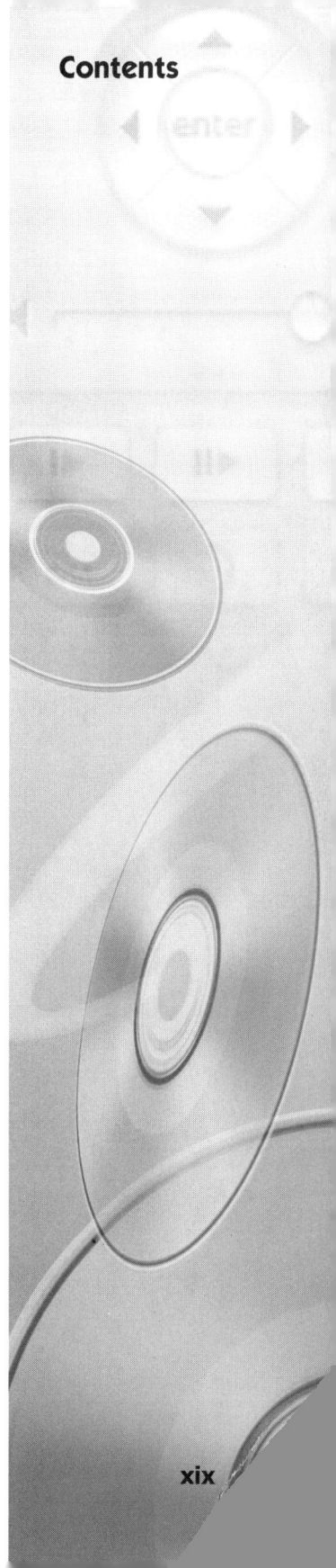

Contents

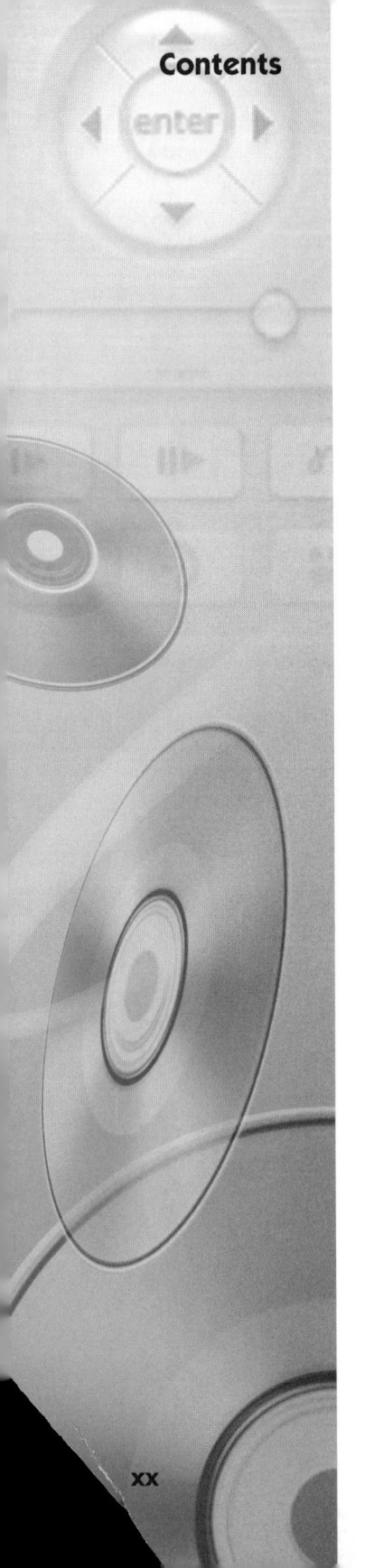

Chapter 9	Authoring Menus	**237**
	Goals	238
	About DVD Menus	238
	About DVD Studio Pro's Menu Editor	239
	Menu Creative Functions Are Built-in	239
	What the New Menu Editor Can Do	239
	Round-Trip Links to Outside Editors	239
	DVD Menu Basics	240
	Understanding Menus in General	240
	Understanding Button "States" in Menus	241
	What Constitutes a "Standard Method" Menu?	241
	Stillframe Menus	242
	Stillframe Menus with Audio	242
	Motion Menus	243
	Adding a Subpicture Overlay	243
	About the Subpicture Overlay File	243
	About Color Settings	243
	Layered Menus	244
	Pros and Cons of Layered Menus	244
	Common Factors	244
	About the Menu Editor	244
	Menu Editor Pop-Up Menus	245
	Menu Editor Settings: Pop-Up	245
	Menu Editor Bottom Tools	245
	Getting Started with Menus	247
	Adding a Standard Menu	247
	Adding a Layered Menu	247
	Adding Submenus	248
	Renaming Menus	248
	Editing a Menu	248
	Adding Assets to Menus	249
	Drop Palettes for Drag-and-Drop Menu Editing	249
	How the Drop Palettes Work	249
	Adding a Menu Template from the Palette	250
	About the Menu Inspector	250
	The Menu Inspector Top Area	250
	The Menu Inspector General Tab	250
	Standard Menu Inspector General Tab Settings	251
	Layered Menu Inspector General Tab Settings	252
	Layered Menu Inspector Settings	253

Contents

Background and Overlay Layer Settings (for PSDs)	254
The Menu Inspector Menu Tab	254
Menu Tab Functions	254
Background Layers	255
Drop Shadow Settings	255
The Menu Inspector Transition Tab	255
The Menu Inspector Color Settings Tab	255
Simple Overlay Colors versus Advanced Overlay Colors	256
Simple Overlay Colors	256
Advanced Colors (Grayscale)	257
Advanced Colors (Chroma)	258
The Menu Inspector Advanced Tab	259
The Simple Steps to a Standard Menu (Still or Motion Menu)	260
The Simple Steps to a Layered Menu	261
Adding a Background Picture to a Standard Menu	261
Using the Menu Editor to Create Buttons	262
Creating a Button	262
Moving and Aligning a Button	262
Resizing a Button	263
Adding More Menu Buttons by Drawing Them	264
Adding More Menu Buttons by Duplicating Them	264
Naming and Renaming Menu Buttons	265
Creating Simple Menu Highlights without Subpicture Overlays	265
Setting the Button's Highlights Using "Overlay Colors: Simple"	265
Help! I Can't See the Selected and Activated Highlights	266
Using Color Set 2 and 3	267
Creating Menu Highlights with Subpicture Overlays	267
Defining the "Subpicture Overlay"	267
How Does the Subpicture Overlay Work?	267
Assembling a Menu with a Subpicture Overlay	268
Specifying the Subpicture Overlay File in the Inspector	269

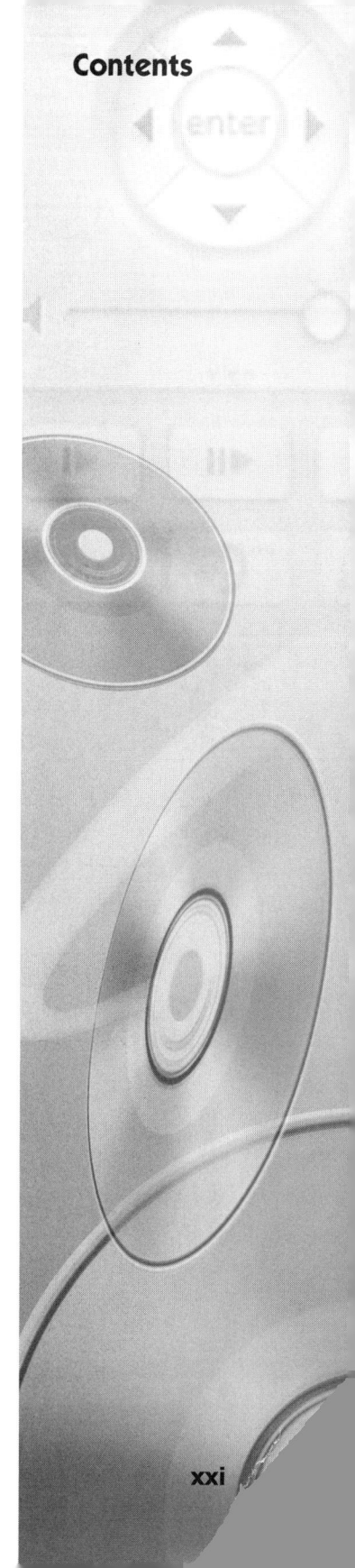

Contents

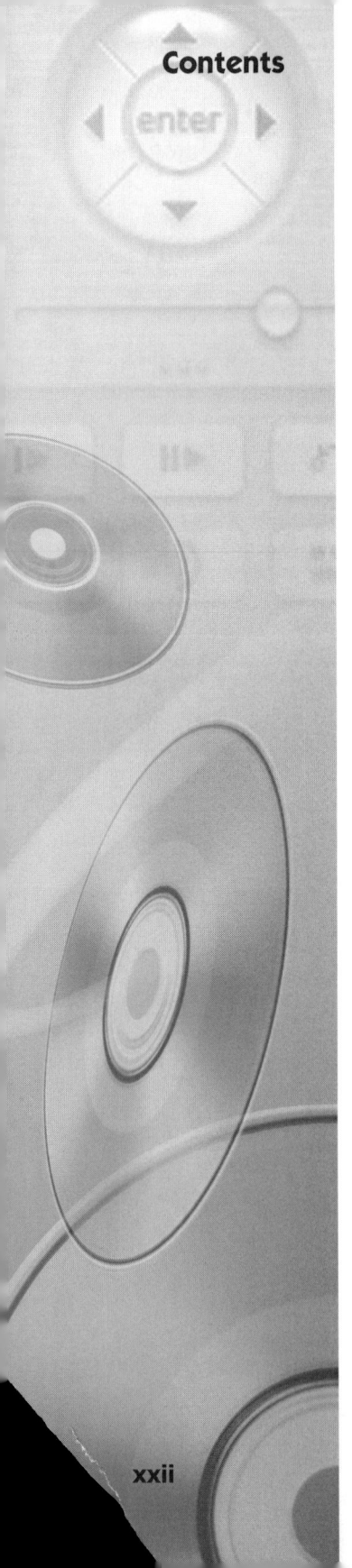

Specifying the Background Picture by Dragging	269
Specifying the Subpicture Overlay File by Dragging	269
Summary of Simple Menu Subpicture Highlights	270
Creating Layered Menu Highlights with Photoshop Layers	271
Displaying the Normal State Using the Menu Background	271
Displaying the Normal State Using Layers	271
Displaying the Selected State Using Layers	272
Displaying the Activated State Using Layers	272
Assigning Highlight States Using Layers	273
Previewing Button Highlight States	273
Examples of Stillframe DVD Menus	273
Making Connections to Buttons	274
Making Connections Using the Button Inspector	274
Working with Templates	275
Summaries of Dragging Assets and Elements into a Menu	275
Dragging Assets and Elements into a Standard Menu	276
Dragging Assets to a Standard Menu's Empty Area	276
Dragging a Video Asset to the Empty Area	276
Dragging Multiple Video Assets to the Empty Area	276
Dragging a Still Asset to the Empty Area	277
Dragging Multiple Stills (or a Folder) to the Empty Area	277
Dragging a Layered Still Image (.PSD) to the Empty Area	277
Dragging an Audio Asset to the Empty Area	278
Dragging a Video Asset with Audio to the Empty Area	278
Dragging Assets to a Standard Menu Button	279
Dragging a Video Asset to a Standard Menu Button	279
Dragging a Still Image to a Standard Menu Button	279

Contents

Dragging Multiple Stills (or a Folder) to a Standard Menu Button	280
Dragging a Video Asset with Audio to a Standard Menu Button	280
Dragging Assets to a Standard Menu's Drop Zone	280
Dragging a Video Asset to a Drop Zone	280
Dragging a Still Image to a Drop Zone	281
Dragging Project Elements to Standard Menus	281
Dragging a Track to the Empty Area	281
Dragging Multiple Tracks to the Empty Area	281
Dragging a Story to the Empty Area	281
Dragging Multiple Stories to the Empty Area	282
Dragging a Slideshow to the Empty Area	282
Dragging Multiple Slideshows to the Empty Area	282
Dragging a Menu to the Empty Area	282
Dragging a Script to the Empty Area	282
Dragging Project Elements to a Standard Menu Button	283
Dragging a Track to a Button	283
Dragging a Story to a Button	283
Dragging a Slideshow to a Button	283
Dragging a Menu to a Button	284
Dragging a Script to a Button	284
Dragging Templates and Styles to Standard Menus	284
Dragging a Shape to the Empty Area	284
Dragging a Shape to a Button or Drop Zone	284
Dragging a Template to the Empty Area	285
Dragging a Template to a Button	285
Dragging a Button Style to the Empty Area	285
Dragging a Button Style to a Button	286
Dragging a Text Style to the Empty Area	286
Dragging a Text Style to a Text Object	286
Dragging a Drop Zone Style to the Empty Area	286
Dragging a Drop Zone Style to a Drop Zone	287
Dragging a Layout Style to the Menu Editor	287
Dragging Assets to Layered Menus	287
Dragging a Video Asset with Audio to a Layered Menu's Empty Area	288

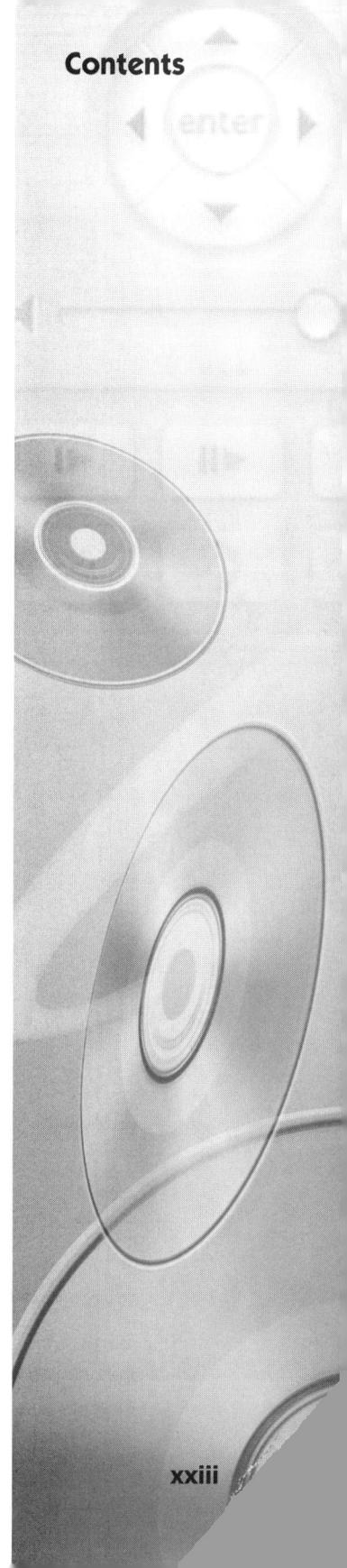

Contents

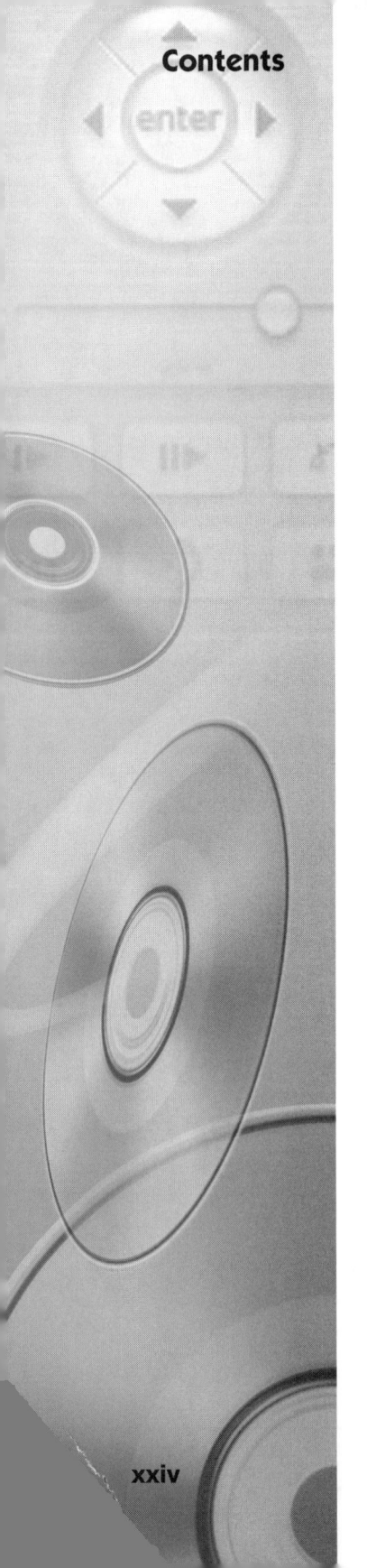

Dragging a Single-Layer Still Image to a Layered Menu's Empty Area	288
Dragging a Layered Still Image (.PSD) to a Layered Menu's Empty Area	288
Dragging Multiple Stills (or a Folder) to a Layered Menu's Empty Area	289
Dragging a Track to a Layered Menu's Empty Area	289
Dragging a Story to a Layered Menu's Empty Area	289
Dragging a Slideshow to a Layered Menu's Empty Area	289
Dragging a Menu to a Layered Menu's Empty Area	290
Dragging a Script to a Layered Menu's Empty Area	290
Dragging Assets to a Layered Menu Button	290
Dragging a Video Asset to a Layered Menu Button	290
Dragging a Video Asset with Audio to a Layered Menu's Button	290
Dragging a Single-Layer Still Image to a Layered Menu's Button	291
Dragging a Layered Still Image to a Layered Menu's Button	291
Dragging Multiple Stills (or a Folder) to a Layered Menu's Button	291
Dragging Project Elements to a Layered Menu Button	291
Dragging a Track to a Layered Menu Button	291
Dragging a Story to a Layered Menu Button	292
Dragging a Slideshow to a Layered Menu Button	292
Dragging a Menu to a Layered Menu Button	292
Dragging a Script to a Layered Menu Button	292
Summary	293

Chapter 10 Authoring Slideshows — 295

Goals	296
Slideshow Basics	296

xxiv

Creative Issues	296
Graphics Formats Usable for Slides	297
Slide Sizes and Aspect Ratios	297
Creating a Slideshow	298
Meet the Slideshow Editor	299
Selecting Slideshow Assets	300
Adding Slides to the Slideshow	301
Need More Slides for the Slideshow?	301
Setting Manual Advance	304
Adding Sound to Your Slideshow	304
Adding an Individual Audio File to a Slide	304
Using an Overall Audio File in the Slideshow	305
Using Fit to Audio	307
Using Fit to Slides	307
Adding Transitions to a Slideshow	307
The Slideshow Inspector	308
The Slide Inspector	310
Some Precautions about Audio for Slideshows	310
Convert to Track	311
Viewing and Simulating Slideshows	312
Simulating a Slideshow	313
Keyboard Shortcuts (Slideshow Editor)	314
Summary	315

Chapter 11 Making Connections   317

Goals	318
Making Connections	318
Making Connections through Inspectors	319
Making Connections from Menu Buttons	319
The Connections Tab	320
Meet the Connections Tab	320
The Connections Tab Tools	321
Connections Tab Layout	321
Using the Connections Tab	323
Basic, Standard, Advanced—the View: Pop-up	323
Next, Previous Jumps—Special Cases	324
All, Connected, Unconnected—the Filter Pop-up	324
The Recipe4DVD Sample Projects	324

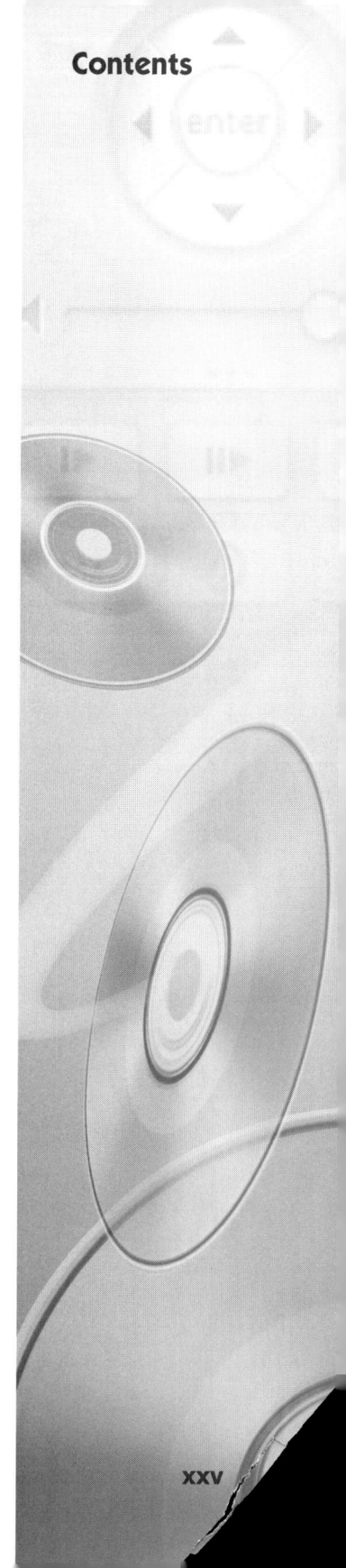

Contents

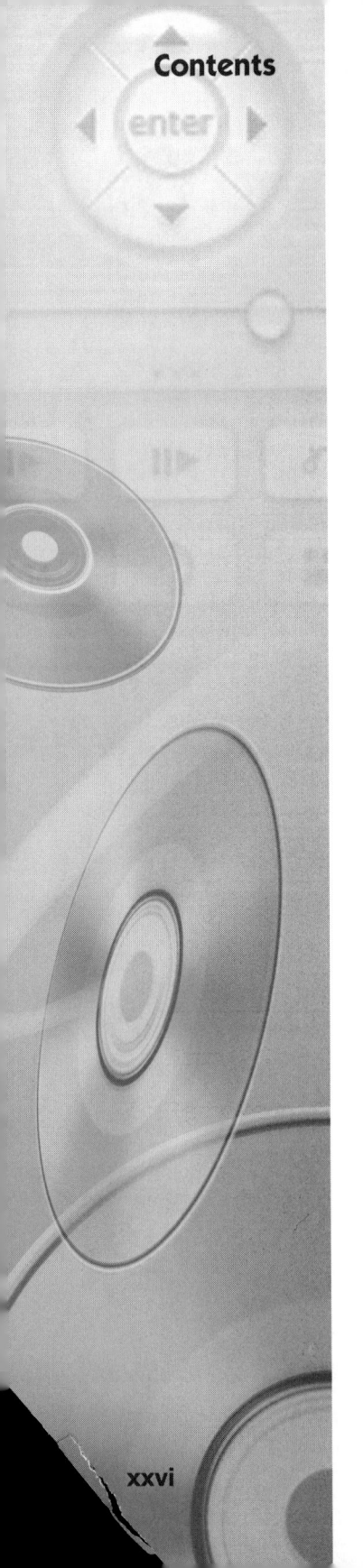

Making Connections	324
Breaking or Modifying Connections	327
Setting the First Play (IMPORTANT!)	328
What Is the First Play?	328
Checking the Connections	328
Avoiding Remote Control Key Problems	329
Understanding Sources and Their Connections	329
Basic Sources in Detail	330
Standard Sources in Detail	331
Advanced Sources in Detail	332
Understanding Targets and How They Work	334
Summary	336

Chapter 12 Creating Subtitles — 337

Goals	338
Some Subtitle Basics	338
Uses for Subtitles	339
Video Commentaries	339
Buttons over Video	339
Some Demonstrations	339
Naming Conventions	339
What Are DVD Subtitles?	339
Closed Captions versus Subtitles	340
How DVD Subtitles Work	341
Subtitling Options in DVD Studio Pro	341
Importing SPUs from DVD Studio Pro 1 Projects	342
Subtitle Tools in DVD Studio Pro 4	342
Checking Subtitles	343
Switching Subtitle Streams in Preview	343
Setting Subtitle Preferences	344
Applying Settings Globally to an Entire Stream	345
About the Subtitle Inspector	345
Top Area	345
General Tab	345
Button Tab	347
Colors Tab	348
Creating a Subtitle	350
Creating the Subtitle Event	350
Working with Text in Subtitles	351

Positioning of the Subtitle Text	353
Using a Graphic File for a Subtitle	353
Advanced Subtitling	353
Summary	354

Chapter 13 Basic Scripting for DVD Studio Pro 4 — 355

Goals	356
Scripting = Enhanced Interactivity	356
What Is Scripting, Exactly?	356
General and System Parameters—Scripting Tools	357
Understanding General Parameters—GPRMs	357
GPRM Issues to Be Aware Of	357
Overflow and Underflow	357
GPRM Volatility	357
Understanding System Parameters—SPRMs	357
Other SPRMs	358
How Do I Use a Script?	359
When to Use Scripting	360
What Kinds of Scripts Are There?	360
Inline Scripts	360
Pre-Scripts	361
Why Use a Pre-Script?	362
General Notes on Using Scripts	362
Making a New Script	362
Renaming a Script	363
Duplicating a Script	363
Saving a Script	364
Loading a Script or Script Description	364
Using the Script Editor	364
Reordering Command Lines	365
Script Command Syntax	365
Using the Script Inspector	365
Editing in the Script Command Inspector	367
The First Pane—Command Selection	367
The Second Pane—Command Options	367
The Third Pane—Compare Command	367
The Fourth Pane—the Comments Field	367
Conditional versus Unconditional Commands	367

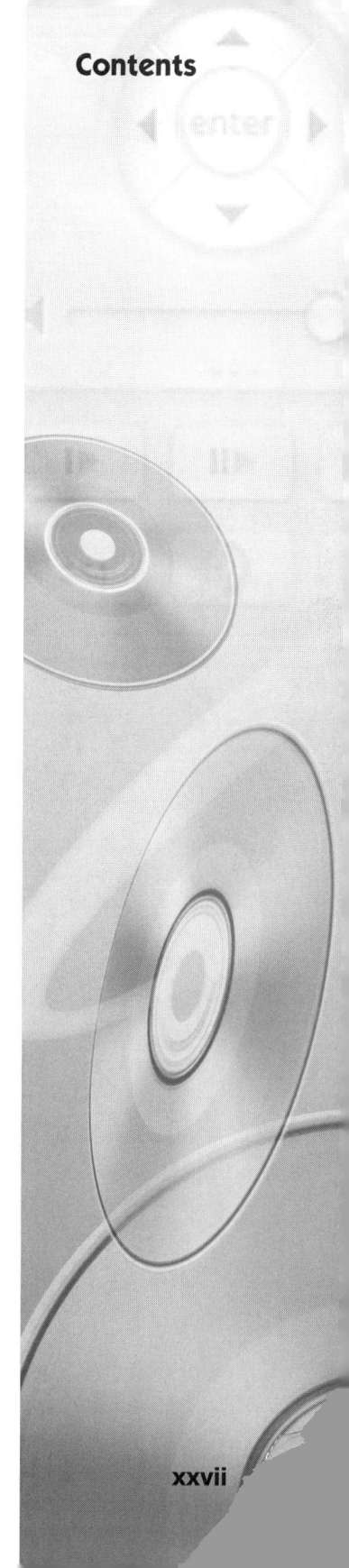

Contents

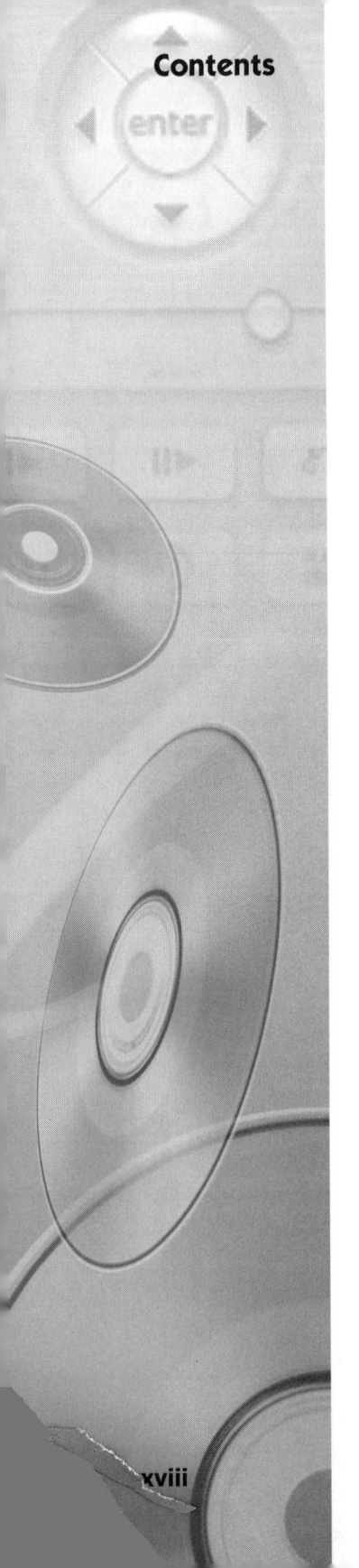

Thinking Logically about Comparisons	370
Scripting Commands in Detail	370
The NOP Command	371
The JUMP Command	371
The Set GPRM Command	372
The Goto Command	373
The Set System Stream Command	374
Resume Command	374
GPRM Mode	374
Exit Command	375
Exit Pre-Script	375
Jump Indirect	375
New Scripting Tool in DVD Studio Pro 4	375
What's a GPRM Partition?	375
Why Use a Partition?	376
Practical Uses of Scripting	377
Anatomy of a "Play All" Script	377
What Does a "Play All" Script Do?	377
The Logic of It	377
Setting up the Script Conditions	378
The "Play All" Script Logic Diagram	378
Summary	381

Chapter 14 Building and Formatting   383

Goals	384
Check Your Project Thoroughly before Output	384
Testing the DVD before Output	385
Using the Viewer	385
Using the Simulator	385
Setting the Simulator Preferences	386
The Simulator Window	387
The Display Window	387
The Remote Control Panel	387
Resolution and Display Mode Selectors	390
Information Drawer	391
The Simulator Display Is "Live" and Active	391
Emulating a Project with Apple DVD Player	391
Build before Emulation	392
Emulating with DVD@ccess Links	392

Setting Disc Properties before Building or Formatting	393
Top Area Settings	394
The Disc Inspector General Tab	394
The Disc Inspector Disc/Volume Tab	395
The Disc Inspector Region/Copyright Tab	396
The Disc Inspector Advanced Tab	397
Outputting a Finished DVD Project	398
Building	398
A Video-only DVD volume	398
A Hybrid DVD—(Video + ROM Data)	399
Formatting	399
Burn, Build, Format, or Build and Format?	400
Burn	400
To Initiate a Burn	400
Build	401
Build Messages	403
Format	403
Format Options in the Disc/Volume Tab	405
Format Options in the Region/Copyright Tab	405
Build and Format	406
Re-Using a Build Folder	406
Creating a DVD Hybrid Disc	408
Making a DVD Hybrid Disc in DVD Studio Pro 4	408
Making a Hybrid Disc Using DVDSP V1.5	408
Creating a DVD-Hybrid in DVD Studio Pro 1.2	409
Making a Hybrid Disc Using DVDSP V1.0-1.2 and Toast Titanium	409
Making a Hybrid Disc Using DVDSP V1.0-1.2 without Toast	409
DVD Hybrid Pitfalls!	410
Joliet Files	410
DVD@ccess Installers	410
DVD@ccess on Windows	410
Hybrid Discs Viewed on Settop Boxes	411
Important Limit on DVD-R Disc Size!	411
Roxio Toast Titanium—the Swiss Army Knife of Optical Media	411
Summary	412

Contents

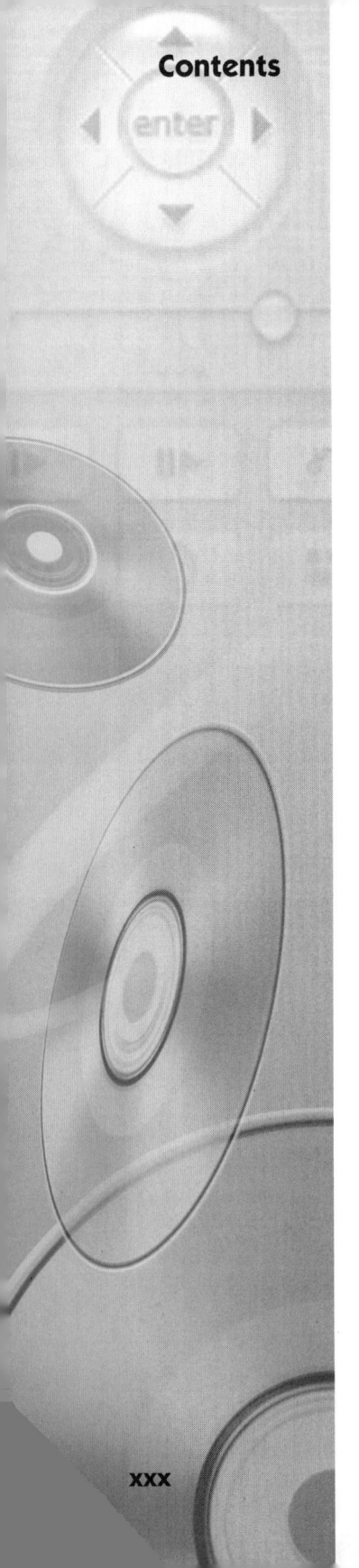

Chapter 15 Graphics Issues for DVD Images 413

Goals	414
Basic DVD Image Concepts (Important Stuff!)	414
Image Size	414
DVD Image Size Basics	415
DVD Image Issues	415
Easy DVD Image Rules	416
Alternate Image Sizes	417
Yet a Third Approach to Image Dimensions!	418
Image Composition Limitations ("Safe Zones")	418
DVD Image Color Rules—RGB versus Video	419
A Global Method to Control Black/White Levels	420
How to Reveal Graphic Image Flaws and Limitations	421
Image Size And Shape Limitations	422
Font Size and Style Issues	422
Using Layer Styles (Layer Effects) with DVD Studio Pro	422
Rendering Layer Styles for Layered DVD Menus	423
Keeping the Colors within Video Standards	424
Using the NTSC Color Filter	424
Completing the Photoshop DVD Graphic Image	425
Motion Menu Basics	425
Motion Image Sizes—720 x 486, 720 x 480	426
Creating Motion Menus in Adobe After Effects	427
To Create and Export a DVD Motion Composition with After Effects	427
About Transitions	428
About Transitions in DVD Studio Pro	428
Creating Motion Menus in Nonlinear Edit Tools	428
Assembling a Motion Menu Sequence in Final Cut Pro	429
Customizing This Simple Menu Sequence	430
Encoding Motion Menus into MPEG Once Completed	430
Motion Menu Overlay Pictures	431
Capturing a Menu Image from a Screen Image	431
Trimming the Screen Capture in Photoshop	432
De-Interlacing	433
Creating the Highlights in the Overlay File	434
Summary	435

Chapter 16 Duplication and Replication   437

- Goals   438
- Duplication and Replication   438
    - "One-offs"—Burning One Disc, or Perhaps a Dozen or Two   439
    - Duplication—Making More than 10, Less than 100   439
    - Replication—When You Need More than a Few Hundred   439
- Duplication versus Replication—How Do I Decide?   440
    - Cost Issues for Replication   440
    - Cost Issues for Duplication   440
    - Compatibility   441
    - Complexity   441
- Duplicators Galore!   441
    - One-to-One Burner—One Drive   441
    - One-to-One Burners—Two Drives   442
    - Single-Burner Automated Duplicators   442
    - Multiple-Burner Straight Duplicators   442
    - Single-Burner Robotic System with Printers   443
    - Multi-Burner Robotic Systems with Printers   443
- Replication by the Numbers   443
    - 1—Delivering the DVD Master—Physical Formatting   444
    - 2—Preparing the "Glass Master"   444
    - 3—Metallizing the "Glass Master"   445
    - 4—Plating—Growing the "Stamper"   445
    - 5—Molding—The "Stamper" Goes to Work   445
    - 6—Sputtering—The DVD Gets a Reflective Layer   446
    - 7—Bonding   446
    - 8—Labeling   446
    - 9—Packaging   447
- Distribution   447
    - Traditional Distribution   447
    - Direct-to-Customer   447
    - Distribution on Demand   448
- Summary   449

Contents

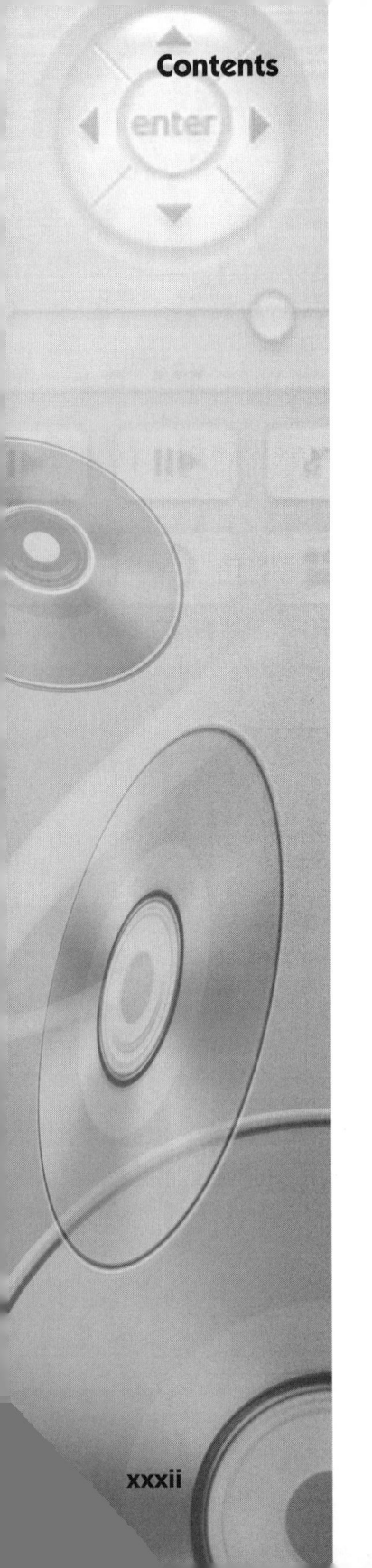

Chapter 17 Advanced Authoring — 451

Goals — 452
DVD@ccess—Interactivity Beyond DVD-Video — 452
  What Does DVD@ccess Do? — 452
  What Do I Need to Use DVD@ccess? — 452
Setting the DVD@ccess Property — 453
  Track Markers — 453
  Menus — 454
  Slides — 454
Setting the DVD@ccess Property — 454
  Proper DVD@ccess Syntax Is Important — 454
Testing Your DVD@ccess Links — 456
  Testing Using the Apple DVD Player — 456
  Testing Using Simulation — 456
  Mounting a .img File on the Mac Desktop — 456
Demo Files and Projects for You to Use in Testing DVD@ccess — 456
Creating DVD Hybrid Discs in DVD Studio Pro — 457
Closed Captioning — 457
User Operations (UOPs) — 458
  About User Operations Properties — 459
  Playback Control — 459
  Stream Selection — 460
Display Conditions — 460
  Setting a Display Condition — 461
  An Example Display Condition: Setting a Language-Based Element — 462
  Conditional Elements and Their States — 462
  Conditional Relationships — 463
  Display Condition versus Pre-Scripts — 464
  About Next and Previous Remote Control Buttons — 464

Chapter 18 Future Developments — 465

So Where Do We Go from Here? — 466

Appendix A Command Reference — 467

Appendix B Alternative Encoders — 469

BitVice from Innobits — 470
   System Requirements for BitVice — 470
   Summary of BitVice MPEG2 Encoder Features — 471
Heuris MPEG Power Professional — 471
   About MPEG Power Professional—DVD — 471
Cleaner 6 from Autodesk — 472
   Cleaner 6 Features — 472
MegaPEG.X Pro HD and MegaPEG.X Pro HD-QT from Digigami — 474
   MegaPEG.X Pro SD and HD Features — 474
Main Concept — 476
Compression Master from Popwire — 479
Look on the DVD for More Information — 480

Appendix C DVD Media Reference Guide — 481

Appendix D DVD Glossary — 485

Appendix E DVD Studio Pro Install/Upgrade — 501

Before You Install—Check Your Current
   System Configuration — 502
   Hardware Requirements — 502
   Software Requirements — 502
   Configuring DVD Studio Pro Software — 509

Appendix F QuickTime™ Installation Guide — 513

Checking the Current QuickTime™
   Version in OS X — 514
   Yes, It's That Simple! — 515

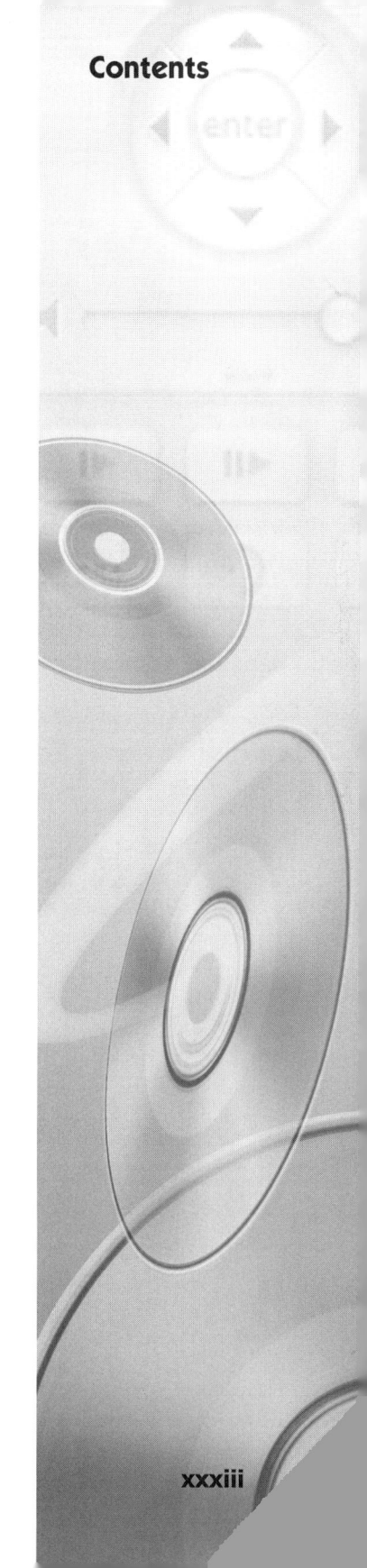

Index — 517

FOREWORD

Becoming an authority on a particular aspect of modern technology is like signing up for indentured apprenticeship—the technology never stands still, so one is obligated to keep up with it or become obsolete and left in the dust. Bruce Nazarian has been serving time in the DVD avocation for many years, long enough to graduate from apprentice to master. And as a master of DVD Studio Pro he's on the hook to keep up with the Jobs-es, so to speak. You hold in your hands the fruit of Bruce's efforts to stay current.

In the early days of DVD authoring, programs such as Scenarist and Sonic DVD Creator required considerable expertise, not to mention considerable cash. Then came products such as DVDit and DVD Studio Pro that did much more for much less. Each succeeding generation has made it easier to produce great-looking DVDs. DVD Studio Pro, like many Apple products, strives to reach a wide audience while staying as simple as possible. It straddles the range above consumers who want to copy home movies to DVD and below the inscrutable specialists who create special-edition movie titles or thousand-menu interactive discs. It hides the full complexity of the DVD format, which makes it usable by mere mortals, but it also hides the full power of the DVD format.

That's where Bruce comes in. He has spent a prodigious amount of time exploring the intricate underpinnings and the murky crannies of DVD. When someone wonders "How'd they do that?" Bruce knows the answer. As anyone who has taught knows, the best way to learn something is to teach it. By this measure, Bruce knows DVD and DVD Studio Pro as well as anyone, since he has taught innumerable training workshops and seminars at conferences such as NAB, EMX, DV Expo, and the like. Bruce's experience at the lectern imbues this book with uniquely relevant detail, since he's been asked the hard questions, the odd questions, and the stupid questions. More importantly he knows how make DVD Studio Pro do its thing: roll over, sit up and beg, and jump through flaming hoops while balancing eight audio tracks and ten subpicture tracks on its nose. This book will give you the inside edge you need to quickly become productive using DVD Studio Pro, and to get more out of it than probably even its creators thought possible.

Jim Taylor
Senior Vice President and General Manager,
Advanced Technology Group, Sonic Solutions
Author of *DVD Demystified* and
Everything You Ever Wanted to Know About DVD

PREFACE

Thanks for purchasing this book.

I sincerely hope you will find it an invaluable reference for not only DVD Studio Pro 4, but all of the other important ancillary information you need to become successful in DVD Authoring, even if your success is only measured by the ability to produce one disc—YOURS!

If you already <u>know</u> DVD Authoring and how to create video and Audio for DVD, but want to start using **DVD Studio Pro 4** right away, read Chapters 2 and 3 to learn the basics of the application, and then go to Chapter 7, "Authoring Tracks." Specific information on using DVD Studio Pro 4 begins here. Remember that to get the best from the entire suite of applications, you should learn **Compressor 2**, for Dolby Digital audio compression and the creation of HD DVD video assets in the H.164 codec. If you require subtitles for your DVD, learning how to use the Subtitle editing features, or to write an STL script will be an important addition to your skill set. This is covered in Chapter 12, "Subtitles."

If you are *completely* new to DVD authoring, I recommend reading the whole book in the order in which the chapters appear. I've designed this book to walk you through the basics of DVD itself, then the basics of DVD Studio Pro 4, and finally the specifics of each of the functions within DVD Studio Pro 4.

Each chapter outlines the specifics for one specific function within DVD Studio Pro. We first cover *Tracks*, then *Markers and Stories*, used within *Tracks*; next, *Menus & Buttons, Slideshows & Slides* and making *Connections* to build navigation. Later chapters cover *Subtitling* using the Track Editor and subtitle files, *Scripting, Building and Formatting* (how to burn DVD-Rs and write DLT tapes), and *graphics issues* for DVD. Finally, we cover *Duplication and Replication* (how to create more than one DVD), and discuss Advanced authoring topics like *DVD@ccess, Hybrid DVDs (containing Video + ROM features)*, and other issues.

An extensive set of *Appendices* add to the reference use of this book, which includes a DVD-centric glossary, Information on third-party encoding solutions, Installation guides for DVDSP and Quicktime Pro, and lots of other good information.

If you purchased the first version of this book, you will notice that Appendix A is no longer the DVD Studio Pro Command library. To be honest, we opted to

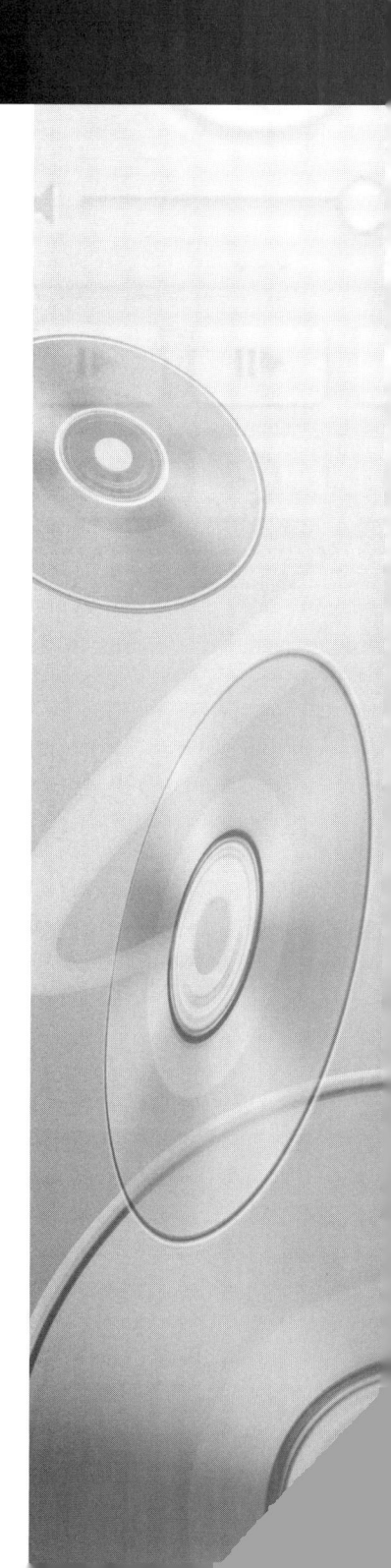

Preface

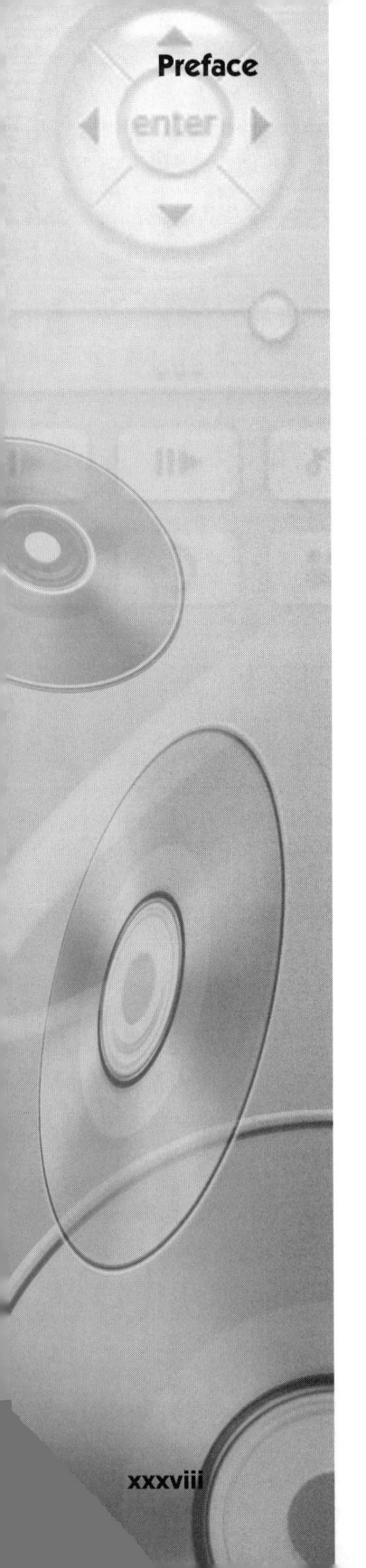

include this in PDF form on the book DVD, as it allowed us to use the printed pages for the additional features that were added in DSP 3 and DSP 4.

So, don't forget to explore the companion DVD included with this book—there's a LOT more information in there to assist you in DVD authoring, as well as assets and projects to practice building DVDs.

Enjoy, and thanks again for your purchase—I welcome your comments, too. I do read all of your emails, and respond to as many as possible. If you are having trouble with your DVD project, I offer consulting by phone and email, and a DLT output service if you wish your projects to be professionally replicated. You can find information on all of these things by visiting my website **www.Recipe4DVD.com**.

You will also find the book companion website there.

ACKNOWLEDGMENTS

First and foremost, allow me to thank Jim Taylor for graciously bringing this book to the attention of McGraw-Hill, and his continuing support since we first met. Jim truly is "Mister DVD" to thousands of DVD people worldwide. Thanks, Jim. I am proud to call you a friend, and have always appreciated your gracious humor ;-).

Special thanks go to Steve Chapman and his crew at McGraw-Hill for providing that all-important interface to the world of the publisher.

Here are important other players in the "Life of the Gnome," without whom, etc…

- My family, all across America—Fred, lotsa nieces and nephews, cousins.
- Roger Williams, Gina Cascone Williams, and Annette Cascone.
- Ralph La Barge, DVD Author "extraordinaire."
- Bernie Mitchell, President, and all of the members of the DVD Association Board of Directors who have consistently lent support.
- Paul Kent of Mactivity, who has believed in DVD from the first day he scheduled it at Macworld SFO 2001—special thanks to IDG's superb staff.
- Mike Evangelist of Wired, Inc., Apple, and Astarte.
- Brian Hoffman, Brian Schmidt, Kirk Paulsen and all from Apple who helped.
- The Apple DVD engineering team, who really built a beauty in DVDSP 2!
- Patty Montesion from Apple Pro Training division.
- Carl White, Vin and Sarah Capone, Tad Shelby, Eric Thomas, Tony Knight.
- The Apple Asia Crew: Phil Hickey, Graham Perkins, Pete Barber, Ian Chapman-Banks, and all who made the Asia trips fascinating and fun besides.

Acknowledgments

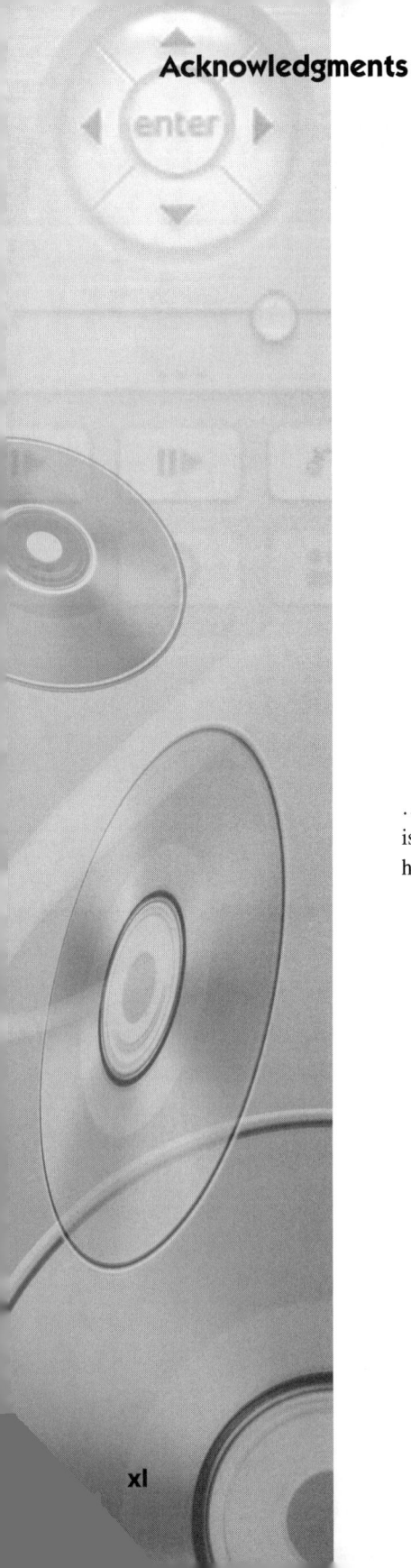

- Philip Hodgetts and Gregory Clarke of Intelligent Assistance, for years of help in developing the often-and-still-delayed DVD Companion. "Back to the drawing board one more time for HD, eh."

- ALL of the dedicated former Apple DE's, including: Joe Meldrum, Mike Descher, Denis Flynn, Sheila Vaccaro, Diana Musto, Barrett Thomson, Mike "Captain Video" Balas, Jack Quattlebaum, Tom Coen, Damon Stewart, and more than I can name.

- Apple SE's everywhere including Brian Snoddy, Bill Lee (Oz), Sammy Huang, Peter Qi and Mr. Mu, Ahmad "Shah," and all of the others who have helped along the way.

- All of the ASE's who are now ACN's—you know who you are.

- The former Apple SDS team: Amy Vivona, Heather Lopez-Cepero, Ann Hahn, Lisa Fortney, Barbara Reukel, Verna Swim. A tip of the Gnome chapeau, ladies.

- Michele Serra, without whom there would not have been a Recipe4DVD.

- Nilceia de Souza—For reasons only she and Jessica know…

… and to all the many others who have helped me along the way—if your name isn't here, it isn't because you didn't count—you know who you are, and you have my thanks as well.

DVD STUDIO PRO 4

This book is designed for enthusiasts who want to learn all the features of DVD Studio Pro 4 in a well-ordered manner. You can build DVD projects that include still menus, chapter points, stories, slideshows, web linking and scripting to familiarize you with the DVD Studio Pro 4 interface. You will also learn the basics about encoding video and audio, creating subtitles, scripting, and so much more. This book gives you specifics on how you can harness the power of DVD.

Contents of the DVD

This book includes a companion DVD which includes many informative documents, PDF product brochures, technical reference documents, and more. In addition, the DVD includes Samplers of the Recipe4DVD productivity tools: Pro-Pack 1 (DVD projects), Pro-Pack 2 (Scripting for DVDSP 2), and Wedding-Pack 1 (menu graphics suitable for use in wedding-themed DVDs).

Additional Folders contain some demonstration DVD projects to assist you in understanding how certain DVD authoring functions work. Finally, there are lots of DVD assets provided to use in building our sample projects—temporary Video Angle MPEG streams, temporary Audio Streams, and building block graphics and MPEG assets for menus and informational screens. You may feel free to continue to use these as you "mock up" your own DVD projects—I am confident you will find them useful, as I have.

The Companion Website

This book features a companion website, where updates, suggestions, errata, and other information will be posted. You can reach that website through our home website:

http://www.recipe4dvd.com

Username: **book** Password: **swordfish**

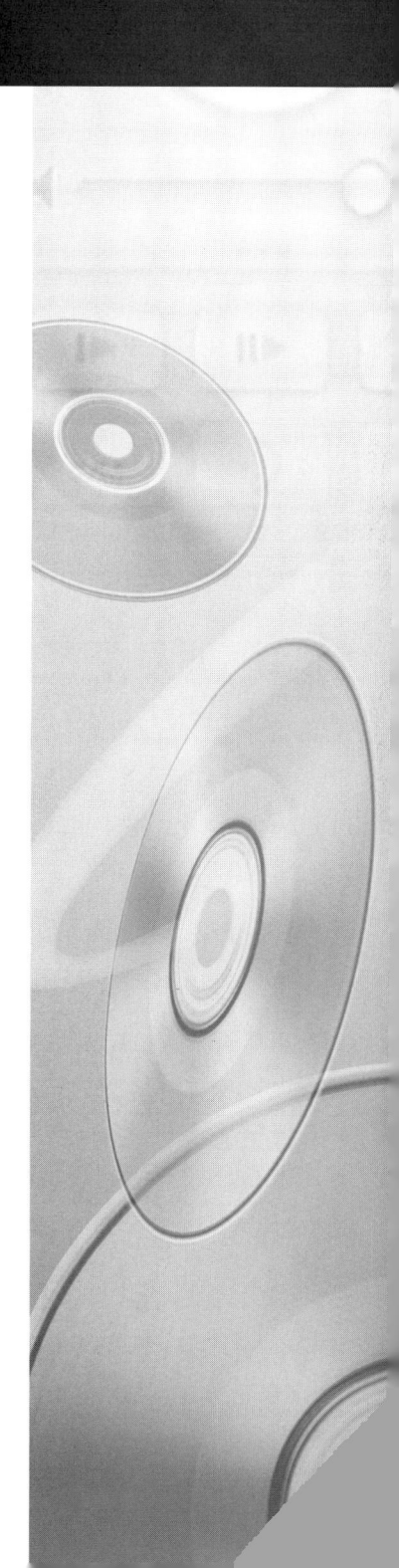

DVD Studio Pro 4

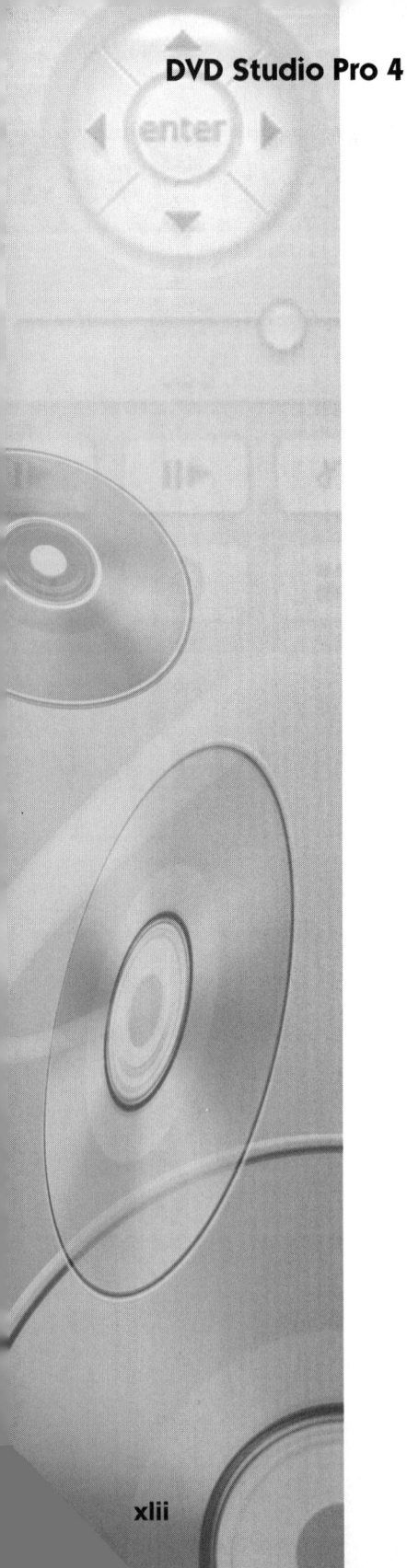

Credits

Written by Bruce Nazarian, Member, Apple Consultants Network

Ideas produced by Gnome Digital Media

Graphic screenshots created using Snapz Pro X

Special Characters Used in This Book

Note!
This icon signifies an *important concept* that should be noted!

Tip!
This icon denotes a *tip* or *idea* to assist you in creating DVDs.

Important!
This icon denotes an *important point* to remember.

This icon denotes a feature of concept specific to the new HD DVD format in DVD Studio Pro 4.

This icon signifies a *new feature or concept in DSP 4 or new implementation* of a previous DSP feature.

INTRODUCTION

Before You Get Started

Goals

By the end of this section, you should:

- Understand the hardware requirements for DVDSP 4
- Understand the software and OS requirements
- Understand how to check your system
- Be familiar with drive storage requirements
- Be familiar with DVD Burners
- Be familiar with DLT drives
- Be familiar with DLT tape cartridges
- Be aware of the difference in system requirements for SD and HD DVD authoring
- Understand the implications of HD-DVD authoring

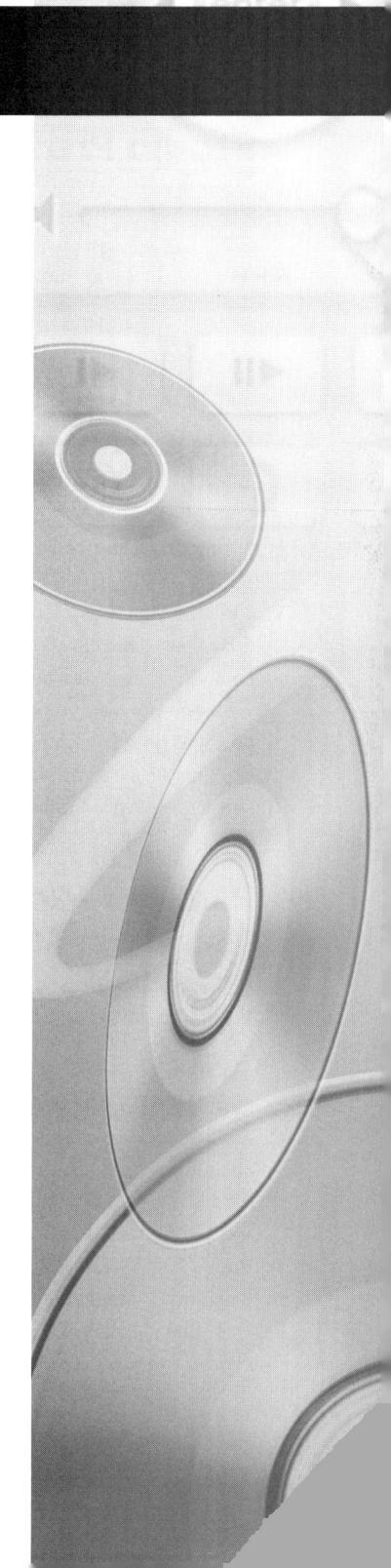

Introduction

Welcome to DVD Studio Pro 4!

Before you rush off and try to start authoring, there are a few things you need to know in order to use DVD Studio Pro 4 effectively.

Operational Requirements

You will need to be sure you have a Macintosh that is properly equipped to take advantage of all that DVD Studio Pro 4 has to offer. This means a fast CPU (or even two!), lots of RAM, scads of hard disk drive space, and the appropriate peripherals (burners, DLT drive, etc.). It also means configuring the software on your Mac properly, so all of the applications you will need for DVD Production will be ready to go! Finally, the right Mac OS update and QuickTime software will need to be installed and/or updated.

Hardware—G4 or G5 Towers, Titanium or Aluminum PowerBooks

DVD Studio Pro is designed to run on any suitable Macintosh that has a G4 or G5 CPU, but be aware that HD DVD authoring has far more stringent requirements (more on this in a second). MPEG 2 encoding algorithms, in particular, take advantage of the Velocity engine (the AltiVec™ Processor) that speeds up the G4 and G5 chips. With DVD Studio Pro 4, a really fast machine will take fullest advantage of everything that the application can do. If you want to do HD DVDs, you will find a dual-processor system a necessity, and not a luxury!

The currently recommended Macintosh for installation of DVD Studio Pro 4 is a G4 with minimum processor speed of 733 MHz and an AGP graphics card. You will find that the additional processing overhead inherent in DVD Studio Pro 4 will cause slower than desirable response on Mac models below this speed. If you are a diehard PowerBook fan, you will find that DVD Studio Pro 4 *will* install on a Titanium or Aluminum PowerBook, but it will run slow on a 667—it's better to upgrade to a faster machine if you can afford it. No sense wasting time waiting for the cursor to stop spinning!

A Superdrive Is Great, Too

Some G4 models released since 2001 have included the Apple SuperDrive as a standard peripheral, while other models offered it as an additional cost option. If you have a previous model Macintosh G4, i.e., a 733, 867, 933, Dual-800, Dual 1GHz, Dual 1.25 GHz or Dual 1.42 GHz, you may already have a model with the Apple SuperDrive installed. Today, all shipping G5 CPUs include a 16X Superdrive with double-layer burning as standard. To take advantage of everything DVD Studio Pro 4 can do, you will want to have a DVD burner available.

No Superdrive? No Problem... Go External

If you have an Apple SuperDrive mounted internally, DVD Studio Pro will recognize it and allow you to burn your own Personal DVDs on it. If you have a system that did NOT come with a built-in SuperDrive, don't despair. Since February 2001, Pioneer Electronics has made available their aftermarket version of this same drive, known as the A03, A04, A05, A06, A07, A08, and most recently the A09. This drive is functionally identical to the Apple SuperDrive, burning both DVD-R (General) and

Introduction

DVD-RW media. The A06 added DVD+R/RW capabilities and the newest drive, the A09, includes 16X burn speed, along with DVD+R DL capabilities (and you can now use this format directly from DVD Studio Pro 4).

Both the Pioneer aftermarket drives and the OEM Apple Superdrive also write CD-R and CD-RW formats. (Why do you think they call it a "Super" drive?) Also since that date, other manufacturers have made available General Media DVD burners. There really are a lot of choices available, but be sure you have a drive that can write DVD media, or DVD Studio Pro 4 won't be able to burn DVDs for you. These days, with burners selling for under $70 online and at retail, the odds of your not having a burner are pretty slim.

Battery-Powered DVD Authoring?? You Bet!

Since the introduction of the Titanium PowerBook with a G4 CPU, portable DVD Authoring with DVD Studio Pro has been possible. For compulsive world travelers like me, this is nothing short of a miracle!

Editing video on my PowerBook while comfortably transiting the Atlantic and Pacific at 35,000 feet is a real treat, but nothing prepared me for the feeling of power that accompanied my first mid-air DVD authoring experience. While flying to Singapore, I successfully completed authoring a DVD using DVD Studio Pro with an external 20 GB buss-powered FireWire hard drive, and upon touchdown in Singapore, emailed the finished DVD project back to the USA as a stuffed (.SIT) DVD project file. I could also have used the "Description" function of DVD Studio Pro to send the DVD project data as plain text, sent within an email message. (See the chapter on Advanced Authoring for more on the Description File function.)

What If I Only Have a G3?—It's Time to Upgrade!

While it was possible to run DVD Studio Pro 1 with a G3 (although you could not encode) you just CANNOT do this anymore with DVD Studio Pro 4. Even if you do have the proper OS X installed on your G3, the installer will not let you install on a G3, and even if you COULD, you would NOT be happy with the performance. Besides, you still wouldn't be able to encode! If you're a diehard G3 owner, get over it—DVDSP 4 IS the reason to convince yourself that it's time to upgrade. You won't regret having a zippy new Mac with DVD Studio Pro 4.

The Need for Speed—G5 CPUs Rule

If you are contemplating the purchase of a new Macintosh for DVD authoring (it's really a great excuse to justify a G5 purchase), allow me to provide a few suggestions to guide your choices:

This is NOT the time to skimp on your system purchase. Think about the intangible issue of what your TIME is worth, and you will immediately see that a faster computer is capable of turning out finished DVDs more rapidly. This will not only save you time, it will also let you make back its purchase price faster.

This Return on Investment (ROI) decision is the kind of thing that keeps high-priced accountants and corporate CFOs up late at night, but you've got it easy. The right decision is: buy the fastest machine you can afford right now, even if it means cringing a bit. The time saved in DVD Authoring is a repetitive saving—meaning, a faster machine will save you time every time you encode, every time you compile, every time you preview, and ultimately, on every build. This is especially true when you build the finished DVD vol-

xlv

Introduction

ume, as DVD Studio Pro is primarily crunching data and doing disk I/O at that point, and a faster CPU gets you to the finish line (or the dinner table) first!

Install Lots of RAM

In fact, buy as much RAM as you can afford (3–4 GB is a good amount). Remember that in Mac OS X, RAM is allocated dynamically as each application needs it, but the system cannot allocate nonexistent RAM! The more real RAM you have, the easier it is to keep your fast system running fast. Think of RAM as the "lungs" of your system—the bigger the lungs, the easier it can breathe.

Experienced DVD hands will also tell you that a fast machine with a ton of RAM means nothing if you are wasting time trying to juggle hard drive space to make room for the next project, so…

Buy Plenty of Hard Drive Storage

Large-capacity hard drives will help keep your DVD workflow streamlined. Not having enough storage will drive you crazy! In DVD Authoring, you are going to generate so many gigabytes of data, through encoding, building, imaging, etc., that you will wish you owned stock in a hard drive company! (Hmmm… now there's an idea!)

Pros and Cons on Storage

While there are a number of different ways to implement large amounts of storage on your system, including centralized methods like SAN (Storage Area Network) and NAS (Network Attached Storage), in my opinion locking your workflow into a centralized storage system is bad for two reasons:

1. Unless you backup religiously (and who does?) you are VERY vulnerable to the possibility of a hard drive crash or directory corruption. If you are using a non-redundant RAID array, and lose a drive, this could be big trouble.

2. It will be more difficult to file transfer those bits around your network via Ethernet or FiberChannel than it will be to move them around on portable drives, so allow me to put in a good word for what I call "Portable Distributed Storage." No, it's not a fancy new digital technology, it's nothing more than some good old FireWire hard drives used properly as SneakerNet.

SneakerNet: Save Time…and Save Money

What's SneakerNet? It's the manual movement of data from one computer to another, using human power! You (and your sneakers) walk a removable hard drive from one system to another. This method of storing your DVD project data has really come into its own since the prices and sizes of high-density portable storage have dropped.

Today, it's possible to purchase an external FireWire hard drive with 80 GB of 7200 RPM ATA/100 storage for under US$150. But why buy only 80 GB? Why not 100 GB or even 200 GB? Portability, that's why! Read on and you'll see why this makes a lot of sense.

Hard Drives and Your DVD Workflow

I can tell you from years of hard-won DVD experience that one of the most frustrating things is to fight your

Introduction

computer system for storage, and lose! Inevitably during DVD production, you'll have a few last minute encodes that come in and oops! There's just no room on your internal hard drives to encode them, or to build the project if you DO encode them. What the heck do you do now? Go buy some new drives!

To avoid running out of space, and also to facilitate moving, holding, and backing up your DVD projects, consider inexpensive, tiny portable FireWire hard drives for your project storage needs. Checking with some well-known street price engines like **www.froogle.com** and **www.pricewatch.com**, I have found FireWire drives from 40–200 GB at ridiculously low prices. Who could have dreamed a few years ago that you would be able to purchase 100 GB of hard drive storage for around $150? It's unthinkable, but it's here; so run—do not walk—to your nearest favorite retailer or Internet site selling reliable hard drive mechanisms with Oxford 911 chipset-equipped FireWire enclosures, and bask in the warmth of your own wisdom.

You might even want to consider the tiny self-powered models. Even though the drive's RPM speed may be a bit slow, these little beauties will work their little hearts out for you, and best of all, you can slip 'em into your pocket, pack 'em into your travel bag, or just disconnect 'em from your desktop computer and put them away on the shelf while the project is under revision. No power supply means "ease of use!"

What Size Drives Do I Need?

The average DVD project is a DVD-5, with a data requirement of perhaps 10–15 GB, worst case—even if you had to completely redo EVERY encode, you'll likely not exceed 20 GB of storage requirements (not counting raw video files). Putting each DVD Studio Pro project on its own 20- or 40-GB drive means you can easily move it from the graphics computer (where the menus are built) to the encoding system (to grab the MPEG files) to an authoring system, then onto a "build" station (to Multiplex the finished project), and eventually to a mastering/backup station where you can unload it to tape (finally!).

If you are going to jump into the realm of HD-DVD authoring, the size drives you will need may depend on which *codec* you are planning to use. Knowing that HD-DVD discs can contain far more gigabytes than SD (Standard Definition) DVDs, you should plan accordingly.

What Speed Drives Do I Need?

We don't often talk about rotational speed of the drive but it can be an issue. Many of the tiniest portable units use drives designed for notebook computers. These tiny (2.5 in) drives generally have speeds in the 4200–4500 rpm range. That's fast enough to retrieve most data, but sometimes not fast enough for certain media tasks—I'd be careful about trying to edit video from one of these, but it should have enough speed to feed DVD data to DVD Studio Pro for previewing or simulation. However, a faster drive (7200 rpm) may be the better choice for both editing and DVD work, since the faster speed means performance won't suffer under demanding tasks. Another factor to consider is that once the drive begins to exhibit fragmentation under use (as the stored pieces of files get scattered), a faster drive may be better able to keep up with DVD previewing and simulation than a slower drive.

In some cases, the decision may hinge on your desire to have a FireWire-powered drive than can be run from your PowerBook while on battery power. In this case, a self-powered drive (like the little FireLite drives from Smart Disk) may be exactly the ticket for you.

In any case, the major advantage of FireWire is the plug-and-play nature of the beast. Hands down,

Introduction

FireWire is the easiest method of adding or moving large amounts of storage for your DVD projects.

Don't Forget to Back Up Your Indispensible Data!

You DO backup your projects don't you?? If not, you should!

FireWire Drives, DVD-R, Streaming Tape formats

If you have properly organized your workflow, you may already have a spare computer with FireWire that you are using to backup your DVD projects when they are finished. If you prefer, small projects can be backed up to DVD-R data discs using Roxio's *Toast Titanium* (**www.roxio.com**). Toast is the INDISPENSIBLE tool you need—I call it the "Swiss Army Knife" of optical disc recording.

Larger projects can be backed up to a wide variety of streaming tape formats. If you have a DLT, your can use that drive to archive your DVD projects as well as create DLT replication masters for your DVD projects. Of course, you could use AIT, or LTO archiving if you prefer.

Software in "Retrospect"

If you are going to use DLT tape as an archival medium (and it's not a bad choice if you have the drive), I'd suggest looking into *Retrospect* from Dantz development (**www.dantz.com**). There are few things in life as indispensable as a rock solid backup program, and *Retrospect* has been a reliable staple for years! As your DVD production workgroup increases from one to many computers, you can expand Retrospect to take care of your entire network, including (ahem) any non-Macs that may find their way into your network.

Peripherals for DVD Production

Many of you will burn your DVDs on a built-in Apple SuperDrive, using inexpensive DVD-R General Media, available from Apple (and many other media manufacturers). But DVD Studio Pro can do much more than just that.

In addition to burning your personal DVD-R, DVD Studio Pro is capable of creating finished DLT (Digital Linear Tape) masters for replication of large-quantity professional DVD projects, as well as burning DVD-R Cutting Master Format (CMF) discs using the Pioneer DVR-S201 and DVD-R (A) [Authoring] media. Note that the DVR-S201 can *only* burn DVD-R *authoring* Media, though, and these discs are not suitable for use in the Apple SuperDrive or any General Media DVD burner.

Replication is the manufacture of hundreds or thousands of identical DVDs, and many if not most replicators still prefer delivery on DLT. (See Chapter 16, "Duplication and Replication" for more on this.)

Using a DLT drive or the Pioneer DVR-S201 DVD-R Authoring Drive requires a PCI SCSI adapter card be installed into your Mac. Be aware that a number of SCSI cards that previously worked fine in OS 9 are NOT supported under OS X. Two popular manufacturers are Adaptec (**www.adaptec.com**), and ATTO (**www.attotech.com**). Of the two, ATTO seems to have the more effective solutions these days—in either case, check the manufacturer's website for current information on OS X compatibility. Especially take note of the fact that higher-end G5 systems

xlviii

Introduction

require **PCI-X** cards, and not the familiar **PCI** cards. The difference is considerable—PCI-X slots do not accept PCI cards, so that older SCSI card you have may not work in a newer G5. Be sure to check your G5's specifications on **www.apple.com/ support** to be certain you are purchasing the proper SCSI card.

DLT Drives

In case you were wondering, **DLT** stands for **D**igital **L**inear **T**ape. It is a robust streaming tape format that is excellent for not only delivery of high-density data like DVD replication Masters, but also extremely well suited to archival uses. I'd venture that there are probably thousands of DLT drives out there busily archiving trillions of bytes of data each and every day in small-, medium-, and large-sized businesses of every stripe. These really are workhorse drives.

Different DLT Drives Do Things Differently!

There are a number of different DLT drives available, so be careful of which one you pick as your choice. Different DLT drives can record on different types of tapes at different speeds and data densities, which accounts for their differences in prices. Many also have different interfaces. This can cause some confusion when you go to buy a drive.

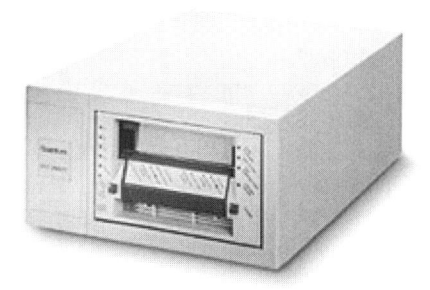

Figure I-1a – DLT 2000 XT

Figure I-1b – DLT 4000

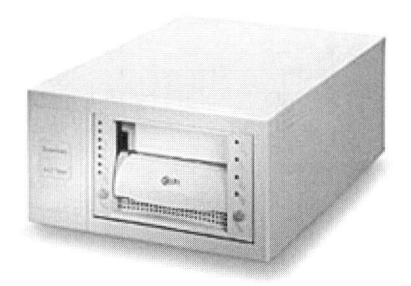

Figure I-1c – DLT 7000

Figure I-1d – DLT 8000

Figures I-1a to I-1d – DLT drives of various models

xlix

Introduction

Which DLT Should I Get?

Any of the DLT models that can successfully record a compatible DVD master, reliably and at the lowest cost, is a good choice. Several DLT drives fit this bill—the "retired" DLT 2000 (10/20 GB) and 2000 XT (15/30 GB), the stalwart, but nonetheless "retired" DLT 4000 (20/40 GB), DLT 7000 (35/70 GB), and DLT 8000 (40/80 GB). The DLT 8000 is no longer a current product, but even used, it may be just a bit pricey for the average desktop DVD producer, with some units commanding prices approaching $1,000. You should be able to find many "lovingly preowned" DLT4000s, 7000s, and 8000s, though, if you search around for them (think "eBay"!).

Beware of the DLT-1 and VS-DLTs—All DLT Drives Are NOT Alike

You may be tempted by the availability of the newer SDLT, DLT-1, or VS-DLT tape drives that you will find shown on the Quantum website (**www.quantum.com**), and at other supply houses. Most of the DLT drive mechanisms are, in fact, made by Quantum, regardless of whose name is on the outside. At this time, SDLT, DLT-1, and VS-DLT drives are not generally usable for DVD Mastering, as many replicators cannot accept or read these master tapes. The best advice is to check with your replicator BEFORE you buy a drive, or need to deliver a DVD project.

Different DLT Media Can Keep You Guessing

DVD Masters are traditionally written on DLT Type III tapes, which hold 10 GB at their native capacity (20 GB if compression is used on the drive—but don't use compression for DVD Masters). Some replicators may accept DLT Type IV tapes (20 GB native, 40 GB compressed) but why waste money on unused capacity?

A DVD DLT Master only includes the data for a single layer, and consequently, uses slightly less than 5 GB, worst case. For Dual Layer DVDs (DVD-9), two DLT Master tapes are written, one for each layer of data. So, why waste a 20-GB Type IV tape on only 5 GB of data?

As you can see from Table I-1 DLT capacities vary with drive model, but one thing is common: All of the above drives can create a DLT Master tape for DVD, using one or more 10-GB DLT Type III cartridges (note the * mark).

A Last Word about DLT Drives

DLT is just about as robust a format as you can get, as long as you don't drop or jar the drive. The delicate tolerances inside this mechanism can be screwed up by a sudden shock, and the heads can get dirty after a certain number of passes. That's one reason to

Table I-1 DLT sizes, capacities, and usability

Drive	Native	Compressed	Media Type
DLT 2000	10GB	20GB	DLT Type III ***DVD**
DLT 2000 XT	10GB		
	15GB	20GB	DLT Type III ***DVD**
	30GB		DLT Type IIIxt
DLT 4000	10GB	20GB	DLT Type III ***DVD**
	15GB	30GB	DLT Type IIIxt
	20GB	40GB	DLT Type IV
DLT 7000	10GB	20GB	DLT Type III ***DVD**
	15GB	30GB	DLT Type IIIxt
	20GB	40GB	DLT Type IV
	35GB	70GB	DLT Type IV
DLT 8000	10GB	20GB	DLT Type III ***DVD**
	15GB	30GB	DLT Type IIIxt
	20GB	40GB	DLT Type IV
	35GB	70GB	DLT Type IV
	40GB	80GB	DLT Type IV

Introduction

always have a DLT Cleaning Cartridge on hand. Be sure to clean the heads regularly, and the DLT should perform reliably for a long time.

DLT Media Types

Remember: DVD Mastering should be done on DLT Type III cartridges, as there is nothing to be gained from using the larger Type IV cartridges, and possibly something to be lost: compatibility! Be sure you confirm which DLT your replicator wants, before you send it.

Figure I-2 – Different colors of DLT Media denote different sizes. Tan, 10 GB; white, 15 GB; dark brown, 20 GB.

It's easy to tell the DLT tapes apart. The format designers were kind enough to design a different color scheme for each size tape (Fig I-2). Remember that for DVD Mastering you want the Tan DLT Type III tapes, not the White Type IIIxt tapes, or the Dark Brown Type IV tapes.

DVD-R (A) Authoring Media—Using the Pioneer DVR-S201 Burner

There are two distinctly different kinds of DVD-R Media, called General (G) and Authoring (A). Your SuperDrive or Pioneer A03–A09 (and some other drives as well) can burn General media discs quite easily, but these discs may be hard to read by certain DVD players (there is a list of suitable players at **http://www.apple.com/dvd/compatibility**). To accommodate this, you can choose to burn DVD-R (Authoring) media, which are available in two different sizes: the usual 4.7-GB, and the original 3.95-GB ones which, in some cases, are more compatible (Figs. I-3a and I-3b).

> **NOTE!**
> The Pioneer DVR-S201 *requires* Authoring media, and cannot write to General media discs.

DVD-RAM

Different from DVD-R, **DVD-RAM** (**R**andom **A**ccess **M**emory) allows read/write capability on the disc without requiring full or partial erasure prior to rewrite. This is different than the re-writeability of

Figure I-3a – DVD-R (A)—3.95GB.

Figure I-3b – DVD-R (A)—4.7GB.

li

Introduction

Figure I-4 – Panasonic DMR-E30K DVD recorder.

DVD-RW, and much closer in function to a "removable hard drive" than an optical disc. Since its original implementation in Mac models in 2000, the DVD-RAM has had strong proponents for data storage uses, but in my opinion, its early usage in DVD-Video applications fell into a gray area. Compromised by slow write speeds, and by being the only "DVD" format that required a hard shell case for its media, DVD-RAM was not the first thought of many (including myself) for use as a DVD-Video distribution format. Interestingly, a lot has changed in a few short years!

DVD-RAM Reborn—Without the Shell

There are now a number of DVD Set-top Disc recorders that can utilize the DVD-RAM format in its new "uncased" form, much simpler to use than the previous hard-to-use "encased" DVD-RAM cartridges (both are shown in Fig. I-5a and I-5b). Notwithstanding this new easier-to-use version, the DVD-RAM format is still a bit harder to distribute video projects on, strictly from a compatibility standpoint. DVD-RAM is not generally compatible with the DVD-ROM drives found in the vast majority of DVD Set-top players, and DVD-equipped computers. In my opinion, it is better to stick with DVD-R ("Dash-R" Recordable) or replicated DVDs, if mass distribution is your end goal.

However, DVD Studio Pro can and will write to a DVD-RAM drive—not a set-top recorder, but an Internal (ATA/IDE) or External (FireWire or SCSI) DVD-RAM drive, if you choose to do so. (But considering the difficulty that still exists in using these discs for distribution, why would you want to?)

DVD+RW, DVD+R

Yet another new format raised its head and, for a moment, caused a bit of confusion in the DVD community—DVD+R/RW ("DVD Plus-R"), promoted by the +RW Alliance (**www.dvdplusrw.org**). This group of manufacturers included Sony, Hewlett-Packard, Ricoh, and other players in the industry who claimed that this new format (while NOT approved or embraced by the DVD Forum) provided greater compatibility and faster rewritable disc burning than the Forum-approved formats (DVD-R, DVD-RW) (See Figs. I6a and I6b).

Figure I-5a — DVD-RAM in a case.

Figure I-5b — DVD-RAM, no case.

Introduction

Figure I-6a — Sony DRU530A Internal DVD±R/RW Burner.

Figure I-6b — Sony DRX510UL External DVD ±R/RW Burner.

One thing that got the DVD+R camp off to a shaky start was the initial release of the DVD+RW-only drives, which were claimed to be upgradeable to include the DVD+R format when it was eventually released. The DVD+R release date came and went and, lo and behold, the original series of +RW drives were NOT upgradeable to include DVD+R as well. This has since been rectified, but not before it caused a bit of discomfort for the +RW Alliance.

The DVD+R/RW format has had its ups and downs, and its alternating proponents and opponents, but an amazing thing happened—instead of taking sides in the war, DVD drive manufacturers soon learned how to put *both* the –R ("Dash–R") and +R ("Plus–R") formats in the same burner! The net result was peace in the valley, and thousands of DVD authors who breathed a huge sigh of relief as they could now use either format they wished to use, without having to choose sides. The great news is that the current version of the Superdrive shipping in Macintosh G5s also supports both formats, so you can choose to write a DVD+R or +RW disc using DVD Studio Pro 4, if you wish.

For a great comparison of DVD+RW and the other formats of recordable and rewritable DVD, be sure to see Ralph La Barge's great comparison study published in the July 2002 issue of DV magazine, and also posted on its website (**www.dv.com**), and on Ralph's own AlphaDVD website (**www.alphadvd.com**).

A Final Thought on Recordable DVD Compatibility

Compatibility between DVD-recordable discs and players is an ongoing topic of frustration for many DVD authors, but with a little research, good encoding practices, and some street smarts gained from DVD websites and DVD forums on Recipe4dvd.com, 2-pop.com, DMN.com, Creative Cow.com, DVDmadeEasy.com, and many more, you can eliminate the frustration and deliver solidly readable DVDs.

IMPORTANT!
A few words of advice:

- Always use trusted, name-brand media.
- Burn your discs at the "native speed" of the media; if the media are 4X, burn at 4X—if 8X, burn 8X.
- Always check your burned discs immediately on a drive other than the drive you burned them on—and preferably, a set-top DVD player.
- Playing back a DVD-R or +R disc on a computer may be convenient, but it can mask playability problems.

Some Assets for Your DVD Learning Projects

During this book we will refer to the DVD Studio Pro tutorial assets, found on the Final Cut Studio Installer

Introduction

discs, but you can also build a typical DVD Studio Pro project using the special Recipe4DVD Demo Assets and Projects provided on the DVD disc accompanying this book.

Since it's not likely that your system will have the "**Recipe4DVDsource" folder already installed, take a moment now and copy that folder from the DVD to your Macintosh hard drive.

The "**Recipe4DVDsource" folder contains a number of prepared assets that allow us to practice using many of the features included within DVD Studio Pro. It also includes some assets that allow you to practice MPEG encoding, as well as some sample assets, graphics, and scripting projects from the DVD Companion Pro-Pack series of products. You will also find more information on the Pro-Packs in their respective folders on the DVD.

DVD Studio Pro 4

The Complete Guide to DVD Authoring with Macintosh

Chapter 1

What Is a DVD?

DVD Studio Pro 4

Goals

By the end of this chapter, you should:

- Know what a DVD is
- Understand how DVDs work
- Understand DVD logical formats
- Understand DVD physical formats
- Understand DVD disc sizes available
- Understand about Binary GB (gigabytes) vs. Decimal GB
- Understand about single-layer DVD formats
- Understand about dual-layer DVD formats
- Understand how the DVD Remote Control Works
- Understand DVD Authoring workflow
- Understand the HD DVD and Blu-ray Disc format

What Is a DVD, Anyway?

DVD is one of the fastest growing technologies around today. Formerly known as either Digital Video Disc or Digital Versatile Disc, a DVD is a great device that can be used to communicate using full motion video and digital sound. By purchasing DVD Studio Pro 4, you've acquired one of the most affordable DVD authoring systems currently available. This software, combined with a PowerMac G4 or G5 and the Apple SuperDrive, provides practically everything you need for a complete, self-contained DVD authoring and burning facility, capable of creating individual "one-off" DVD-R discs for distribution and playback on DVD set-top players, or professional DLT masters for replication of thousands of DVDs. All you need to do is create or obtain the video content you want to put on your DVD and you'll be set! Have you started editing yet? Well, get going!

What Is an HD DVD?— What Is a Blu-ray Disc?

If you're asking about these formats, chances are you already know the basics of DVD. For the answers to these two important questions, skip ahead to to the section "What Is HD DVD and Blu-ray Disc?" on page 18.

What Makes Up a DVD?

A DVD-Video disc is an optical disc-based media delivery system, specifically engineered to provide for the playback of high-quality video material, in conjunction with digital audio soundtracks, and graphical bitmap subpictures ("subtitles") in a digital, nonlinear, random-access delivery format.

Every DVD-Video disc is potentially able to deliver a combination of content consisting of:

- Up to nine video programs ("Video Streams" or "angles"),
- Up to eight audio programs ("Audio Streams"), and
- Up to 32 graphic subtitles ("Subtitle Streams").

This specific list of Streams refers to authored elements called "Tracks" in DVD Studio Pro. The precise combination of these streams available to the viewer is dependent on which assets are assembled into the Tracks during Authoring. The available

What Is a DVD?

streams can be selected from interactive menus or controlled in real time by the DVD user using the DVD player's remote control. The other elements in DVD Studio Pro (Menus and Slideshows) combine DVD assets in different ways.

> **NOTE!**
> In DVD Studio Pro, only *Tracks* can combine the 9 angles, 8 audios, and 32 subtitles listed above. Other elements combine different combinations of assets. See Chapter 3, "Workspace and User Interface" for more information on these "Elements."

Video Streams are delivered from digitally compressed MPEG video files. Properly preparing the video for DVD (encoding) is a critical step. (We'll discuss this more in Chapter 5, "Making Video for DVD.")

Audio that accompanies the video programs is delivered from digitized audio programs, which are aligned with the video during Authoring. The Audio may also be compressed using Dolby Digital AC-3 encoding to save space and reduce the playback data rate. When Audio is used in DVD Video, it must always have an image of some form associated with it—either moving video or a still image. (See Chapter 6, "Making Audio for DVD.")

Subtitles are displayed by the DVD player in real time, from compiled bitmap instructions assembled into the DVD during the authoring stage of DVD workflow. Subtitles are deliberately designed to take very little room on the DVD, so there can be many of them available. (See Chapter 12, "Creating Subtitles.")

Menus in the DVD allow for the interactive presentation of the available program choices, which the user can select from a Button on the menu. High-quality graphics design in Menus is an important element in making an enjoyable and professional-looking DVD. (See Chapter 9, "Authoring Menus.")

Slideshows are another way of assembling assets, and in DVD Studio Pro, Slideshows are assembled from still images and audio files. Think of a Slideshow as being a sequence of pictures, with individual sound files or an overall soundtrack, and you've got the right idea. (See Chapter 10, "Authoring Slideshows.")

How DVDs Work

DVDs work their magic by playing back the properly digitized and prepared video, audio, and subtitle media from the tiny optical disc that contains the finished DVD files. If you examine a DVD in your computer, and open the disc to explore it, you will find a world of files exists on the DVD that is invisible to you when watching a DVD on a set-top DVD player. It is these files that make up the DVD you experience on that disc.

The files in a finished DVD are first organized into a specific folder (directory), called the "VIDEO_TS," then further divided into groups called *Video Title Sets* (VTS), each of which contains three specific types of files to organize the media and the navigation instructions you will create in your DVD. Some DVDs also contain an additional folder, called "AUDIO_TS," containing DVD-Audio materials. (The new HD DVD standard adds more to this subject; we'll cover it in a later chapter.)

Each set of VTS files contains one or more **.VOB** (**V**ideo **OB**ject) files to hold the prepared media, plus a **.IFO** (**InFO**rmation) file with the navigation info, and a **.BUP** (**B**ack**UP**) file, providing a duplicate of the IFO file, used if the IFO becomes damaged or unreadable. The .IFO tells the DVD how to playback the .VOB files.

The trick to making a DVD work properly is to *Author* it properly—that is, to properly prepare the media and correctly establish the navigation structure. DVD Studio Pro makes this easy with its intuitive graphical user interface and simple to use drag-and-drop, point-and-click authoring. It has also

3

DVD Studio Pro 4

implemented many new ideas to make authoring even easier than before.

One Finite Limit to Your DVD's Size

The exact number of VTS file groups that are included in your finished DVD depends on how you organize the media and the navigation while creating your DVD. Due to limits built into the DVD Video Specification (the "DVD Spec"), a maximum of 99 VTS groups can be created. Since each Track, Story, or Slideshow in DVD Studio Pro can create a VTS group, this amounts to a practical limit of 99 Tracks, Stories, and Slideshow elements *combined*. Don't despair, though—within that 99 VTS limit, you can create a LOT of DVD! Another limit is disc size, which we will explore a little later in this chapter. (We'll discuss build issues in greater detail in Chapter 14, "Building and Formatting.")

Making a DVD Play Back

DVDs require a specific tool, a DVD Player, in order to decode the digitized and encoded video and audio files, and to reproduce the prepared subtitles (if you used any). There are two different mechanisms that do this, one to decode the Video and Audio and another to reconstruct the graphical bitmaps that are superimposed on the MPEG video to create the running subtitles or graphics images. DVDSP allows you to create text subtitles using the Subtitle function within the Track Editor. It is also possible to create graphical overlays, like watermarks, using Subtitle events. These can be created while authoring Tracks. (See Chapter 12.)

The Remote Control Is the Key to DVD Navigation

The DVD Remote Control is the key to the user's ability to navigate your DVD project, and its functions should guide you in laying out your DVD. Understanding the function of the DVD remote will enable you to easily and quickly plan your DVD project and its navigation.

Transport Controls

The DVD Remote control manages both the Transport functions of the DVD Player—typically Play/Pause, Stop, Skip Back, and Skip Fwd (see Fig. 1-1). Set-top DVD remotes typically also have adjustable speed playback, both in forward and rewind.

Figure 1-1 — Apple DVD Player Remote Control Transport functions.

Presentation Options

The DVD Remote Control also can control the Presentation Options of the current DVD—in which Video Angle is visible, Audio Stream is audible, and Subtitle Stream, if any, is being displayed, as shown in Figures 1-2, 1-3, and 1-4.

4

What Is a DVD?

Figure 1-2 — Subtitle Button.

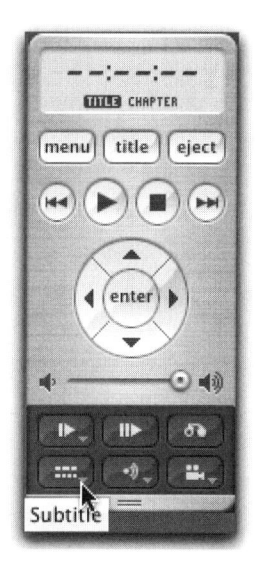

Figure 1-3 — Audio Button.

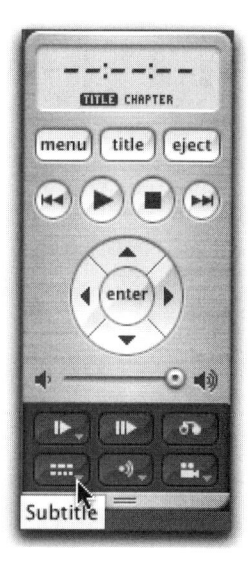

Figure 1-4 — Angle Button.

Menu Navigation Arrows, and Enter Key

To access the DVD Menus, a quadrant of arrow keys on the face of the remote allow for up, down, left, and right navigation through the available Menu Buttons while the "ENTER" key (usually centered within the Menu Arrows) allows activation of the *currently* selected and highlighted menu button.

Additionally, most remotes include numbered keys for entering numerical data or selecting buttons directly by number (if you have authored the DVD Menu to allow the buttons to be numerically selectable). These numerical keypad buttons are not shown in our simplified remote graphic (Fig. 1-5) because on a computer, the keyboard or keypad keys are used for this function.

Navigating to and from Menus: Using TITLE, MENU; RETURN

The *Menu* button (Fig. 1-6), the *Title* button (usually called *Top Menu* on new DVD players) (Fig. 1-7), and the *Return* button (Fig. 1-8) complete the typical DVD remote control.

The *Title* button navigates to the Top Menu of the DVD project, as determined by the DVD author.

The *Menu* button usually returns to the Menu presentation previously viewed or the Menu element associated with that video presentation, but again, this is determined by the Author during authoring. In DVD Studio Pro, this can be programmed in many different ways.

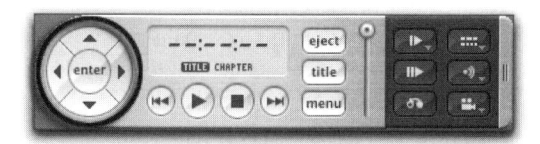

Figure 1-5 — Menu Navigation Buttons.

DVD Studio Pro 4

Figure 1-6 — Menu Button.

Figure 1-7 — Title Button.

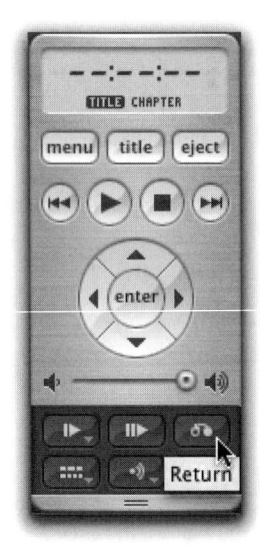

Figure 1-8 — Return Button.

The *Return* button provides yet a third level of navigation capability, but unlike the *Title* or *Menu* buttons, has no preset behavior built into the DVD player. How and *if* it functions is determined during authoring. DVD Studio Pro gives you lots of options for creating functions for this button.

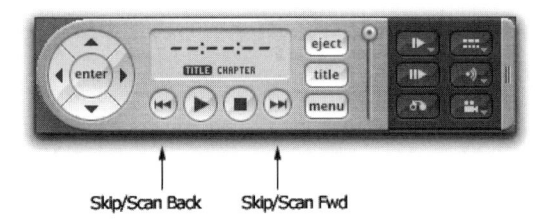

Figure 1-9 — Apple DVD Player Remote Chapter Skip Controls.

Using the Remote Control or Mouse for DVD Navigation

Knowing how the DVD user will interact with the DVD remote is a good first step in understanding how to layout the navigation of your DVD and knowing what you can control within the Presentation.

The easiest controls to understand are probably *Skip Forward* and *Skip Back* (see Fig. 1-9). The DVD viewer will use these to move forward and backward through the Chapter Markers (max. 99 per Track) that can be programmed into each of the Video Tracks (max. 99 per DVD). In the Apple DVD Player, clicking and holding either of these will play the DVD in fast play mode, either forward or backward (see Fig. 1-10).

Keep in mind also that most computers these days provide for playback of DVD Video discs as well, but the method of navigation and menu interaction is different: A mouse frequently takes the place of, or augments the use of a typical DVD remote. In fact, few if any computers actually *have* a stand-alone remote

What Is a DVD?

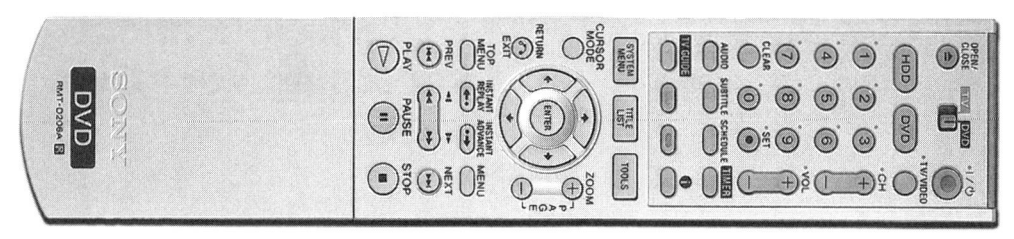

Figure 1-10 — A typical settop DVD player remote control.

control for the DVD player (although this functionality can be added using devices like the Keyspan Media Remote). The mouse allows the user to directly navigate over the menu image, selecting a button by entering that button's "hot zone" (the rectangle created for that button during the Menu Editing phase of DVD Authoring).

DVD Disc Sizes (How Much Video Will My Disc Hold?)

Single-Sided Discs

Many commercial DVD discs are single-sided discs, 12 cm in size (the same size as a CD), and may contain up to 8.5 GB*, anywhere from 120–260 minutes of good quality video, depending on the MPEG *video encode bitrate* (see Table 1-1). Because DVDs are made up from two pieces of media bonded together, disc capacities larger than CD's 650–800 Megabytes are easily created using multiple-sided or multiple-layered discs, although some of these formats are only available through Replication. (See Chapter 14.)

> **NOTE!**
> *Be certain to read the section on GB (gigabytes) versus BB (billions of bytes) disc sizes—understanding this difference is crucial to being a good DVD author.

Table 1-1 Single-Sided DVD Configurations

DVD-5	4.70 GB	Single Side/Single Layer	SS/SL
DVD-9	8.54 GB	Single Side/Dual Layer	SS/DL

Because of the two-piece construction of DVDs, single-sided discs may contain either one or two layers of data, as shown in Figures 1-11 and 1-12. Dual-layer DVD-9 discs are quite the norm these days for commercial movie releases, as that format easily accommodates a large movie and some special features. But even those discs don't hold the amount of data some DVD authors would like, leading to the next generation of DVDs—HD DVD and BD.

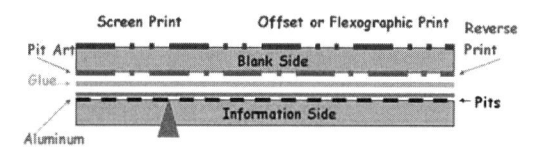

Figure 1-11 — A DVD-5: Single-sided disc with one layer of DVD data. (Illustration courtesy of DVD Today website. Original artist unknown.)

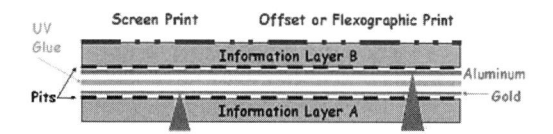

Figure 1-12 — A DVD-9: Single-Sided disc with two layers of DVD data. (Illustration courtesy of DVD Today website. Original artist unknown.)

Double-Sided Discs

When additional media are needed to be stored on a DVD, it is possible to create double-sided discs that can store the digital info on each side of the DVD (see Table 1-2). Standard configurations called DVD-10 and DVD-18 are created by making a two-sided disc with either one or two layers on each side (see Fig. 1-13). A special disc (DVD-14) can be created by making a disc with both a dual-layer side and a single-layer side. While DVD-14 discs are a little tricky to manufacture, and possibly a bit pricey, they do exist and are being manufactured today, although in recent years the demand for these discs has slowed, as DVD9s became the dominant format in commercial releases.

Table 1-2 Double-Sided DVD Configurations

DVD-10	9.40 GB	Dual Side/Single Layer	DS/SL
DVD-14	12.32 GB	Dual Side/Mixed Layer	DS/ML
DVD-18	17.08 GB	Dual Side/Dual Layer	DS/DL

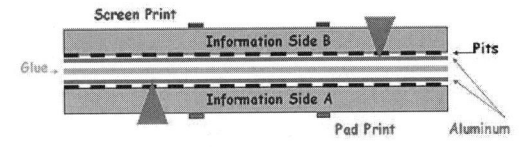

Figure 1-13 — A DVD-10 or DVD-18: Double-sided DVDs. (Illustrations courtesy of DVD Today website. Original artist unknown.)

Mini-DVDs

There are four DVD discs in a smaller size (8 cm), called **DVD-1 (SS/SL), DVD-2 (SS/DL), DVD-3 (DS/SL), and DVD-4 (DS/DL)** as well as DVD Business cards that can contain tiny 350- or 650-Mbyte DVD projects. Many replicators can produce the DVD business card sizes for you, if this format is needed. Currently, just a few DVD-R discs are available in the 8-cm size.

A Quick Note about Recording Your Own DVDs

Just a little over a year ago, I wrote:

> "DVD-9 discs CANNOT be burned on *any* DVD-Recordable or DVD-ReWritable disc by *any* DVD Burner."

Well, a LOT has changed since then! DVD-9 Double-Layer burners and media are now pretty commonplace—in fact, the current generation of Macintosh G5s ship with 16x double-layer SuperDrives as standard equipment. And while the double-sided DVD-10, -14, and -18 discs must *still* be manufactured by replication, it is now common for DVD authors to use the double-layer DVD-9 recordable discs as a way to create "test discs" of their projects for clients. There is much more on this in Chapter 14. Please read this chapter carefully, as the burning of double-layer discs should not be done without some precautions and some knowledge of the possible pitfalls.

Making DVD Discs with Data— DVD-ROM, DVD-Hybrids

DVD Studio Pro is specifically built to author DVD-Video discs, but can also be used to create *DVD-Hybrid discs*, where Video and Data are combined on the same DVD. DVD Hybrid discs are a powerful combination of video media and computer files on the same optical disc. DVD data discs without Video, called *DVD-ROM* discs, can be burned with the SuperDrive using the Mac OS, or applications like Roxio Toast Titanium. These data discs are perfect for storing large archives of files, but will NOT play in a DVD player unless they contain a properly built DVD volume—i.e., a VIDEO_TS folder, and usually an AUDIO_TS folder as well.

What Is a DVD?

HD Because the new HD DVD specification allows for a new kind of DVD Volume, an HVDVD_TS may be built instead. We'll cover the HD DVD changes in DVD Studio Pro in each chapter as we go along—look for the HD Icon in the margin.

What DVD Formats Are Available?

When we open a discussion of DVD formats, we should be careful to separate the Logical Formats (what the DVD disc *does*) from the Physical Formats (what *type* of disc media the DVD is delivered on).

This part of the discussion will expand dramatically in the near future as the HD DVD and Blu-ray Disc (BD) formats become available to us for use in DVD video authoring. But let's begin with the tried and true Standard Definition (SD) DVD items.

DVD's Logical Formats

At the heart of every DVD-Video disc is the DVD-ROM format. A DVD-ROM disc is essentially a computer data disc, in many ways just like a CD, but much larger in capacity (think "CD on Steroids"). *DVD-Video* discs are created by properly preparing the video and audio content, authoring the navigation structure (Menus and such), and then formatting it to be delivered on a disc formatted as a *DVD-ROM*. Because DVD-ROM is underneath the DVD-Video content, this also means your DVD-Video disc can simultaneously deliver computer-readable data to DVD users who have a DVD-equipped computer!

The beauty of the logical format system though, is that it means DVD-Video discs are readable by practically every DVD-ROM–equipped computer, and practically every other DVD format drive as well, because most DVD drives can not only read their own format, but DVD-ROM as well (see Table 1-3).

Table 1-3 Current Standard Definition (SD) DVD Logical Formats

DVD-ROM	Data	Computer Software, data
DVD-Video	Video	Movies, Industrial, Corporate
DVD-Audio	Music	High-resolution Audio, Surround Sound

DVD-Audio is another logical format you may run across, but DVD-Audio discs are VERY different from DVD-Video discs, and desktop authoring of DVD-Audio is still not yet anywhere near as well-developed (or as inexpensive) as DVD Video authoring, nor as popular. Regardless, DVD Studio Pro is an authoring application for DVD-Video and doesn't create the DVD-Audio format.

DVD's Physical Formats

DVDs not only come in different logical formats, they are also different in their Physical format as well. Today, you have a bewildering choice of drives available in these formats: DVD-R (record once), DVD–RW (ReWritable), DVD+R (record once), DVD+RW (ReWritable), DVD+R DL (double layer), and even DVD-RAM (random access read/write), as well as the traditional replicated DVD-ROM disc.

Arguably the most compatible DVDs are those that are manufactured in *replication*, but the fuel that has been powering the explosion of personal DVD authoring in the past few years would have to be the *recordable* DVD formats. Part of the problem for new DVD authors has been to cut through the many different recordable and re-writable formats that exist, and to make a decision as to which one to use.

Luckily, certain standards have emerged that make this decision a bit easier to make. DVD Studio Pro can access devices that can write to both DVD-R (General) and DVD-R (Authoring) media directly from the authoring environment, but you need to use the correct media in each.

DVD Studio Pro 4

Since the last edition of this book, great strides have been made in DVD recordable technology, and today the Apple SuperDrive can write to practically any format media—DVD-R (G) media, DVD+R, DVD-RW, and +RW media, and even the newest format, DVD+R DL, the new double-layer recordable format. Nothing is yet on the horizon for DVD-R DL, but that should surface soon.

Burning to DVD-R (A)

Authoring media still requires a Pioneer burner (DVR-S201) designed specifically for this format. Fortunately, both the (G) and (A) formats can be read by most, if not all, DVD players, once they have been written (burned). Discs burned on DVD-R (G) or DVD-R (A) can also be read by most, if not all, personal computers.

Please be aware that there may still be a compatibility issue with the DVD+R/RW format ("DVD plus R"), which cannot be written or read on a DVD-R/RW ONLY ("DVD dash R") drive. These days, practically every burner you can buy is a multi-format burner, which writes and reads all DVD-recordable formats (but not DVD-RAM), and CD-R and CD-RW as well! DVD Studio Pro does now support the DVD+R/RW part of these drives, and the newest SuperDrives can write to double-layer +DL media as well.

You can usually tell which formats a particular drive accepts by looking at the logos on the faceplate. Drives that support CD-R and CD-RW formats will have the familiar Compact Disc Logo, drives that support the DVD-R/RW formats will have the DVD Forum logo with R and/or RW underneath, and drives that support the DVD+R/RW formats will have the +RW alliance logo with the word *recordable* or *rewritable*. (See Table 1-4.)

This is the original write-once DVD-R media format, but this can only be used with DVD Studio Pro if you have the proper DVD-R burner. These media may be written to, but not erased. It uses a different laser wavelength than General Media, and only the Pioneer DVR-S201 (or 101) can write to it.

There are two sizes of DVD-R (A) media—the original 3.95 GB, and Version 2.0 media, which can hold 4.7 GB. In either case, you will find these media to be more expensive by far than DVD-R General Media.

> **REMEMBER!**
> You will require a SCSI adapter on your Macintosh, if you want to use the Pioneer DVR-S201 or S101 burners. If using a newer G5, you will need a PCI-X compatible card. Check www.attotech.com.

Table 1-4 Current SD DVD Physical Formats

Format Description	Method	Max Cap.	
DVD-ROM	Manufactured	18.0 GB	read-only
DVD-R (A)*	Burned	4.7 GB	SS/SL write-once
DVD-R (G)	Burned	4.7 GB	SS/SL write-once
DVD-RW	Burned	4.7 GB	SS/SL rewriteable
DVD+R	Burned	4.7 GB	SS/SL write-once
DVD+RW	Burned	4.7 GB	SS/SL rewriteable
DVD+R DL	Burned	8.5 GB	SS/DL write-once
DVD-RAM	Burned	4.7 GB	SS/SL rerecordable
DVD-RAM	Burned	9.4 GB	DS/SL rerecordable

*Creating DVD-R (A) Authoring media requires a special Pioneer burner.

What Is a DVD?

DVD-R (A)

- 4.7 GB/3.95 GB SS/SL
- DVD-R Authoring Media—Not suitable for SuperDrive
- Use the Pioneer DVR-S201 or S101
- Can be written using DVD Studio Pro 4

DVD-R (G)

- 4.7 GB SS/SL
- DVD-Recordable—known as "dash-R"
- Can only be written once—cannot be erased
- Uses DVD-R (recordable) General Media
- Usable in the current Apple SuperDrive
- Can be written using DVD Studio Pro 1, 2, 3, or 4

This is the standard write-once media used with DVD Studio Pro. This media may be written to, but not erased. It uses a different laser wavelength than the DVD-R Authoring Media and requires a General Media-capable DVD Burner, which includes all SuperDrives built since 2001. This is the BEST overall media for predictable compatibility with the majority of DVD set-top players and DVD-enabled computers.

DVD-RW

- 4.7 GB SS/SL
- DVD-Re-Writable—known as "dash-RW"
- Can be erased and re-written
- Uses DVD-RW re-writable Media
- Usable in the current Apple SuperDrive
- Can be written directly with DVD Studio Pro 2, 3, or 4, or
- Build DVD with DVDSP, then use Toast Titanium with SuperDrive

DVD+R

- 4.7 GB SS/SL
- DVD+Recordable—known as "plus-R"
- Can only be written once—cannot be erased
- Uses DVD+R (recordable) General Media
- Requires DVD+R capable drive (not older SuperDrives)
- Usable in the current Apple SuperDrive
- Can be written directly with DVD Studio Pro 3 or 4, or
- Build DVD with DVDSP, then use Toast Titanium with SuperDrive

DVD+RW

- 4.7 GB SS/SL
- DVD+Re-Writable—known as "plus-RW"
- Can be erased and re-written
- Uses DVD+RW re-writable Media
- Usable in the current Apple SuperDrive
- Can be written directly with DVD Studio Pro 3 or 4, or
- Build DVD with DVDSP, then use Toast Titanium with SuperDrive

DVD+R DL

New!

- 8.5 GB SS/DL
- DVD+Recordable Double Layer Media—"double-layer"
- Requires DVD+R *DL-capable* drive—very recent

DVD Studio Pro 4

- Can be written directly with DVD Studio Pro if SuperDrive is +DL, or

- Build DVD with DVDSP, then use Toast Titanium 6.0.7 or later with DL burner Toast will automatically determine the Layer Break, if using Toast 6.07 or later. Check **www.roxio.com** for version updates.

DVD-RAM

- 4.7 GB SS / 9.4 GB DS
- DVD-Read/Write
- Not usable in Apple SuperDrive, period!
- Build DVD with DVDSP, then use Toast Titanium with DVD-RAM drive

Disc Size—Gigabytes (GB) versus Billions (BB)

Many newcomers to DVD are puzzled about why they cannot fit 4.7 GB of data on a DVD disc advertised as having 4.7 GB of capacity. The answer to this puzzle is: "It depends on the meaning of *gigabyte*."

DVD Disc Sizes are expressed as GB, but are really measured in decimal Billions of Bytes. DVD GB are smaller than an equal number of real (i.e., Binary) GB. Bottom line? *Your 4.7 GB DVD only holds 4.37 GB of hard disk data!* If you wish to go into the details of why, read on, but if you just want to take this fact as gospel, move on to the summary at the end of this chapter.

In DVD usage, the capacity of a 4.7-GB disc isn't calculated using the traditional binary Gigabyte (1 GB = 1,024 Megabytes) used to represent computer hard drive and memory capacities. Instead, DVD disc sizes are expressed in decimal Billions of Bytes, rather than binary "or powers of two." This means a 4.7-GB DVD contains 4.7 billion Bytes and is smaller than 4.7 GB of hard drive storage. Confused? Don't worry—let's decipher this for you:

Remember that computers are binary counting machines that can only think in two states—on and off, one and zero. Counting in computer terms means 1, 2, 4, 8,16, 32, 64,128, 256, 512, and ultimately 1,024. 1,024 Bytes of RAM memory would be called a *kilobyte* (KB—thousands); 1,024 kilobytes is called a *megabyte* (MB—millions); 1,024 megabytes equals a *gigabyte* (GB—billions), and so on up the binary food chain. Notice that each time we are multiplying by 1,024, and not by 1,000.

All these KB, MB, and GB have one very important common factor: They are all multiples of 1024, and not 1000! This means that the more of these we add up, the greater our error between the Binary value and the Decimal value of the same number of decimal places. (See Fig. 1-14.)

We have been using the *kilo-*, *mega-*, and *giga-* references ever since computers first walked the earth, and we all knew exactly what they meant...But somehow...(big sigh here)...the billions of bytes "GB" reference managed to sneak its way in when DVD came about.

While we were busy relating to our hard drives, memory chips, and other units in binary KB, MB, and GB, the DVD revolution happened and, before we knew it, manufacturers were reckoning their capacities not in the familiar MB and GB, but in DECIMAL BILLIONS.

Figure 1-14 will put it in better perspective for you.

Luckily, our bit budget spreadsheet does all of the appropriate size calculations for you in decimal billions, properly. All you need to do is enter the proper number of audio streams, subtitle streams, total run-

What Is a DVD?

DVD File sizes: Gigabytes vs Billions of Bytes

Name	Billions of Bytes	Gigabytes	diff. (decimal)	difference (KB)	difference (MB)
DVD-1	1,460,000,000	1,357,489,362	102,510,638	100,108.05	97.76
DVD-2	2,650,000,000	2,463,936,170	186,063,830	181,702.96	177.44
DVD-3	2,920,000,000	2,714,978,723	205,021,277	200,216.09	195.52
DVD-4	4,120,000,000	3,830,723,404	289,276,596	282,496.68	275.88
DVD-5	4,700,000,000	4,370,000,000	330,000,000	322,265.63	314.71
DVD-9	9,400,000,000	8,740,000,000	660,000,000	644,531.25	629.43

Figure 1-14 — GB versus BB in DVD Disc sizes.

ning time, and the proper disc size, and you're on your way.

(*You will find the Bit Budget spreadsheet "MakeABitBudgetV23.xls" in the Tidbits folder of the DVD accompanying this book.)

A note—in dealing with the new topic of HD DVD, we will encounter some new video codecs, which I will cover in the appropriate chapters. HD-specific paragraphs will always display this HD alert icon.

DVD Workflow

Before we get into the inner workings of DVD Studio Pro, let's review the typical method of DVD production, called "DVD Workflow." (See Figure 1-15a.)

There is a definite *workflow* used in creating a DVD—this is the order in which the DVD production steps normally take place. It begins with the presumption that you will have already created or licensed the video assets for the DVD, or will be creating them shortly.

1. **PLAN** the Project, **ACQUIRE** video, audio, graphics assets.
2. **ENCODE** the video program material as appropriate.
3. **AUTHOR** the navigation and interactivity.
4. **BUILD** the DVD (*mux* the file/folder data or a disc image).
5. **DELIVER**—BURN a DVD or WRITE a DLT tape master.

In each one of these steps, an important part of the DVD Production process takes place—when this complete workflow is finished, the DVD has been completely premastered, and this finished DVD master (as it is known) may either be burned to some flavor of DVD-recordable or re-writable disc, or delivered to a DVD replicator on DLT (Digital Linear Tape) for the mass production of large quantities of DVD discs (called *replication* of discs). Let's outline what happens in each stage of the DVD workflow:

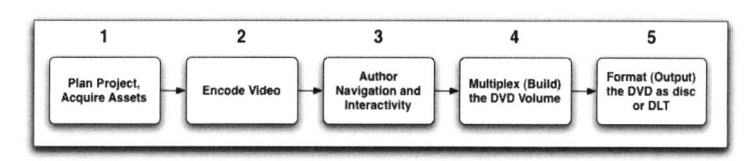

Figure 1-15a — Five Steps of DVD Workflow.

13

1—Plan the Project, then Acquire the Assets You Need

Planning a DVD Project should begin by developing an idea of what you wish to present and how DVD can be used to present it. Planning is helped immeasurably by understanding a bit about how DVDs work—the basic building blocks of DVD as used in DVD Studio Pro are *Tracks* that deliver the Video, *Menus* that present interactive choices, *Slideshows* that display linear sequences of graphics, and *Scripts* to control enhanced navigation.

Planning should define the structure of the DVD project and the relationships of all of these elements. During the planning stage, you will need to define the length of video program including any additional audio or subtitle tracks desired, as this will determine the video encoding bitrate required to make the project fit on the desired DVD disc size. This calculation is generically called *bit budgeting*, reflecting the need to carefully plan the usage of the limited number of bits available on the DVD disc. Proper bit budgeting will outline the specific requirements for not only video (MPEG) encoding rates, but also how best to take advantage of audio compression (typically Dolby Digital in most DVD regions), and also how to allow bandwidth for any subtitle streams that may be needed.

We have included a working bit budget spreadsheet in the *Tidbits* folder on the accompanying DVD. Open this sheet in Microsoft Excel, and work with it while reviewing the bit budgeting instructions found in Chapter 2, "What is DVD Studio Pro 4?" This will make your bit budgeting simpler, and faster.

Once the bit budget has defined the asset encode rates, **Acquiring the Assets** consists of gathering the video, audio, graphics, and any ROM data required for the DVD. In Chapter 6, "Making Audio for DVD," we will discuss in detail how to properly create audio programs for DVD using **A.Pack** (for any of you still using DVD Studio Pro 2 and 3) and **Compressor 2** (which is new with DVD Studio Pro 4 and Final Cut Studio). And in Chapter 15, "Graphics Issues for DVD Images," we will outline methods used in correctly creating graphics assets for DVDs—this is primarily of interest for making menus, but also relevant for slides and generic info screens. While there are a lot of new features in DVD Studio Pro that allow you to create Menus directly within the program, it is invaluable to know these DVD graphics basics.

2—Encode the Video

Once the list of video presentations (Tracks and Motion Menus) that will be needed has been determined during planning, and the video programs created or acquired, the video must eventually be converted into compressed digital files using the MPEG-2 Codec. (MPEG-1 also works, and is currently implemented in DVD Studio Pro, but has some restrictions on its use. Consult DVD Studio Pro help.)

This process is usually called *encoding* if it means transferring programs from Videotape, or *transcoding* if it means converting an existing digital file into an MPEG file. In either case, the end result is an MPEG-compressed video bitstream (sometimes just called a Stream). This MPEG Stream must contain ONLY the video program for that particular presentation. DVD Authoring requires this type of Stream, called an *elementary MPEG Stream*, which is different from an MPEG-2 *Program Stream* or *Transport Stream*, which some encoders can create.

DVD Studio Pro installs a new software encoder called *Compressor 2*, as well as a new background encoder that can create MPEG-2 streams from QuickTime movies while authoring in DVD Studio Pro. While DVD Studio Pro 1, 2, and 3 added encode components into the Macintosh QuickTime engine to enable MPEG-2 encoding functions in QuickTime,

What Is a DVD?

DVD Studio Pro 4 relies on Compressor 2 exclusively. The QuickTime MPEG-2 Export Component (a.k.a. MPEG-2 encoder) facilitates the MPEG export/encode function in any QuickTime application that can use *export components*.

Apple's own Final Cut Pro is capable of exporting (encoding) directly into MPEG-2 either through Compressor or QuickTime. The QuickTime Pro Player can also export to MPEG-2 directly. There are several other software MPEG-2 encoders that can be used with DVD Studio Pro. (See Appendix B, "Alternative Encoders.")

3—Author the Project

Authoring the DVD involves assembling the content (the assets) while adding the navigation and interactivity that will control the presentation of the content created in the encoding stage. Authoring defines the functionality of the DVD, and what the DVD user will experience when this particular disc is viewed. During Authoring, you will use Previewing and Simulating to verify the proper operation of the various DVD functions that have been created during Authoring. DVD Studio Pro is the Authoring program, and includes an integral preview solution that allows simulated playback of the DVD in progress. (We will cover each aspect of Authoring in subsequent chapters as we go forward.)

4—Build the Project

Building (a.k.a. *multiplexing*, or *mux-ing* for short) means reassembling the raw DVD data from the original asset files (the *elementary streams* used for authoring) into the finished interleaved data file format, called the *DVD Volume* that all DVD players require. DVD Studio Pro can build this DVD Volume to your Macintosh hard drive, or to any hard disk volume attached to the host Macintosh. FireWire Drives have made this a very inexpensive and trouble-free process. Building the DVD is absolutely essential to creating a working DVD. Until you *build*, you just have a DVD Project, not a real DVD!

5—Deliver the Project—Burn a DVD or Write a DLT

The final output of a DVD project is generally a recorded DVD Disc (of any format), or a DVD *replication master* written on DLT (*digital linear tape*).

Final distribution of the DVD can take one of three possible forms:

A "ONE-OFF" DISC on DVD-R/RW, DVD+R/RW, or DVD+R DL

A DVD disc may be written on any acceptable recordable DVD media using a suitable DVD Burner (like the Apple SuperDrive). These kinds of discs are created with DVD-R or DVD+R discs, typically DVD-R. Typical DVD burners are manufactured by Pioneer (DVD±R/RW +R/RW), Panasonic (DVD-R/RAM), Ricoh and H/P, (DVD+RW/+R), Sony (DVD±R/RW), and many, many others. Many new drives (including the newest generation of Apple SuperDrives) offer the capability for burning the new double-layer recordable media (generally +DL). Be careful to use the correct drive for the chosen media!

Duplication

In the event that more than one disc is needed for final delivery, a DVD *duplicator* may be used to create multiple copies of the original disc (see Chapter 16, "Duplication and Replication"). These disc copies are typically created with DVD-R discs, but may be created on any recordable or rewritable format, including the new +DL. From a cost-effectiveness standpoint, duplication loses its advantage when the quantity of discs required exceeds 50–100 discs or more. As the cost of recordable DVD media continues to plummet, this number will continue to

DVD Studio Pro 4

increase, making duplication more cost-effective for even larger runs.

Replication

When more than 100 discs are needed (or more typically 1,000 or more), DVD *replication* is used to create large quantities of identical discs. These discs are the most compatible DVDs you can make. Their manufacture is accomplished at a DVD replication facility, and prior to manufacture, the manufacturing parts (called *stampers*) must be created through a process known as *glass mastering*.

If you are going to replicate your DVDs, the most accurate delivery method is the DLT, although not all replicators require DLT any more. Many can use the DVD-R (G) media that you can burn on your own SuperDrive using DVD Studio Pro. Be careful, however, that you deliver the highest quality master you can. If you are going to deliver DVD-R (G) instead of a DLT, take care to deliver two discs, and do NOT play them more than once after they have been burned. Believe it or not, playing the discs may compromise the data recorded on the discs, making the mastering process more complicated. If you are going to deliver a DVD-9 dual-layer project, you *should* deliver on DLT, as this is the method generally used by replicators to create your DVD's glass master(s).

Today's Workflow Is Different

Of course, new developments mean sometimes we have to update our old standards and dogma. DVD Studio Pro has brought new tools and methods to the DVD party, so it is no longer necessary to create encoded video before authoring, and it is now possible to create a menu entirely within DVD Studio Pro's very powerful Menu Editor! The DVD Studio Pro 3 update added video and still transitions so, in effect, you now have access to a powerful compositing engine within DVD Studio Pro—ah, you've got to love progress! (See Figure 1-15b.)

OK, I'm Going to Replicate—What Now?

If you are going for replication, prepare your master DVD disc or DLT master using the methods we will outline in Chapter 14, and then deliver it to the replication facility. (Be sure to verify that they will accept something other than DLT, though.) The replicator will then prepare the DVD manufacturing parts by the method we outline in Chapter 16.

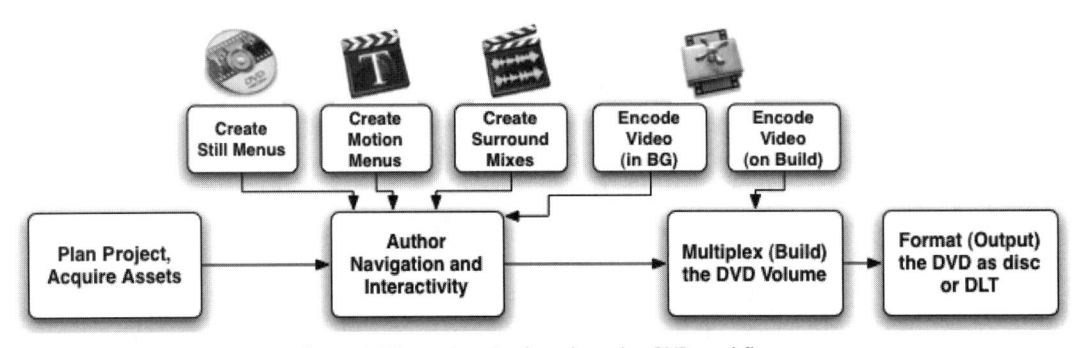

Figure 1-15b — A revised modern-day DVD workflow.

What Is a DVD?

How Do I Plan My DVD?

Planning a DVD requires a combination of authoring insight, graphics design, and technical number crunching. Planning a DVD begins by designing the structure and layout of the disc, based on the navigation and project needs, and then determining the disc size desired, and video encoding parameters. This last part is called *bit budgeting*.

Creating a bit budget involves determining three important things:

1. Determine the audio and subtitle encode rate overhead required for the DVD project. You'll do this by adding up all of the bitrates for the audio and subpicture (subtitle) Streams you need for your DVD project. Subtracting that total value from 9.8 (for a one-angle DVD) will give you a video "maximum bit rate" value. You will use this value if you are performing a VBR (*variable bit rate*) encode, but may not need it for a CBR (*constant bit rate*) encode.

2. Determine the *average bitrate* of the project. This can be done easily by dividing the total bit size of the disc by the number of seconds of video program you wish to include.

3. Subtract **1** from **2** to determine the *video-only encode rate*. The video-only encode rate determined in this last step will be used to program the encoding software that you choose to use in creating your MPEG video files. (See Chapter 5 for more information.)

To make life easy for you, the DVD with this book includes a copy of the "Make a Bit Budget" spreadsheet that I designed a few years ago, and that has been downloaded by DVD authors around the world. (See Figures 1-16 and 1-17.) It requires a recent version of Microsoft Excel to run (although I heard that several users have successfully loaded it into Appleworks!).

Helpful Data for Use While Calculating a Bit Budget

Format	Nominal encode rate
Dolby Digital Audio (Stereo)	192 kbps
Dolby Digital Audio (5.1)	384 kbps, MAX 448 kbps!
MPEG-1 Layer II Audio (Stereo)	192 kbps (DVDSP 1.5 +)
PCM Audio (AIFF/WAV)	1.6 Mbps (yow!)
Subtitle streams (each)	.04 Mbps

Figure 1-16 — Table of Nominal Encode/Data Rates.

Disc	Billion Bytes	Gross Megabits	usable (less 4%)
DVD-1	1.46	11,680	11,213
DVD-2	2.66	21,280	20,429
DVD-3	2.92	23,360	22,426
DVD-4	5.32	42,560	40,858
DVD-5	**4.70**	**37,600**	**36,096**
DVD-9	**8.54**	**68,320**	**65,587**
DVD-10	9.40	75,200	72,192
DVD-14	13.24	105,920	101,683
DVD-18	17.08	136,640	131,174
HD DVD SL*	15.00	120,000	unknown
HD DVD DL*	30.00	240,000	unknown
HD DVD 3L*	45.00	360,000	unknown
BD SL*	25.00	200,000	unknown
BD DL*	50.00	400,000	unknown
DVD-RAM I	2.58	20,640	19,814
DVD-R (1.0)	3.95	31,600	30,336
DVD-R (2.0)	4.70	37,600	36,096
DVD-RW	4.70	37,600	36,096
DVD+R	4.70	37,600	36,096
DVD+RW	4.70	37,600	36,096
DVD-RAM II	4.70	37,600	36,096
DVD+R DL	8.54	68,320	65,587

Proposed capacities

Figure 1-17 — Table of DVD Disc size equivalents and usable Megabits.

17

DVD Studio Pro 4

DVD Disc Run Times (minutes-to-seconds)

Minutes	Seconds
10	600
20	1,200
30	1,800
60	3,600
90	5,400
100	6,000
110	6,600
120	7,200
130	7,800

Figure 1-18 — Disc Running Times in seconds.

Table 1-5 SD Encode Rate Benchmarks for Reference

Disc Size	Prgm Length	Encode Rate Mbps*
DVD-5	<60 Minutes	6 Nominal 8 Max
DVD-5	60 Minutes	6 Nominal 8 Max
DVD-5	90 Minutes	6 Max
DVD-5	120–130 Minutes	4.5 Max
DVD-9	120–130 Minutes	6 Nominal 8 Max**
DVD-9	180 Minutes	6 Nominal and Max
DVD-9	240–260 Minutes	4.5 Max

* Assumes one Dolby Digital Soundtrack at 192 kbps
** Why would you do this? Make a DVD-5!

Significance of Encode Rates

The MPEG video encode rate, more than anything else, is the key to good-looking video on your disc. (See Figure 1-19 and Table 1-5.) If you use the proper encode rate for your proposed project (and also assuming you have good-looking source video, and a decent software encoder), you should be able to create good-looking MPEG files for your DVD. Remember, though, that you should create a bit budget for each DVD project to verify the proper encode rates required. You are also bound by this relationship, which I always preach in my encoding seminars and classes:

"Running time and encode rate are interrelated!"

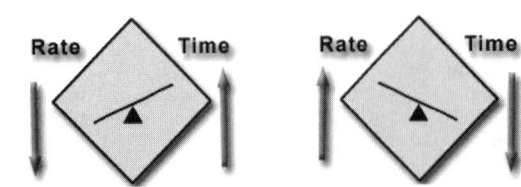

Figure 1-19 — Lower encode rate = longer run time; higher encode rate = shorter run time.

What Is HD DVD and Blu-ray Disc?

What could be better than one new format for next-generation DVDs? TWO formats, right? (Well, not really—actually two formats are a huge accident waiting to happen, but I digress....)

HD DVD and Blu-ray Disc (BD for short) represent two different approaches to extending the current DVD specification to take into account the newer high-definition video formats, and the desire for more storage options and interactive features to go with the improved video.

While both formats are attempting to accomplish similar goals, the methodology used by the two different camps hasn't been resolved into a single, unified format (yet—there's always hope, but the odds are getting slimmer).

If there is a format compromise, we will all be the better for it. But frankly speaking, it looks like the next version of DVD is going to go forth into the world with two different and not necessarily compatible formats... I guess it will remain to see what happens when the consumers (that's you!) make their judgment with their checkbooks and credit cards.

What Is a DVD?

In the interim, we can discuss what we do know about the HD DVD format, since that format is included in the DVD capabilities of DVD Studio Pro 4.

First on the Block for HD DVD Authoring

DVD Studio Pro 4 is the first application to be able to create discs that conform to even a small part of the new HD DVD specification. Apple has taken the 0.9 version of the HD DVD specification (the part that was stabilized at the prerelease point) and has implemented those capabilities within DVD Studio Pro 4.

This means you can make widescreen DVDs that utilize many of the great newest video formats, like DVCPRO HD and HDV. If you have Final Cut Pro, you can even capture and cut all the way up to 1920 × 1080i (you'll need a pretty powerful system for that, though—not a Powerbook).

This might be a good time to outline what these new video specs are, so we're all on the same page.

> **NOTE! ****
> These formats are not allowed in HD DVD using the MPEG 1 codec.

NTSC for HD DVD

Rates with "i" indicate interlaced; Rates with "p" indicate progressive scan.

Resolution	Frame rate	Aspect ratio	Notes
352 x 240	29.97i	4:3	a.k.a SIF format **
352 x 480	29.97i	4:3	a.k.a 1/2 D1
704 x 480	29.97i	4:3, 16:9; 16:9 is anamorphic	a.k.a. Cropped D1;
720 x 480	29.97i	4:3, 16:9; 16:9 is anamorphic	a.k.a. Full D1;
720 x 480	59.94p	16:9 HD only,	a.k.a. 480p; is anamorphic
1280 x 720	59.94p	16:9 HD only,	a.k.a. 720p
1440 x 1080	29.97i	4:3, 16:9 HD only;	16:9 is anamorphic
1920 x 1080	29.97i	16:9 HD only	a.k.a. 1080i

PAL for HD DVD

Rates with "i" indicate interlaced; Rates with "p" indicate progressive scan.

Resolution	Frame rate	Aspect ratio	Notes
352 x 288	25i	4:3	a.k.a SIF format **
352 x 576	25i	4:3	a.k.a 1/2 D1
704 x 576	25i	4:3, 16:9; 16:9 is anamorphic	a.k.a. Cropped D1
720 x 576	25i	4:3, 16:9; 16:9 is anamorphic	a.k.a. Full D1;
720 x 576	50p	16:9 HD only,	a.k.a. 576p; is anamorphic
1280 x 720	50p	16:9 HD only,	a.k.a. 720p
1440 x 1080	25i	4:3, 16:9 HD only;	16:9 is anamorphic
1920 x 1080	25i	16:9 HD only,	a.k.a. 1080i

DVD Studio Pro 4

There is more to cover on the HD DVD, but we will cover those topics as we go along in subsequent chapters. Each reference of interest for HD will have the *HD alert* logo that is shown here.

Some Notes on How This Book Is Arranged

Beginning with Chapter 3, each chapter of this book deals with one specific topic regarding DVD Studio Pro, and covers it in depth.

The book is best experienced when used with the demonstration project contents from the accompanying DVD, and also in conjunction with the tutorial media from DVD Studio Pro itself.

Shipping with every copy of DVD Studio Pro is a folder of tutorial project media. This media includes pre-encoded MPEG files, prepared PCM Audio Files, as well as menu graphics and MPEGs and a host of other media used to build the DVD Studio Pro tutorial projects and explore the operation of the application programs.

Added to this, we have provided you with a second complete project and one full-length video clip to use as source media for MPEG encoding using the DVD Studio Pro encoding system.

We have also included a number of folders of additional source material on the DVD accompanying this book and some great resources for learning more about available enhancements for your DVD Studio Pro authoring system.

Please note that Apple's QuickTime MPEG-2 encoder requires a G4 or G5 processor in order to encode—this is due to the encoder's use of the Velocity Engine (the AltiVec instruction set) available only on G4 or G5 chips.

In cases where our tutorial materials are used, the book will refer you to the proper asset contained in the DVD that accompanies this book. If you wish, you can build our tutorial project along with us.

Our goal has been to provide you with an authoritative guide to DVD Authoring, as well as the ancillary issues you will need to know to produce and effectively distribute a DVD. We sincerely hope you will find the remainder of this book useful and enjoyable.

What Is a DVD?

Summary

DVD is a digital delivery medium for Video, Audio, and Data, and the fastest growing Home Entertainment medium ever introduced. There are a wide variety of DVD formats, both *logical* (DVD-ROM, DVD-Video, DVD-Audio), and *physical* (DVD-ROM, DVD-R/RW, DVD+R/RW, +DL, DVD-RAM).

DVD-Video discs combine digitally compressed MPEG Video files (streams) with digitized Audio programs (usually PCM or Dolby Digital AC-3) and graphical Subtitle streams to provide a high-quality nonlinear, random-access delivery format.

To properly create a DVD requires the right software tool (an Authoring application), and to properly burn a DVD requires the proper hardware drive, and software to drive the burner. DVD Studio Pro provides this built-in burning function, and can also connect to a Digital Linear Tape (DLT) tape drive to create a DVD Tape Master.

Understanding the navigation of a DVD using the DVD remote control is a key to creating good DVD projects, as is understanding the proper creation of video, audio, and graphics Assets used in the making of a DVD. Planning a DVD project requires knowing the structure of the disc, as well as calculating the various encoding parameters that will be used. This is called *Bit Budgeting*, and must be done carefully to avoid wasting time or money by re-encoding assets.

Running time, encode rates, and disc space are all interrelated, and must be balanced properly for the DVD to be technically successful.

DVD Studio Pro 4 running on a PowerMac G5 or G4 with the Apple SuperDrive gives you a complete self-contained DVD-Video Authoring station, capable of encoding, authoring, and burning professional-quality DVD-Video discs in both the Standard Definition (SD) and High Definition (HD) DVD formats. If you have a G5, you can be one of the first to create the new format HD DVD discs and play them back with the Apple DVD player on any suitable Mac running OS 10.4 Tiger.

If you still have a G4, trust me, you'll have a lot of fun making DVDs, but you won't be able to play back HD DVDs. Is it upgrade time?

Chapter 2

What Is DVD Studio Pro 4?

DVD Studio Pro 4

Figure 2-0 — Welcome to DVD Studio Pro 4.

Goals

By the end of this chapter, you should:

- Understand the DVD Studio Pro 4 application

- Understand how the individual DVD Studio Pro **functions** are applicable to the various stages in the DVD workflow

- Understand the role of the **DVD Studio Pro 4** application

- Understand the role of the **Compressor 2** application

- Understand the role of the **DVD@ccess** feature

- Understand differences between **DVD Studio Pro 3 and 4**.

- Understand the role of **Apple Qmaster** in encoding

- Understand the role of **Apple DVD Player** in emulation

Using Earlier DVD Studio Pro Projects in DVD Studio Pro 4

To use a project begun in earlier versions of DVD Studio Pro in version 4, it's really as simple as opening the file into DVD Studio Pro 4. You will get an update alert dialog (see Fig. 2-1) that reminds you not to save over the original, as doing so will PERMANENTLY change the file, and prevent it from ever being reopened in DVD Studio Pro 3.

Figure 2-1 — There's no going back to DSP3 if you ignore this warning.

> **IMPORTANT!**
> IMMEDIATELY perform a "Save As..." as soon as you import an old DVD Studio Pro 3 project file. Give it a name with something specific to indicate it's a project for SP4.

What Is DVD Studio Pro 4?

DVD Studio Pro 4 is a full-featured application for making professional DVD-Video discs that can be played on almost any consumer DVD player. Like its predecessors, DVD Studio Pro 1, 2, and 3, DVD Studio Pro 4 is a simple, yet professional tool. DVD

What Is DVD Studio Pro 4?

Studio Pro 4 also includes some great features not previously available in DVD Studio Pro 3. It is an excellent complement to Final Cut Pro, Apple's popular desktop video editing application, adding DVD output to the Final Cut workflow.

In point of fact, DVD Studio Pro 4 is actually a combination of several applications and supporting software that work together as an integrated system to help you author DVD videos quickly, and easily.

When DVD Studio Pro 4 is fully installed, you will have:

- *DVD Studio Pro 4*—the Authoring application.
- *Compressor 2*—the video and audio compression application.
- *Apple QMaster*—to control distributed encoding with Compressor 2.
- *QuickTime MPEG-2 decoder*—to facilitate previewing your DVD.

For those readers who are familiar with DVD Studio Pro 3, this chapter will outline some of the many new features of this version, with a goal of getting you on your way to authoring in this new environment with a minimum of confusion and delay.

If you have just acquired DVD Studio Pro 4, and want to start fresh, perhaps it may be better for you to skip to "Getting Started with DVD Studio Pro 4," later in this chapter, and begin from there. The first part of this chapter will deal with what's new in DVD Studio Pro 4, and in many cases what's different (and sometimes similar) between this version and previous versions. If you haven't owned DVD Studio Pro before, this may not be of interest to you. On the other hand, we will cover the new features in DVD Studio Pro 4, so it might be a good place for you to get a quick overview.

What's New in DVD Studio Pro 4?

How is DVD Studio Pro 4 different from DVD Studio Pro 3? Well, many very cool features have been added, and all of the many good features from previous versions remain, some of which have been enhanced.

The first one is the biggest breakthrough:

New!—HD DVD Authoring— First in the World!

DVD Studio Pro 4 adds support for using new high-definition video formats in DVDs that comply with the new HD DVD standard. You now can make widescreen DVDs with stunning content from HDV, DVCPRO HD, or other formats of High Definition Video. Best of all, DVD Studio Pro 4 allows use of the next generation video codec H.264 (AVC) for enhanced quality or extended running time. These are major breakthroughs, especially since at the current time, there are no other commercially available products that will enable H DVD authoring. You will find updated info on HD capabilities throughout the chapters of this book—just look for the HD alert icon.

New!—24P Compatibility

DVD Studio Pro 4 allows use of video content encoded at 24.fps (23.98 fps in reality). Progressive 24-frame content is a natural for DVD, as you will see in your exploration of this new format.

> **IMPORTANT!**
> DOWNLOAD DVDSP 4.0.3 FROM APPLE.COM TO GET ENHANCED HD DVD FEATURES.

DVD Studio Pro 4

New!—VTS Editor in the User Interface

If you have bemoaned the limit of 1 GB in each Menu domain in the past, you will find the new VTS editor to be a welcome addition. Not only does it allow you to redistribute Menus into different VTSs, breaking free of the previous 1 GB total limit, but you can also reallocate DVD elements to different VTSs. This can help with scripting, as reallocating scripts can speed up the execution of the script code in the finished DVD. (See Chapter 3, "Workspace and User Interface.")

New!—Templates for HD

DVD Studio Pro 4 now supports templates for the HD format as well as the tradition SD formats. (See Chapter 9, "Authoring Menus.")

New!—Integrated Encoding for HD MPEG2

DVD Studio Pro 4 continues the tradition of MPEG background encoding by adding MPEG2 encoding for HD DVD formats. This is a real breakthrough. You can now use HD formatted movie files (of certain codecs) without encoding them into MPEG files first. As with previous versions, you can defer encoding until the build process. (See Chapter 5, "Making Video for DVD.")

New!—Integrated Encoding for AC3 in Compressor 2

A.Pack has been replaced by the new AC3 encoding capabilities in Compressor 2. This allows you to convert your video and your Dolby Digital audio in the same encode job. (See Chapter 6, "Making Audio for DVD.")

New!—Dual-Layer Burning Support (Sans Hassles)

OK, OK. This feature was really introduced in DVD Studio Pro 3.02, but since all G5s now ship with a double-layer enabled SuperDrive, burning +DL discs out of the box is a reality. (See Chapter 14, "Building and Formatting.")

New!—Support for Blue Laser Burners

DVD Studio Pro 4 includes support for both red laser burners (up to 8.5 GB), and the upcoming blue laser burners (15–50 GB). Of course, we'll need to wait for the actual release of some burners that use blue lasers in this country—some do exist in Japan at this time, but we will begin to see them in the USA in 2006. (See Chapter 14, "Building and Formatting" for more.)

New!—Simulation to Digital Cinema Desktop

You will welcome DVD Studio Pro 4's revised *Simulator*—the updated version will now allow you to preview your DVD projects on a Digital Cinema Desktop (or any other widescreen digital monitor with enough screen resolution to accommodate the video resolution you are using). Wow—what a difference a big screen makes! (See Chapter 14, "Building and Formatting" for more.)

New!—Simulation to an External Video Monitor

If you've been longing to have a true *What You See Is What You've Made* display, it's finally here! Simulator now allows you to route the simulation display to an external video display. You can't click

What Is DVD Studio Pro 4?

on menu buttons in that window, though—you'll still need to use the Simulator display or keyboard shortcuts. Previewing on an external video monitor should end any questions you've had about the quality of your video encodes. (See Chapter 14, "Building and Formatting" for more on the new features in the simulator.)

New!—Support for the DTS 6.1 Audio Format

Building on the DTS capability originally added in DVD Studio Pro 3, DTS support is now expanded to include 6.1 format files, as well as audio in the 96-kHz, 24-bit format.

New!—Simulation to an External Audio Monitor

You can now connect your Macintosh to an external audio system via the optical digital, FireWire, or USB output. This will allow you to preview surround sound digitally even via a Dolby or DTS decoder.

New!—GPRM Partitioning for Scripting

Scripting enthusiasts will appreciate this: DVD Studio Pro 4 adds *partitioning* to the 8 GPRMs, so you may no longer need to do complex binary bit-manipulation to access a single bit. Each GPRM may be used as a 16-bit whole, or divided into up into 8-bit, 4-bit, 2-bit, or 1-bit partitions. Each partition may be named uniquely, as well. (See Chapter 13, "Basic Scripting for DVD Studio Pro " for more on GPRM partitions.)

New!—Dual-Layer Breakpoints Can Be Set in the DVD-ROM Zone

Dual-layer DVDs that include larger amounts of ROM data than video can now set the breakpoint within the DVD-ROM material.

New!—Better Access to Menu Loop Points

There are a number of additions to the Menu loop point usage, including the ability to script a jump directly to the Menu Loop point.

Getting Started with DVD Studio Pro 4

Once you have installed DVD Studio Pro 4, you should find the following applications, which comprise the DVD Studio Pro application suite. It takes a number of Gigabytes to install this package, but there are good reasons why… there are a lot of media included with this application and they have to go somewhere!

Figure 2-2a — DVD Studio Pro 4: The DVD Authoring application.

DVD Studio Pro 4

Figure 2-2b — Compressor 2 (MPEG-2, H.264, and AC-3) Encoder. Compressor encodes the MPEG-2, AVC, and AC3 Streams for your DVDs.

Which Applications Are Installed

The applications provided with DVD Studio Pro 4 provide tools to cover every aspect of DVD authoring. With the onboard Menu Editor, it is quite possible to reduce or eliminate reliance on external graphics applications to create menus. Of course, if you own Final Cut Studio, you may find an irresistible temptation to learn Apple's *Motion* application to create your motion menus. You may still use other motion graphics applications, however, quite easily—the image and movie files you create will work quite nicely in DVD Studio Pro 4 (especially if you follow the guidelines set forth in Chapter 15, "Graphics Issues for DVD Images").

The DVD Studio Pro 4 installer installs these applications:

- **DVD Studio Pro 4** is the DVD authoring application for assembling the finished product (Fig. 2-2a shows its icon). You will do most of your DVD work in this application. (Starting with Chapter 3, features are detailed in depth.)

- **Compressor 2** can be used to encode MPEG-2 files from your video assets, and AC-3 Dolby Digital files from your audio assets. It provides high-quality video and audio encoding and batch mode processing. Fig. 2-2b shows its icon. (Learn more in Chapter 5, "Making Video for DVD" and Chapter 6, "Making Audio for DVD.")

- **QuickTime MPEG encoder** is no longer used by, or installed with, DVD Studio Pro 4 or Final Cut Pro 5. If you installed it previously, perhaps with Final Cut Pro 3 or 4, or DVD Studio Pro 2 or 3, it may likely still be accessible from within Final Cut Pro or QuickTime.

Documentation Files Installed

The new *DVD Studio Pro 4 User Guide* is close to 700 pages and is viewable in PDF form while you are authoring. It's *very* complete!

Extras Installed

DVD Studio Pro 4 also installs the *Apple Qmaster* distributed rendering software to allow Compressor 2 to use the power of all suitable Macs attached to your local area network (LAN) for increased encoding horsepower. The *Apple Qmaster Node* installer for the slave Macs is included in the *Extras* folder.

Templates, Shapes, Patches, Fonts

The DVD Studio Pro 4 application bundle weighs in at a hefty 3+ GB on disk. It isn't because the application is *bloatware*—on the contrary. The DVD Studio Pro 4 bundle contains several GB of prebuilt DVD Menu templates, plus transition movies, shapes, patches, and fonts that enable the Menu Editor to do all of the really incredible new things it can do. Think of it as an "art department in a folder."

What Is DVD Studio Pro 4?

Launching DVD Studio Pro 4 for the First Time

Setting the Default Configuration

When you launch DVD Studio Pro 4 for the very first time, you will notice something new and different: A dialog box will appear asking you to select a "default" configuration—that is, the configuration your workspace windows will assume each time you launch DVD Studio Pro 4. This is very different from DVD Studio Pro 1 where all the windows were preset, and they would resume those preset locations each time the application launched. Certain standard arrangements of the display are preset in DVD Studio Pro 4, called "Default Configurations." (This is just the beginning of new features you will encounter in this version.)

The Configuration Dialog (Fig. 2-3a) only appears the first time you launch DVD Studio Pro 4 after installation. Once made, this selection will govern all new launches of the application, but NOT launches initiated by double-clicking an existing project file—that project will reload with the same configuration it was last saved with.

Three workspace defaults are available in DVD Studio Pro 4: *Basic*, *Extended*, and *Advanced*. Each configuration has differing levels of detail available in the displays, and each has its own particular authoring benefits, as you will see in Chapter 3.

If you are new to DVD, or coming to DVD Studio Pro 4 from iDVD, select the Basic configuration. If you are an experienced DVD Studio Pro author, or have experience authoring on another platform, try the extended or advanced configuration to see which is more comfortable. Remember that any configuration can be highly customized at any time.

Getting Acquainted with Your Options

DVD Studio Pro 4 has a large array of operating options—so many that it takes 10 preference panels to list and organize them all. Before you dash off and start trying to author, it might be a good idea to

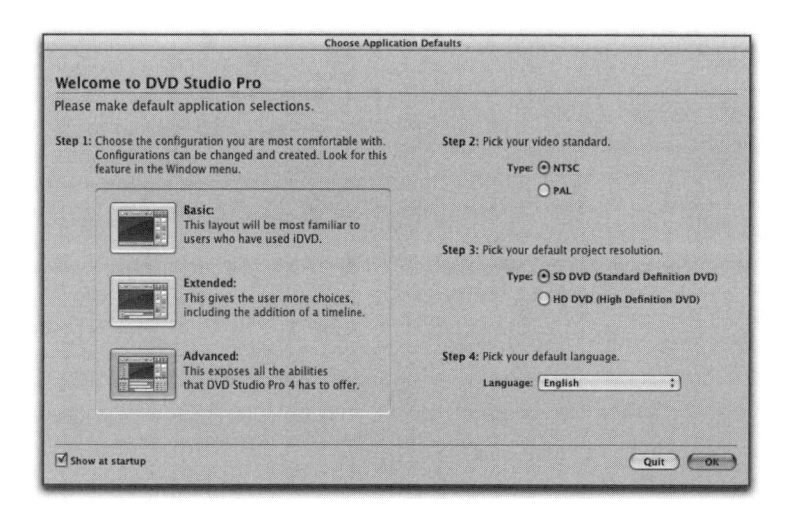

Figure 2-3a — The DVD Studio Pro 4 Configuration Dialog.

DVD Studio Pro 4

review the preferences you have control over, and perhaps to take a run through the Preferences and set things the way you would like them to operate.

Establishing Project Basics

Upon launching DVD Studio Pro 4, or creating a new project, you will have an empty workspace available to you. At this point, the project has some basic settings that should be set before proceeding further:

- **The Disc Name**—DVD Studio Pro 4 gives the name UNTITLED_DISC to every new project by default, but you should modify this early on, and then save the project with the disc name established. (See Fig. 2-3b.)
    - Click on the UNTITLED_DISC icon in the Outline Tab.
    - If the Inspector isn't visible, select **View>Show Inspector** in the menu bar.
    - Select the *General* Tab.
    - Click in the *Name:* field.
    - Double-Click or Click-Drag to select the default name.
    - Type a new name—hit *Return* or *Enter*—it's that simple.
- **The Video Standard**—DVDs can be made with NTSC or PAL video assets, and DVD Studio Pro 4 needs to know which system you will use.

You can set a default preference in the *General Preferences* Pane (see later in this chapter) and you can change it directly in the *Disc Property Inspector*:

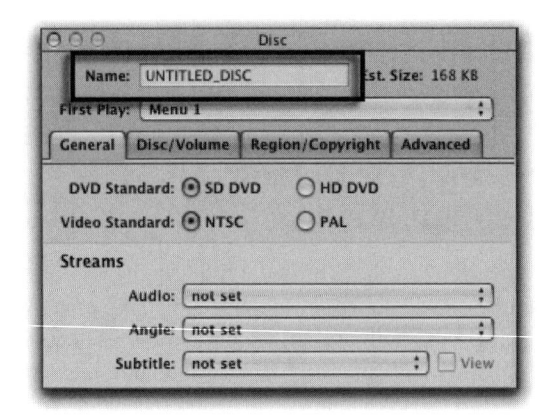

Figure 2-3b — Setting the Disc Name.

- Click on the UNTITLED_DISC icon in the Outline Tab.
- If the Inspector isn't visible, select **View>Show Inspector** in the menu bar.
- Select the General Tab.
- Select your project's desired TV standard.

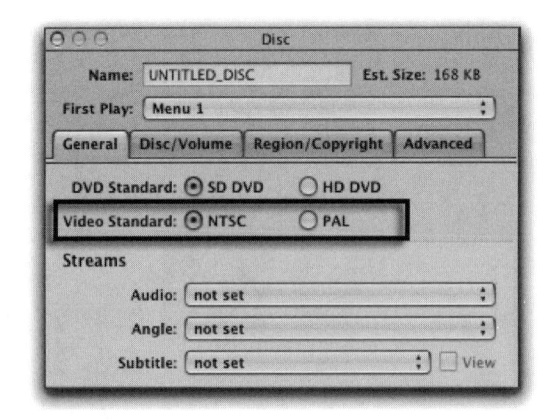

Figure 2-3c — Setting the Video System Standard.

What Is DVD Studio Pro 4?

> **NOTE!**
> Switching to the PAL standard means using PAL video assets as well as graphic images (Slides, Menus) that conform to the PAL image size of 720 x 576. (See Chapter 15.)

If you are not certain which standard you need, here are some general guidelines:

- *North America*—supports the NTSC standard (29.97 Frame rate, 525 lines, 60 Hz sync). In general, countries that use 60 Hz power use this standard for DVD, but exceptions exist, like Brazil, which uses Pal-M.

- *Europe*—In general, countries using 50 Hz power will support the PAL standard (25 fps, 625 lines, 50 Hz sync).

> **TIP!**
> For a complete list of the countries of the world and the video standards they support, see the DVD Studio Pro 4 Help Manual, installed with the application.

- Select **Help>DVD Studio Pro Help** from the menu bar, or

- Type *Command-Question Mark* (⌘-**?**).

- **The DVD Standard**—in DVD Studio Pro 4, DVDs can be made to the SD or HD DVD standard and DVD Studio Pro 4 needs to know which system you will use. You can set a default for this in the *Project Preferences Pane* (see later in this chapter), and you can change it directly in the *Disc Property Inspector* (See Fig. 2-3d.)

  – Click on the UNTITLED_DISC icon in the *Outline* Tab.

  – If the Inspector isn't visible, select **View > Show Inspector** in the menu bar.

  – Select the *General* Tab.

  – Select your project's desired TV System.

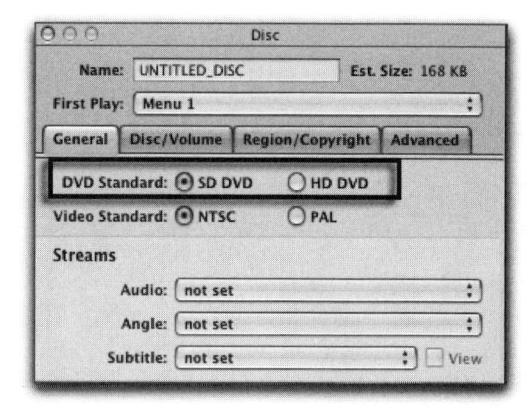

Figure 2-3d – Setting the DVD Standard.

> **NOTE!**
> Switching to the PAL standards means using PAL Video as well as Graphic Images (Slides, Menus) that conform to the PAL image size of 720 x 576 (see Chapter 15).

- **The Disc Size**—DVD Studio Pro 4 defaults to a DVD-5 size project, but you can modify this in the Disc/Volume Tab of the Disc Property Inspector (Fig. 2-3e).

  – Click on the UNTITLED_DISC icon in the *Outline* Tab.

  – If the Inspector isn't visible, select **View > Show Inspector**.

  – Select the *Disc/Volume* Tab.

  – Select your project's desired size and layer configuration.

Other options can be set in the *Disc Inspector* as well, like number of sides, layer options, track direction, break point, etc. (See Chapter 14, "Building and Formatting" for more on disc options.)

31

DVD Studio Pro 4

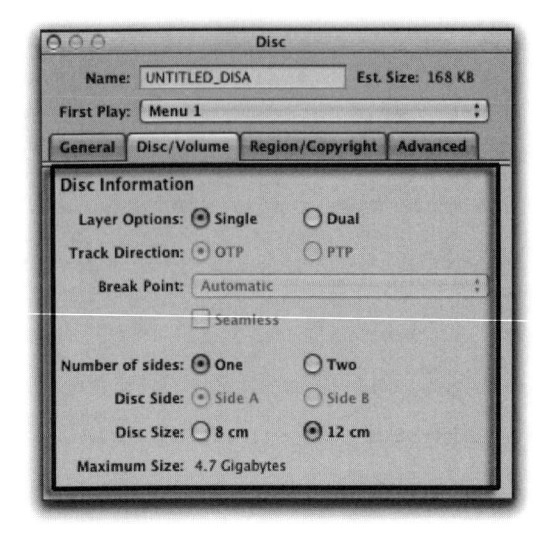

Figure 2-3e — Setting the Disc Size.

Figure 2-3f — Accessing the DVD Studio Pro 4 Preferences panes.

DVD Studio Pro 4 Preferences

Many aspects of the operation of DVD Studio Pro 4 are controlled or selectable by setting options in the Preferences panes. There are now nine of these, and they control everything from the background color of the Menu window to the encoding rates and details of the new embedded MPEG-2 encoder.

To Access the DVD Studio Pro 4 Preferences:

- Select **DVD Studio Pro>Preferences**... in the menu bar, or
- Type *Command-Comma* (⌘-,)

Once opened, the *Preferences Panel* appears as a single window (see Fig. 2-3f for an example), with 10 selection tools across the top, each of which can select a specific group of preferences.

To use the Preferences window (see Fig. 2-3g), do the following:

- Click an icon along the top to open a specific preference pane.
- Click *Apply* to apply the current settings and leave the Preferences window open.
- Click *OK* to apply the current settings and then close the Preferences window.
- Click *Cancel* to close the Preferences window without applying the current settings.

Project Preferences

Project Preferences control defaults used at the origination of a new DVD project.

DVD Standard (pop-up menu)

- *SD DVD:* selects Standard Definition DVD standard for new projects.
- *HD DVD:* selects High Definition DVD standard for new projects.

What Is DVD Studio Pro 4?

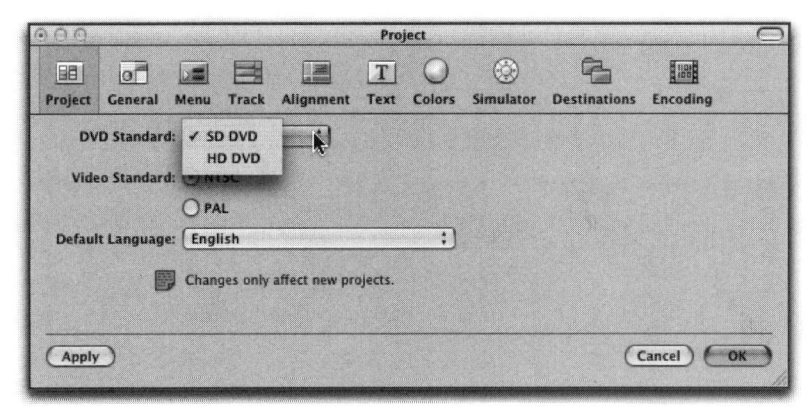

Figure 2-3g —Project Preferences Pane.

Video Standard

- *NTSC:* selects NTSC video standard for new projects.
- *PAL:* selects PAL video standard for new projects.

Default Language (pop-up menu)

- *Default Language:* Click to choose a default menu language for new projects.

General Preferences

General Preferences control defaults most likely used day-to-day:

Slideshows and Tracks

- *Default Slide Length:* This will be the default duration of images added to slideshows or added to a track's video stream.
- *Background Color:* You can set the color used as a background by the Slideshow Editor. Normally, this color should be covered by the slide. If the slide image is not the correct size to fill in the video frame, this color fills the background gaps that exist. This color is also used by the Track Editor if you add still images that are not the correct size to fit in the video frame. (See Fig. 2.4.)

> **IMPORTANT!**
> *This setting is not saved as part of the project.* It is a Global Application Setting. If you change this setting, it will affect all projects, including previously saved ones, that have this color visible.

SD DVD Menus, Tracks, and Slideshows

- *Display Mode:* This will be the default aspect ratio that is used to create Menus, Tracks, and Slideshows in Standard Definition DVD projects. The available options are: 4×3, 16×9 Pan-Scan, 16×9 Letterbox, and 16×9 Pan-Scan & Letterbox.

33

DVD Studio Pro 4

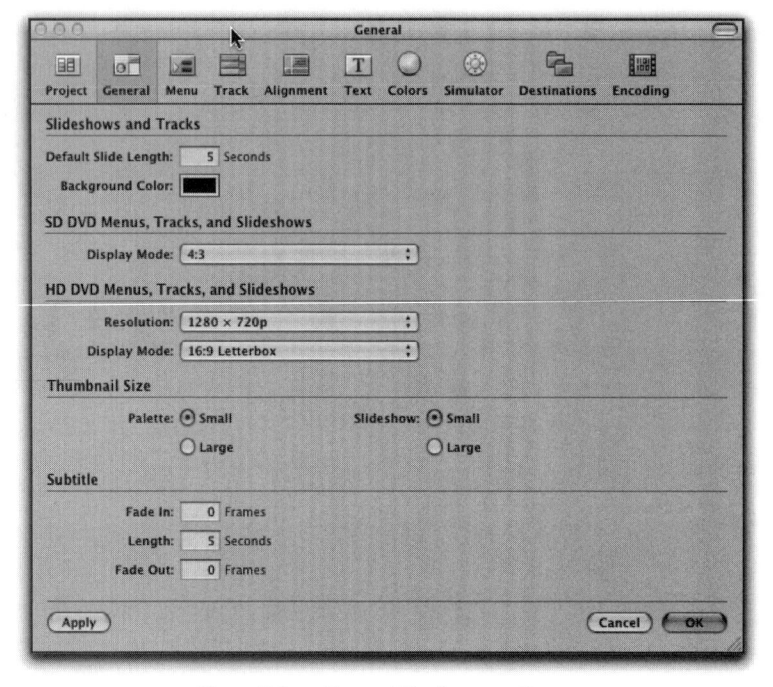

Figure 2-4 — General Preferences Pane.

HD DVD Menus, Tracks, and Slideshows

- *Resolution:* This will be the default image resolution that is used to create Menus, Tracks, and Slideshows in High Definition DVD projects. The options available are: 720 × 480i, 720 × 480p, 1280 × 720p, 1440 × 1080i, and 1920 × 1080i.

- *Display Mode:* This will be the default aspect ratio that is used to create Tracks and Slideshows in High Definition DVD projects. The available options are: 4 × 3, 16 × 9 Pan-Scan, 16 × 9 Letterbox, and 16 × 9 Pan-Scan & Letterbox.

> **NOTE:**
> 4 x 3 is only available in 720 x 480i mode.

Thumbnail Size

- *Palette:* Choose Small or Large thumbnails to use in the Palette.

- *Slideshow:* Choose Small or Large thumbnails to use in Slideshows.

Subtitle

- *Fade In:* Enter 0 for "instant on" or a value in frames to determine the subtitle fade in time.

- *Length:* Enter a value, in seconds, that controls the default length of new subtitles you create.

What Is DVD Studio Pro 4?

- *Fade Out:* Enter 0 for "instant off" or a value in frames to determine the subtitle fade out time.

Menu Preferences

The *Menu Preferences* pane (Fig. 2-5) controls Menu-related default options:

- *Motion Duration:* Enter a default duration, in seconds, to use when you create a motion menu (can be overriden in the Menu Inspector).

- *Auto Assign Button Navigation:* Choose whether the Auto Assign feature wraps the button navigation for left-to-right or right-to-left reading.

- *Final Rendering:* Because menus can contain a variety of elements layered over the background, such as buttons, drop zones, and text, they must be rendered into a single layer. Still menus render quickly, but motion menus can be a lengthy process. The Final Rendering settings allow you to choose how the menus are rendered:

  – *Hardware based:* This setting provides the fastest rendering. However, because the quality is dependent on your system's video card, the quality might not be as good as when using the "Software based" setting, and may not be consistent between different systems.

  – *Software based:* This setting provides consistent, good-quality rendering on all systems. However, depending on your system, it may take substantially longer than "Hardware based" renders.

- *Drop Palette Delay:* Position the slider to control how long of a delay there is before the Drop Palette appears over the Menu Editor when you drag an asset to it.

- *Video Background Color:* You can set the color used as a background by the Menu Editor when no asset has been assigned as the menu's background. Normally, this color

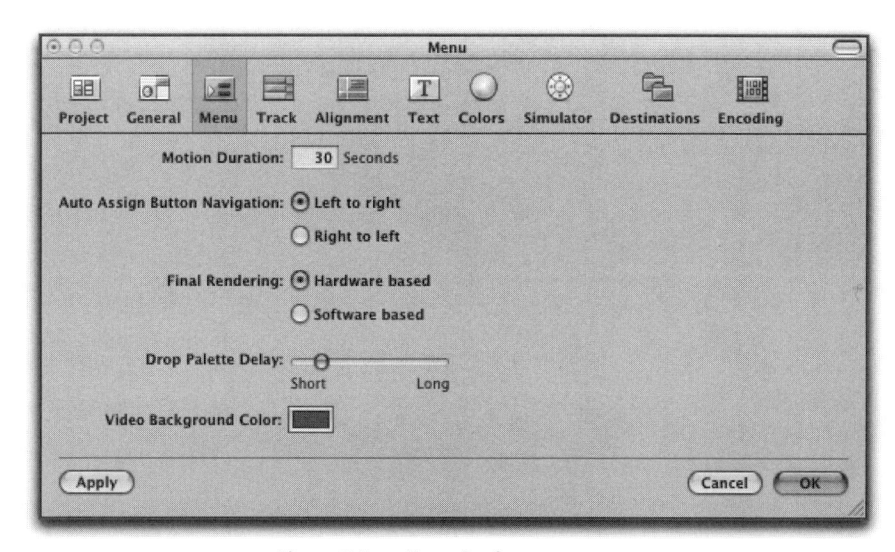

Figure 2-5 — Menu Preferences pane.

is not seen because it is covered by the menu background. It will be seen if the background image is not the correct size to fit in the video frame. In that case, this background color fills the gaps that exist.

> **IMPORTANT!**
> This setting is not saved as part of the project. If you change this setting, it will affect all projects, including previously saved ones, that have this color visible.

Track Preferences

Track Preferences (Fig. 2-6) control options encountered while creating Tracks:

- *Marker Prefix (Root) Name:* Enter the name that all new markers use as their prefix. The Generate Marker Names setting, below, controls whether this root name is followed by a number of timecode value.

    – *Check for unique name:* Select this checkbox to ensure you will not have two markers with the same name within a track.

- *Generate Marker Names:* You can choose how markers are named when you create them.

    – *Automatically:* Names new markers with the (root) name prefix (above) followed by a number that increments each time you add a marker. An advantage of this option is that you can easily tell how many markers have been added (although the number does not take into account any markers that you may have deleted). A disadvantage is that, since you can add markers between existing ones, the numerical order of the markers is not necessarily the same as their order in the timeline—the numerical order is based on the order the markers are created, not on their position.

    – *Timecode based:* Names new markers with the (root) name prefix (above)

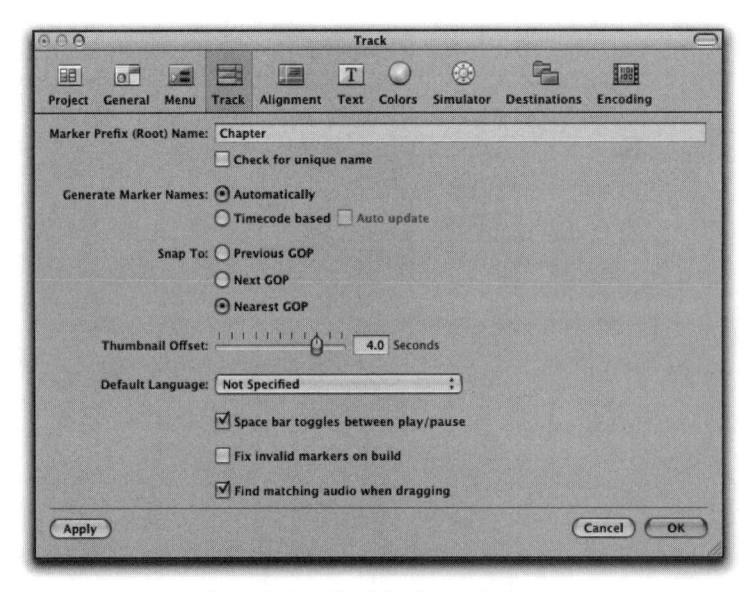

Figure 2-6 — Track Preferences Pane.

What Is DVD Studio Pro 4?

followed by the timecode of the video asset. An advantage of this option is that the marker names are always in the order they appear in the timeline. A disadvantage is that you cannot easily tell how many markers you have in the track.

> **NOTE!**
> You can rename the markers in the Marker Inspector.

- *Auto update:* When you select "Timecode based," the "Auto update" checkbox becomes available. Selecting this checkbox causes the timecode value assigned to a marker's name to update if you move the marker.

• *Snap To:* Controls where a marker will be placed in the timeline. Markers can only be placed on I-frame headers, which occur at the beginning of a Group of Pictures (GOP). A GOP is typically 12 to 15 frames long, which means that when you place a marker, you may not be able to place it on the exact frame you would like. This setting determines how DVD Studio Pro chooses the frame to use.

- *Previous GOP:* Places the marker on the GOP occurring before the selected location.
- *Next GOP:* Places the marker on the GOP occurring after the selected location.
- *Nearest GOP:* Places the marker on the GOP closest to the selected location, either before or after it.

• *Thumbnail Offset:* A thumbnail image of the video asset, representing its first frame, appears in the Video tab of the Palette, in the Video Asset Inspector that appears when you select a video asset in the Assets tab, and at the first frame of the video stream in the Track Editor. This setting allows you to determine whether the thumbnail used in those places is based on the video frame at the start of the video asset or on a frame up to five seconds later in the asset. Because video assets often start at black and fade up, this setting allows you to set the thumbnail to a frame that has video that better represents the asset.

• *Default Language:* You can choose a language that is automatically assigned to all audio and subtitle streams in new tracks. Choosing *Not Specified* leaves the stream language tag unassigned.

> **NOTE!**
> You should always set the audio and subtitle stream tags!

• *Space bar toggles between play/pause:* This setting controls what happens when you press the Space bar while viewing a track. When you play a track, the timeline's playhead follows along. With "Space bar toggles between play/pause" checked, the playhead in the track's timeline remains at its current position when you press the Space bar. When the checkbox is not selected, the playhead jumps back to where you started playing from when you press the Space bar.

Note that this setting does not affect the pause and stop controls in the *Viewer* tab—only what happens when you use the Space bar while playing a track.

37

DVD Studio Pro 4

- *Fix invalid markers on build:* While creating your tracks, it is possible to end up with markers that either are not positioned on GOP boundaries or fall outside of the V1 stream. When this checkbox is not selected, a build operation will abort if any invalid markers are detected. With the checkbox selected, a build operation will automatically reposition any markers that are not placed on GOP boundaries and will remove any markers that are outside the V1 stream.

- *Find matching audio when dragging:* When selected, DVD Studio Pro will automatically try to locate an audio file with the same name as the video file you have dragged to an element of your project. DVD Studio Pro only checks the folder the video file was dragged from. DVD Studio Pro does not try to find matching audio if you drag a mixture of video and audio files. You can press the Command key while dragging a video asset to temporarily override this setting.

Alignment Preferences

Alignment preferences (Fig. 2-7) control the measurement units and guides used in the Menu and Subtitle Editors.

Rulers

These settings allow you to customize the Menu Editor's rulers.

- *Show and Hide:* Select one of these options to set whether the rulers are displayed.

> **NOTE!**
> Alignment guides can only be dragged onto the Menu Editor if the rulers are visible.

- *Units:* Choose the units for the rulers from the pop-up menu.
  - *Pixels:* The rulers will measure in pixels.
  - *Centimeters:* The rulers will measure in centimeters, based on 28.35 pixels per centimeter.

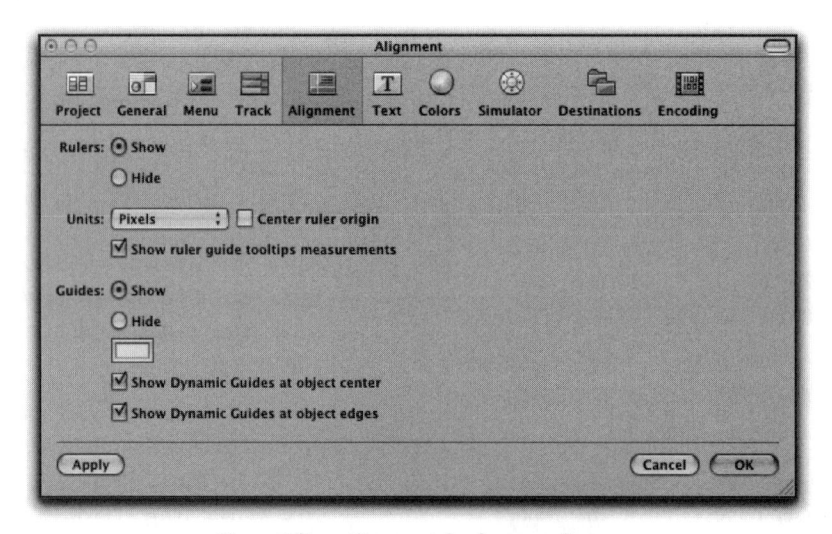

Figure 2-7 — Alignment Preferences Pane.

What Is DVD Studio Pro 4?

- *Inches:* The rulers measure inches, based on 72 pixels per inch.
- *Percentage:* The rulers measure percentage of the frame.

* *Center ruler origin:* Select this checkbox to have the rulers start at the center of the frame. This places "0" at the center of the rulers, with the units counting up from there in both directions. When this checkbox is not selected, the ruler's origin is the frame's upper-left corner.

* *Show ruler guide tooltips measurements:* Select this checkbox to have the pointer's coordinates appear when dragging an alignment guide. The values are in pixels, based on a position of 0,0 for the upper-left corner.

Guides

These settings apply to the guides that appear in the Menu and Subtitle Editors.

* *Show and Hide:* Select whether menu guides are displayed. You create menu guides by dragging in from the ruler.

> **NOTE!**
> You cannot add Menu guides in the Subtitle Editor.

* *Guide color:* Set the color for the menu guides and the dynamic guides that appear when you move an item in the Menu or Subtitle Editor. Opens the Colors window.

* *Show Dynamic Guides at object center:* Select this checkbox to have the dynamic guides that appear when you drag an item in the Menu and Subtitle Editors show lines referenced to the item's center.

* *Show Dynamic Guides at object edges:* Select this checkbox to have the dynamic guides that appear when you drag an item in the Menu and Subtitle Editors show lines referenced to the item's edges.

Text Preferences

Text Preferences (Fig. 2-8) control how DVD Studio Pro 4 handles text elements in the various editors where text is used:

* *Show:* Choose which text to configure (subtitle, menu, or menu button). Your selection determines the other settings in this pane.

* *Font:* Shows the current settings.

* *Font Panel:* Click to open the Fonts window to configure the font.

Subtitle Text Settings

> **NOTE!**
> You cannot set the subtitle text color as you can for the menu button and menu text items.

* *Horizontal and Vertical:* Use to set the default positioning of text-based subtitles you create in DVD Studio Pro. See Chapter 12 for more information.

* *Offset:* Enter values to modify the horizontal and vertical settings. For example, a horizontal offset value will move the text towards the center when "left justified" is the horizontal setting. Negative horizontal offset values move the text to the left and positive values move it to the right. Negative vertical offset values move the text up and positive values move it down.

39

DVD Studio Pro 4

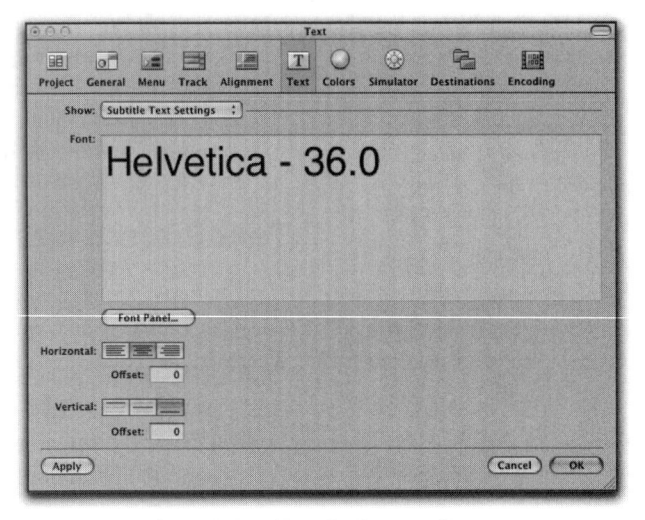

Figure 2-8 — Text Preferences Pane.

Menu Button Defaults

- *Color Panel:* Opens the Colors window for configuring the text color.

- *Position:* Choose the default position for the button's text (Bottom, Top, Right, Left, or Center).

- *Include text in highlight:* Select this checkbox to have the text included as part of the button's highlight area.

- *Motion enabled:* Select this checkbox to set the button's asset to appear as full motion, if the asset supports it.

Menu Text Settings

- *Color Panel:* Opens the Colors window for configuring the text color.

Colors Preferences

Colors preferences (Fig. 2-9) control Menu and Subtitle editor color options:

- *Show:* Choose whether to show the settings for the Menu Editor or Subtitle Editor. This affects all other settings in this pane.

- *Mapping Type:* Choose the type of overlay mapping, Chroma or Grayscale, you are using. This only applies if you are using advanced overlays.

- *Selection State:* Choose which of the three selection states (Normal, Selected, or Activated) to configure.

- *Set:* Choose which of the three color mapping sets to configure. These settings are only active when the Selection State is set to Selected or Activated.

- *Key, Color, and Opacity Settings:* Choose the color and opacity setting for each of the

What Is DVD Studio Pro 4?

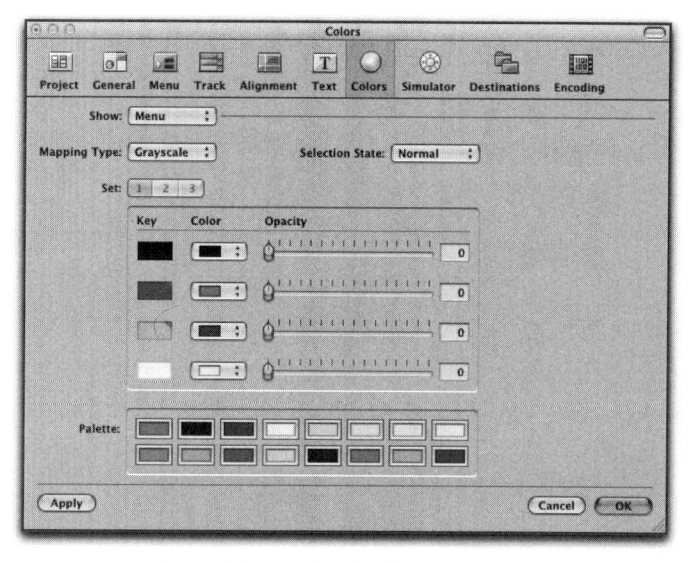

Figure 2-9 — Colors Preferences Pane.

overlay's four colors (shown in the Key column) for each selection state and set.

> **NOTE!**
> Shapes and simple overlays use the black key color (the top one) for their highlights.

- *Palette:* Shows the colors in the color mapping palette. You can change a color by clicking it and selecting a new color from the Colors window.

Simulator Preferences

Simulator preferences control options that are used to set the default Simulation environment:

- *Default Language Settings:* Choose the languages to use as the Simulator's default for audio, subtitles, and DVD Menus. This simulates the language settings in a DVD player. If you want any of these elements to use the stream settings in the Disc Inspector in place of the language settings, choose *Not Specified*.

- *Features:* When *Enable DVD@ccess Web Links* is checked, the Simulator is allowed to process any *DVD@ccess* links it encounters in the project. This is useful to verify that email and web links work correctly. See *Chapter 17* for more on DVD@ccess.

- *Region Code:* Choose the region code to simulate. You can choose All regions or one specific region.

- *Playback Output:* Controls the various new options for routing the simulation display to internal or external video or audio destinations.

    - *Video:* Choose simulator window, or Digital Cinema Desktop—yes, this means you should finally be able to preview and Simulate to an external Video Monitor, with the proper hardware.

41

DVD Studio Pro 4

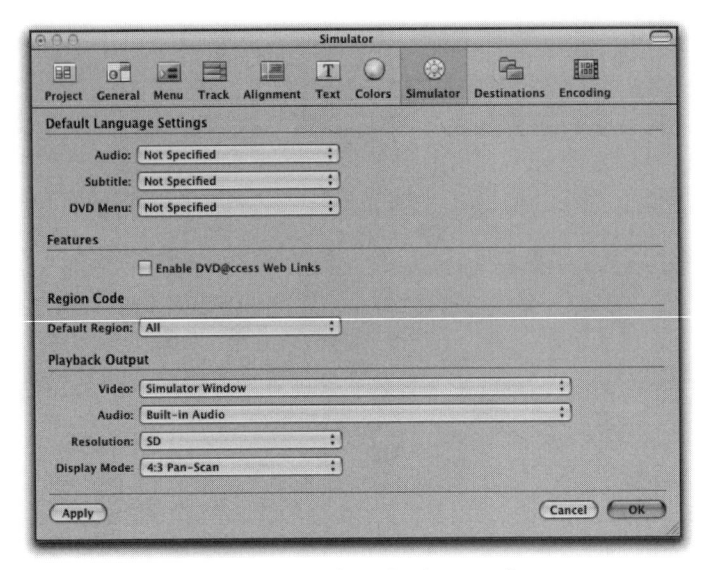

Figure 2-10 — Simulator Preferences Pane.

- *Audio:* Select Built-in analog, or Built-in Digital (S/PDIF)

- *Resolution:* Select SD, HD 720, or HD 1080. Settings of this parameter also affect:

- *Display Mode:* Select from, when available,

  - 4:3 Pan & Scan: Simulates a 4:3 aspect ratio monitor with a DVD player configured to use the pan-scan method when showing 16:9 content.

  - 4:3 Letterbox: Simulates a 4:3 aspect ratio monitor with a DVD player configured to use the letterbox method when showing 16:9 content.

  - 16:9: Simulates a 16:9 aspect ratio monitor.

Destinations Preferences

Destinations preferences (Fig. 2-11) are used to route files created during various file-generating operations in DVD Studio Pro:

- *Show:* Choose the process whose path you want to configure:

  - *Encoding:* Specifies the location for the video and audio files that are created by the embedded QuickTime MPEG Encoder when you import QuickTime files into your project.

  - *Image Encoding:* Specifies the location for the MPEG files that are created when you use non-MPEG still images in your menus, slideshows, and within a track.

  - *MPEG Parsing:* Specifies the location for the parse files that are created when you import MPEG files.

42

What Is DVD Studio Pro 4?

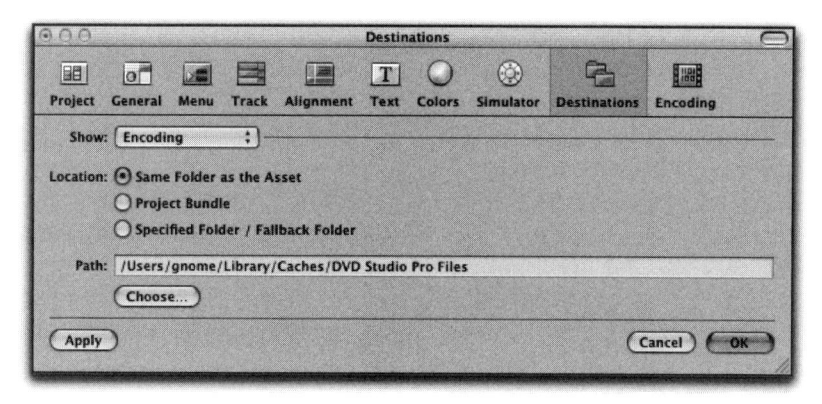

Figure 2-11 — Destinations Preferences Pane.

- *Build/Format:* Contains settings that define the default location to which the AUDIO_TS and VIDEO_TS folders, created during the build process, are saved.

- *Location:* Choose the location to save files to:

  - *Same Folder as the Asset:* This is the default setting. It places the files into an MPEG or PAR subfolder (depending on the type of file being created) in the original file's folder. For QuickTime assets, an MPEG folder is created, with a PAR subfolder. The encoded file is placed in the MPEG folder and, if necessary, the parse file is placed in the PAR folder. For assets that are already MPEG-encoded, a PAR folder is created and the parse file is placed there. If the original files are on a volume that can't be written to, such as a CD-ROM disc or a disk you do not have Write privileges for, DVD Studio Pro automatically writes to the Specified Folder/Fallback Folder location.

  - *Project Bundle:* This setting saves the files to the project file. To see the contents of the project file (the file created when you saved your project), locate the file in the Finder, Control-click it, then choose Show Package Contents from the shortcut menu. If you have not saved your project yet, the files are saved at your Specified Folder/Fallback Folder location.

  - *Specified Folder/Fallback Folder:* This setting saves the files to a disk and folder you choose. You can either enter the path directly or click Choose to open a dialog to choose the folder to use. It is also used when DVD Studio Pro is unable to write to the Same Folder as the Asset or Project Bundle locations. The default path is in your home folder at /Library/Caches/DVD Studio Pro Files. If you specify a location that cannot be written to, this default path is used in its place.

43

DVD Studio Pro 4

Encoding Preferences

The Encoding panes (Figs. 12-12a and 2-12b) contain the settings that control the internal MPEG-2 Encoder, which is used when you import a QuickTime asset.

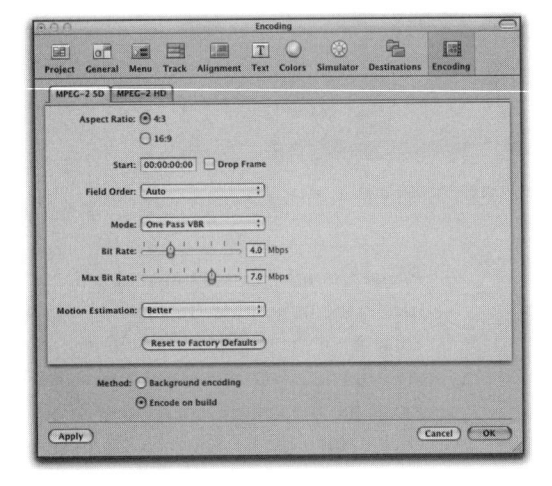

Figure 2-12a — Encoding Preferences Pane SD.

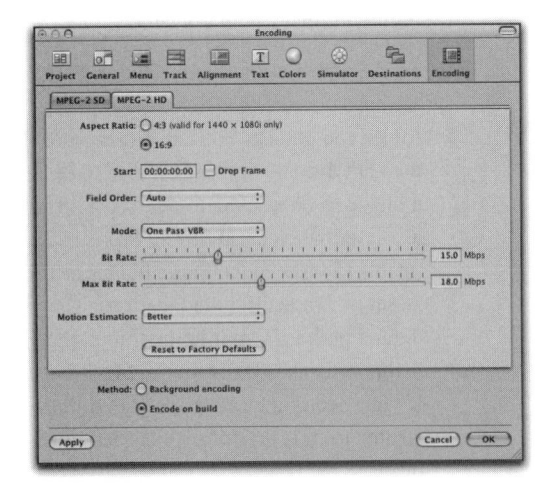

Figure 2-12b — Encoding Preferences Pane HD.

DVD Studio Pro 4 now contains an encoder to create MPEG-2 from HD resolution content. (See Chapter 5 for more information on MPEG Encoding and details on these settings.)

- *Aspect Ratio:* Select 4:3 or 16:9 as appropriate for the video content used in the project.

- *Start:* Select a beginning timecode value for the encoded file.

  – *Drop Frame:* This checkbox will force DFTC values in the stream.

- *Field Order:* Select from Auto (lets DVD Studio Pro make the determination), or Top or Bottom field first. Leave at Auto if you're not certain.

- *Mode:* Selects the encoding process desired:

  – *One Pass*—uses the Bit Rate value to encode a CBR stream.

  – *One Pass VBR*—uses both rate parameters to create an expedited VBR stream.

  – *Two Pass VBR*—uses both rate parameters to create a high-quality MPEG bitstream (this takes longer to encode).

> **NOTE!**
> Be sure to calculate the proper Bit Rate needed by Bit Budgeting FIRST!

- *Bit rate*—sets the Target or Average Bit Rate for all encodes.

- *Max Bit rate*—sets the maximum permissible encode rate for VBR encodes. Note that this parameter must be set properly for all encodes and have some rate limitations for multi-angle content!

What Is DVD Studio Pro 4?

- *Motion estimation*—sets the tracking level for motion estimation, which affects encode time somewhat. Options: Good, Better, Best.

- *Method:*
  - *Background encoding*—allows MPEG stream encoding to occur during authoring.
  - *Encode on Build*—postpones encoding until the Build operation is selected.

DVD Studio Pro 4

Summary

DVD Studio Pro 4 contains all of the features expected in a professional application. It allows you to create DVDs using many of the features of the DVD Video Specification, including multiple video Angles, alternate Audio Programs, and Graphical Subtitles.

DVD Studio Pro is not just one application, but a *suite* of applications, each of which performs some function in the DVD **workflow**. The suite includes DVD Studio Pro and Compressor 2, as well as the Apple Qmaster application to control Distributed Rendering/Encoding.

DVD Studio Pro 4 also includes a scripting language, which extends the navigation and interactivity of your DVDs, as well as a Story capability, allowing playlists of video segments to be created. The Compressor 2 application installed by DVD Studio Pro 4 allows you to control encoding of both video and audio, including Dolby Digital AC3. You can create DVDs with up to about 2 hours of video program or more, depending on MPEG encoding format and rate. You can also create DVD-9 Dual-Layer discs of up to 8.5 GB size (on DLT).

DVD Studio Pro 4 adds HD DVD authoring to its formidable arsenal of talents. Compressor 2 allows encoding in the new H.264 (AVC) format to take advantage of recent improvements in codec technology.

All SD DVD video discs are based on the MPEG-2 format (or, to a lesser extent, MPEG-1), which allows video data to be encoded and compressed at different data rates from 1.0 Mbps to 9.8 Mbps. For best results, MPEG-2 encode rates should be kept between 3.5 Mbps and 6–8 Mbps depending on the desired DVD player type. This allows for large amounts of data to be put on small 4.7-GB discs.

HD DVD discs may use MPEG-2 or the newer H.264 Advanced Video Codec (AVC), or VC-1, for which there is not currently a Mac encoding solution. In order to accommodate the larger screen resolutions, encode rates of up to 29 Mbps may be used in HD encoding.

DVD Studio Pro builds your DVD using these project *Elements*:

- **TRACKS**—These are the containers for the MPEG Video files, with Audio Streams, and optional Subtitle streams. (See Chapter 7.)

- **MENUS**—Menus are the containers that display the choices available for the user to select. Menus also display button highlights, to give the user feedback. (See Chapter 9.)

- **SLIDESHOWS**—Slideshows contain sequences of graphics images that may or may not have audio tracks accompanying them, or an overall audio file. (See Chapter 10.)

- **SCRIPTS**—DVD Studio Pro has a proprietary scripting language to give you access to the DVD Command set, which uses System Parameters and General Parameters to control the DVD player. (See Chapter 13.)

All in all, DVD Studio Pro 4 is an amazing system for a modest investment.

In the coming chapters, we will look at each of these *Elements* in greater depth as we learn about DVD Studio Pro Authoring methods.

Chapter 3

The Workspace

DVD Studio Pro 4

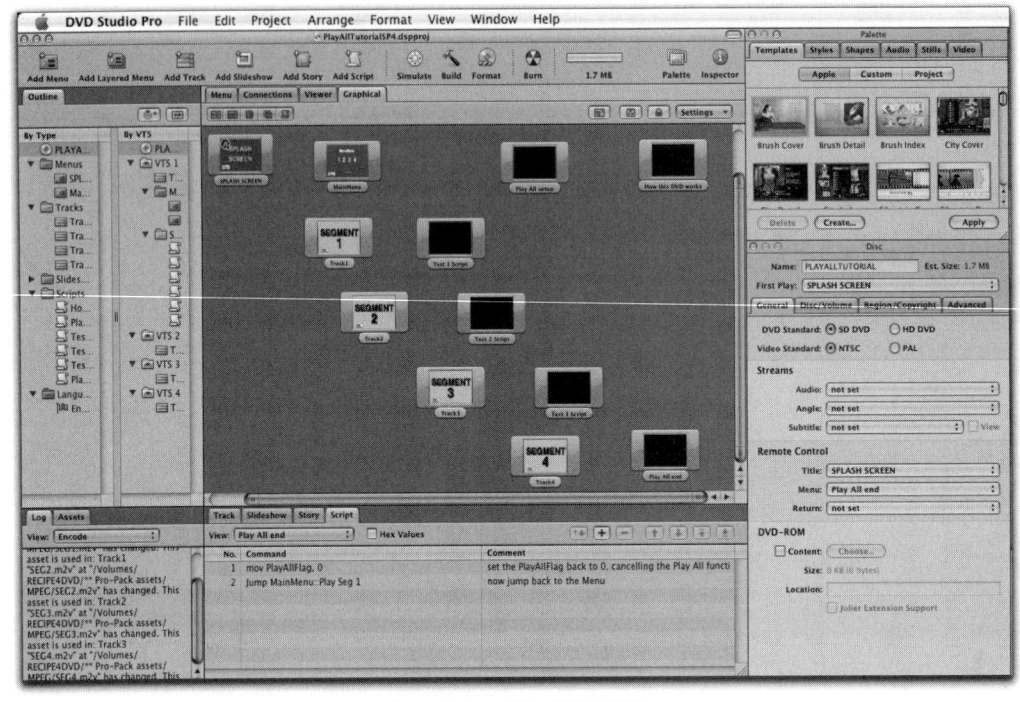

Figure 3-1 — The DSP 4 GUI.

Goals

By the end of this chapter, you should be able to:

- Understand the DVD Studio Pro 4 workspace
- Configure and customize the workspace layout
- Navigate throughout the workspace
- Understand *Quadrants* and *Tabs*
- Understand how *Quadrants* and *Tabs* interact
- Understand how to resize *Quadrants*
- Understand how to move *Tabs*
- Understand how to tear-off *Tabs* into *Windows*
- Understand how to restore *Tabs* to their *Quadrants*
- Understand the role of the *Palette*
- How to locate assets in the *Palette* and Author with them
- How to add and remove folders and assets in the *Palette*
- Understand the new role of the *Inspector*
- View Properties in the various *Inspector* displays
- Re-order Items in the *Outline View Window*

The Workspace

- View item allocation in the VTS display
- Name and rename Items using the *Property Inspectors*

The DVD Studio Pro 4 Workspace

If you are, for whatever reason, only just now upgrading from DVD Studio Pro 1, you're in for a big readjustment. In that version, the workspace had a fairly standard window configuration and, although the windows could be moved around on the desktop and resized, there was little in the way of serious customization that could be done.

This changed dramatically in DVD Studio Pro 2. With the exception of the *Inspector*, which remained an independent display, and the addition of a similar independent display, the *Palette,* all of the other display functions were combined in a Main Window. This window can be divided into 1 to 4 segments, called *Quadrants*. Within each *Quadrant, Tabs* identify selectable subdisplays, which can further organize the data being displayed, allowing a great deal of information to be made available in a small but controlled screen space. The entire workspace is organized as a *Configuration*, of which there are infinite possible variations. This layout has remained consistent since then, although I was pleased to note the re-emergence of the *Graphical View* window in DVD Studio Pro 3, a very useful authoring tool.

The biggest change in DVD Studio Pro 4 is the addition of the *VTS Editor* to the outline View tab. We will cover the VTS Editor in due course in this chapter.

Using the Interface Configurations

Each screen Configuration in DVD Studio Pro 4 uses a combination of *Quadrants* and *Tabs* to organize the authoring data in the Main Window, and to show or hide the *Inspector* and *Palette*. The beauty of this approach is DVD Studio Pro's ability to completely customize the workspace display to each user's exact specifications, save it that way, and easily recall it at any time—even at the touch of an F-key!

Of note is the ability to resize each *Quadrant* within the window and rearrange the order and placement of Tabs contained within each *Quadrant*. Complete personalization is easily possible with this much flexibility, and especially for multiple users of the same Macintosh. We'll cover *Quadrants* and *Tabs* in depth once we have covered the *Default Configurations— Basic, Extended*, and *Advanced*.

Sizing the Main Window for Best Effectiveness

The DVD Studio Pro workspace is best viewed at a screen resolution of at least 1024 × 768 (or better). Using a smaller sized screen will cramp the display, cutting off some important features. Users with Studio Displays, Cinema Displays, or even PowerBooks will benefit from the ability to create configurations specifically designed to take advantage of their particular screen sizes.

Users with two displays (especially two widescreen displays), will be able to take special advantage of the ability to preview and simulate HD content in true widescreen resolution, or preview to an external video monitor (see "Setting the Simulator Preferences," in Chapter 14 on page 387).

DVD Studio Pro 4

All users will benefit from the ability to "tear-off" *Tabs* from their original *Quadrants* to create free-standing windows, moveable around the desktop.

The *Basic Configuration* is a good place to start tinkering, especially if you have just upgraded from iDVD to DVD Studio Pro, because this Configuration most resembles iDVD's workspace and workflow. If you are familiar with previous versions of DVD Studio Pro, you may find the *Extended* and *Advanced* configurations to be a little more familiar territory.

The Basic Configuration

In the *Basic Configuration* (Fig. 3-2), the workspace defaults to a unified view containing one window (showing the Menu Editor Tab) on the left side, and the *Palette and Inspector* displayed on the right (we'll have more specifics on the *Menu Editor*, *Palette*, and *Inspector* later in this chapter). If you look closely at Figure 3-2, you will see that at the top of the Menu Editor window are *Tabs* (Menu, Slideshow, Viewer), similar to those found in a number of other recent Apple applications. These Tabs are used to organize different data displays, which use the same Window space, conserving the screen space. The media *Palette* and context-sensitive *Inspector* are displayed on the right side.

The Extended Configuration

In the *Extended Configuration* (Fig. 3-3), the workspace becomes a lot more informative and complex. This is the first display in which the *Quadrants* are clearly visible.

Figure 3-2 — The DVD Studio Pro 4 Basic Configuration.

The Workspace

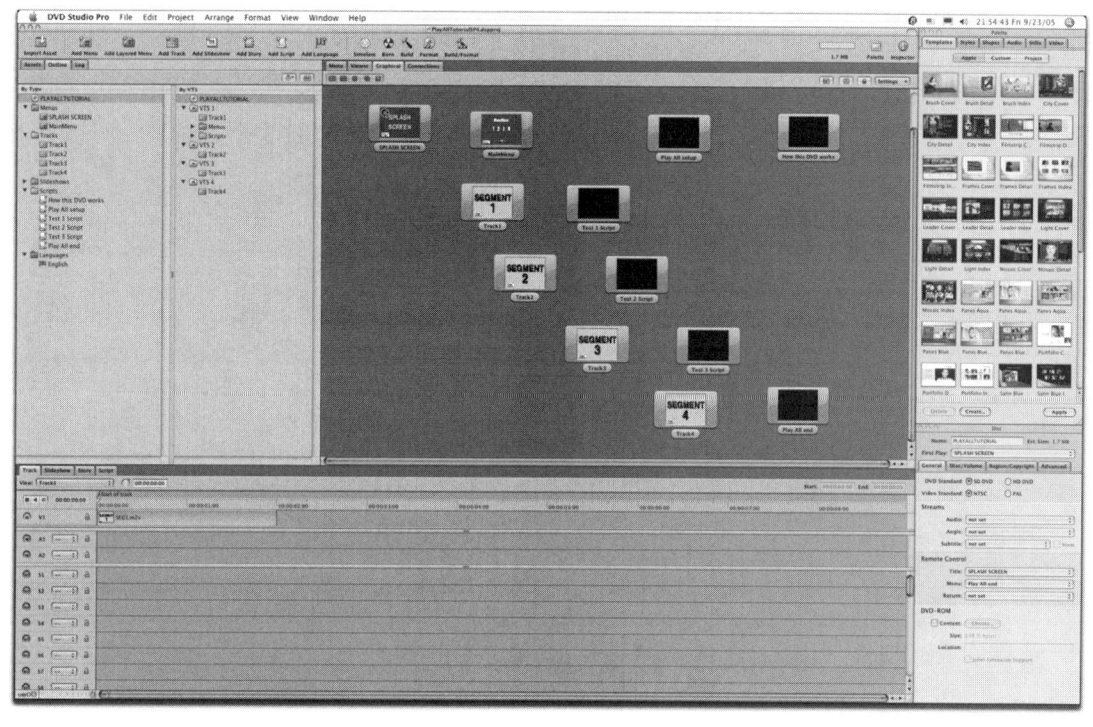

Figure 3-3 — The DVD Studio Pro 4 Extended Configuration.

In this configuration's default setup, the Main Window is on the left of the screen, with three *Quadrants* visible. The upper left Quadrant shows the *Assets, Outline,* and *Log Tabs*, while the upper right Quadrant shows the *Connections, Menu,* and *Viewer* Tabs. The bottom half of the window shows only the lower Right Quadrant, which contains the Tabs for the *Track, Slideshow, Script,* and *Story* editors, respectively in the default configuration.

In the default Extended view, the Lower Left Quadrant is hidden, but it can be revealed at any time. The media *Palette* and context-sensitive *Inspector* are displayed on the right side. The Track Editor Tab works with a timeline, and looks similar to edit timelines in Final Cut Pro (looks can be deceiving, though!). In the *Track Editor,* you have access to areas for up to 9 Video Angles, 8 Audio streams, and 32 Subtitle streams. You can also create *Markers* for *Chapters, Cells* and the *Layer Break,* and *Subtitle* events directly in this Editor. We'll cover the *Track Editor Timeline* and the other Tabs in the Lower Quadrants (*Script, Story,* and *Slideshow* editors) in greater detail further on in this chapter, and in specific detail in other chapters.

The Advanced Configuration

In the *Advanced Configuration* (Fig. 3-4), the bottom half of the window splits to reveal its two *Quadrants*, retaining the *Track, Slideshow,* and *Script* Tabs in the lower right Quadrant while the lower Left Quadrant adds the Tab for the *Log* window and relocates the *Assets* Tab down from the upper left Quadrant. The *Story* Tab is moved to the upper left Quadrant to

51

DVD Studio Pro 4

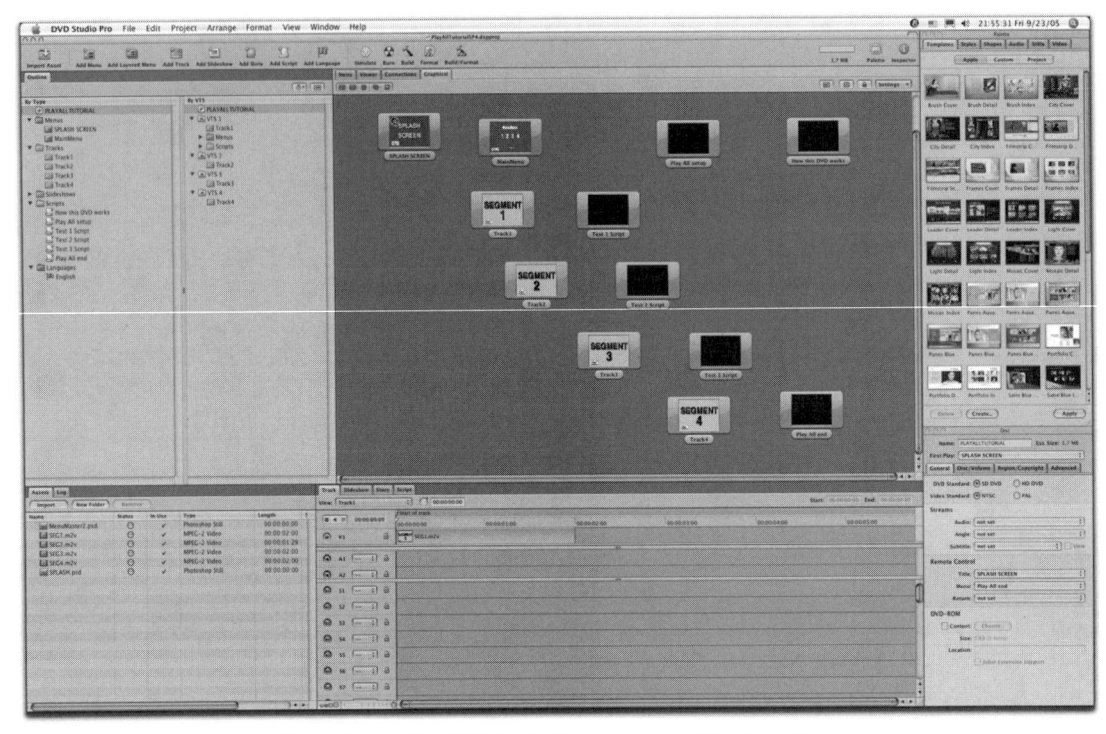

Figure 3-4 — The DVD Studio Pro 4 Advanced Configuration.

replace the now-relocated *Assets* Tab (although I personally prefer moving it to the upper right Quadrant for better readability). Once again, the *Palette* and context-sensitive *Inspector* are displayed on the right side.

The *Advanced Configuration* is the most elaborate of the default displays, but you may still perform further customization of the workspace and save highly personalized configurations as you see fit.

The Inspector

The *Inspector* in DVD Studio Pro is by now a familiar friend, being a much-improved version of DVD Studio Pro 1's *Property Inspector*. The *Inspector's* function is to give you detailed information on any project element or asset that is currently selected.

The Inspector is quite powerful and very informative. The Inspectors now have Tabs as well, to allow for even greater detail of display. We'll cover the Inspectors (and there are lots of them) in greater detail a bit later on. (See Fig. 3-5.)

Showing and Hiding the Inspector

The *Inspector* is an independent display, and has a higher display priority than the authoring window—when it is actively displayed, it will always be visible on top of the window. It may be revealed or hidden in any Configuration at any time, as follows:

- Select **View > Show Inspector** in the menubar (it might say **Hide Inspector**), *or*

52

The Workspace

- Type keyboard shortcut *Command-Option-I*: (⌘-**Opt-I**).

Both the menu command and keyboard shortcut operate as flip-flops; if the Inspector is hidden, the menu command will say **Show Inspector** and selecting it will reveal it; if it is visible, the command will say **Hide Inspector** and selecting it again will hide it.

Styles, etc.) and your own DVD assets, whether they are already encoded, or in their natural pre-encoded state (movies, audio files, pictures, etc.). The Palette can organize your DVD authoring assets into one convenient location, if you choose to use it that way. We'll cover the Palette in detail a bit further on in this chapter, but right now let's see how to show it and hide it.

Showing and Hiding the Palette

Like the *Inspector*, the *Palette* is also an independent high-priority display. (See Fig. 3-6.) It may also be revealed or hidden in any Configuration at any time. Here's how:

- Select **View > Show Palette** in the menubar (it might say **Hide Palette**), *or*

- Type Keyboard shortcut *Command-Option-P* (⌘-**Opt-P**).

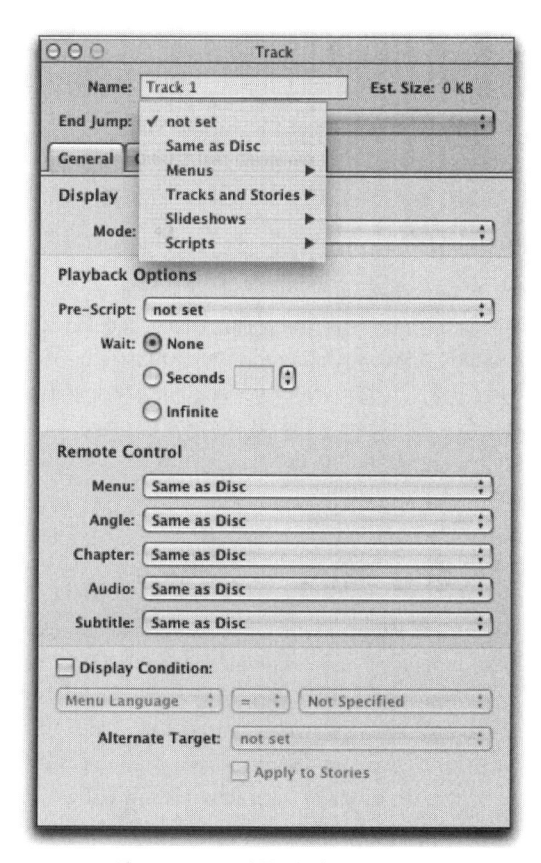

Figure 3-5 — A Typical Inspector.

The Palette

The *Palette* was a new concept with DVD Studio Pro 2, a sort of über-browser and collecting place for both the Apple-supplied media (Templates, Shapes,

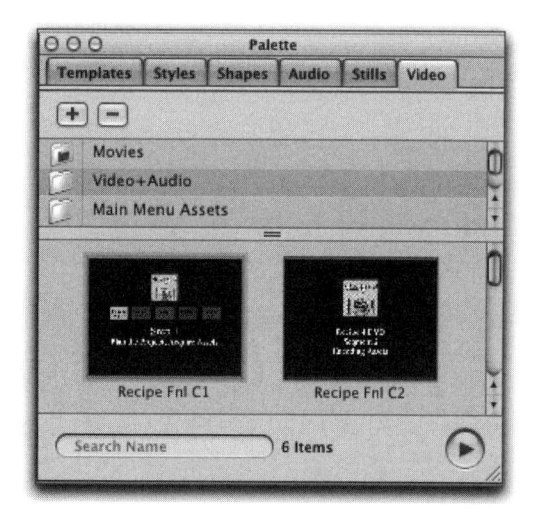

Figure 3-6 — The Palette.

As with the Inspector, the menu command and keyboard shortcut both operate as flip-flops; if the

53

DVD Studio Pro 4

Palette is hidden, the Command will say **Show Palette** and selecting it will reveal it; if it is visible, the Command will say **Hide Palette** and selecting it again will hide it. If both the Inspector and Palette are visible, they may overlap each other, and may require rearrangement for each to be clearly visible. Both the Inspector and Palette have higher viewing priority than the main window and will always be visible on top.

Using Windows, Quadrants, and Tabs

One of the great strengths of DSP's approach to the workspace is the degree of total customization that is now possible. The default configurations only hint at this, so let's explore it in greater depth.

(This section will be easier to understand if you are looking at the Advanced configuration, since it already uses a four-quadrant display. If you don't have your display in the Advanced configuration, please select that default configuration now.)

- Select **Window > Configurations > Advanced** in the menubar, or
- Type **F3**

Reducing or Enlarging the Main Window Size

The Main Window has a traditional size tool at the lower right corner (the crosshatched edge—see Fig. 3-7a). Click there and you can drag the window to scale its size larger or smaller, as you desire. Note that the *Quadrants* inside the window will automatically scale in size as the window resizes.

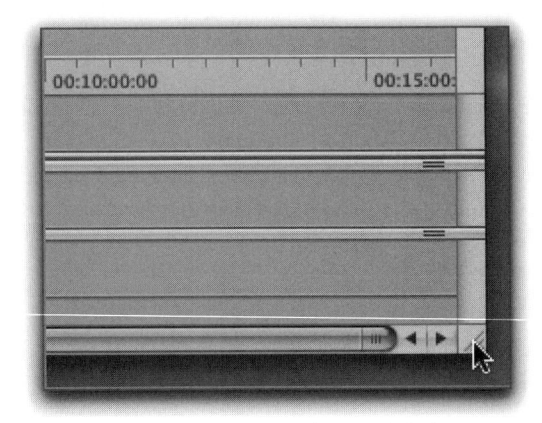

Figure 3-7a — Resizing the Main Window.

Quadrants Divide the Main Window

The Main Window can be divided into as many as four *Quadrants*, with both Horizontal and Vertical boundary lines separating the *Quadrants* (refer back to Fig. 3-4). *Quadrants* are not fixed in size, and may be resized by placing the cursor in the correct location, and dragging up, down, left, or right. There are three possible methods to resize:

Resizing Window Halves

- **Left/Right:** Place the cursor on the vertical boundary and it will change into the left-right cursor (see Fig. 3-7b), indicating resizing may be done left to right. Drag left or right as needed to resize the relative widths of the two vertical window halves.

- **Up/Down:** Place the cursor on the horizontal boundary and it will change into the up-down cursor (see Fig. 3-7c), indicating resizing may be done up and down. Drag up or down to resize the relative heights of the upper and lower halves of the window.

The Workspace

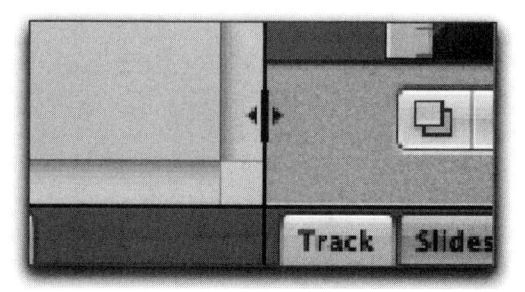

Figure 3-7b — The 2-way Left/Right cursor.

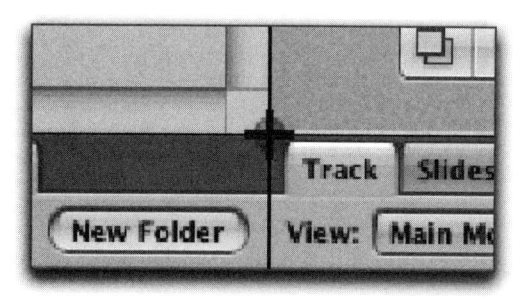

Figure 3-8 — The 4-way cursor.

Figure 3-7c — The 2-way Up/Down cursor.

Resizing All Quadrants Simultaneously

Place the cursor on the center boundary of all four *Quadrants* and it will become the four-way cursor (see Fig. 3-8), indicating resizing of all four Quadrants is possible. Drag up, down, left, or right, or a combination of these directions to resize all four Quadrants simultaneously. Of course, as some Quadrants grow larger, others will shrink to compensate. The overall size of the Window will not change.

Resizing Quadrants Independently of Each Other

You've probably noticed that if you move Quadrants up-down or left-right, ALL of the Quadrants in the Main Window are being resized at one time. Sometimes, it may be desirable to unlink the Quadrants from each other, to allow for independent resizing. This is easily done.

Unlinking the Quadrants from Each Other

Place the cursor on any Quadrant's horizontal or vertical boundary with its neighbor, and move it while holding the Option key; the selected boundary will unlink and can be moved independently from the other Quadrants. Once unlinked, you need not use the Option key to move it again. For example, using the Advanced configuration:

- Hold the Option key down and place the cursor on the horizontal boundary between the Upper-left and Lower-left Quadrants (see Fig. 3-9a). You will be able to move that boundary up and down without affecting the boundary between the Upper-right and Lower-right Quadrants. (See Fig. 3-9b.)

55

DVD Studio Pro 4

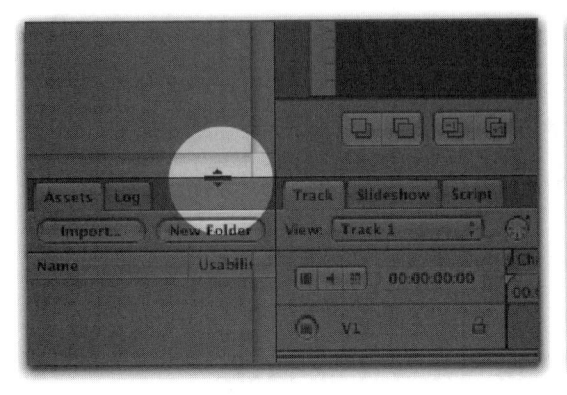

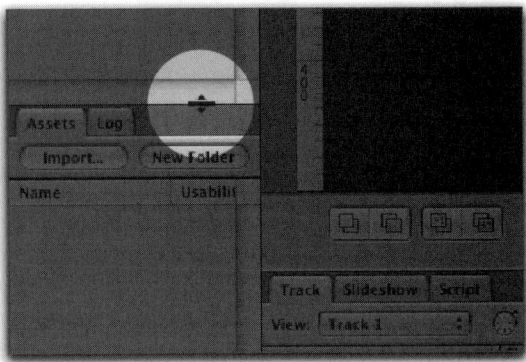

Figure 3-9a and 3-9b — Before and After: Unlinking the horizontal boundary.

Once unlinked from each other, the two horizontal boundaries will now work independently of each other, and you can freely adjust the sizes of the Quadrants on the right without changing the size of the Quadrants on the left.

Dragging an "Unlinked" Quadrant Boundary

Place the cursor on the Quadrant boundary between the upper-right and lower-right *Quadrants* (containing the Menu and Track editors) and notice that the boundary between them may now be moved without affecting the size or position of the Quadrants on the left side (see Fig. 3-10).

Hiding and Revealing Quadrants

Whether or not you have "unlinked" the Quadrants, you will find that dragging the boundary lines will allow you to resize the Quadrants to make them disappear from the Window (temporarily, if you wish).

Dragging to Hide a Visible Quadrant

Place the cursor on the boundary line between the upper-left Quadrant and lower-left Quadrant and drag all the way down to the bottom of the Window. The lower Quadrant will disappear, leaving only the Upper Quadrant (usually showing the Outline Tab—see Fig. 3-11a).

To restore visibility of the now-hidden Assets and Log Tabs, you can drag the Quadrant boundary in the reverse direction (see Fig. 3-11b).

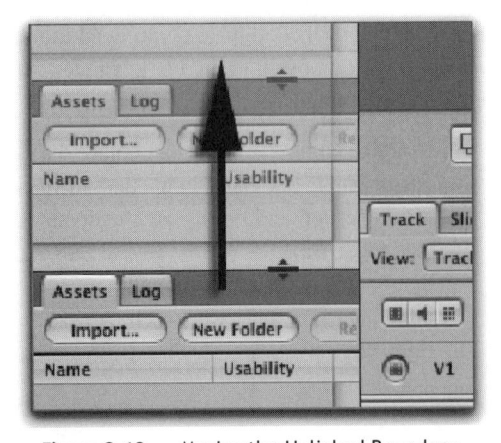

Figure 3-10 — Moving the Unlinked Boundary.

The Workspace

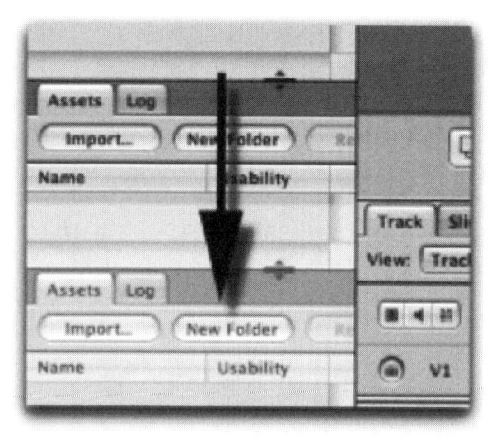

Figure 3-11a — Dragging a Quadrant to hide it.

Dragging to Reveal a Hidden Quadrant

If you are in doubt as to which side is hidden, put the cursor on an edge, *any* edge, and check to see if the cursor changes shape. If it does, drag towards the opposite side of the window to reveal the hidden Quadrant.

When only one Quadrant is shown, placing the cursor on the edge of the Quadrant window will determine if an invisible Quadrant exists at that edge. If you see an arrow cursor, a hidden Quadrant is waiting there to be revealed.

Figure 3-11b — Dragging a Quadrant to reveal it.

> **NOTE!**
> Tabs will not reveal themselves when the Quadrant is hidden.

While in this hidden state, the Command-Key shortcuts for the Track (⌘-**9**), Slideshow (⌘-**7**), and Script Tab (⌘-**6**) will *not* force the Quadrant to open to reveal the selected Tab. This works the same for all hidden Tabs, at all times.

Figure 3-12 — Top Left Quadrant waiting to be revealed (basic view, cursor at left edge).

- If the vertical Quadrant boundaries are unlinked, you will have independent control of this function in each half (top and bottom) of the window.

57

DVD Studio Pro 4

- If the top half of the window is currently visible, place the cursor at the bottom edge of the window. The cursor will change to the Up/Down cursor, and you can push up to reveal the hidden Quadrant containing the Track Tab. If the bottom half of the Window is visible, put the cursor at the top edge of the Window, just below the Toolbar. The cursor will change to the Up/Down cursor, and this time you can push down to reveal the hidden Quadrant containing the Menu Editor Tab.

- If the horizontal Quadrant boundaries are unlinked, you will have independent control of this function in each half (left and right) of the window. In these cases, you will be moving the cursor left or right to reveal the hidden Quadrant.

Unlinking One Boundary Doesn't Unlink the Other!

If you have unlinked the horizontal boundaries, as in our example above, you will notice the vertical boundaries are still linked, and any left-right moves will affect all *Quadrants* at the same time. This also works in the same manner if you have unlinked the vertical boundaries—the horizontal boundaries will remain linked, and all *Quadrants* will adjust simultaneously to any up-down moves you make. (See Figs. 3-13a and 3-13b.)

Dragging "Linked" Quadrants

- Assuming you still have the horizontal boundary unlinked from our previous demonstration, place the cursor on the vertical Quadrant boundary between the upper-left and upper-right Quadrants (between the Outline View and the Menu Editor)—the left-right cursor will appear—notice that this boundary may only be

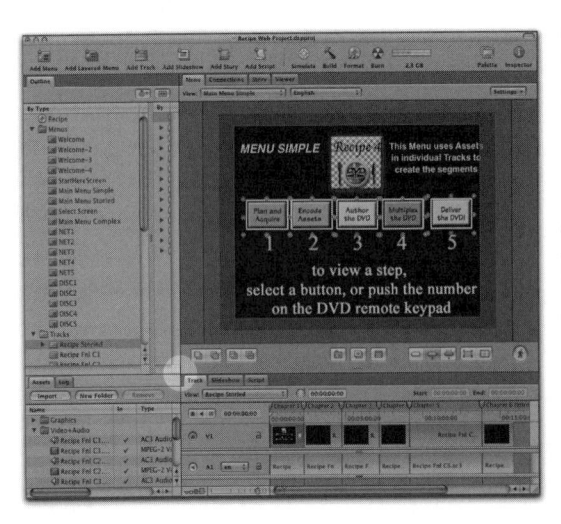

Figure 3-13a — Resizing all quadrants—before.

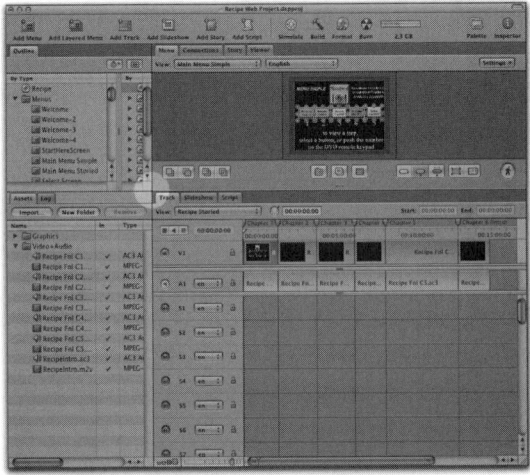

Figure 3-13b — Resizing all Quadrants—after.

The Workspace

moved left or right, affecting both halves of the Window simultaneously.

Re-Linking the "Unlinked" Boundaries—Returning to Normal

Should you need to restore the Quadrants to their original linked configuration, simply drag the unlinked boundary back into alignment with the other similar boundary and they will snap back together again. If you experience any difficulty in this, try moving the unlinked boundary all the way down or up, left or right so as to hide that Quadrant, then reveal it again by dragging in the opposite direction. Once the Quadrant is again visible, it should snap in to place.

Moving Tabs around in the Workspace

The Default Configurations are just that—defaults. These layouts are not etched in stone, by any means. And that goes for both Quadrants and Tabs. Tabs exist in an area of the Quadrant called the *Tab Bar*, along the top edge of the Quadrant. Tabs can easily be moved from Quadrant to Quadrant, or even moved within *Quadrants*, allowing you to put the functions exactly where YOU want them to be. Tabs can be moved by Dragging between *Quadrants*, and also by Control-Clicking in the destination Tab Bar. Control-Click is a context-sensitive function that behaves differently depending on where you Control-click in the Main Window.

Moving Tabs between Quadrants by Dragging

Dragging a Tab from the Tab Bar of one quadrant to another will relocate that Tab into the selected Quadrant. As an example, try this:

- Using the default Advanced configuration, place the cursor on the Story Tab (upper-left) and drag it to the Tab Bar of the Lower-right Quadrant (put it to the right of the Tabs for Track, Slideshow, Script).

- Once you are within the target zone, a highlight rectangle will appear, and you can release the cursor, dropping the Tab into that Quadrant's display, joining with the existing Tabs. (See Fig. 3-14.)

- You will find that you can place the Tab exactly where you want it to land, and the existing Tabs will adjust to make way for the incoming Tab.

Figure 3-14 — Dragging a Tab to a new home.

DVD Studio Pro 4

Moving Tabs between Quadrants Using Control-Click

- You can force the relocation of any Tab by Control-clicking in the Tab bar of the Quadrant you wish to move it to. Because the Control-Click function is context sensitive, a pop-up dialog will appear, listing all Tab names—you may select the Tab you wish to relocate to that Quadrant. That Tab will relocate to that Quadrant's display, taking its place to the right of the existing Tabs. This is also a very efficient way of quickly relocating a Tab to a location in a Quadrant after it has been operating as a freestanding Window. (See Fig. 3-15.)

Figure 3-15 — Moving or revealing Tabs using Control-Click.

Turning Tabs into Freestanding Windows

Tabs are useful devices—they can organize and display data from a variety of screens and editors within SP2. Occasionally though, it's better to access a Tab in its own freestanding Window (the Assets Tab comes to mind here). This is very easy to accomplish, using one of the following methods:

Tearing Off Tabs to Become Windows

- Click and hold on any desired Tab.
- Drag the Tab away from its home in the Quadrant and drop it anywhere on the window <u>except</u> in the Tab bar of another Quadrant.
- The torn Tab is now a freestanding Window of the same display.

Or, as an **alternative**,

- Control-Click on a Tab and select "Tear Off Tab" in the Menu.

Restoring a Torn-Off Tab to a Quadrant

- Click on the desired Tab in its current Window location, and drag it back to the Tab bar of any desired Quadrant.
- Release the Tab, and it will relocate to that Quadrant.

Or, as an **alternative**,

- Control-Click in the Quadrant's Tab bar and select the Tab you wish to place in that Quadrant.
- Release the mouse and the selected Tab will appear there.

The Workspace

Using Keyboard Shortcuts to Select/Display Tabs/Windows

While Tabs are at home in their Quadrants and visible, these Keyboard Shortcuts can be used to bring them to the foreground. (See Table 3-1.) The Command-Key + number syntax is simple and consistent. Executing a Command-Key combination will bring that Tab to the foreground in its Quadrant. If the Quadrant containing the desired Tab is NOT currently visible, nothing will happen when the Command-Key is used.

Table 3-1 Command-Keys combinations and Their Tab Equivalents

- ⌘-1—displays Assets Tab
- ⌘-2—displays Connections View
- ⌘-3—displays Log Tab
- ⌘-4—displays Menu Editor Window
- ⌘-5—displays Outline View
- ⌘-Option-5—displays Graphical View
- ⌘-6—displays Script Editor
- ⌘-7—displays Slideshow Editor
- ⌘-8—displays Story Editor
- ⌘-9—displays Track Editor
- ⌘-0—displays Viewer

Using the Window Buttons—Close, Minimize, Maximize

A torn-off Tab acts like any other window under OS X—using the window controls (the Red, Yellow, and Green dots in the upper left corner), you may close (•) the window, minimize (–) it down into the Dock, or maximize (+) the window to fullscreen.

There are some rules to these functions:

- A *minimized* window will go to the Dock immediately. It will return to its original location and size in response to its Command-Key.

- A *maximized* window will expand to the entire size of the screen, including the space occupied by the Inspector and Palette. Maximizing it a second time will return it to its former size and location.

- A *closed* window is just that—closed. It is not forgotten, though—if you use the Command-Key shortcut of that Window, it will reappear in the foreground. If you look behind the Main Window, you will usually find it hiding there, waiting to pop up and be of service again.

When Tabs are torn-off from their original Quadrants and acting as freestanding windows, the Keyboard Shortcut keys can bring that window to the foreground if it is hidden. Keyboard Shortcuts can also bring a Minimized window up into the foreground from the Dock, if it has been parked there.

> **POWER USER TIP!**
> If wrangling Tabs is driving you crazy, here's an alternative that gives you access to large windows:
> - Tear-off a Tab to make a freestanding Window of that function.
> - Maximize it with the (+) button.
> - Immediately minimize it into the Dock using the (-) button.

Now when you use the Command-Key Shortcut for that window, it will immediately pop up from the Dock, and appear full screen. Work to your heart's content, then minimize it again to put it to sleep.

DVD Studio Pro 4

> **BE CAREFUL!**
> Be aware that while the Close Button behavior will work normally for torn-off tab windows, and will usually move them to the background behind the Main Window, the *Main Window Close Button* will try to **close the current DVD Project**! Luckily though, it will give you a Save dialog if your project isn't currently saved.

Using Workspace Configurations

As you create or modify your own workspace configurations, you may wish to save them for future use. Once you have suitably modified a configuration, use the `Window > Save Configuration` command to save it. In the dialog box that appears, enter a new or modified name for this configuration, and click Save. The modified configuration will be added to the existing configurations and you can easily assign an F-Key to that configuration for easy access.

Managing Configurations

You may also manage and organize your configurations with the `Window > Manage Configuration` command. Once the dialog box has been opened, do the following:

- To add a configuration, Click the Add (+) icon.

- To delete a configuration, select it and click the Delete (–) icon.

- To rename a configuration, double-click it, and enter a new name.

- To apply a configuration, select it and click the *Apply* button.

- Click *Done* when finished, to close the dialog box.

Recalling Workspace Configurations Quickly

Both default and customized Configurations may be recalled using the F-keys (F1 through F15) located along the top of the Mac keyboard. You can even do this while you are working on a project, and the new workspace configuration will appear with your current project alive and well!

> **NOTE!**
> Do not confuse F-keys with Command-Key shortcuts [⌘-1, ⌘-2, etc.], which are used to show or hide the various Tabs (or their torn-away Windows).

If you're just getting to know to DVD Studio Pro 4, you will want to spend a minute becoming familiar with the operation of these two sets of keystrokes—while the Command-Key shortcuts are familiar to users of DVD Studio Pro 1, their function in DVD Studio Pro 4 is somewhat different, and could cause a bit of confusion. Specifically, if a *Cmd+Number* (like ⌘**-1**) combination is pressed, and the Window Quadrant containing that Tab is not currently open, neither the Menu Command or the Keyboard Shortcut will force the visibility of that Tab. If it is invisible, it will remain invisible, unless the Quadrant is opened to display that Tab.

Like Final Cut Pro, DVD Studio Pro's user interface has been designed so that everything is accessed from these convenient Quadrants and Tabs in the Workspace. Almost everything you need to do to author a professional DVD can be done with simple drag-and-drop, or point-and-click ease. Well, ALMOST everything…once in a while you may have to type in a name or a few characters! With a little forethought, you can design your ideal workspace configuration and each user can customize his own "perfect world" without affecting any other user's world.

The Workspace

The Palette—A Virtual Media Browser

The Palette is a tool that was new in DVD Studio Pro 2, so if you already know DVD Studio Pro 2 or 3, you can skip ahead—if not, let's take a moment and get acquainted with its function—many of your DVDs can begin life with elements found here.

The Palette (Fig. 3-16) is a window used to organize and access both presupplied media and your own DVD authoring assets. It also features the now-familiar Tabs, each one giving access to one of six "media organization areas," if you will.

Figure 3-16 — The Palette's Tabs.

Templates, Styles, and Shapes

The first three of these areas, *Templates*, *Styles*, and *Shapes*, contain many preset elements provided by Apple to get you going making DVDs quickly.

- *Templates* are similar to Themes in iDVD, but with much more potential. *Templates* contain prebuilt background graphics, buttons, shapes, drop zones, and can even have button commands and linkages prebuilt. You can access Apple's supplied templates, and can also save your own Custom Templates. Apple and Custom Templates are globally available for any new DVD project. Project Templates are only available within the project file where they were saved (or copies). The installed DVD Studio Pro 4 bundle contains lots of

beautiful, professionally created menu template designs, in thematically related sets. This allows you to create professional-looking themed DVDs quite easily. The inclusion of customizable graphics Drop Zones and buttons means you can take a prepared template design as a working beginning, and add your own graphics, text, and motion elements (see Fig. 3-17).

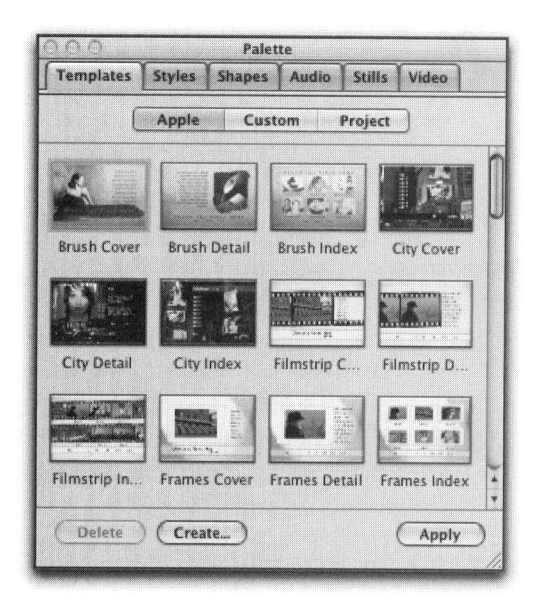

Figure 3-17 — The Templates Palette, showing small thumbnails.

- *Styles* give you a quick way to select from a large selection of preset *Buttons*, *Type Styles* (Fonts with attributes), *Drop Zones*, and *Layouts*. Used within the Menu Editor, these style building blocks allow you to develop a graphically rich menu without depending on an external graphics application (see Figs. 3-18a and 3-18b).

63

DVD Studio Pro 4

Figures 3-18a and 3-18b — The Styles Palette—Buttons (L), Layouts (R).

- The *Shapes* Tab (see Fig. 3-19a) provides a convenient location to organize all of the Shapes and Patches items you can use in DVD Studio Pro. *Shapes* are a library of prepared lines, arrows, colors, frames, and other graphics "tidbits" that can be used in the Menu editor. You can create your own Shapes using Photoshop or any similar layer-based pixel graphics application. *Patches* are Shapes that have visual effects, blurs, color tints, and even motion elements or effects built in, easily giving rich visual effects to a menu.

Video, Audio, and Stills

The three remaining Tabs give you access to your *Video, Audio*, and *Stills* files—DVD asset elements that can be stored on and made available from your Mac. To make these asset elements easy to find and

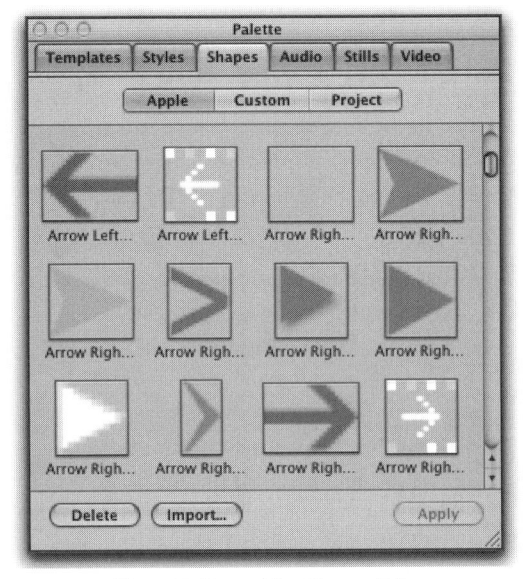

Figure 3-19a — The Shapes Tab.

The Workspace

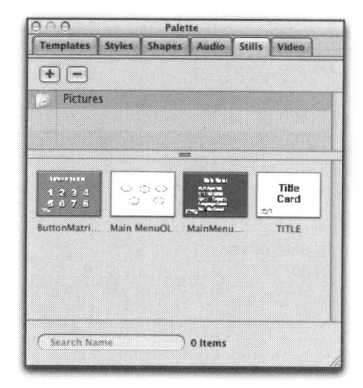

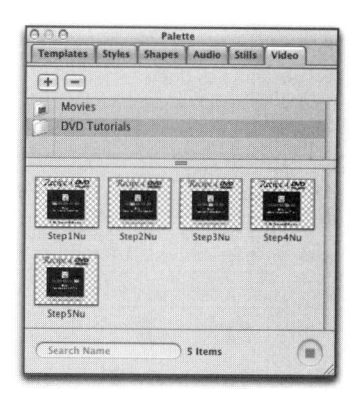

Figure 3-19b, c, and d — The Audio, Stills and Movies Tabs.

quick to use, the first folder in each tab defaults to the Current OS X User's Home folders:

- The *Audio* tab gives access to the User's *Music* folder (see Fig. 3-19b).
- The *Stills* Tab gives access to the User's *Pictures* folder (see Fig. 3-19bc).
- The *Video* tab gives access to the User's *Movies* folder (see Fig. 3-19d).

We will cover these palettes in detail in this chapter, just ahead.

Palette Default Entries Are Tied to the Current User

Since the Palette keeps its Preferences in the user area of the currently logged-in OS X user, the contents of the DVD Studio Pro Palettes will change from user to user, and users can customize the Palette as they see fit. Folders visible in one user's Palette will NOT be visible in another's, unless they both have added the same folders to their respective Palettes.

Adding Your Own Elements in the Palette, or Removing Them

Additional folders can be added to the Video, Audio, and Stills Tabs using the *Add* (+) Icon, while existing folders can be removed from those Tabs using the *Delete* (−) Icon (see Figs. 3-19b, c, d).

To Add a Folder to the Palette:

- Click the *Add Icon* (+) in the Palette, **or**
- *Control-Click* in the Tab's Folder List, and select *Add Folder*.
- In the Dialog box that opens, select the folder to add;
- Click *Add* to complete this operation.

To Remove a Folder from the Palette:

- In the Palette, click on the folder selected for removal;
- Click the *Delete* Icon (−) in the Palette; **or**
- *Control-Click* on that Folder, select *Remove Selected Folder*.

65

DVD Studio Pro 4

> **NOTE!**
> This operation is instantaneous, with no confirmation dialog! Click Undo immediately if the wrong folder is removed.

> **IMPORTANT!**
> While adding asset files to the Palette can help organize your asset collection for authoring, it does NOT add them to a DVD project! You will actually have to Author an asset into the DVD for it to become a part of the project, or move it into the Asset Tab of DVD Studio Pro in anticipation of using it.

Changing the Size of the Thumbnail Images

The General Preferences Pane allows selection of small or large thumbnail images in the Palette and the Slideshow Editor.

Palette Entries Stay in the Palette

Quitting DVD Studio Pro does NOT remove your user-added folders from the Palette. The Palette contents remain in the Palette after quitting the application, and will reappear there when you again launch the application.

> **NOTE!**
> You may find that folders added from FireWire hard drives may "get lost" from time to time, necessitating the removal of the broken icon and the addition of the desired folder again. On rare occasions, this may also occur to a folder added from a local hard drive.

The Video Tab

To display the contents of any folder in the Tab, click on the folder to select it. Its contents are then displayed in the lower area of the Palette (see Fig. 3-21). If more than one folder is highlighted, the contents of all highlighted folders will be displayed.

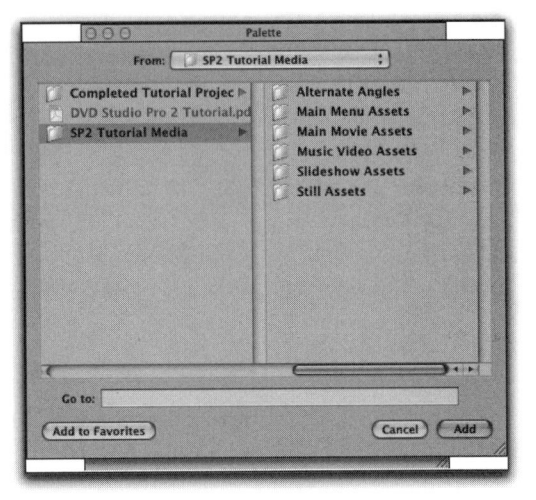

Figure 3-20 — The Palette Import Dialog.

Searching through the file names is possible by using the Search Name field at the bottom of each Tab. In the Video tab shown in Figure 3-21, notice that within the selected folder is a file called "Main

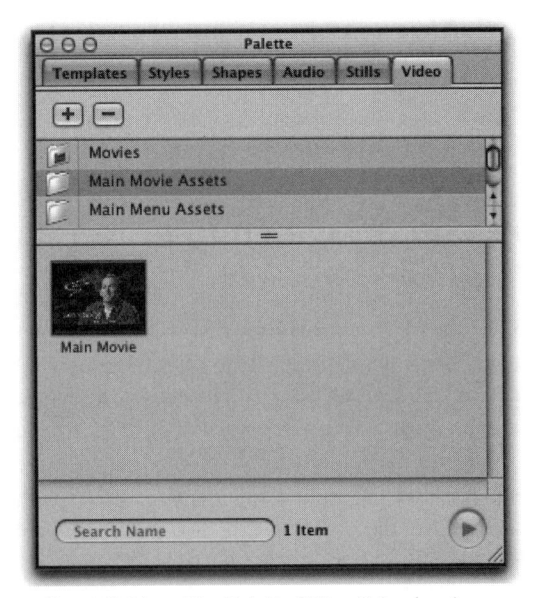

Figure 3-21 — The Palette Video Tab, showing a thumbnail.

The Workspace

Movie"—we can see a thumbnail image of this movie—better still, if you click on that movie to select it, you can then click on the Play triangle at the bottom right of the window, and the movie file will play in the Palette! This could be just the thing for sorting out similar-looking movie files before authoring them into your DVD. Clicking that Icon a second time will stop the preview (the PLAY triangle will turn into the STOP square after you begin playing the movie—not shown here).

Using Movie Files in Your DVD

DVD Studio Pro 4 allows you to use Movie files (QuickTime movies) before they are encoded into MPEG2. You can control the encoding parameters using the *Encode* Pane of the Preferences. You may also still use MPEG-2 and MPEG-1 video assets directly.

NEW—For HD DVD, you may now also use H.264 (see Chapter 5).

The Audio Tab

The *Audio* Tab defaults to showing the User's Music folder, but additional folders can be added, as outlined previously in the Video Tab section. In the Audio Tab of our example (see Fig. 3-22), the Music folder has been selected, and an iTunes music file selected from within that folder. Clicking on the Play triangle at the bottom of the window will cause the selected audio file to play from within the palette, useful for auditioning your music files before using them in the DVD. Clicking the Icon a second time (when it is a square) will stop the preview.

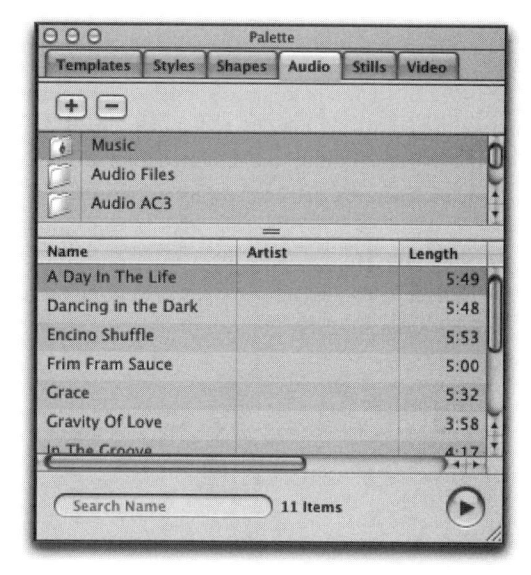

Figure 3-22 — The Palette Audio Tab.

Using Music and Sound Files in Your DVD

DVD Studio Pro 4 allows you to import many different kinds of audio files and, while they may NOT technically be legal to use in DVD, MP3 files from your iTunes library can be selected, and DVD Studio Pro 4 will convert them to usable DVD sound assets while you author, or when you build the DVD.

Creating Dolby Digital (AC3) Files

Beginning with DVD Studio Pro 4, Apple no longer includes the **A.Pack** application in the DVD Studio Pro suite of programs. To prepare your own Dolby Digital AC-3 files prior to authoring, use Compressor 2. (See Chapter 6 for more on making Dolby Digital audio bitstreams).

67

DVD Studio Pro 4

The Stills Tab

The *Stills* Tab defaults to the User's Pictures folder, but additional folders may be added (see Fig. 3-23). In the Stills Tab example, the imported tutorial folder called "Slideshow Assets" has been selected, and displays some older DVD Studio Pro tutorial slides from within that folder. Note that while you can still search, there is no Play triangle at the bottom of the window because stills are just that—still images without motion.

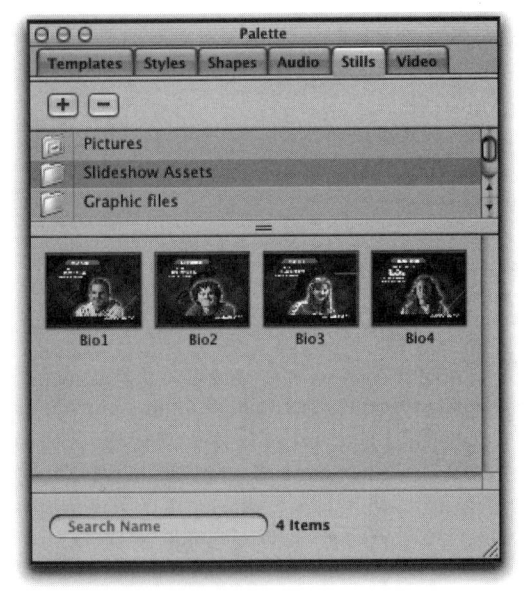

Figure 3-23 — The Palette Stills Tab showing tutorial media Slides.

What Kind of Graphics Stills Can You Use?

DVD Studio Pro is very forgiving in the kind of graphics you can use for stills. Almost *any* graphic format can be used in Studio Pro 2, including:

- Adobe Photoshop PSD files using the 8-bit RGB mode
- PICT format files
- BMP format files
- JPEG format files
- QuickTime Image files
- Targa (TGA) format files
- TIFF (TIF) format files

Although DVD Studio Pro will convert any of these file types into usable assets, you'll find that better results will be achieved using uncompressed files (TIFF, PICT, BMP, TGA) instead of compressed files (JPEG).

Organizing Assets Using the Palette

We've been discussing the Palette before moving on to other topics because DVD Studio Pro 4's authoring methods enable you to pre-organize your collection of movies, pictures, and sounds for more efficient authoring. Taking a little time to appreciate this feature, and organizing your DVD files in advance of authoring will save you lots of authoring time in the long run.

Configurations

Now that you have organized your DVD assets (and raw movie, sound, and graphic files), let's get further into the configurations you are likely to use.

The Basic Configuration Close-Up

One of the authoring features new in DVD Studio Pro 4 is the ability to manipulate live elements like text, Buttons, and Shapes within the DVD authoring appli-

The Workspace

cation, instead of having to rely on an external graphics application like Adobe Photoshop. DVD Studio Pro 4's new Menu Editor owes a lot to the design and success of iDVD's drag-and-drop authoring approach. The intuitive nature of designing Menus in a WYSIWYG (*what-you-see-is-what-you-get*) environment makes DVD authoring as simple as possible.

In the Menu Editor window, the largest component of the Basic Configuration, you can work directly with many of the DVD elements you may already be familiar with—MPEG streams, Audio streams, and Graphics files. DVD Studio Pro also gives you the ability to work with movie files before they are encoded into MPEG, and graphics elements smaller than the full 720 × 480 (NTSC) or 720 × 576 (PAL) sizes ordinarily required in DVD Studio Pro 1 (these are the shapes and patches we referred to earlier).

Because DVD Studio Pro's menu editor allows you to assemble a finished rendered menu from movies, tiny pieces of graphics, and Type elements added directly onto the menu, the flexibility and speed of menu design is truly dramatic. This directly translates to improved authoring efficiency, and faster DVD completion. This also means that DVDs previously completed can be modified and re-mastered faster, without delays, while graphics or motion menus are being redone in an external application or on a different workstation.

The Basic Configuration is an interesting workspace, because it has been deliberately kept simple, allowing you to create your DVD by creating the Menu, which automatically links its Buttons to the Menus, Tracks, Slideshows, and Scripts created in your project.

Let's take apart the various elements of the Basic Configuration and see how they fit together.

The Menu Editor Window

Like many other innovations in DVD Studio Pro 4, the *Menu Editor* window (see Fig. 3-24) includes UI

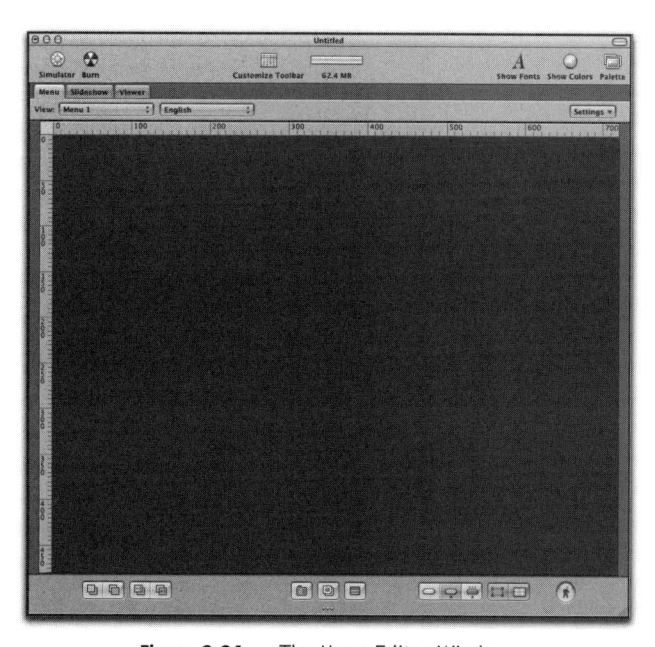

Figure 3-24 — The Menu Editor Window.

features developed in other Apple applications, like iDVD and Keynote, and OS X in general. The drag-and-drop asset linking feature from iDVD now appears in DVD Studio Pro, which can automate the creation of Buttons and speed their linkage to their assigned assets. One amazing new feature is the ability to create, link, and name the buttons for a Chapter menu with ONE mouse drag. Once you have created a video stream with embedded Chapter markers (using Final Cut Pro, Final Cut Express, or iMovie), you can experiment with this feature. The *auto-alignment* guides feature of Keynote has been incorporated to make working with Menu elements a snap! The Menu editor will be covered in depth in Chapter 9, "Authoring Menus."

The Menu editor window occupies most of the Basic Configuration in DVD Studio Pro 4. On it, you can author iDVD-style within DVD Studio Pro 4. By dragging Templates, Assets, Shapes, Buttons, Graphics, and Text elements directly onto this window, you can quickly and easily create a very high-quality Menu. This Menu can contain motion graphics backgrounds (with audio), Drop Zones (to add stylized graphic elements or movies), Shapes, Buttons, and links. The drag-and-drop style of authoring in the menu editor is a very easy step up for users who have learned DVD authoring using iDVD. The Recipe4DVD Basic Tutorial project relies on the capabilities of the Menu Editor window to ease your entry into the use of DVD Studio Pro. The Menu Editor also allows editing of Buttons, Button highlights, and Button commands—the basics of DVD navigation.

Other interesting features in the Menu Editor window include the *View*: and *Menu Language*: pop-up selectors (on the left), and the *Settings*: drop-down dialog (on the right), as well as the (optional) rulers and static and dynamic alignment guides.

The Toolbar

The *Toolbar* is located across the top of the Menu Editor window (see Fig. 3-25a), and contains a customizable interface displaying click-on tool icons

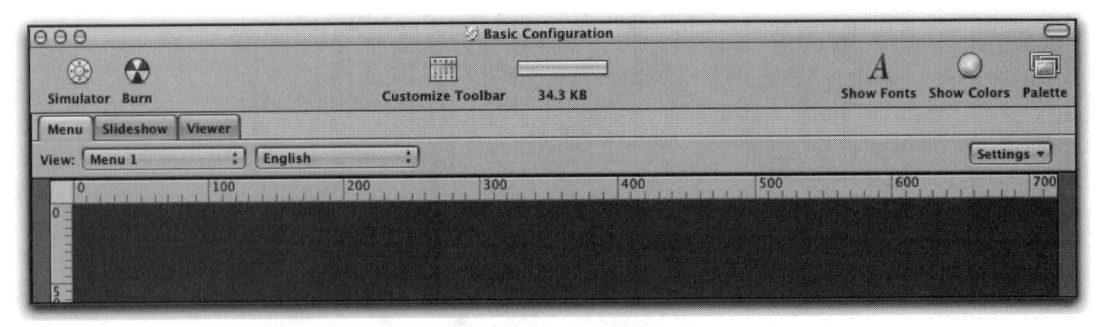

Figure 3-25a — The Menu Editor Window Top close-up showing Toolbar.

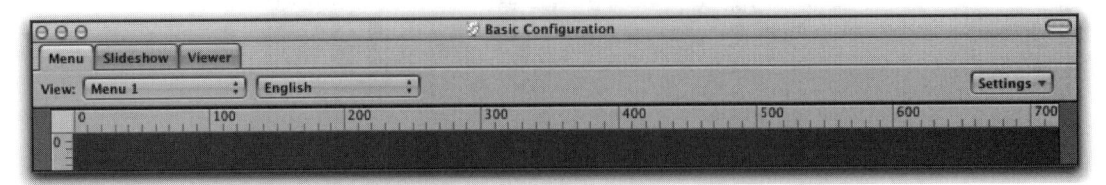

Figure 3-25b — The Menu Editor Window Top, with Toolbar hidden.

The Workspace

Figure 3-25c — The Menu Editor Window Top, with Toolbar Button featured.

used to invoke commands (e.g., Show Fonts, Add Track), start processes (e.g., Simulate, Burn), or display info (Disc Size) during authoring. The Toolbar is a great addition, but if for some reason you don't want to display it, use the **View > Hide Toolbar** menubar command (Fig. 3-25b shows it closed). You can also click the Button in the upper right corner of the Project Title bar, which will quickly hide the Toolbar, or restore it to visibility, if hidden (see Fig. 3-25c).

More about The Toolbar

The Toolbar is completely configurable to your personal taste, using the *Customize Toolbar* menu command (**View > Customize Toolbar...**) or the *Customize Toolbar* Tool (in Fig. 3-26, it's just to the left of the Bit Budget display).

Clicking on the Customize Toolbar Tool or using the Customize Toolbar command opens the Tool center (see Fig. 3-26), where your choice of new tools can be selected to add to the toolbar in any configuration. You can also use the Customize Toolbar tool to remove unwanted tools, or just to unlock the Toolbar to rearrange the existing tools in a more desirable layout.

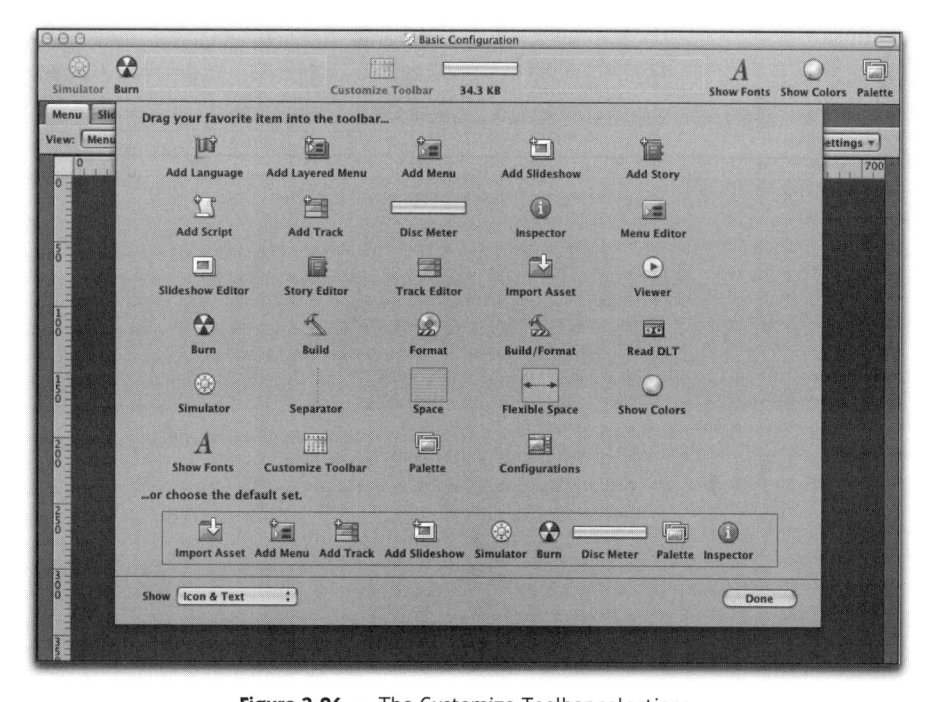

Figure 3-26 — The Customize Toolbar selections.

71

DVD Studio Pro 4

Customizing the Toolbar

Why customize the Toolbar? Because you can! Seriously, the Toolbar gives you a central location to select many of the common authoring commands with your mouse. The sheer beauty of the current DVD Studio Pro is that so much of the interface is customizable, it's like having your very own application, tailored just to your working method. Of course, you may prefer to use Keyboard Shortcuts, and you won't be disappointed there—there are dozens of such Shortcuts programmed into DVD Studio Pro 4, and they are readily available for use, regardless of which window configuration you are showing.

Some commands require complex keyboard shortcuts, or are used so often that they are good candidates to be placed in the toolbar. One such command is **Show/Hide Inspector** [⌘-**Opt-I**]—the Inspector window is such an important part of the authoring interface that I think you will want to keep it handy all the time. Luckily, it's a snap to add it to the toolbar in any configuration that lacks it.

Here's how to do it:

Adding Tools to the Toolbar (or Removing Them)

To add the Inspector Tool (or any other tool):

- Click on the Customize Toolbar Tool (if visible), **or**

- Select **View > Customize Toolbar...** in the menubar.

When the drop-down Palette opens to display the available tools, click on the desired tool (in this case, the circle-I Icon which represents the Inspector Tool) and drag it to the toolbar (see Fig. 3-27a). I like to place it right next to the Palette Tool, which is already located at the right end of the Toolbar, but feel free to drop it wherever you please. When you have finished dragging it to the toolbar, release the mouse and the tool will "land" in the Toolbar in the location you have selected. Other tools will "make way" to allow a space for the incoming new tool. Remove a tool by dragging it away from the Toolbar—release the Mouse and "Poof"—the tool will disappear (see Fig. 3-27b).

When you have finished editing the Toolbar, you'll have to click on the *Done* button at the bottom of the

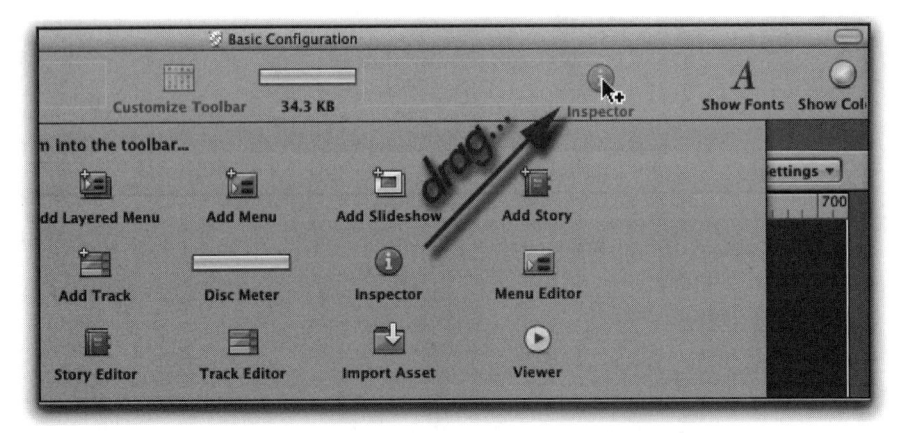

Figure 3-27a — Adding the Inspector Tool to the Toolbar.

The Workspace

Figure 3-27b — POOF! Deleting a Tool from the Toolbar.

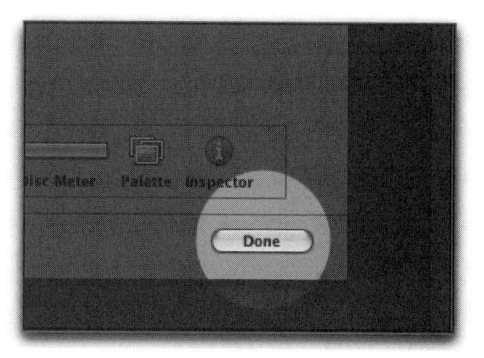

Figure 3-27c — Click here to close the Tool Bin and lock the Toolbar.

Customize Toolbar window (see Fig. 3-27c) to exit Toolbar Edit mode. This will return the Toolbar to proper functioning, and lock all of the tools in their current locations.

Moving Tools around in the Toolbar

If you decide you wish to move any of the tools around again, just invoke the *Customize Toolbar* Tool again, move the tools around, and again click *Done*, locking the Toolbar once more.

Figure 3-28 shows the Inspector Tool being moved to the right, after landing in the Toolbar, and Figure 3-29 shows it locked in its final resting place next to the Palette Tool.

I like having the Inspector Tool handy on the toolbar, especially if I am using the mouse heavily, as I do when in the Menu Editor. You will find this especially useful because the mouse plays a large part in editing in the Menu Editor window.

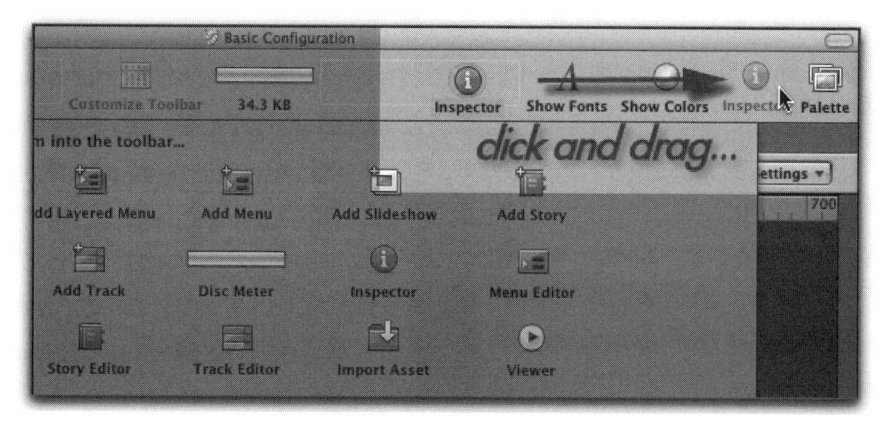

Figure 3-28 — Moving the Inspector Tool.

73

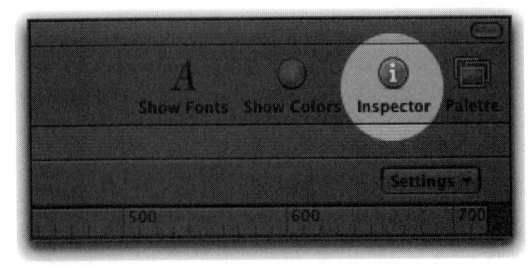

Figure 3-29 — The Inspector Tool in its final location.

Saving the Toolbar Configuration

The Toolbar configuration will be saved with your project, so it is possible to create a highly customized template even without creating a separate configuration. Just save a project to use as a template and use it repeatedly at the start of your new project. Be sure to immediately save your project with a different name each time you use the template to avoid overwriting your template project and ruining it.

To Save the Toolbar Setup as Part of a New Configuration:

- Select `Window > Save Configuration` in the menubar.

Available Tools for the Toolbar:

One of each of the following tools can appear in the Toolbar, with the exception of *Space, Flexible Space*, and *Separator*. They are listed in the order in which they appear in the Toolbar Palette:

- **Add Language:** Adds a new Menu Language to the project. The new Menu Language can be selected in the Outline tab.
- **Add Layered Menu:** Adds a new layered Menu. The new Menu can be selected in the Outline tab or in the Menu Editor *View*: pop-up.
- **Add Menu:** Adds a new standard Menu to the project. The new menu can be selected in the Outline tab or in the Menu Editor *View*: pop-up.
- **Add Slideshow:** Adds a new Slideshow to the project. The new Slideshow can be selected in the Outline tab or in the Slideshow Editor *View*: pop-up.
- **Add Story**: Adds a Story to the currently selected track. If none is selected, it will add a Story to the first track in the Track list of the Outline Tab. It will add a new Story in numerical order for each click.
- **Add Script**: Adds a new Script to the project. The new Script can be selected in the Outline tab or in the Script Editor View: pop-up.
- **Add Track**: Adds a new Track to the project. The new Track can be selected in the Outline tab or in the Track Editor View: pop-up.
- **Disc Meter**: Displays the Disc Meter Bar display to indicate the relative amount of disc space used and still available. This display is aware of the Disc Media setting in the Disc/Volume tab in the Disc Inspector. The number underneath the progress bar shows disc space used in the current project.
- **Inspector**: Displays the Inspector in its last position.
- **Menu Editor**: Displays the Menu Editor.
- **Slideshow Editor**: Displays the Slideshow Editor.
- **Story Editor**: Displays the Story Editor.
- **Track Editor**: Displays the Track Editor.
- **Import Asset**: Opens the Import Asset dialog, allowing you to select assets to import into the project's Assets Tab.

The Workspace

- **Viewer**: Displays the Viewer tab.
- **Burn**: Builds the DVD Volume and writes it to your DVD burner using the current Disc Inspector settings.
- **Build**: Multiplexes your project to create a DVD Volume folder (containing the VIDEO_TS or HVDVD_TS folder) of your project.
- **Format**: Writes a built DVD Volume folder to a DVD burner, DLT (*Digital Linear Tape*) drive, or as a disk image to a hard disk.
- **Build/Format**: First builds the DVD Volume, then writes it to a DVD burner, DLT drive, or as a disk image to a hard disk. (If any assets need to be encoded first, it will do so prior to attempting to Build.)
- **Read DLT**: If a DLT drive is connected and a DLT tape inserted, reads the tape contents and writes them to the designated hard drive as a Disc Image file (.IMG). This can be played in Apple DVD player, but does NOT contain the DVD project—only the finished DVD—re-authoring is not possible without additional work.

> **IMPORTANT!**
> Save your project and asset files if you expect to return to this project again! The master DLT is NOT editable.

- **Simulator**: Opens the project Simulator, allowing you to try out your project before actually building it.
- **Separator**: Inserts a vertical bar into the toolbar, to group tools.
- **Space**: Inserts a fixed space into the toolbar. This helps you to group and arrange tools by adding a fixed space between them.
- **Flexible Space**: Inserts a space into the toolbar that automatically expands to fill any existing empty space.
- **Show Colors**: Opens the system Colors window.
- **Show Fonts**: Opens the system Fonts window.
- **Customize Toolbar**: Opens the toolbar palette ("Tool Bin").
- **Palette**: Displays the Palette in its last configuration.
- **Configurations**: Opens the Configuration Manager dialog so that you can manipulate and manage the configurations.

Extra Menu Editor Tools

At the bottom of the Menu Editor, groups of additional icons give you access to more functions used while authoring in the Menu Edit window. The bottom of the Menu Editor window normally defaults to showing the Menu editor tools (see Fig. 3-30a), but these can be hidden using the three dots tool, bottom center (see Fig. 3-30b).

From the left, these additional tool groups are:

- **Arrange Controls**: Buttons in DVD Studio Pro 4 are prioritized by their button number; these tools control these functions by allowing the selected button to be moved forward (to a higher priority) or backward (to a lower priority).

75

DVD Studio Pro 4

Figure 3-30a — The Menu Editor Window bottom showing tools.

Figure 3-30b — The Menu Editor Window—bottom without tools.

- **Add Submenu, Add Slideshow, Add Track**: These tools allow you to add a new button to the current menu and at the same time make a new Element to connect the button to. New Buttons created will assume the current menu's default Button style. Add Submenu will create a new Submenu based on the current menu template.

- **Button State Selections**: These tools control the display state of the currently selected button(s)—Normal, Selected, Activated.

- **Button Outlines**: This button switches the button outline display on or off, so buttons may blend into the Menu Background.

- **Guides button**: This button turns the alignment guides on and off in the menu, and also enables or disables the guide creation function.

- **Motion button**: The "running man" Icon enables or disables the menu motion, if there is a motion asset in the background.

Context-Sensitive Drop Palettes

Dragging files to the Menu Editor window and pausing before releasing the mouse button will display a Context-Sensitive "Drop Palette," another cool feature of DVD Studio Pro. These multifunction palettes display the most likely connections, based on *what* you have dragged, and *where* you have paused (see Figs. 3-31a and 3-31b).

There are literally dozens of drop palette functions, which we will cover in more depth in Chapter 9, "Authoring Menus."

You will find the DVD Studio Pro 4 Help Manual has quite a complete chart of Drop Palette connections available. Refer to the Index.

Hey! The Graphical View Window Is BACK!

If you are a veteran of DVD Studio Pro 1, you know that the Graphical View was a large part of the User Interface in that version (see Fig. 3-32a). By adding, selecting, and clicking on the *Item Tiles* (*the Icons*

The Workspace

Figure 3-31a — Drop Palettes Video.

Figure 3-31b — Drop Palette Still.

*that represent Track*s, etc.) and working with the various Property Inspectors, you made your way around the DVD project, authoring and connecting the Items—Menus, Tracks, Scripts, Slideshows, and Menu Languages—that made up the DVD.

Luckily, as of DVD Studio Pro 3, the Graphical View window has returned! (See Fig. 3-32b.) You can once again arrange logical storyboards for your DVD projects and, in addition, use the Outline Tab and Connections Tab to organize the display and interconnections of Element data in your project. (We'll cover the Connections Tab and other methods of linking things in Chapter 11.)

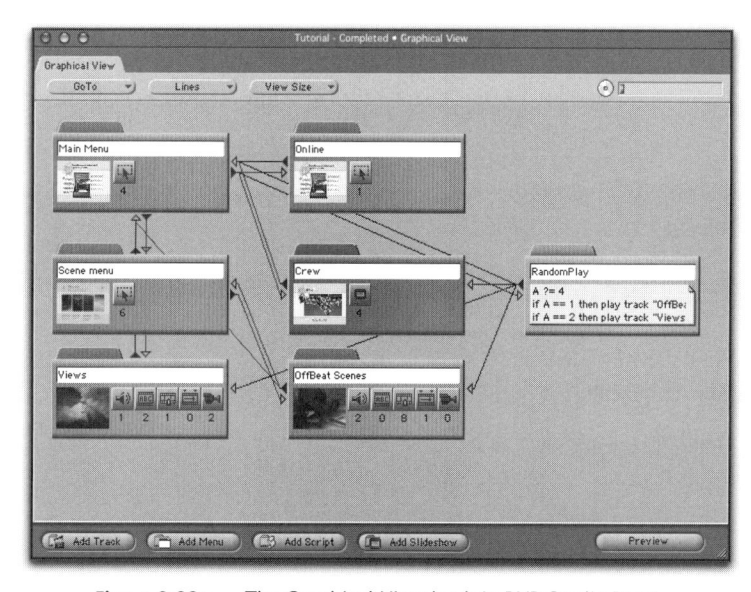

Figure 3-32a — The Graphical View back in DVD Studio Pro 1.

77

DVD Studio Pro 4

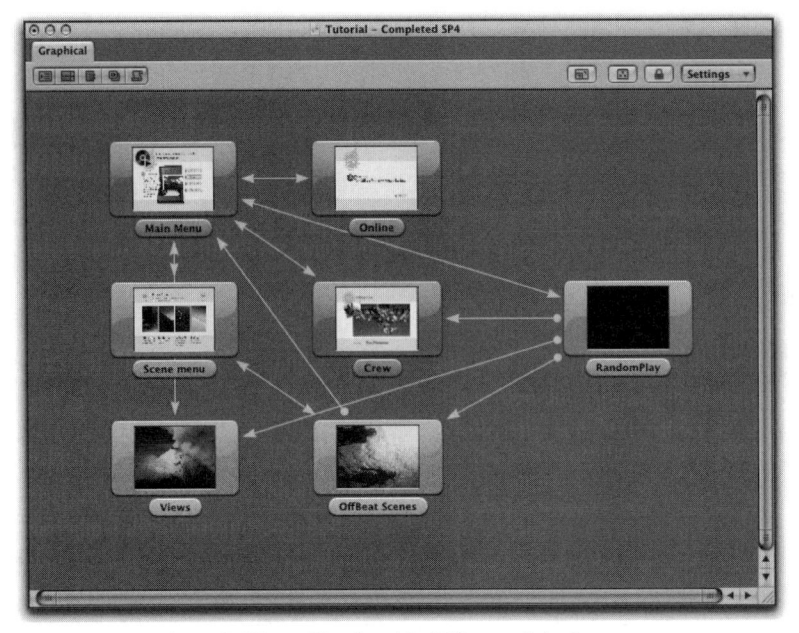

Figure 3-32b — The Graphical View as it looks now.

Workspace Tabs Close-Up

By name, in alphabetical order, the available Tabs are:

- **Assets** listing your DVD's assets currently in use
- **Connections** showing interconnections of Elements
- *New!* **Graphical View** showing Element interconnections graphically
- **Log** tracks Simulation, Building, Encoding events
- **Menu** contains the Menu Editor
- *New!* **Outline View** a hierarchical view of project Elements
- **VTS Editor** allows redistribution of project assets in the finished DVD's Title Sets
- **Script** contains the Script Editor
- **Slideshow** contains the Slideshow Editor
- **Story** contains the Story Editor
- **Track** contains the timeline-driven Track Editor
- **Viewer** allows you to preview your DVD's Elements

In the following paragraphs, we'll use the Advanced Configuration as the basis of our exploration of the individual Tabs and Windows of the Workspace. The default Advanced Configuration contains all 11 of the Tabs available in DVD Studio Pro 4's arsenal.

The Workspace

The Outline Tab

The *Outline Tab* (see Fig. 3-33) originally replaced the Graphical View of your project in DVD Studio Pro 2, but since that's come back, it's become a great companion for it. Outline View is one of the central "overviews," if you will, of the structure of your DVD, regardless of how complex or simple that DVD may be. As you build your project, Outline View will show the Elements you create. Unlike the Graphical View, the Outline Tab is organized hierarchically in *Folders*, one for each type of Element. New Elements are always created within the Folder they belong in. It is possible to move Elements around in their own Folder, but Elements cannot be moved into an unrelated Folder. The Element Folders will always remain in alphabetical order in this Tab. The Disclosure Triangle to the left of each Folder allows you to close or open each one, as needed, to manage the display. Selecting an Element in the Outline Tab will bring it into the Inspector for further examination. It will also be visible in the Connections Tab for investigation of how that Element interconnects with other Elements in the project. One other impressive sign of how Apple really thinks about its software design is reflected in the simple inter-relationship of Elements in the Outline View and the Graphical View—if an element is selected in one view, it is highlighted in the other, so you can see exactly where you are in the project at all times!

Adding New Elements to the Outline

Elements may be added to your DVD project as follows (see Fig. 3-34):

- Select **Project > Add to Project** in the menubar, **or...**
- Use *Keyboard Shortcuts* (⌘ + a modifier key), **or...**
- Using *Toolbar Tool Icons* (if visible), **or...**
- *Control-Click* in the Outline Tab to reveal the pop-up, then select **Add >.**

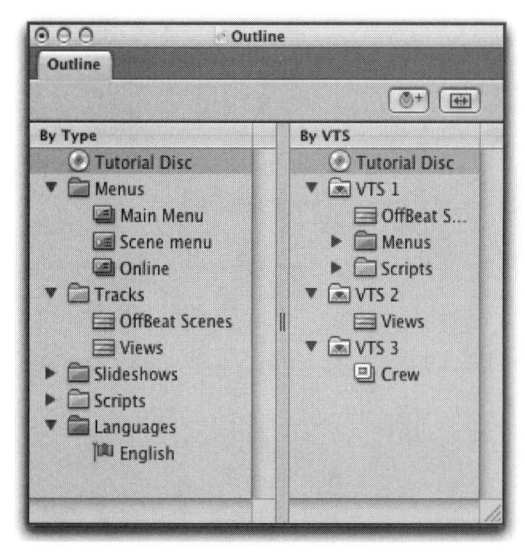

Figure 3-33 — The Outline Tab, with VTS Editor opened.

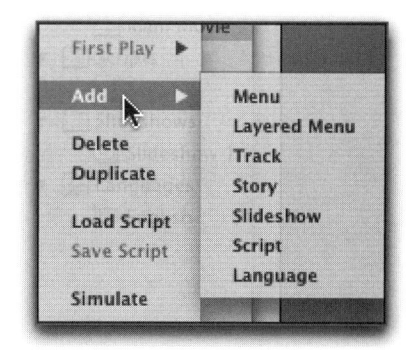

Figure 3-34 — Control-Click in the Outline Tab to add Elements.

DVD Studio Pro 4

The Story Tab

A Story is DVD Studio Pro's method of defining a content segment *within* a Track Element. You could think of it as a subclip, in editorial terms. A Story can define the entire Track it is in, or it can define just a small segment of that Track. Unlike a Chapter, which only tells you where to begin playback, a Story allows you to define a starting point and an ending point for a specific segment within a track; in other words, where to jump in, and where to jump out. Each story can also have an *End Action*, an instruction of where to navigate when that story has completed playback. Stories may have duplicate sets of story markers, but different end actions, if you need this feature. (See Chapter 8, "Creating Markers and Stories.")

Placement of the Story Tab can affect its usefulness. If you open the Story Tab inside of the upper left Quadrant (see Fig. 3-35a), you will find it is cramped and difficult to read and use. It is much easier to use if you move this Tab into the upper right quadrant. You can also tear-off this Tab at any time to turn it

Figure 3-35a — The Story Tab can be more readable in the right Quadrant.

into a Window, and then revert it into a Tab easily (see Fig. 3-35b). Review "Turning Tabs into Freestanding Windows" earlier in this chapter for more on this.

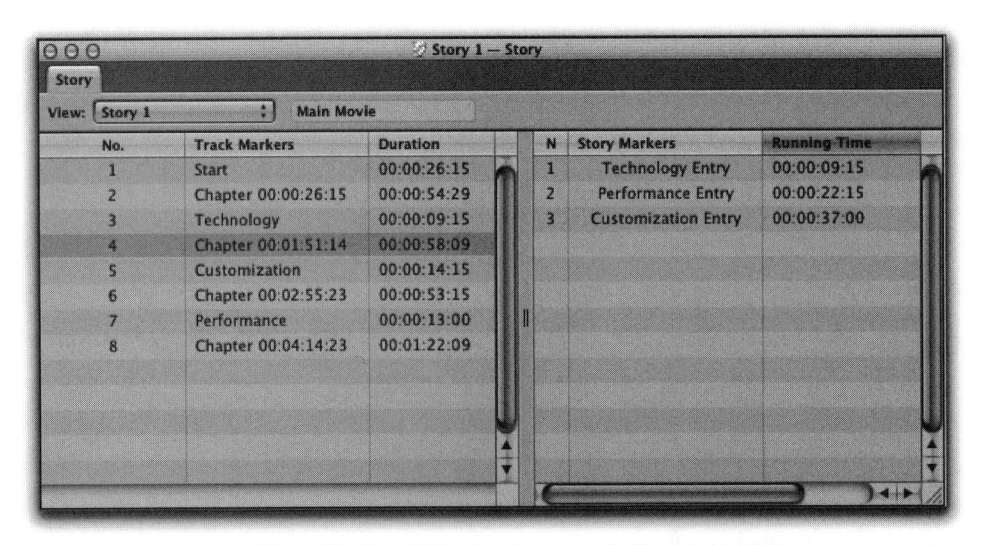

Figure 3-35b — The Story Tab, showing a sample Story in the Editor.

The Workspace

Adding a New Story

To add a new Story to a project, do one of the following:

- Click the *Add Story* Toolbar Icon, **or**
- *Control-Click* **on** a Track in the Outline Tab, select *Add Story*.

If a Track is <u>currently selected</u> in the Outline view, you may also:

- Select **Project > Add to Project > Story**.
- Type *Command-Shift-T* (⌘-**Shift-T**).

The Menu Tab

The *Menu Tab* is home to one of the most powerful features of DVD Studio Pro, the Menu Editor Window (see Fig. 3-36). This editor is dramatically different from the Menu Editor in Studio Pro 1. Taking a large creative nudge from iDVD, the Menu Editor can be used to create a WYSIWYG Still or Motion Menu, including things that just couldn't be done in DVD Studio Pro 1, and still can't be done in some other authoring applications or other platforms. It can be the authoring nerve center for many DVD projects. It is so powerful you may *never* have to visit the Connections Tab to link things.

The Menu Editor contains pop-up dialogs along the top edge, which access Menu and Language selections, and the Settings pop-up, which displays various menu authoring and display preferences. Along the bottom of the Menu Editor are tools to control various Menu and Button authoring functions. We will cover the Menu editor in great detail in Chapter 9, "Authoring Menus."

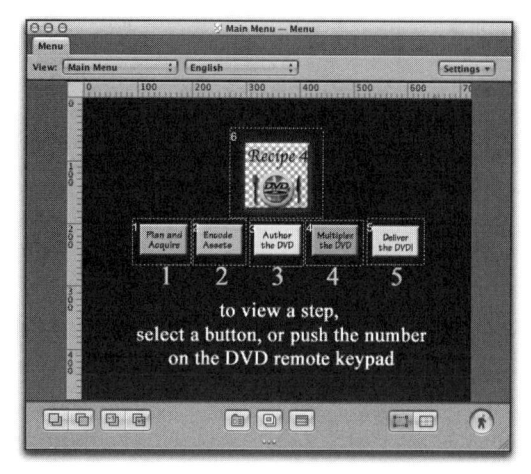

Figure 3-36 — The Menu Tab (seen as a window).

The Connections Tab

The *Connections* Tab is the place to check or define an Element's link destinations with the most in-depth detail (see Fig. 3-37). You may rarely visit this Tab in your simplest DVDs, if only because making linkages between Elements is so amazingly simple to do in DVD Studio Pro's Menu Editor. In many cases, you can build an entire DVD and never use this Tab; however, Connections is the best place to get an overall in-depth view of the connections in your DVD. Connections, together with the Outline Tab and the Graphic View, provides all of the information you need.

Similar to the Menu Tab, the Connections Tab has pop-ups along the top. The *View:* pop-up allows selection of three different levels of depth in the view—Basic, Standard, and Advanced. The adjacent pop-up allows three levels of filtering—*show All*, *show only Connected*, and *show only Unconnected*. Adjusting this pop-up to the proper level of filtering shows you what's connected, or what's left to connect! Further to the right there is a Button to *Connect* or *Disconnect* a selected linkage. At the far right,

81

DVD Studio Pro 4

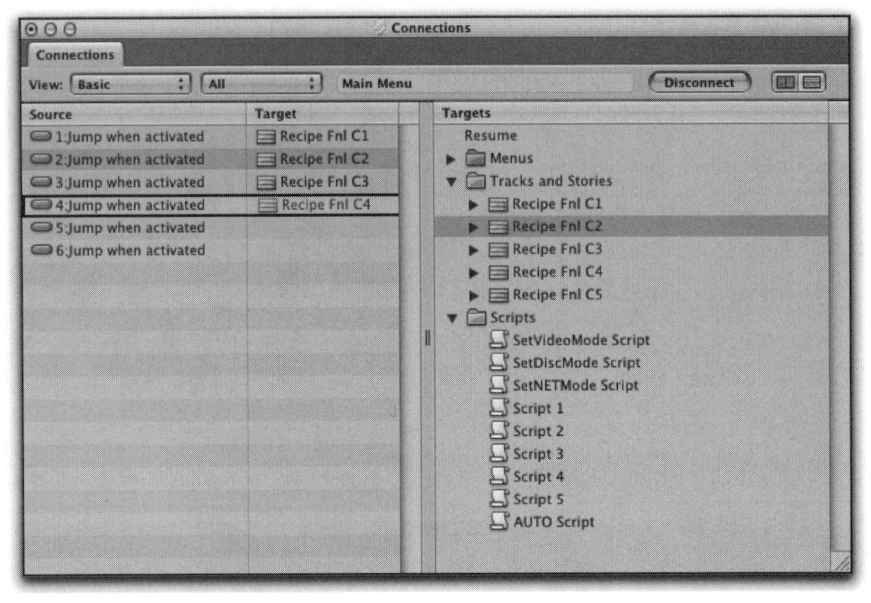

Figure 3-37 — The Connections Tab—make or check links here.

there are window control tools, which can split the Connections Window Horizontally, listing Sources in the Top pane, and Targets in the bottom pane. (See Chapter 11, "Making Connections" for a far more detailed review of this tab.)

The Viewer Tab

The *Viewer* is an interesting feature of DVD Studio Pro (see Fig. 3-38). This Tab spends most of its time patiently waiting for you to need to preview something, and then it springs into action, displaying the requested preview of a Track, Asset, or Slideshow. The Viewer is good for quick things, but the *Simulator* window gives a far more refined preview of the finished DVD, and also gives you complete navigation control, real-time parameter information, and an extensive Log window.

Display in the Viewer can be a bit raggedy from time to time, as it is using un-multiplexed assets to approximate what the DVD will look like when finished. In general, however, the Viewer will give a good rough approximation of what a finished Track or Slideshow will look like. When you need navigation control, use the Simulator.

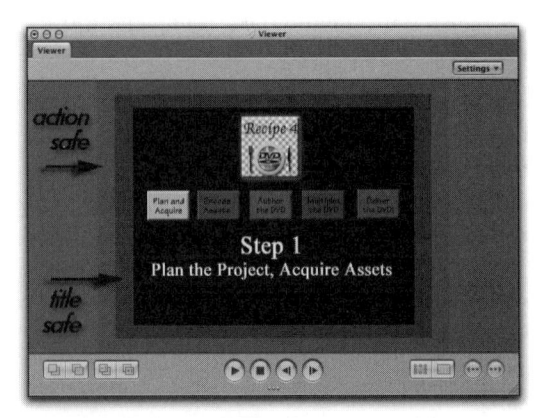

Figure 3-38 — The Viewer Tab—in 4 x 3 format with Action and Title Safe Zones overlaid on the video.

The Workspace

Another function of the Viewer window is to visualize Subtitles as they are being created. Some of the tools at the bottom of the Viewer are helpful while creating Subtitles. We'll cover those in detail in Chapter 12, "Creating Subtitles."

The Assets Tab

The Assets Tab (see Fig. 3-39a) displays files, which are actually in the Asset Bin and are either *available for use* or *already in use* in the current DVD project.

> **NOTE!**
> Files displayed in the *Palette* are available to be added to the Project, but may not already be contained in the Asset Tab, or in use in the DVD.

Shown in Figure 3-39b as a torn-off window, the Assets Tab uses columns to organize the asset data that are available to be displayed. Files in this Tab may be organized using Folders (a handy feature),

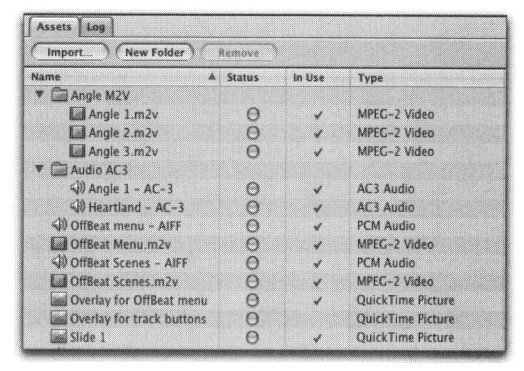

Figure 3-39a — The Assets Tab docked—showing assets and asset folders.

and unused files may be deleted from this Tab using the *Remove Button*. As in Final Cut Pro, these data columns may be used as sort criteria, and rearranged in the order desired. (See Chapter 4, "Assets and Elements" for a detailed review of this Tab.)

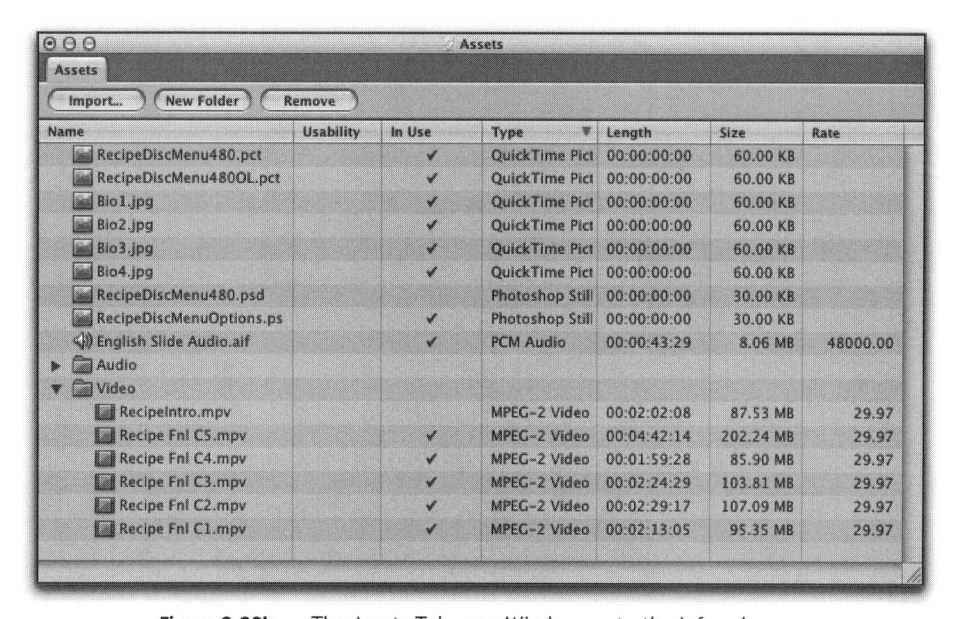

Figure 3-39b — The Assets Tab, as a Window—note the Info columns.

83

DVD Studio Pro 4

The Log Tab

The *Log Tab* is three displays in one (Fig. 3-40). Again using a *View:* pop-up, this Tab may selectively display events relevant to either Simulation, Encoding, or Building.

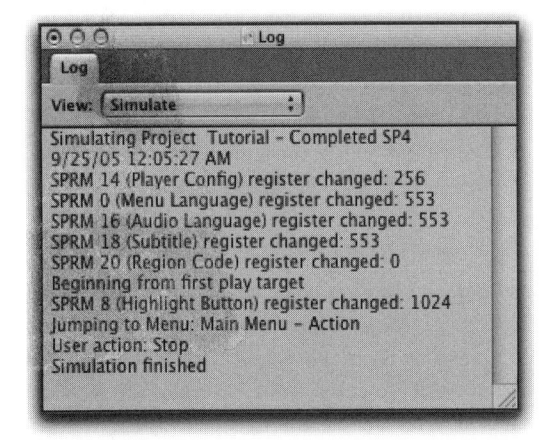

Figure 3-40 — The Log Tab, showing *View:* pop-up and Simulation events.

The Track Tab

The *Track Tab* (see Fig. 3-41) contains the timeline-based Track Editor, a tool used to assemble the various Video, Audio, and Subtitle streams that make up a Track container. There may be a combined total of up to 99 Tracks, Stories, and Slideshows in a project.

The Track Editor can be also used to create and edit Markers of all kinds, as well as creating Subtitle events in their Stream areas. Streams can be trimmed as to length, moved about in time, and multi-angle and nixed-angle tracks can be constructed in this Tab.

You will find detailed information about the Track Editor in Chapter 7, about Markers and Stories in Chapter 8, and about Subtitles in Chapter 12.

The Slideshow Tab

The *Slideshow Tab* (see Fig. 3-42) contains the Slideshow Editor, used to organize and create Slideshows. Each *Slideshow* is a linear sequence of

Figure 3-41 — The Track Editor Tab—Video, Audio, Subtitles, and more.

84

The Workspace

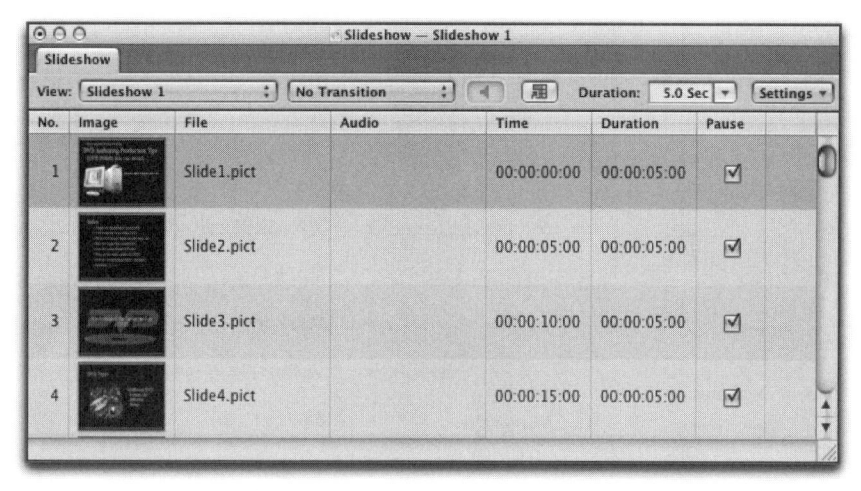

Figure 3-42 — The Slideshow Tab—an Editor for graphic event playlists.

up to 99 still image Slides, each with or without audio, or with a continuous single audio program underneath all.

The Slideshow Editor contains a *View*: pop-up to quickly switch from one slideshow to another during editing. In DVD Studio Pro 3 Transitions were added, and there is a pop-up to select them, as well as a tool to allow for the creation of an *overall audio track*—one audio file playing continually underneath all of the files in the slideshow. A *Settings*: pop-up controls how these functions are applied. We will cover this in detail in Chapter 10.

The Script Tab

The *Script Tab* contains the Script Editor window that, in conjunction with the Script Inspector, allows you to create the Scripts that tap into the control commands and variables available in every DVD player (see Fig. 3-43).

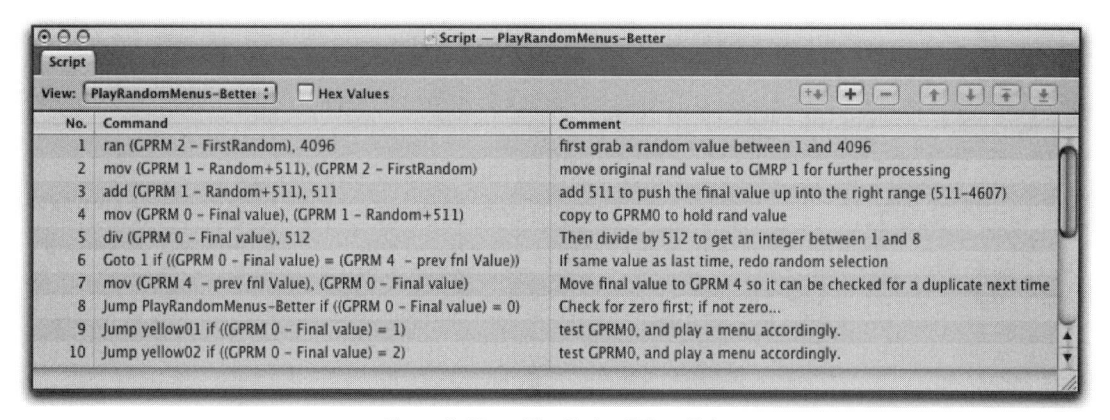

Figure 3-43 — The Script Editor Tab.

85

The Script Editor places these tools across the top:

- The *View*: pop-up, which selects which script to edit.

- The *Hex Values* checkbox, which, when checked, displays numerical script values in their Hexadecimal (base 16) form, instead of the Decimal (base 10) default.

- Three tools to *Insert, Add*, and *Delete* commands.

- Four tools to move command lines: *Move Up, Move Down, Move to Top,* and *Move To Bottom*. These allow reordering of the script command lines within the script.

See Chapter 13, "Basic Scripting for DVD Studio Pro 4," for details on the Script Editor functions, and how to script your DVD. You will see the order of the command lines is very important to the function of the script.

In the DVD accompanying this book, you will find several samples of Recipe4DVD's *Pro-Pack 2* Scripting training CD-ROM. Several projects are provided to help you learn more about Scripting in DVD Studio Pro 4, as well as versions for DVD Studio Pro 3, 2, and even 1.

About the Inspectors

We briefly mentioned it earlier, but let's have a closer look at the Inspector—I mean Inspectors—because there are lots of them.

The Inspectors are an integral part of authoring in DVD Studio Pro, providing detailed information about a project Element or Asset file.

Inspectors are available for any currently selected project Element:

- Disc
- Menu
- Menu Button
- Menu Drop Zone
- Menu Text element
- Track
- Markers
- Video Clip
- Audio Clip
- Subtitle Event
- Script
- Script Command Line
- Slideshow
- Slide
- Language (Menu display Language)

A New Inspector exists in DVD Studio Pro 4:

- VTS

Inspectors are also available for any currently selected project Asset:

- Video Asset
- Audio Asset
- Picture Asset

Because there are frequently several Inspectors for each project element, we'll detail the Inspectors in their respective chapters—Menus and Buttons, Tracks, Markers and Stories, Scripts, and Slideshows.

The Workspace

To Select a Project Element for Inspection

To easily select any project Element for inspection, click on it in either the *Outline View* or the *Graphical View*. If the Outline Tab isn't visible, select it and make it visible, then select the Element you wish to Inspect.

The Outline Tab normally lives in the upper left Quadrant of the Main Window, if you haven't moved it from its default position. The *Outline Tab* gives us the complete rundown on all Items created in a DVD project. You'll note that it contains Folders, which organize all of the project Elements. Each folder contains the items created in that specific type—Menus, Tracks, Scripts, Slideshows, and Menu Languages.

The disclosure triangle to the left of each folder is the Macintosh tool used to open the folder to reveal its contents. Clicking on a Folder will do nothing—you must open the folder to reveal the Elements you wish to access (see Fig. 3-44).

Figure 3-44 — The Outline Tab—with folders closed.

- The *Menus Folder* will display Menus in the Current Project. To inspect a Menu, click the disclosure triangle to open the Menus folder, then click on the desired Menu. To inspect a Button, Drop Zone, or Text element, you must first open the desired Menu into the Menu Editor Tab, then click on the desired element within the Menu where it exists. See Chapter 9 for in-depth information on Menu, Button, Drop Zone, Shape, and Text Elements, and their Inspectors.

- The *Tracks Folder* shows Tracks and their Stories (if any Stories have been created). To Inspect a Track, open the Tracks Folder then click on the desired Track within the Folder. To access a Marker of any kind, first select the desired Track, double-click it to reveal the Track Editor Tab, then select the Marker from within the Track Editor. To Inspect a Story, open the Tracks Folder then click on the desired Story within the appropriate Track. See Chapter 7 for more information about the Track Element itself, and the Track Inspector. See Chapter 8 for more information about Marker and Story Elements and their Inspectors.

- The *Scripts Folder* shows the Script elements only, not their Commands. To inspect a Script, click the disclosure triangle to open the Folder, then click on the desired Script from within the Folder. To Inspect a Script Command Line, first reveal the Script in the Script Editor, then click on the desired Command Line within the Script. See Chapter 13 for more information about the Script and Script Command Elements and their Inspectors.

- The *Slideshows Folder* shows Slideshows only, not Slides. To inspect a Slideshow, click on the Slideshow from within this display. To Inspect a Slide, reveal the Slideshow in the Slideshow Editor, then select the Slide from within the Slideshow Editor. See Chapter 10 for more information

87

about Slideshow and Slide Elements and their Inspectors.

- The *Languages Folder* shows Menu Visual Languages available within the DVD project.

> **NOTE!**
> Menu Languages should not be confused with the Spoken or Subtitle languages used in Tracks. Language, used here, refers to the Menu Visuals, which can be "tagged" with a particular language code, to allow the DVD player to choose from among multiple menu languages that may be available on a specific DVD.
>
> If you add additional Languages in this Tab, you may need to provide a new menu picture for each Language. Be sure to read Chapter 9 for more information.

To Select a Project Asset for Inspection

The *Assets Tab* shows all project assets used in the current project, and those that have been imported but not yet used or assigned. To inspect an Asset file that is in the root level of this Tab (i.e., not inside of a Folder), just click on the desired Asset within the Tab. To inspect an Asset file that is inside of a Folder, first open the Asset Folder containing that asset, then click on the desired Asset within the folder. See Chapter 4 for more information about Assets and their Inspectors.

The Workspace

Summary

The DVD Studio Pro Workspace contains all of the information displays required for you to author your DVDs—the Main Window, which organizes the project Elements as they are created into Tabs for the *Outline, Story, Connections, Menu, Viewer, Asset, Log, Track, Slideshow,* and *Script Tabs*. The workspace also contains *The Palette*, which organizes and displays Templates, Styles, and Shapes that are installed with DVD Studio Pro 4, and Audio, Still, and Movie asset files that are resident on the Authoring system. There is also an additional display, the *Inspector*, that provides more detailed information.

Clicking on an *Element* (Track, Marker, Story, Menu, Button, Slideshow, Slide, Script, Asset File) will cause the display of the appropriate *Inspector* for that Element. The Inspector displays and allows setting of the specific properties of each Element created or Asset used in the Project, or loaded into the Asset Tab. It is in the Inspector that many settings are made that define the operation, navigation, and interaction of the various Items.

The *Viewer* Tab is used to roughly preview Video and Audio Assets, as well as project elements like Tracks, and Slideshows. It does not provide the sophisticated navigation control available in the Simulator.

The *Simulator* allows you to simulate the DVD project you are creating, to test its functionality at every step of the way. The Simulator display also includes a selectable *Info Pane*, which reads out the current Title and Chapter numbers, the Title Absolute and Chapter Relative running times, and the numbers of the currently selected Angle, Audio, and Subtitle Streams, as well as System and General Parameters.

New in DVD Studio Pro 4, the simulator can output video to a Digital Cinema Desktop (or other secondary display monitor), and surround audio can be routed out digitally or optically to a multi-channel decoder for Dolby Digital or DTS decoding.

Chapter 4

Assets

DVD Studio Pro 4

Figure 4-0 — Assets are essential to your DVD.

Goals

By the end of this chapter, you should be able to:

- Understand **Assets**
- Understand how to **Import Assets**
- Understand how to use **Assets without importing them first**
- Understand **Video Assets**, and how to create them
- Understand **Audio Assets**, and how to create them
- Understand **Graphics Assets**, and how to create them
- Understand **Subtitle Assets**, and how to create them
- Understand about **ROM Assets**
- Understand how **DVD-Hybrid discs deliver assets**

What Are Assets?

Assets are the essence of every DVD—its basic building blocks. The generic name "Asset" is assigned to any video, audio, graphic, or subtitle element required in the creation of a DVD. To author a DVD, we need to design and create (or acquire) the DVD project's various Assets, and make them available for use inside of the Authoring environment.

Starting with DVD Studio Pro 2, you no longer need to Import asset files into the Assets Tab *before* they can be used for Authoring. Assets may now be dragged directly into the Main Window from the Palette, or even from the Finder, *prior* to being encoded into MPEG! All that is required is that the file is of a type that is suitable for DVD authoring, and that suitability depends on what kind of asset we are describing.

What Kind of Assets Are There?

DVD assets include the following kinds of media files: Video Assets, Audio Assets, and Still Assets (graphics images). Other assets are graphics or text files for Subtitles, and computer-readable files used in the DVD-ROM zone of Hybrid discs. (See Chapter 17 for more on Hybrid DVD discs.)

Video Assets

Video Assets can be used in Tracks and Motion Menu elements. They cannot be used in Slideshows in DVD Studio Pro.

Video Assets for DVD can include the following:

- **MPEG-2** files already encoded from video files or videotape.

Assets

- **MPEG-1** files already encoded from video files or videotape.
- **Movie Files**—QuickTime Movie files not yet encoded.
- **H.264 (AVC)** files encoded in next-generation HD DVD Codec.

DVD Studio Pro can use both MPEG-2 and MPEG-1 files, movies, and some newer codecs, as you see. MPEG-2 files must comply with either the NTSC (*National Television System Committee*) video format at 720 × 480/30 fps or the PAL (*Phase Alternating Line*) video format—720 × 576/25 fps to be compliant with DVD Studio Pro. MPEG-1 files have similar requirements. They may be either NTSC at 352 × 240/30 fps, or PAL at 352 × 288/25 fps.

New to DVD Studio Pro 4 is the use of the *Advanced Video Codec* (known as AVC, or H.264). This codec is used in conjunction with the creation of HD DVD discs, the cutting edge of next-generation DVDs. See Chapter 5 for more on AVC, and Chapter 17 for more on HD DVD authoring.

> **NOTE!**
> Video assets for DVD may be either NTSC or PAL, but it isn't possible to use both NTSC and PAL on the same side of the same DVD. Plan accordingly! By the way, you will note there is NO entry for the SECAM system, as PAL is used instead of SECAM.

Using Video Files as DVD Assets

Video Assets for DVD can begin as video content on videotape or as files created in a nonlinear video edit system like Final Cut Pro. This content may be captured into QuickTime through a variety of means, both hardware and software. One very common method is capturing through FireWire from a FireWire-equipped camera or deck. Video Assets may have already been encoded into MPEG-2 or AVC from videotape or transcoded into MPEG-2 or AVC files from digital video files. Video Assets may also be used in their pre-encoding state, as QuickTime movie files, in a number of different codecs.

New to DVD Studio Pro 4 is the ability to use movies in the HDV codec natively. Note that they will still need to be encoded into a DVD compliant form.

On the Macintosh, MPEG-2 encoding is available in software through Apple's *Compressor 2* Application, which is installed with DVD Studio Pro 4. The QuickTime MPEG-2 Export Component, previously used in DVD Studio Pro 2 and 3 to allow for direct export to MPEG-2 from the QuickTime Pro Player, is no longer installed as part of DVD Studio Pro 4 or the Final Cut studio.

In addition to direct encoding in Compressor 2, DVD Studio Pro continues to offer background MPEG-2 encoding internal to the application. As you work, the DVD Studio Pro internal encoder can be converting your spare CPU cycles into MPEG-2 encoding time. This feature allows movie files to be used as authoring assets in DVD Studio Pro before being encoded into MPEG-2, just like in iDVD.

Third-party Software MPEG encoding is also available through applications from Innobits (BitVice), Main Concept (mccoder), Digigami (MegaPEG Pro), Popwire Technologies (Compression Master), and Heuris (MPEG Power Professional). Software encoding is almost always used on a video program that has been captured as a file, since it obviates the need to export to tape and capture again.

Software encoders for the new AVC (H.264) codec are still a bit hard to find, but you will find one of them already installed—that's right: Compressor 2 is a dandy AVC encoder, and will create HD DVD compliant assets directly from Final Cut Pro or from practically any QuickTime compatible movie of the proper resolution.

93

DVD Studio Pro 4

For real-time encoding from videotape, hardware encoding cards are available for the Macintosh from Optibase, Wired, Inc., and Sonic Solutions. Other hardware cards may be used to capture video in a form other than MPEG-2, but rely on software encoding to convert the captured video content into MPEG-2 after the capture has been completed.

You will find lots of additional information on software and hardware encoding in the DVD accompanying this book, and in Appendix B. We will cover MPEG Encoding in greater depth in Chapter 5, "Making Video for DVD." Because AVC (H.264) is so new, we'll cover it it more depth on the recipe4dvd website, since new developments will be happening in the coming years.

Audio Assets

Audio Assets for DVD can take a few different forms:

- **AIFF**—Mac PCM (*Pulse Code Modulation*) files in *Audio Interchange File Format*).
- **WAV**—PC PCM files in the *WAV Audio File Format*.
- **AC-3**—*Dolby Digital* bitstream files containing from 1.0 to 5.1 channels, compressed in size and reduced in data rate.
- **DTS**—(*Digital Theatre Systems*) beginning with DVD Studio Pro 4, you may use DTS audio files in formats up to 6.1, encoded in formats that are compressed in size and reduced in data rate, and compliant with DVD authoring. DTS compatibility was added in DVD Studio Pro 3.

Audio Assets can be used as soundtrack elements in Tracks, as Soundtracks for Menu elements (both Still menus and Motion Menus) and as individual or overall audio files for Slideshows.

About Audio Files

AIFF files are created anytime a QuickTime movie file containing audio is exported into MPEG-2 using Compressor or the QuickTime MPEG encoder (if you still have that component installed). *WAV* files may be created on a PC, and imported onto the Mac for use with DVD Studio Pro. AIFF or WAV files may be used directly in DVD Studio Pro, or may be further encoded into the *Dolby Digital* (AC3) format using the Compressor 2 application, a part of the DVD Studio Pro 4 application suite. (Earlier versions of DVD Studio Pro provided a separate application, A.Pack for this task.) It is recommended to use Dolby Digital (AC3) files because of the advantages of smaller file size and reduced data bandwidth.

Other formats are also available for audio in DVDs: *MPEG-1 layer 2* (a.k.a. MP2) is a spec-legal format of audio for DVD, but not in Region 1 (North America). It is usually coupled with PAL video, when making assets for non-NTSC regions. While MP2 was originally part of the DVD Specification, this format has lost favor in many regions where its use is still permitted, with Dolby Digital (AC3) becoming the more often used format.

We will cover audio for DVD in greater depth in Chapter 6, "Making Audio for DVD."

Still Assets

Still Assets are graphic bitmaps images used to create DVD Still Menus, Slides for Slideshows, and Subpicture Overlays of many different types, including Menu Highlights, Watermarks, and Buttons over Video highlights.

Menus are the interactive presentations that provide choices for DVD users, to allow selection from among the many assets available on the DVD. *Menu Highlights* inform the user as to which Menu Button

Assets

is currently selected, and also confirm activation of a Menu Button. These Highlights are frequently created using *Subpicture Overlay* files to provide a *Highlight Map* for the Menu. Still images are also used in a DVD Slideshow, essentially a playlist of from 1-99 still images, with or without audio.

Still Assets for DVD are typically prepared using Adobe Photoshop or another pixel-based graphics application, but DVD Studio Pro has a built-in Menu Editor that allows creation of compliant still menus by compositing still backgrounds, additional graphics images, and type elements. When the project is built, the completed menu is rendered into the required MPEG still image. In DVD Studio Pro, Highlights may also be created by information contained within the *Button Shapes* used in the Menu Editor.

> **NOTE!**
> DVD Menu assets may be prepared in vector format, but *must be exported* in pixel format to be usable. Graphics formats usable in DVD Studio Pro include .PICT, .TIFF, .TARGA, .BMP, and Photoshop layered PSD files. DVD Studio Pro even allows usage of the JPEG format, but it is recommended to use the highest quality image possible as the original source for any DVD image.

We will cover Graphics preparation in great depth in Chapter 15, "Graphics Issues for DVD Images." Menu creation will be discussed in Chapter 9, "Authoring Menus."

Subtitle Assets

Subtitles are Graphical bitmaps that have been prepared to display dialog, stage direction, or commentary information as text info superimposed over the MPEG background stream in the finished DVD.

In DVD Studio Pro authoring, Subtitle events are created in the desired Subtitle stream of the selected Track, using text or prepared subtitle files. Many times, Subtitles are created by hand in the Track Editor, by creating a Subtitle event and adding text and style attributes as needed.

Creating subtitles manually is a process that isn't really practical for long-form content with hundreds of subtitle events. In these cases, Subtitle events may be created from an external edit list containing either the text entries themselves, or a list of event times (very similar to an EDL) and the prepared graphics files that are to be displayed during those timed events. Subtitles using this last method are usually prepared by a third-party subtitling company. (Searching *Google* for "DVD Subtitling Company" will find many of these.) A large number of these companies can also do translations for alternate languages, as well.

In DVD Studio Pro, industry standard Subtitle files may be imported in these formats:

- *STL*—The Spruce Technologies subtitle format
- *SON*—The Sonic Solutions bitmap-based format
- *SCR*—The Daiken-Comtec Labs Scenarist bitmap format
- *TXT*—A plain text file

We will cover Subtitle preparation in depth in Chapter 12, "Creating Subtitles."

Other Assets—Hybrid (DVD-ROM) Deliverable Files

Hybrid Discs are those DVDs that contain both DVD-Video data and computer-readable data in the *Others* or *ROM* zone (typically at the Root level of the DVD Disc). Computer-readable assets destined for the ROM zone of the DVD do not need to be imported into the Asset Tab of DVD Studio Pro. Instead, they should be collected into a single folder so the location of that folder can be set in the "DVD-ROM" property fields in either the *Disc Property Inspector*, or the **File > Advanced Burn > Format** dialog. ROM assets do not get imported into the Assets Tab, and will not be found there. We will cover Hybrid discs in more detail in Chapter 17, "Advanced Authoring."

Organizing Your Assets

Regardless of the method you choose to import your DVD assets, you must first locate the assets you would like to import. To make this easier, organizing your DVD assets into folders that correspond to Asset Type is something to consider. Take a quick glance at the Recipe4DVDsource media folder (Fig. 4-1) for an example of asset management. Because there are advantages to locating both Audio and Video from your movie encodes into one folder, notice that this technique has been used for the Tutorial media. (You'll learn more about Dragging Matching Audio in Chapter 7, "Authoring Tracks.")

Organizing Assets in Folders before Importing

Figure 4-1 shows a typical DVD Project folder. Note the various folders organizing the different types of assets for easy access. Proper DVD project management includes Asset File Management—meaning, learning how to organize your asset files so they can be found easily, and more importantly, so they are difficult to misplace! Separating by type is easy and efficient.

Organizing Assets Using the Palette

Another good location for Asset organization is the Palette. Adding Folders into the Audio, Video, and Stills Tabs of the Palette keeps them readily available for Authoring. The Palette Tabs can also be searched, making it easy to find a specific file, even in a folder containing a large number of files.

The Assets Tab

The *Assets Tab* (see Fig. 4-2) is the place where all Movies and Video Streams, Audio Files and Streams, Graphics images, and Subtitle Streams will be imported when they have been used for Authoring. In DVD Studio Pro, Assets may be used for Authoring without having been imported first—just drag them in from the Palette or the Finder! But if desired, Assets may still be imported before Authoring.

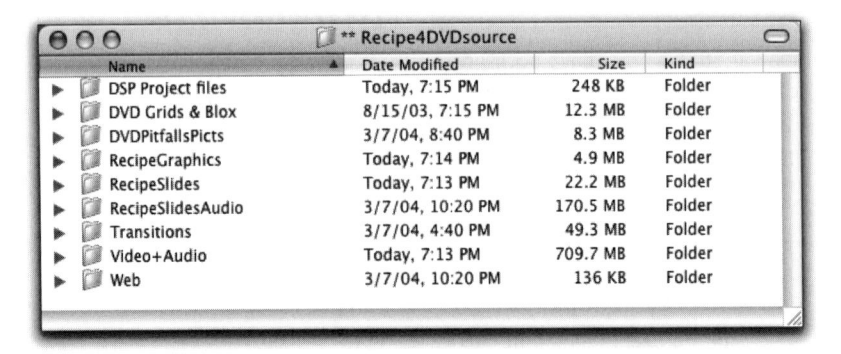

Name	Date Modified	Size	Kind
DSP Project files	Today, 7:15 PM	248 KB	Folder
DVD Grids & Blox	8/15/03, 7:15 PM	12.3 MB	Folder
DVDPitfallsPicts	3/7/04, 8:40 PM	8.3 MB	Folder
RecipeGraphics	Today, 7:14 PM	4.9 MB	Folder
RecipeSlides	Today, 7:13 PM	22.2 MB	Folder
RecipeSlidesAudio	3/7/04, 10:20 PM	170.5 MB	Folder
Transitions	3/7/04, 4:40 PM	49.3 MB	Folder
Video+Audio	Today, 7:13 PM	709.7 MB	Folder
Web	3/7/04, 10:20 PM	136 KB	Folder

Figure 4-1 — The Recipe4DVD Demo Project folder.

Assets

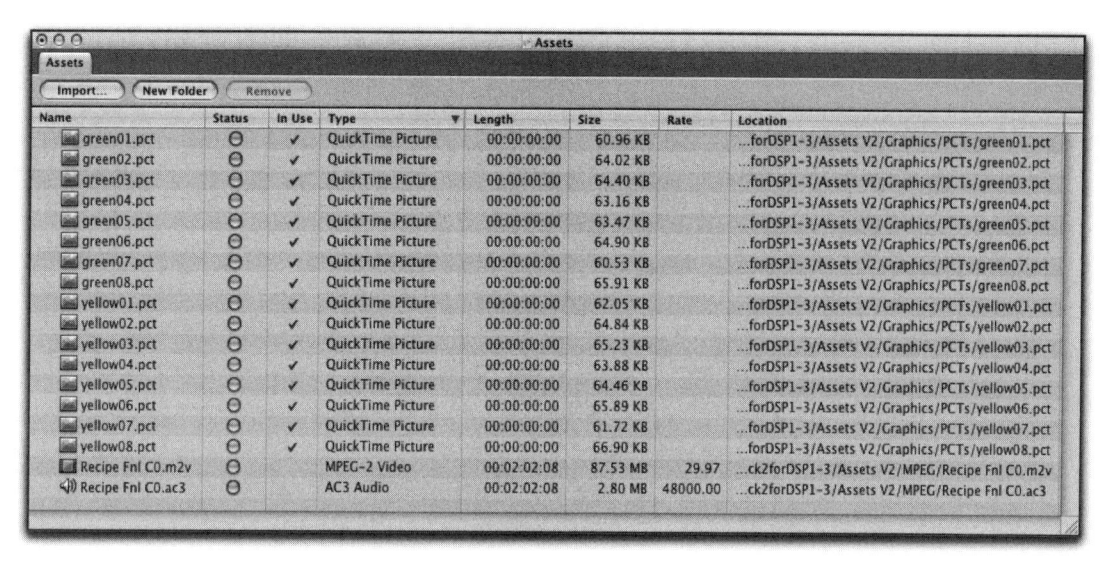

Figure 4-2 — The Assets Tab.

When an asset file is imported, DVDSP creates an entry in the Assets Tab, which points to the file's actual location on the storage attached to the Authoring system. This storage may be Internal Hard Drives, FireWire Drives, SCSI Drives, or Network Attached Volumes.

> **IMPORTANT!**
> **Be aware** of using ANY storage device that may become unattached or dismounted! If an authored asset goes missing due to this, DVD Studio Pro will not be able to finish the project until the asset has been restored, or a substitute asset assigned that *is* currently available.

Where Do the Assets Go?

Ordinarily, imported assets will land in the root level of the Assets Tab. This can create quite a long list of files, as you would expect. DVD Studio Pro adds an additional level of organization to the Assets Tab: Folders and Subfolders may now be created within the Tab and used to group similar or related assets, as you choose (see Fig. 4-3).

Before you click the *Import...* button to initiate an asset import, select the proper Folder, or select the Root level of the Assets Tab. Proper placement of the selection cursor in the Assets Tab is important. Depending on where you place the cursor, your imported files will either be placed into that selected Folder or at the root level in the Assets Tab. Once the Import Assets dialog has been used to initiate an

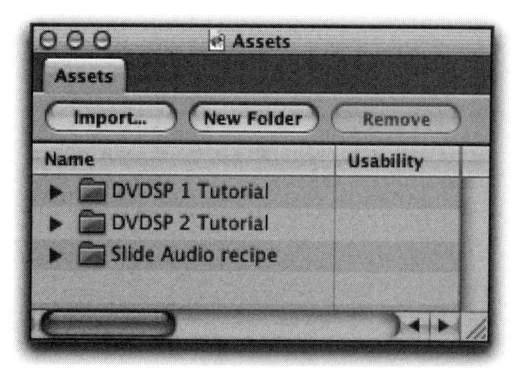

Figure 4-3 — The Assets Tab, shown after Importing some Folders.

DVD Studio Pro 4

import, the selected files will appear in the Assets Tab at (hopefully) the chosen location.

If you would like the imported files to be organized in a folder within the Assets Tab, you can just Import a folder of files, or you can create a folder within the Assets Tab to organize files. This folder may be created before or after importing. Figure 4-3 shows the Assets Tab using folders to organize imported files.

Creating a New Folder within the Assets Tab

To create a new folder within the Assets Tab, use one of the following methods:

- Click the *New Folder* Button in the Assets Tab, **or**
- *Control-Click* in the Assets Tab, or on a Folder within the Assets Tab, and select *Create New Folder* in the pop-up, **or**
- Select **Project > New Asset Folder** in the menubar, **or**
- Type *Command-Option-B* (⌘-**Opt-B**) on the keyboard.

A New Folder (called "Untitled Folder") will be created in the Root level of the Assets Tab if no folder is selected, or it will be created within the folder that was selected at the time the command was executed.

Adding Assets to the Assets Tab

There are two ways to import asset files or folders into DVD Studio Pro's Assets Tab:

- Use any of the many available *Import Commands*, **or**
- *drag-and-drop* directly into the Assets Tab.

To Add Assets Using an Import Command

First, let's look at importing using *Import Commands*. There are several of these, but they all result in the same operation:

- Select **File > Import > Asset** from the menubar, **or**
- Type *Command-Shift-I*, (⌘-**Shift-I**), **or**
- Click the *Import Asset* Tool in the toolbar (if visible), **or**
- Double-click the mouse inside of the Assets Tab, **or**
- *Control-Click* in the Assets Tab, select *Import Asset...*, **or**
- Click the *Import...* button in the Assets Tab (Fig. 4-4).

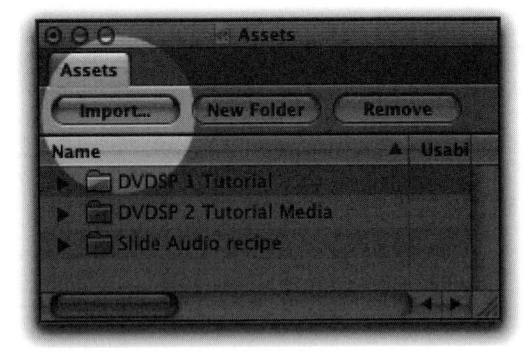

Figure 4-4 – The Assets Tab Import Button

Using any of these options will open the *Import Assets Dialog* box (see Fig. 4-5), which will invite you to locate and select the asset(s) you wish to import.

The Import Assets Dialog box allows you to navigate around your Mac and import DVD asset files or folders (even folders containing other folders) from any currently accessible storage Volume. You may select one or more files, one or more folders, or a combination of files and folders. Of course, you may invoke this command as many times as needed to gather in all of the needed assets.

> **NOTE!**
> DVD Studio Pro 4 does not have a method to collect files in an Import List before importing. If you need to gather files from more than one folder location, do several Imports.

Figure 4-5 — The Import Assets dialog box.

> **NOTE!**
> If your assets exist on a CD, FireWire drive, or any other removable storage media, it's essential to copy them to your project drive before importing them. We'll discuss this more in the section titled Missing Assets, later in this chapter.

Initiating the Asset Import

Once you've selected the desired assets in the Import Assets dialog, clicking the *Import* button begins the import process (Fig. 4-6). During the import, you will likely see a progress dialog box (Fig. 4-7).

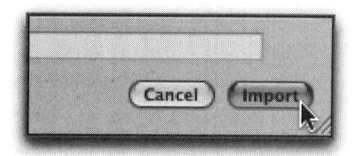

Figure 4-6 — The Import Button in Import Assets Dialog ready to go.

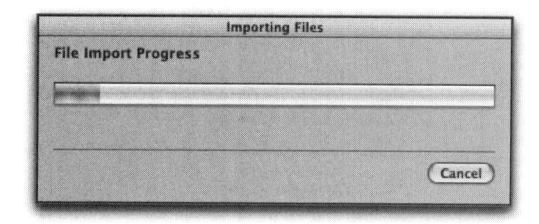

Figure 4-7 — File Import Progress dialog.

Once the assets have been completely imported, they will appear in the chosen location within the Assets Tab—either at the root level, or within the selected Folder. If you have imported a Folder, the entire Folder will appear, containing the files you imported.

Adding Assets by Drag and Drop

You can also drag-and-drop files into the Assets Tab from the Mac desktop or the Palette. The only prerequisite is that you must be able to access the Assets Tab while you are dragging—so be certain it is open first.

Importing Files or Folders by Drag-and-Drop

To import files using drag-and-drop, select the desired files or folders, and then drag them directly into the Assets Tab, or onto a Folder within the Assets Tab.

99

DVD Studio Pro 4

They should immediately be visible in the Asset list, and usable for authoring. For video assets that require encoding, the status of the encoding will be indicated in the Usability Column, in color coding: red, yellow, or green. Learn more in Chapter 5, "Making Video for DVD."

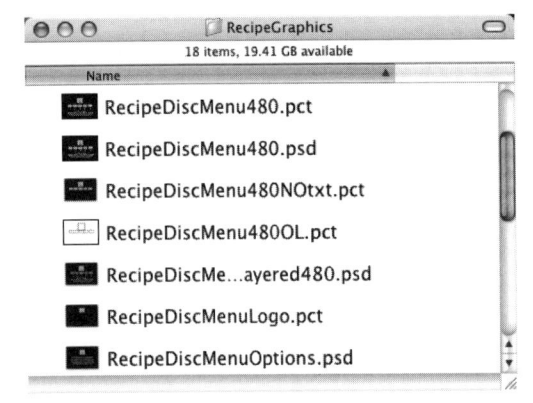

Figure 4-8 — The Recipe Assets folder.

How DVD Studio Pro Organizes Assets

Files imported into the Assets tab may be organized and sorted by practically any of the columns of criteria displayed in the tab. Click on the criteria name to sort. (Also see Fig. 4-8.)

Checking Imported Asset Files for Usability

DVD Studio Pro will check every file imported into the Assets Tab, to be sure that it is either usable for .DVD Authoring as-is, or can be encoded into a usable format. If not, an error message will display (see Fig. 4-9).

This dialog will also display if, for example, you include a folder with DVD Project files in the folder containing Asset files as well. Only usable Asset files are recognized during Import.

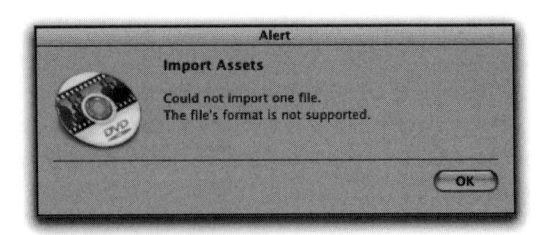

Figure 4-9 — The Alert Dialog showing Import error message.

To prevent problems while importing, it's always best to start with the compliant asset file formats we discussed earlier in this chapter.

What Does Importing Really Do?

Importing assets into DVD Studio Pro does NOT move the actual asset file into the Application—instead, the Import function creates a symbolic link (like an *alias*) between DVD Studio Pro and the actual physical location of the asset file.

> **IMPORTANT!**
> When it creates the symbolic link to an asset, DVD Studio Pro records the current drive volume and all folders in the file path to an imported asset. This is important because moving a file, or removing the CD or DVD it is on once it has been imported, can and will cause trouble. Moving projects from one system to another may also cause some difficulty. Read more in the section titled Missing Assets.

Locating an Asset File in the Finder

The imported assets are listed in the Assets Tab, and their location can be determined by displaying the Location Column (see Fig. 4-12b) or by Control-Clicking in the Assets Tab and selecting *Reveal in Finder*. You can also locate a selected asset file by choosing **File > Reveal in Finder** from the

Assets

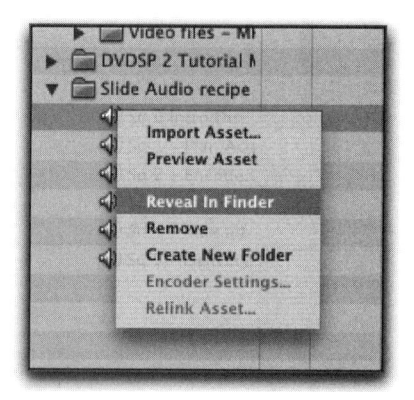

Figure 4-10 — Revealing an Asset in the Finder.

menubar. This command is only available when an individual asset file has been selected (see Fig. 4-10).

Seeing More Details in the Assets Tab

The Assets Tab is a columnar display, with the default displaying six of the available 15 columns (Fig. 4-11). The Tab can reveal much more detail about a file than the default six-column display shows.

Displaying More Data in the Assets Tab

To display more data columns in the Assets Tab, do this: *Control-Click* on the Column Title bar, and select a column in the pop-up. To add more than one column, repeat the command. Only the Columns that are checked (✓) will be displayed in the Tab. (See Fig. 4-12a.)

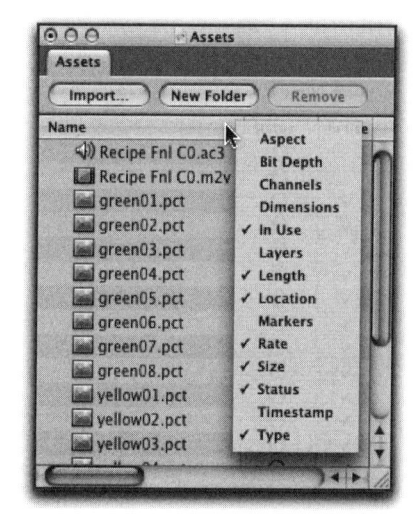

Figure 4-12a — The Assets Tab displays its available headings.

Figure 4-11 — The Six-column Asset Display.

101

DVD Studio Pro 4

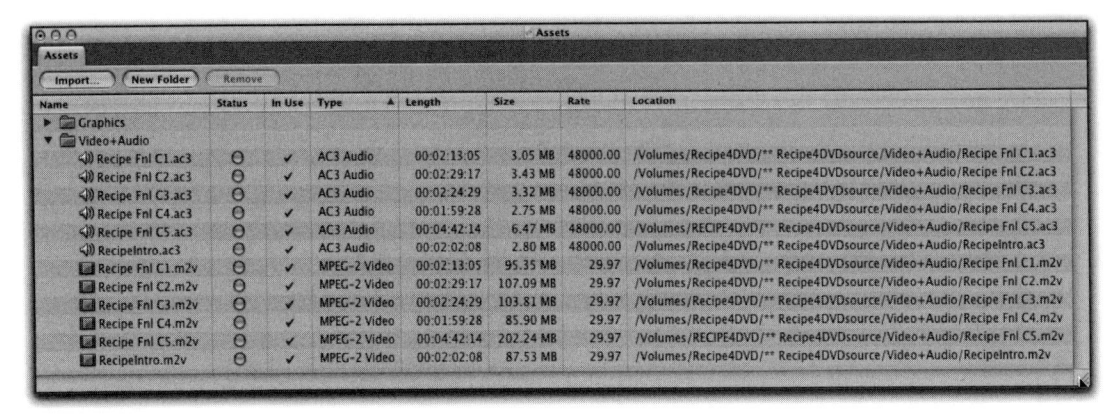

Figure 4-12b — Asset Tab with additional Columns displayed.

> **TIP!**
> If the Location Column is displayed, the physical location of the asset on the host Macintosh can be easily seen.

Missing Assets...

DVD Studio Pro keeps track of every imported asset file by its name and its complete location. If the asset file is moved or renamed once it has been imported (or the project is copied to a new hard drive), the symbolic link to the file may accidentally break. In this case, the asset file will not be found when the project is next opened, and this missing file will disrupt the normal operation of DVD Studio Pro.

When an asset file goes missing, the Assets Tab will display the corresponding entry in RED (Fig. 4-13). This file <u>must</u> be located or a substitute asset assigned before the current DVD can be built.

Fixing Missing Assets

If an asset file goes missing while you are working on a DVD, there are several possibilities why: it may be that the FireWire drive or network volume it is on

Figure 4-13 — Missing Assets Display Sq0 to Sq 5 are listed in red on screen, but the red color displays as a light grey in our Greyscale images.

has dismounted, or perhaps a CD or DVD containing the asset has been removed. It may be possible for you to relink the missing file by repairing its asset file path. This operation is shown below.

102

Assets

Relinking One Missing Asset File

To relink a single missing asset file, use one of these methods:

- Select the asset you need to relink by clicking on it, then
- Select **File > Relink Asset...** from the menubar, **or**
- Control-Click ON the missing Asset, select 'Relink Asset' (see Fig. 4-14).

Figure 4-14 — The Relink Asset... command opens a Relink dialog box.

Relinking an Entire Folder, or More than One File

If more than one asset within a folder is missing, DVD Studio Pro will recognize that fact. When a file within that folder is relinked, an *Alert* will offer to relink the remaining files (see Fig. 4-15).

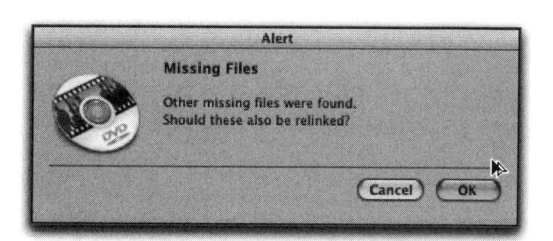

Figure 4-15 — Relink remaining missing files?

Asset File Properties

Selecting an individual asset file in the Assets Tab will also show that Asset's details in the Inspector, if the Inspector is visible. These details generally include the asset's name, its composition (type, length, encode rate, and so on) and for Video assets, a browse pane, which allows scrolling through the asset (see Fig. 4-16 and Fig. 4-17 for Audio Assets).

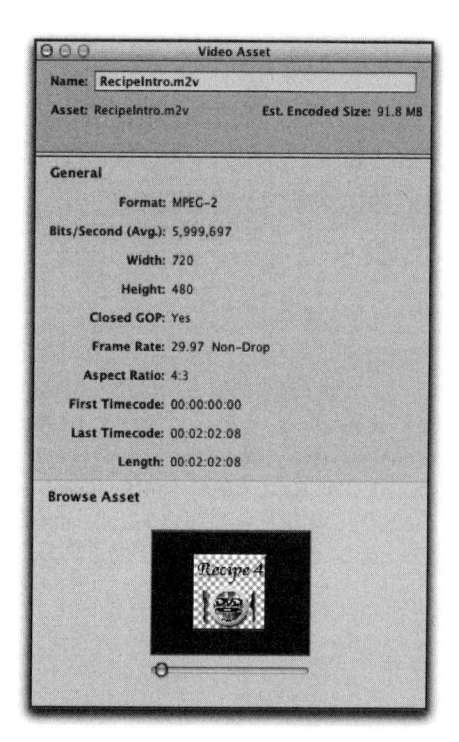

Figure 4-16 — Video Asset Inspector.

103

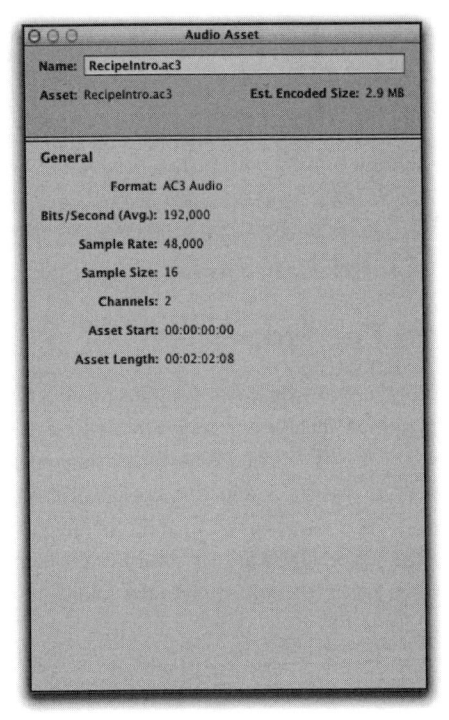

Figure 4-17 — Audio Asset Inspector.

Removing a File or Folder

If there is ever a difficulty with an import, or if you have mistakenly imported files you do not need, it's best to remove them from the Assets Tab, so they will not be cluttering up your list. Entries within the Assets Tab *will* be considered as Project Assets, and DVD Studio Pro will expect to find them the next time the project is opened. This could be a problem if you decide to delete those unneeded files at some future time—DVD Studio Pro will still be looking for them!

To Remove a Folder or File from the Assets Tab

Click on the desired file(s) or folder(s) to select, then use one of the following methods:

- Click the *Remove* Button in the Assets Tab, **or**
- Hit the *Delete* Key, **or**
- Select **Edit > Delete** from the menubar, **or**
- Select **Edit > Remove Asset** from the menubar, **or**
- *Control-Click* on a File or Folder and select *Remove* from the pop-up.

Adding Assets by Dragging Directly into Elements

DVD Studio Pro has a great philosophy regarding assets—you don't *need* to import them prior to using them. Simply dragging an Asset file into *any* project element (Track, Menu, or Slideshow) or into the Graphical View will add that asset into the Assets Tab of that project.

Adding a file that is not yet encoded into a DVD-compatible form will require the built-in MPEG encoder to convert that file into usable form either while you're authoring, or at the time of the Build. We'll cover this in Chapter 5. Also see Figure 2-12 regarding the Encoding preferences pane.

Dragging an Asset into the Track Editor

Video files and Streams, audio files and Streams, and still images may be dragged into the proper Stream areas of the Track Editor. If video files have not yet been encoded into MPEG-2, the settings for the background encoder may be set in the Encode preferences pane (see Fig. 2-12). Still images will be imported in at the default duration set in the General Preferences Pane (see Fig. 2-4). Their duration may be easily edited in the Track Editor once they have been placed.

Dragging an Asset into the Slideshow Editor

Still images may be dragged into a Slideshow element and will be placed in the slideshow at the default duration set in the General Preferences pane (see Fig. 2- 4). Slide assets can be of many different graphic formats. More detail can be found in Chapter 10, "Authoring Slideshows."

DVD Studio Pro 4

Summary

Assets are the building blocks of your DVD project.

Assets are the video, audio, and graphics files needed to complete your DVD project. Assets may also include subtitle text files and DVD-ROM deliverable files. Assets may be imported into the project or used without being imported. Assets that are used directly in project elements will be imported automatically. Assets may be imported manually using one of the many import commands or by drag-and-drop.

Imported Video, Audio, and Graphics Assets are organized and stored in the Assets Tab. Other assets, like Subtitles and files for the DVD-ROM zone are not stored in the Tab. The actual Asset file is not stored in the Asset Tab; instead, a pointer is created that links DVD Studio Pro to the file's location on the host system. If the asset file is moved to a different location, removed from the system, or even renamed, the link to that asset may get broken and need repairing.

Assets with broken links will not perform properly in the DVD project until they are relinked to their original file. This is done with the relink command.

Chapter 5

Making Video

DVD Studio Pro 4

Figure 5-1 — Compressor 2.

Goals

By the end of this chapter, you should:

- Understand **what kinds of video you can use** in DVD authoring
- Understand **what encoding is**, and what it means
- Understand what **MPEG-2, MPEG-1 and AVC (H.264)** are
- Understand what **Compressor 2** is, and the role it plays
- Understand **how to set up an encode** in Compressor 2
- Understand **how to set up an MPEG-2 encode in DVD Studio Pro**
- Understand **how to set up an MPEG-2 encode in QuickTime** (even though this is no longer a preferred encoding tool)
- Understand how to **export to Compressor 2** from Final Cut
- Understand about the **old QuickTime encoder** (no longer used)
- Know **how to use movies** in DVD Studio Pro
- Know **how to use MPEG-2 Video Streams** in DVD Studio Pro
- Know **how to use MPEG-1 Video Streams** in DVD Studio Pro
- Know **how to use AVC Video Streams** in DVD Studio Pro
- Understand **how to optimize your encoding** for DVD
- Understand about **hardware encoders versus software encoders**
- Be familiar with the **software encoders** available for Macintosh

Making Video

Before You Author, You Need Video Content

You certainly can make a DVD without using Video (you could use still images and audio), but overall, video is the driving force behind creating a DVD—the *raison d'etre*, if you will. This chapter does not deal with the creative aspects of conceptualizing your video content, or editing; there are plenty of great books on how to use nonlinear video edit systems and the esthetics of editing a great program. What we *are* concerned with is what happens *after* you have the video content created and are ready to make a DVD using that content. With that in mind—let's look into the role that encoding plays in DVD.

Before You Author, Do You Have to Encode?

That *used* to be true, but DVD Studio Pro 2 changed that a while back. One of the breakthrough changes in DVD authoring with all versions of DVD Studio Pro is that you can use practically *any* QuickTime-compatible video movie as content for your DVD, without first having to encode it into MPEG before authoring*. While it sounds simple, this was a very revolutionary step in DVD production. DVD Studio Pro can handle MPEG-2 encoding for you in the background, while you work. Of course, you *still* can encode your content before you author if you want to, but you don't have to. Today, there are any number of great MPEG-2 encoding solutions, both software and hardware. That's what this chapter is about—encoding your video content successfully.

(*If you want to use the H.264 format, however, you DO need to create those encodes first.)

Recent developments in computers have made even more powerful MPEG encoding applications available, which do more and streamline the DVD workflow even better. Meanwhile, Apple's been making its product line simpler—today, Compressor 2 is the answer for encoding.

MPEG-1, MPEG-2, MPEG-4—What Video Can I Use?

The developers of the DVD specification were thoughtful enough to allow for the use of either MPEG-1 or MPEG-2 files in SD DVD authoring.

DVD Studio Pro 4 brings us the advent of HD DVD authoring and, with it, the ability to use HDV content natively and H.264 (AVC) encoded assets for High Definition DVD authoring.

In the previous edition of this book, I said:

> "MPEG-4 is a different format, for a different form of video distribution, and is not a part of the DVD specification. It's possible the DVD Specification may be updated or enhanced some day to include High-Definition Television (HDTV), Interactive TV, and maybe even MPEG-4."

Today that sounds strangely prescient because—that day has arrived! Today, DVDs may be made with High Definition content encoded using the H.264 (AVC) codec (basically MPEG-4 Layer 10), or MPEG-2, as well as Standard Definition video encoded into elementary streams using the MPEG-2 or MPEG-1 codec. Most DVD authors have chosen MPEG-2 over MPEG-1, because of the superior picture quality. DVD Studio Pro can also use MPEG-1, although with certain restrictions. (We'll discuss more about MPEG-1 later in this chapter.) So the

109

answer to the question is—you can use whichever format of video best suits your purpose.

Elementary Streams, My Dear Watson

Regardless of whether you use SD or HD, or encode using software or hardware, make sure your encoder creates *Elementary* (separate) video-only and audio-only stream files that are DVD-compliant. Many MPEG encoders can create MPEG forms known as *System Streams, Program Streams*, or *Transport Streams*. These kinds of files are not immediately usable for DVD authoring, as they are already *multiplexed*—that is, the video and audio components have already been married together. You shouldn't really need to worry about this issue because Compressor 2 is built to create elementary streams for DVD, the presets are clearly labeled, and the DVD Studio Pro encoder can only create elementary Streams. Final Cut Pro sequences are easily exported into elementary stream formats through Compressor 2. Of course, in Final Cut, you could always elect to output a QuickTime movie, and encode it using an alternative encoder! We will provide a quick overview of what's available in alternative encoders at the end of this chapter, and have devoted all of Appendix B to these software and hardware encoder options.

About Encoding

Before you can burn your DVD to disc, or create a DLT (*Digital Linear Tape*) master for replication, you *will* need to encode your video content into a DVD-compliant form—in DVD Studio Pro 4, that is either MPEG-2 or MPEG-1, or H.264 (AVC).

Hardware Encoders

In the early days of DVD production (way back in the mid-'90s), the standard method of encoding was to create MPEG files directly from videotape, using a *hardware encoder board*. This was generally a peripheral card that was installed into a desktop Macintosh or PC (or even an SGI machine!) that contained specialized chips to facilitate the creation of MPEG files directly from videotape in real time. These early DVD production systems contained proprietary software to control the hardware encoder boards, and additional software to author and build the DVD using the encoded MPEG assets.

A lot has changed since then…what *hasn't* changed is that hardware encoders are predictable and generally fast. On the simplest encodes, a 1-pass CBR (*constant bit rate*), you can count on a predictable 1x encode time—one pass through the tape and you're done. On VBR (*variable bit rate*) encodes, perhaps 2 or 3 passes may be needed, but back then, hardware encoding was still a lot faster than any software encoder.

Which Encoding Approach to Use?

Your ultimate choice of hardware or software encoding may be guided by the library format of the content you need to process, and perhaps your budget. In my opinion, and that of others in the DVD business, hardware encoders excel at streamlining the DVD production workflow for libraries comprised largely of videotape, since there is no "capture-then-encode" penalty to pay. Software encoders, on the other hand, are well suited to video content created in nonlinear editors, since software MPEG or AVC encoding from existing digital video files is a natural extension of the editing workflow.

Making Video

In fact, "camera-to-edit-to-disc" is an accurate description for today's digital video content creation cycle. It's fast, it's easy, and these days, it's very affordable. Did I mention that it's really easy? Oh, yes—so I did.

Software Encoders

Today's faster computers, more powerful processors, and cheaper mass storage devices have brought software MPEG-2 encoding into its own, and enabled convenient H.264 encoding as well.

Software encoding dovetails very nicely with desktop nonlinear video editing—a large part of today's content creation methodology. You can see this quite clearly in *Final Cut Studio*, where Apple has created a streamlined content workflow that features software encoding. Edited output from Final Cut Pro is generally encoded using Compressor 2, but QuickTime movies can also be used for authoring and encoded directly inside of DVD Studio Pro. Connectivity is seamless.

New! This connectivity is also evident in another new feature of DVD Studio Pro 4, the ability to author using HDV content natively—edited HDV material from Final Cut can be brought directly into DVD Studio Pro and encoded automatically as appropriate.

Both Compressor 2 and DVD Studio Pro have powerful encoding capabilities, albeit with some small variations between them. Both contain enhanced control of the encoding parameters, including true 2-pass VBR setups, motion estimation, and other great features. Compressor 2 has a particularly rich feature set, and offers detailed control of encoding parameters, pre-encode filtering (see Fig. 5-2), batch processing for unattended encoding, access to the H.264 codec (not found in DVD Studio Pro's internal encoder), AC3 audio encoding, and much more.

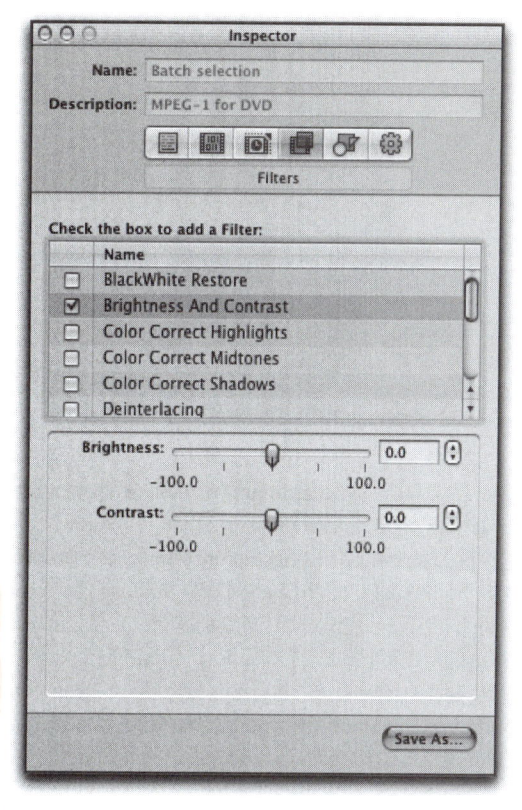

Figure 5-2 — The Compressor 2 Filters—an array of problem solvers.

Software Encoding Options

If you are familiar with earlier versions of DVD Studio Pro (2 and 3, specifically), you will want to make note that things have changed with time and the old standby QuickTime MPEG encoder is no longer supplied, installed, or required by DVD Studio Pro 4. In its place, Apple relies heavily on a new version of an old favorite—so we lead off our encoder quick tour with *Compressor 2* (see Fig. 5-3a).

DVD Studio Pro 4

Compressor 2

Figure 5-3a — Compressor 2 logo.

The icon may look familiar, but there's a lot more under the hood in Compressor 2 than in Compressor 1. Apple's revamped flagship encoder gets installed with both Final Cut Pro 5 and DVD Studio Pro 4, and is also an integral part of *Final Cut Studio*. Right out of the box, it's ready for action, making improved MPEG-2 streams. It may not be able to leap tall buildings in a single bound, but the quality of Compressor's encodes can be superior to other software encoders, especially at the critical low bitrate encode points (4–4.5 Mbps).

(In a recent encoder shootout conducted by noted compression gurus, Ben Waggoner and Darren Giles, published in *DV Magazine*, Compressor was "best of breed" for the Macintosh.)

New to this version of Compressor is the ability to create AC3 streams and H.264 (AVC) streams for HD DVD authoring.

Compressor 2 continues the batch processing model for its encoding (see Fig. 5-3b), and allows fairly complete control over each detail and parameter of each encode. It is a 2-pass VBR encoder when you need it to be, but can also be run in 1-pass CBR mode.

Compressor comes with a vastly expanded set of presets for not only MPEG-2 encoding but practically every other format of media. These presets include both *fast* and *best quality* MPEG-2 encodes for 4 × 3 and 16 × 9 video content at 90, 120, and 150 minute running times. Each preset includes a video setting, which is very customizable, and an audio setting that defaults to the de-facto standard 16-bit 48kHz PCM and a new addition, Dolby Digital 2.0 and 5.1 (Stereo and Surround AC3 streams). (See Fig. 5-3c.)

In lieu of *A.Pack,* Dolby Digital AC3 encoding is now standard in Compressor 2, as are settings for H.264 (for HD DVD), 3G Mobile Devices, and an

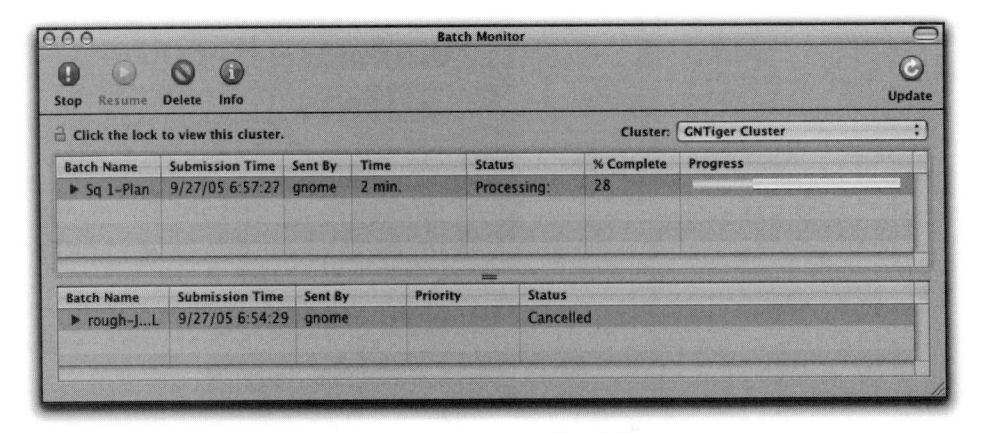

Figure 5-3b — Compressor 2 Batch Monitor.

Making Video

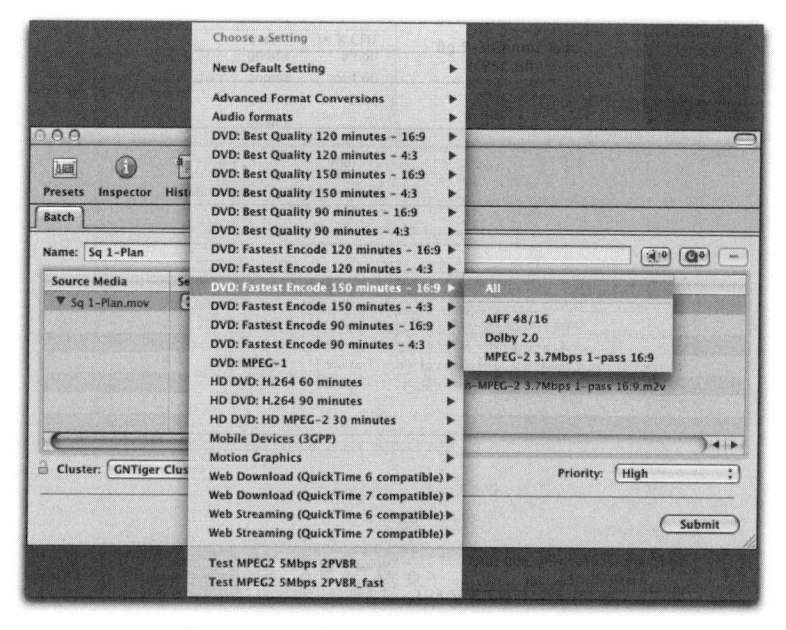

Figure 5-3c — Compressor 2 Encode Settings.

incredible array of Advanced Format Conversions ranging from DV and DVCPro to HDV and uncompressed High Definition 1080p. Man, this baby's loaded for bear! (See Fig. 5-3d.)

NTSC-to-PAL and PAL-to-NTSC Conversions Now Possible

Compressor 2 adds video format standards conversion to its impressive arsenal of tricks—you can now convert your NTSC content into PAL, and vice versa. That should end those inevitable questions about how to accomplish this previously very tricky conversion.

Qmaster for Distributed Encoding

An additional and significant new development in Compressor 2 is the addition of Apple's *Qmaster* distributed processing software. You can now harness the processor horsepower of multiple Macs on your network for the encoding your DVD assets. You may need them too. The additional features in Compressor 2, including the new Optical Flow Analysis, can add lots of time to encoding chores. For this incarnation of DVD Studio Pro, a very fast processor (or a dual) isn't a luxury—it's a necessity!

A Qmaster node can be installed on suitable Macintoshes on your local area network, and the cluster, as it is called, can be controlled from the designated *Qadministrator* machine.

You can find out more about using Qmaster and how to set up your own cluster on your Macintosh network by reading the Qmaster and Qadministrator help file (see Figs. 5-3e and 5-3f).

113

DVD Studio Pro 4

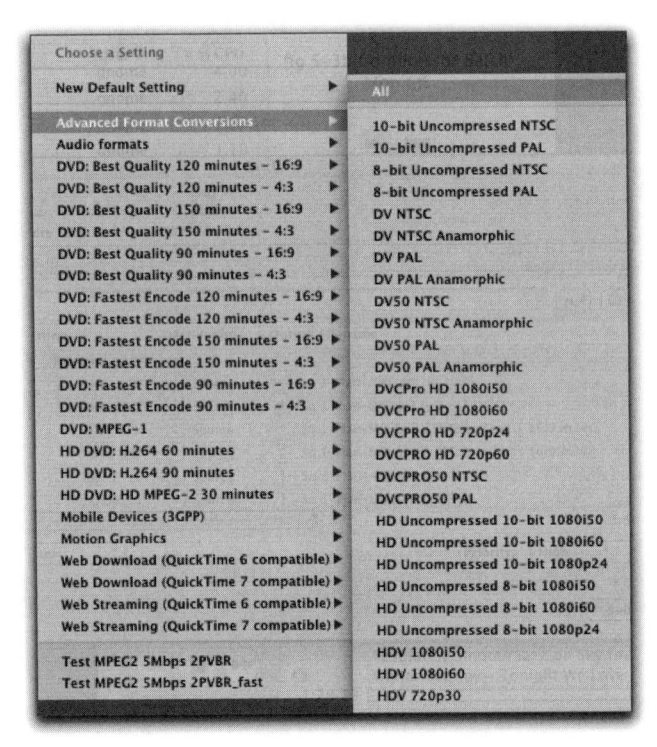

Figure 5-3d — Compressor 2 Advanced Format Conversion settings.

Figure 5-3e — Qmaster 2.

Figure 5-3f — Qadministrator 2.

Making Video

DVD Studio Pro 4 Onboard MPEG-2 Encoder

DVD Studio Pro 4 still has the built-in MPEG-2 encoder that can run in the background as you work, or at build time (see Fig. 5-4). There's a little difference in its implementation in DVD Studio Pro than in Compressor 2. The Encoding Preferences panes (see Figs. 2-12a and 2-12b) in the DVD Studio Pro onboard encoder give you nominal control over the encoding parameters for both SD and HD, but not nearly as deep a level of control and filtering as exists within Compressor 2. Still in all, it's great to be able to work quickly in DVD Studio Pro and have MPEG encoding taking place in the background while you are authoring with QuickTime movies.

Figure 5-4 — DVD Studio Pro 4 logo.

> **NOTE!**
> The DVD Studio Pro onboard MPEG encoder does not create Dolby Digital audio bitstreams—see Chapter 6, "Making Audio," for info on how to use **Compressor 2** to create AC3 bitstreams. The A.Pack application is not installed by DVD Studio Pro 4, and no longer part of Apple's encoding strategy.

Final Cut Pro—Export to MPEG-2

Figure 5-5 — Final Cut Pro logo.

It is possible for you to Export MPEG-2 files directly from Final Cut Pro (see Fig. 5-5), using either Compressor 2 or any QuickTime MPEG *Export Component* that may be installed in your system, like Digigami's MegaPEG Pro QT (see Appendix B, "Alternative Encoders"). Final Cut Pro can also export a QuickTime reference movie, which can be converted into MPEG-2 or H.264 (AVC) by using Compressor 2 or any other software encoder.

Final Cut Studio will install Final Cut Pro 5, DVD Studio Pro 4, Compressor 2, and Qmaster 2 to create a complete video editing and DVD workstation.

QuickTime Pro Player

Sadly, DVD Studio Pro 4 no longer installs the *QuickTime MPEG-2 export Component* in QuickTime Pro (see Fig. 5-6a) as it did in previous versions, but if you had installed it previously, it may not make it go away (there are conflicting reports on this phenomenon). This component enabled the QuickTime Pro Player to encode (export) MPEG-2

115

DVD Studio Pro 4

Figure 5-6a — QuickTime Player logo.

streams suitable for DVD authoring. (You needed QuickTime Pro for this to work.)

Despite the fact that Apple no longer supports this method, there is at least one other component manufactured that still works, even with QuickTime 7 and Tiger. We'll cover the Digigami line of products in Appendix B (along with others), and learn more about MegaPEG Pro HD and its cool QuickTime export component (see Fig. 5-6b). You should also know about Mpressionist Pro, a great analysis MPEG tool for both SD and HD streams!

Hardware Encoder Options

Hardware MPEG-2 encoders for Macintosh are manufactured by Optibase, Inc.; Sonic Solutions, Inc.; and Wired, Inc. Hardware encoders in general are used with videotape decks, to encode MPEG directly from tape using RS-422 transport control.

Optibase, Inc. (www.optibase.com)

The Optibase MPEG Master Encoder Family includes the MPEG Master Basic and Publisher for Power Mac G4 and G5 models running Mac OS-X as well as PC versions. These professional half-size PCI real-time encoder boards for MPEG-1 and MPEG-2 offer a full range of video and audio interfaces, video resolutions, audio encoding formats, and MPEG multiplexing capabilities.

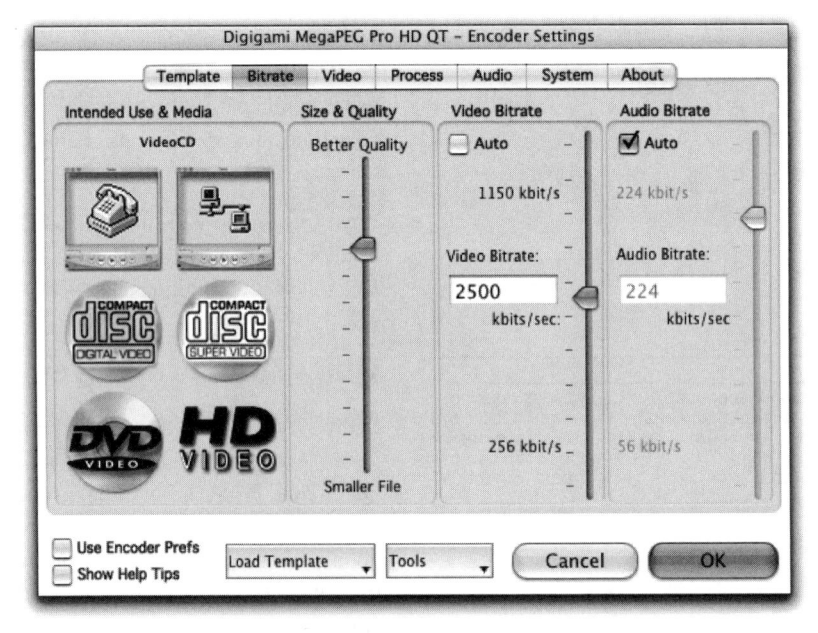

Figure 5-6b — Digigami MegaPEG Pro QT export component in action.

116

Making Video

Optibase encoders for Mac include:

- MPEG Master Basic FD1
- MPEG Master Basic VBR 720
- MPEG Master Publisher SDI
- MPEG Master Publisher DE

Wired, Inc. (www.wiredinc.com)

There are now two Wired MPEG Encoder boards available for Mac:

- Mediapress X
- Mediapress LE
- Mediapress Pro SDI boards may still be available refurbished.

The Wired Mediapress X and Mediapress LE encoders are compatible with Macintosh G5 systems, but the older Mediapress Pro SDI only works on G4 systems. There is Wired encode software available for Mac OS X. Check the Wired website for more information and prices.

Sonic Solutions, Inc. (www.sonic.com)

Sonic offers two versions of its MPEG Encoding product for Mac:

- SD-2000 (Mac)
- SD-1000 (Mac)

Sonic Solutions encoder boards can be found in a large number of Sonic DVD Creator and DVD Fusion systems operating on Macintosh G3 and G4 systems. The Mac software that drives these encoders is currently compatible only with Mac OS 9, with no plans to update them for OS X.

For Mac, Sonic Encoders can make DVD Studio Pro-compatible encodes as long as the "Large File Support" preference has been turned ON before encoding. This causes the encoder to write one large contiguous file instead of the normal 1 GB segments. Normally, Sonic encoders make 1 GB segmented MPEG files, to work around file size limitation present in older versions of the Mac OS.

DVD Studio Pro Helper

If you do have a Sonic Solutions-created encode that is segmented, it might be possible to use it in DVD Studio Pro as long as certain precautions are followed. Sonic encodes are segmented in a manner that prevents them from just being spliced together head to tail and expected to work. Instead, using the Sonic Imager software to copy the files to a separate folder will allow the Sonic Header information to be stripped out. Files may still be segmented, but can then be used in DVD Studio Pro by splicing the segments together with John Brisbin's handy utility MPEG Append. Get more information from: **http://homepage.mac.com/dvd_sp_helper**.

About MPEG-2

So, what makes MPEG-2 such a big deal, anyway? Efficiency of disc use, combined with superb video quality (if you encode properly!).

For one thing, the compression ratios MPEG-2 can achieve are astounding! Consider this: an uncompressed video stream requires about 25 Megabytes (Mega*BYTES* not Mega*BITS*) per second to capture. At that data rate, it would require close to 180 GB of hard drive storage to hold the file created for a typical feature film. Obviously, we don't yet have a DVD disc that can hold 180 GB, let alone one that can hold 80 GB, but they may appear some day given the recent developments in Blu-Ray and other Laser technologies, with HD DVD and BD disc capacities now up to 30 GB and climbing.

DVD Studio Pro 4

Video File size: 120 minutes at selected bitrates			
	Data Rate Megabits/Sec	Digital File Size MB	GB
25 MB Native	200	180,000	175.8
50 Mb DVCPRO	50	45,000	43.9
25 Mb DVCAM	25	22,500	22.0
15 Mb MPEG B'cast	15	13,500	13.2
10 Mb MPEG B'cast	10	9,000	8.8
8 Mb High BR DVD	8	7,200	7.0
6 Mb Med BR DVD	6	5,400	5.3
5 Mb DVD	5	4,500	4.4
4 Mb DVD	4	3,600	3.5

Figure 5-7 — Video File Sizes at various Data Rates.

Figure 5-8 — MPEG-1 vs. MPEG-2 image sizes—NTSC.

Our needs have been met up to today by using the MPEG-2 codec to compress the video data into a more easily stored and retrieved file size. A glance at the table (Fig. 5-7) will show you the dramatic results achieved by MPEG-2 compression rates.

For comparison purposes, the table includes the DVCAM and DVCPRO camera data rates, although very few of these cameras create digital files—yet! We're starting to see this, though, with cameras like the Panasonic AG-HVX200, which shoots to both DVCPRO tape and P2 digital memory cartridges. Most importantly, look at the significant difference in file size between Native uncompressed (175 GB) and 5 Mbits MPEG-2 (4.4 GB). These file sizes are in real computer GB, not billions (1024, not 1000). It would be impossible to make a DVD without some kind of significant data compression, hence MPEG.

There's a Lot of Data in That Tiny MPEG File

An important factor to remember is that while MPEG-2 can create these incredibly tiny files sizes, it is doing so while capturing much more data than its predecessor codec, MPEG-1.

In Figure 5-8, you will note that compared to its predecessor, MPEG-1, MPEG-2 captures a larger screen area. It can also decode a more faithful reproduction from that encoded stream.

> **NOTE!**
> In DVD Studio Pro, MPEG-1 is an option, and projects built with MPEG-1 can be previewed in DVD Studio Pro Simulator and emulated in the Apple DVD Player.

Many factors contribute to a good or great MPEG encode, not the least of which is the one that is most frequently overlooked—starting with a great-looking master image! It has often been said that one cannot "put a tux on a pig," and nowhere is it more true than in MPEG encoding. A low-quality video master, or one with inherent picture problems, will *not* likely provide a satisfactory MPEG master once encoded, although there are a few image enhancement and repair processes that can pay off handsome dividends in the quality of the finished MPEG encode. Apple's new Compressor 2 encoding application has provision for using pre-encode filters, which lends lots of possibilities for video optimization. (Figure 5-2 shows just a few of these filters.)

Making Video

Image Enhancement before or during Encoding

The power of the MPEG-2 codec is a big factor in the success of DVD. The quality of the digitized video in DVD can be spectacular, if care is taken during the encoding of that video.

Post-production processes like **DVNR** (**D**igital **V**ideo **N**oise **R**eduction) can make the MPEG-2 encoder more efficient by reducing noise in the video image. Noise is a big impediment to a successful encode, as the encoder will have a hard time determining what is noise and what is content. An MPEG encoder trying to compress a smoke-filled scene cannot determine that the smoke is an important part of the director's vision for that scene. On the contrary—the encoder interprets smoke as video noise impairing its ability to process the scene. To the encoder, each frame is just around 400,000 pixels of video information, to be properly encoded into its component details—color, movement, contrast, brightness, etc. Now add to this *your* desire to have it finished as fast as possible, and you will see why encoding is often the whipping-boy of DVD creation—there's so much riding on it, and yet many people don't take the time to learn a few basics about how to make their MPEG encodes look better.

Some Hints for Successful MPEG Encoding

1—Start with a great master.

The best video master you can get is the right place to start. VHS is a common format, but would NOT produce the best high-quality MPEG encode. A better choice, by far, is Beta SP or Digital Betacam.

In fact, any digital videotape format is better than VHS. You do need to be careful of using DV, however, as DV has its own particular quirks in terms of color saturation (reds tend to saturate easily in DV), chroma keying (keys tend to be difficult to pull effectively, and sometimes are noisy around the edges), and excessive luminance (DV content can have a tendency to exceed 100 IRE, making it difficult to properly encode and keep within NTSC tolerances).

2—Use the best format you can find/afford/rent/borrow.

Digital Betacam—top of the line choice for encoding—SDI is the BEST format for feeding an Encoder, in my opinion.

- Betacam SP—Analog, but not shabby—using component YUV is a lot better than encoding from a Composite Video Source.
- DVCAM deck with SDI—a good choice if you don't have DigiBeta.
- DV deck with FireWire—good for capturing to your edit system, but may not give the best MPEG encode.
- VHS—is this really ALL you have to work with?
- Web video formats—Real, Flash, etc.—fageddaboutit.

3—Pay attention to technical standards.

While you can argue that your DVDs are never going to be broadcast on air, and may not need to meet network standards, at some point, they WILL be seen on an NTSC or PAL video monitor or television set, and that will be the point where adherence to standards will make all the difference in the world.

Don't forget to align your tape deck, if you are using a hardware capture board. Even a Digital Betacam deck will benefit from a proper alignment, using color bars from the delivered video master as the alignment reference. If you are not familiar with the concept of aligning your video deck to color bars, you can search the Tektronix website for this PDF booklet: **25W_7247_1.pdf**. You can also search for "NTSC Video Measurement."

The booklet was last found at this URL (it may have moved): **http://www.tek.com/Measurement/App_Notes/NTSC_Video_Msmt/25W_7247_1.pdf**

This is an excellent introduction to Video test signals and their importance in proper playback.

4—Use a great encoder.

You can find hardware encoders from $300 to $25,000 that will run on a Macintosh—which one you acquire will depend on your sense of quality, your clients' needs, your turnaround time requirements, and of course, the size of your checkbook.

A small $300 desktop MPEG converter isn't going to make an encode that competes with a $100,000 hardware system with DVNR and all of the attendant digital signal processing bells and whistles.

But *do* remember that a digital video path is going to net you a better encode than an analog video path, in pretty much every case. This means an encoder with an SDI (Serial Digital Interface) input will have a better chance of creating a great encode than one that processes analog. In the analog video world, a properly aligned Component video input should create an encode superior to a *Composite* video.

5—Have realistic expectations.

If your original master is fast-paced, "cutty," highly saturated, effected, and subject to wild swings of brightness and contrast, that's fine—MTV makes this kind of video as a matter-of-fact, and it's all around us in the cinema and on television. But, if you are hoping to encode it at 3 Mbps and make it look fantastic, you may need an MPEG reality check! It takes a pretty amazing encoder to do that at 3 Mbps.

6—Maximize your available video encode bitrate.

Audio is a big part of DVD—using PCM audio can be a mistake if you are looking to maximize the quality of difficult video content, because it will reduce the available video encode rate. By choosing to use Dolby Digital (AC3) compressed audio instead of PCM uncompressed digital audio, the difference in encode rate can be passed along to the MPEG-2 video encoder for use in maximizing the quality of the picture. How much of a difference can it make? A simple stereo soundtrack requires 1.6 Mbps of bandwidth as PCM, but less than 200 kbps as Dolby Digital. The difference, about 1.4 Mbps, can be added directly to your video encode rate. That is a *significant* addition to the MPEG encode rate, and can make a huge difference in the quality of the final encode.

7—Encoding is a two-part equation.

We can discuss encoders all day long, but the *content* being encoded is an equal or perhaps more important part of the equation. Encode rate or encoder quality alone aren't enough to guarantee a successful, artifact-free encode. The very nature of the video itself can determine if an encode at a given rate will be successful or not. To this end, it is helpful to understand some things that drive MPEG encoders crazy and can cause less than acceptable MPEG-2 results.

Making Video

Fast motion content, with lots of cuts and saturation, is difficult for many MPEG-2 encoders to handle, because of the nature of MPEG encoding itself. Without going into a lengthy technical explanation, MPEG encoders are looking to remove redundant pixels in the stream of pictures, so they can compress the video program down into the smallest, most effective file size at a given encode rate. Fast motion reduces redundancy in the video frames and pixels, making encoding more complex and consequently more difficult to do well.

Other potential problems include large areas of solid color with objects moving over them, such as a person walking in front of a solid color wall, as well as large areas of trees blowing in wind, grassy fields, or water in motion. All of these seemingly innocent pictures can create havoc for an MPEG-2 encoder.

In cases of problematic content, selecting VBR encoding is a good possible solution. The 2-pass nature to VBR encoding means that the encode rate can be optimized to bring more bits to the areas of image complexity that need them, while maintaining the proper average bitrate to ensure the file will fit in the finished DVD. We'll cover this a bit more when we discuss encoding from within Final Cut Pro.

> **TIP!**
> On the accompanying DVD, we've included a very nice high-Def clip of fast-moving waves, courtesy of our friends at Artbeats. This clip should be an ideal torture test for any encoder.

8—A really high encode rate isn't necessarily a good thing.

Although a higher MPEG encode rate is generally preferable to a lower one (assuming your content is short enough to allow for it) at a certain point, a high MPEG encode rate can become your enemy.

Conventional wisdom on this topic indicates that you should be careful of selecting an MPEG encode rate in excess of 6 Mbps, as that seems to be a threshold point at which playback difficulties can be caused that can overshadow the quality of the MPEG encode. In general, this issue relates to DVD-Video–capable computers with lesser-powered CPUs, which may have some difficulty dealing with high encode rates, and especially high encode rates combined with PCM audio. Higher bitrates can also cause problems on portable computers and portable DVD players when powered by batteries.

This problem does not seem to surface up as much in set-top DVD players, as they are built specifically to process DVD Video encoded at higher bitrates.

Encoding for DVD in Apple's World

Many of you will encounter DVD authoring as an adjunct to your editing endeavors in Final Cut Pro. Apple has included export options in Final Cut Pro that allow you to export an edited sequence directly into MPEG-2 files from your Final Cut Pro edit timeline.

In Final Cut Pro 5, these export options include exporting through *Compressor 2*, which was not available in earlier versions of Final Cut Pro. Exporting using *QuickTime Conversion* is still available in Final Cut Pro, but Apple no longer provides or supports the QuickTime MPEG2 encoder component for QuickTime 7. The preferred route is now through Compressor 2.

Prepping for Export from Final Cut Pro

Not much is really required for a successful MPEG-2 export from Final Cut Pro. A few standard things need to be addressed, including the following:

DVD Studio Pro 4

Choice of Video Resolution

MPEG-2 encoding for DVD presumes you are using a full-resolution video codec, like DV, Motion JPEG, or Uncompressed that contains a complete 720 × 480 (or 720 × 486) NTSC image at 29.97 fps (or a full 720 × 576 PAL image at 25 fps). Smaller multimedia codecs or picture sizes do not provide the proper initial image for a good MPEG-2 encode, and should not be attempted. Frame rates that are not 29.97 or 25 can cause problems.

Rendering

Recent developments in Final Cut Pro have reduced the number of renders that need to be done in order to preview in real time while editing, but exporting for MPEG-2 is a different animal. To properly export to MPEG-2, all renders need to be completed prior to attempting to export. Failure to completely render will cause failure of the export operation.

Setting In and Out Times

On rare occasions, an MPEG-2 encode of Final Cut Pro content might have one white frame at the end of the encode. This intermittent glitch is usually solved by setting the "out" pointer one frame before the end of the sequence, rather than right at the very last frame.

Setting DVD Markers in Final Cut Pro

Know Your Marker Types

Starting in Final Cut Pro 3.02, Apple added the ability to create DVD-related Markers in a Final Cut sequence timeline, offering the ability to set frame-accurate *Chapter* or *Compression* markers into the sequence before encoding (see Fig. 5-9). These markers carry through the export process and are embedded in the finished MPEG-2 stream complete with their names, ready to become visible and usable in DVD Studio Pro.

Markers prepared in Final Cut Pro 3.02 and 3.04 were compatible with DVD Studio Pro 1.5. Final Cut Pro 4 and above offer the same function, but make Chapter Markers and Compression Markers that are compatible with DVD Studio Pro 2 and above. (The internal coding is different between the two applications, and QuickTime encodes with "embedded markers" that work in DVD Studio Pro 1 may not work in DVD Studio Pro 2 and above.)

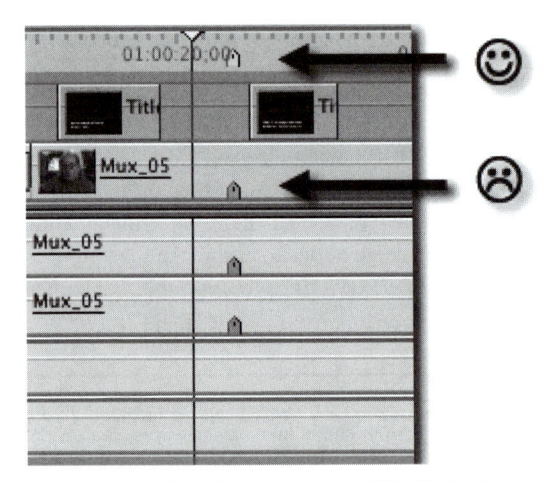

Figure 5-9 — Timeline Markers are OK; Clip Markers are not.

Be certain to set DVD markers into the *Timeline* (see Fig. 5-10 at top), where they appear green, and not into the clip itself (see Fig. 5-10 middle), where they appear pink. Only markers in the *Timeline* can be exported into the MPEG encode.

The following *two* distinct types of DVD markers can be created by this process, and this can tend to create some confusion:

Making Video

Compression Markers

Compression Markers are meant to indicate a location in the sequence timeline where the MPEG-2 encoder should pay special attention to a frame, perhaps because of a scene change or other significant change in image composition. This will force the MPEG-2 encoder to generate a new I-Frame (key frame) at that point. But some things are important to note:

- A Compression I-frame is *not* an I-frame Header frame;
- It cannot be used to establish a DVD Chapter point.

To add Compression Marker status to the current Marker,

- Open the *Marker Edit Panel* (see Fig. 5-10a).
- Click the *Add Compression Marker* button in the Final Cut Pro Edit Marker pane.
- Click *OK*.

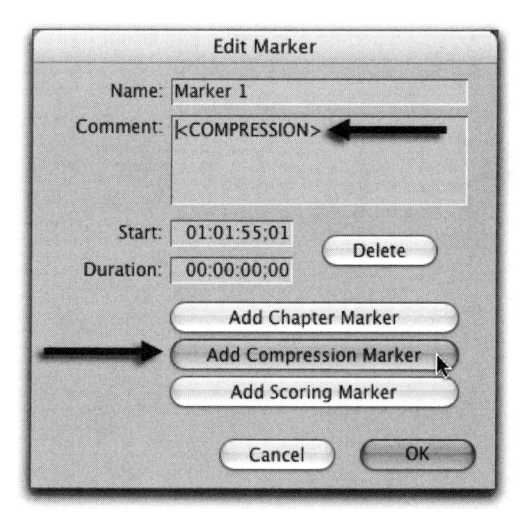

Figure 5-10a — Final Cut Pro Compression Marker.

Chapter Markers

Chapter Markers, on the other hand, are meant to establish not only a Compression event, but also a navigable DVD Chapter point. While sometimes a point of confusion, the difference can be summed up as follows:

- A Chapter Marker is of higher precedence than and contains an inherent "Compression Marker."
- A Chapter Marker signals the encoder to restart the MPEG encoding sequence at that frame, creating an I-frame header, and thus creating a navigable DVD chapter marker.
- If in doubt, use a "Chapter Marker," but remember you can only have a maximum of 99 Chapter Markers per Track in DVD Studio Pro. The Marker at the beginning of the encoded Track is always there, and counts as the first marker in the Track, so only 98 additional markers are allowed. (The Track's End Marker does not count as one of the 99.)

Figure 5-10b — Chapter Marker.

123

DVD Studio Pro 4

To add Chapter Marker status to the current Marker,

- Open the *Marker Edit Panel* (see Fig. 5-10b)
- Click the *Add Chapter Marker* button in the Final Cut Pro *Edit Marker* pane.
- Click *OK*.

Be sure to include DVD Studio Pro Markers in your QuickTime movie exports from Final Cut Pro if you want to use them with DVD Studio Pro. See Chapter 8, "Creating Markers and Stories" for information on how to create DVD markers within DVD Studio Pro.

Exporting into MPEG-2 from Final Cut Pro Using Compressor

To export from Final Cut Pro 4 or 5 using Compressor 2, do this:

- In Final Cut Pro, either select the sequence icon in the Bin or activate the sequence tab in the Timeline that you wish to export to MPEG-2 or H.264 (the subsequent steps are the same for either codec).
- Select **File > Export > Using Compressor...** (Fig. 5-11).
- Compressor will now launch, with the selected clip placed in the Batch List, ready for you to set the encode parameters and destination. (See Fig. 5-12a.)

Using Compressor to Encode (Quick Start)

- If you have just placed a clip in Compressor 2's Batch List, you need to adjust the clip's encode settings, destination, and output

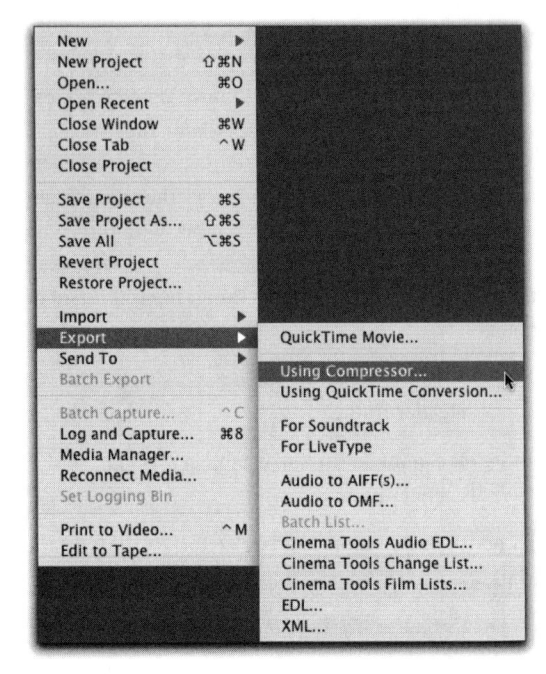

Figure 5-11 — Export from Final Cut Pro 4 using Compressor.

filename. You should also take a moment to name the Batch. (See Fig. 5-12a.)

- Select and fine-tune the encode settings for this encode. (See Fig. 5-12b.)

IMPORTANT!
You will want to be sure you have calculated your Bit Budget before setting an encode rate!

- Set the encoded file's destination—be careful not to overwrite existing files in "Source" if you select that choice (see Fig. 5-12c). Your options are:

- **Source**—puts the encode in the same folder as the source file.
- **Desktop**—puts the encode on your Desktop.

Making Video

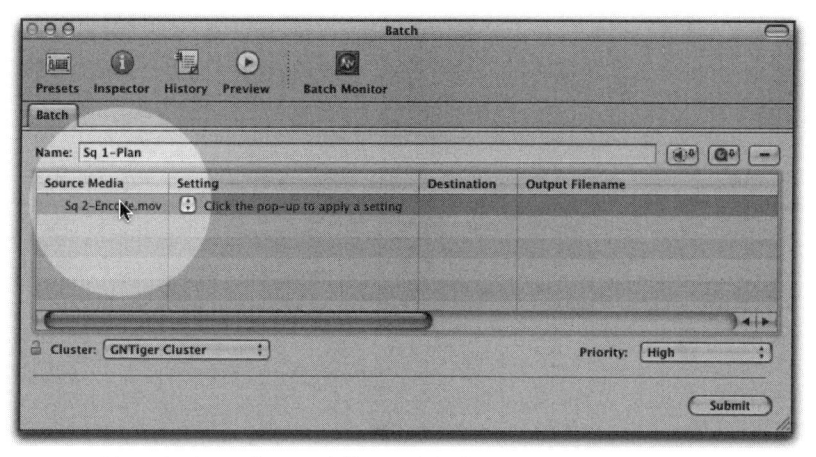

Figure 5-12a — Exported file ready in Compressor batch window.

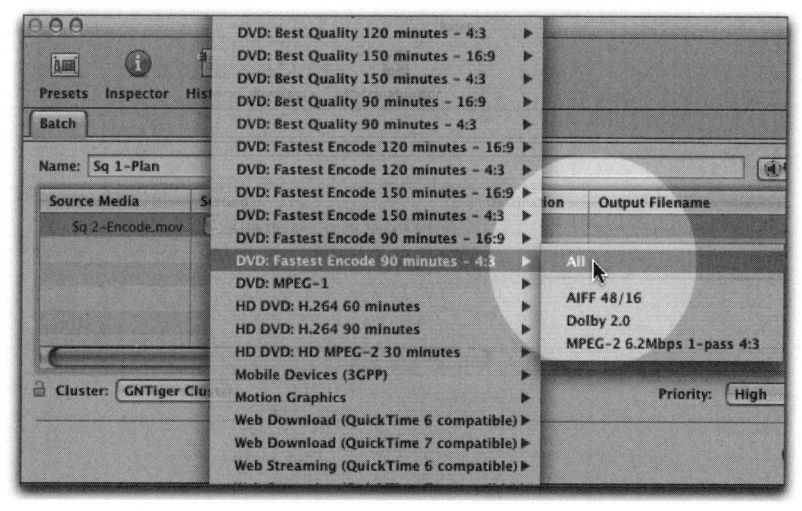

Figure 5-12b — Selecting the Compressor encode settings.

- **User's Movies Folder**—saves encodes in the Movies folder for you, the current user. This is NOT globally accessible if another user signs in instead of you!

- **Cluster Storage**—destination is dictated by the Qmaster cluster settings (see Qmaster User Guide).

- Set the desired filename (see Fig. 5-12d, Step 1).

- Finally, click the *Submit* button (see Fig. 5-12d, Step 2).

125

DVD Studio Pro 4

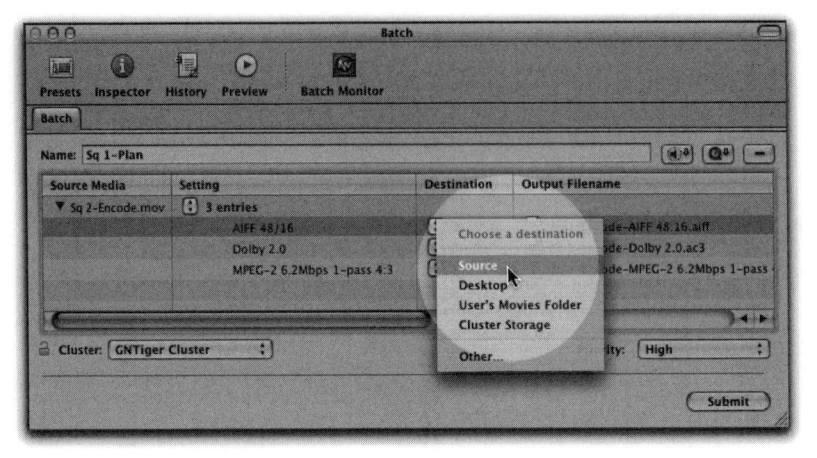

Figure 5-12c — Setting where to save the encoded file.

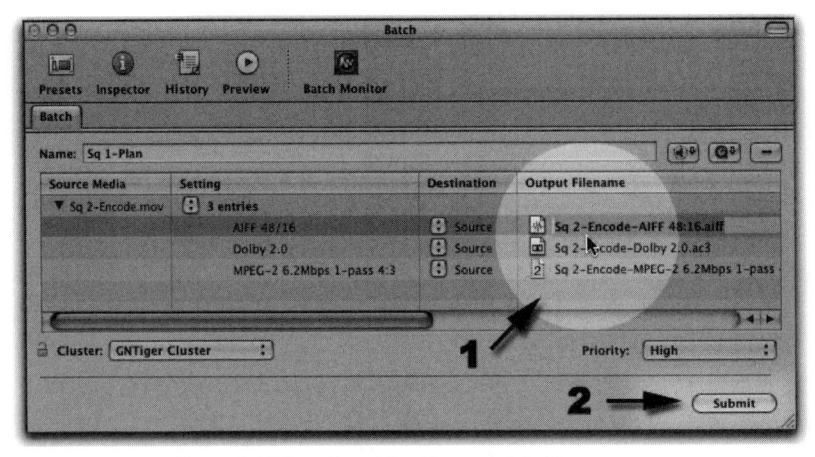

Figure 5-12d — Tweaking the encode's file name.

There is more information for you in the Manuals

This has been a very simplified overview of Compressor 2, designed to get you going. Compressor 2 has a lot of capabilities and possibilities. I would recommend that you:

- Consult the Compressor 2 User Guide for in-depth instructions on how to use Compressor's many features.

- Consult the Apple Qmaster and Qadministrator guide for information on how to implement distributed encoding if you have more than one Mac on a local area network.

Making Video

Encoding Using DVD Studio Pro's Onboard MPEG Encoder

DVD Studio Pro has a unique onboard encoder that can encode MPEG-2 in the background while you do your authoring work in the foreground. This feature uses the same MPEG encoding engine as Compressor. The options made available within DVD Studio Pro are similar to those found in Compressor 2, except DVD Studio Pro offers none of the options available for filtering and image adjustment in Compressor 2. It does allow you the luxury of doing MPEG-2 encoding in the background, while you continue to author your DVD. This maximizes the use of your Mac's CPU cycles, without causing any significant drain on your authoring power.

Creating an MPEG-2 File Using DVD Studio Pro's Encoder

If you elect to author using QuickTime movies, you can create an MPEG-2 video asset in DVD Studio Pro by following these steps:

- First, set the encode parameters in the *Encoding Preferences* pane (see Fig. 5-13).

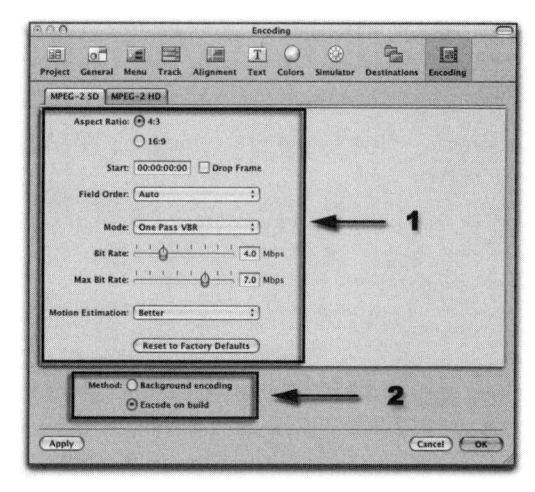

Figure 5-13 — Setting Onboard MPEG Encoding Preferences.

- Select *Background Encoding* if you want the MPEG encodes to be created while you work on authoring.
- Select *Encode on build* if you wish to defer encoding until the end of the authoring phase, during project Building.
- Set the encoded file destination in the Encoding pop-up of the Destinations Preference Pane (see Fig. 5-14).

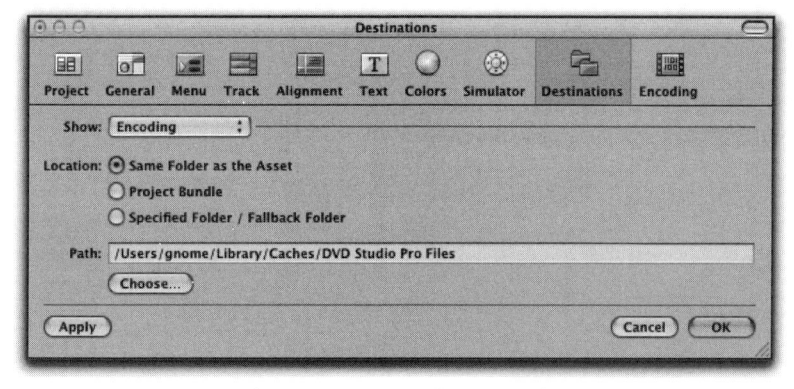

Figure 5-14 — Setting Destination for Encoded MPEG Streams.

127

Once the Preferences are set, you are free to begin using non-MPEG-2 QuickTime movies in your DVD.

To prepare a file for MPEG-2 Encoding in DVD Studio Pro:

- Import a suitable QuickTime movie into the Asset Tab, or
- Author a suitable QuickTime movie in a Track or Menu.

If you have set the *Method:* encoding preference to "Background Encoding," the movie will immediately begin being encoded upon import. "Encode on Build" will defer the encoding until you tell DVD Studio Pro to *Build* the project (see Chapter 14).

Encode Status Indicators

The progress of the Encode can be monitored in the Assets Tab, where the usability indicator will change color to indicate the status of the encoded file (see Fig. 5-15).

- Red—The asset is encoded, but is being parsed.
- Yellow—The asset is being encoded, not ready yet.
- Green—Asset is ready to use—encoding complete.

Exporting from Final Cut Pro Using QuickTime

Exporting Requires QuickTime "Export Components"

Although this was supported in previous versions of DVD Studio Pro, beginning with Version 4, this method is no longer used. Encoding to MPEG-2 through QuickTime uses the same encoding engine as Compressor, but offers a simpler interface and exposes fewer encoding options than using Compressor. To gain more control over your MPEG-2 encode, export *"Using Compressor…"*, as described in the previous paragraphs.

Compressor 2 is the preferred method, but if you still have the older QuickTime MPEG2 export component installed and visible, this is how to use it:

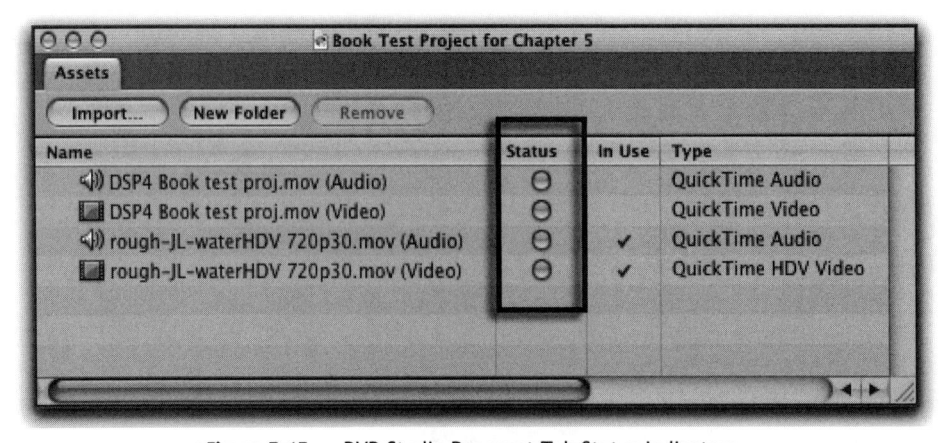

Figure 5-15 — DVD Studio Pro asset Tab Status indicators.

Making Video

- In Final Cut Pro, select the sequence icon in the Bin, or activate the sequence tab in the Timeline that you wish to export.

- Select **File > Export > Using QuickTime Conversion...** (see Fig. 5-16).

- Continue on and follow the instructions in the next section, "MPEG Encoding in the QuickTime Player."

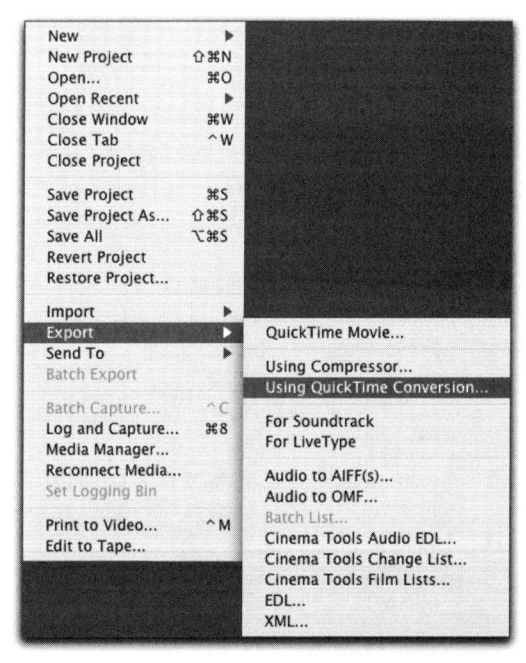

Figure 5-16 — Exporting to MPEG-2 using QuickTime's encoder.

MPEG Encoding in the QuickTime Player

If you still have a QuickTime-compatible MPEG Export Component in your install of QuickTime (like Digigami's *MegaPEG Pro QT*), you can still use this technique.

To encode a QuickTime-compatible movie into MPEG-2, do the following:

- In the Macintosh Finder, locate the video file to be encoded. This can be any NTSC or PAL QuickTime Movie; you can also select a Final Cut Pro reference movie.

> **NOTE!**
> YOUR CONTENT MUST ALREADY BE IN A QUICKTIME COMPATIBLE FORMAT! If you haven't captured already, you can do so in iMovie, Final Cut Pro, Avid, Media100, Premiere, even Cleaner.

- Drag the file on to the *QuickTime Player* Icon (see Fig. 5-6, earlier), or launch *QuickTime Player* and then select **File > Open Movie in New Player** to open the movie you wish to encode.

- Select **FILE > EXPORT** from the menubar*.

> ***NOTE!**
> If you do not see the "**Export...**" command in the QuickTime **File** menu, please review **System Preferences > View > QuickTime** to verify that the QuickTime Pro Player upgrade has been applied to your installation of QuickTime. Check carefully to be certain of which application is launching, especially if you are using an alias to launch the QuickTime player.

- In the "*Save exported file as...*" dialog (see Fig. 5-17), click on the *Export:* Format selector and select the *Movie to MPEG2* format if available, or whatever export component gives you access to MPEG encoding.

- Click the *OPTIONS...* button to open the MPEG Encoder Options window.

129

DVD Studio Pro 4

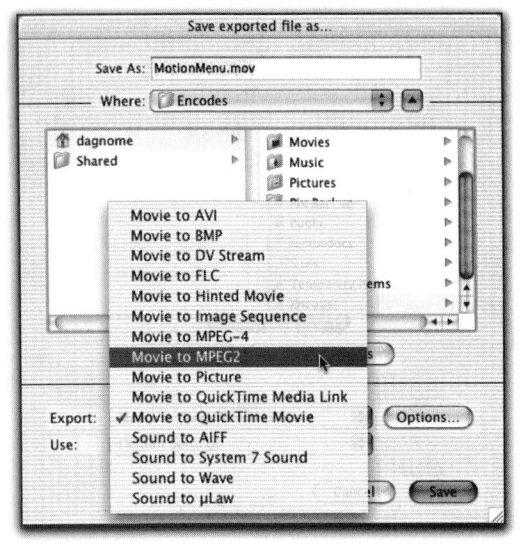

Figure 5-17 — QuickTime File Export Dialog box.

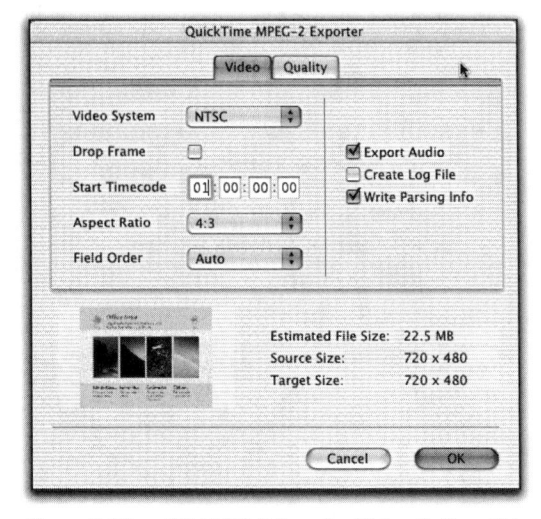

Figure 5-18 — QT MPEG-2 Encoder Video Options Tab in OS X (QT6).

Setting up the QuickTime MPEG Encoder "Video" Options

The QuickTime encoder has two Options Tabs that contain the different parameters to be set. Based on settings you have calculated when you generated a Bit Budget (or using the quick guide settings below), set the MPEG encoder options as needed. Click *OK* when finished, to close the options dialog and prepare to encode that file.

On the left side of the QuickTime Exporter display (Fig. 5-18):

1. Set *Video System* to NTSC or PAL, to match the source.
2. Check (✔) *Drop Frame* if you want Drop Frame code.
3. Set *Start Timecode* as needed, to set a specific value. HH:MM:SS:FF = Hours, Minutes, Seconds, Frames.
4. Set *Aspect Ratio* to 4:3 or 16:9, to match your source.
5. Set the source's *Field Order* (DV typically uses "Auto"). In all formats, you may choose Upper or Lower field.

On the right side of the display:

6. Check (✔) *Export Audio* "On" to save the Audio.
7. Checking *Create Log File* will write an encoder log file.
8. Check *Write Parsing Info* to prewrite the MPEG file parse information that DVD Studio Pro needs to properly use this file.

> **NOTE!**
> Setting this switch will write parse info into the MPEG file itself, making this encode ONLY useable with DVD Studio Pro 2 and above.

130

Making Video

Figure 5-19 – QT MPEG-2 Encoder Quality Options Tab in OS X (QT6).

9. Choose the *Encoding format* desired: One pass = one pass Constant Bit Rate—good, fast. One pass VBR = one pass Variable Bit Rate—better, slower. Two pass VBR = two pass Variable Bit Rate—best, slowest.

10. Set the *Target Bitrate* as needed: The average bitrate for your DVD, as calculated in the Bit Budget, or determined by the guidelines below.

11. Set the *Max Bitrate* as appropriate, but only for VBR; the Maximum allowable bitrate for your DVD, as calculated in the Bit Budget, or determined below.

12. Set the *Motion Estimation* quality desired.

13. The *Info Panel* displays the details of your encode.

Note carefully the Source Size (this should be 720 × 480 for NTSC, 720 × 576 for PAL)—be sure it matches the Target Size—if not, you will experience encoder SCALING, which will slow down the encode a lot!

14. When finished, click "*OK*" to return to the Encode Dialog.

15. Click "*Save*" to begin the Encode.

Quickie MPEG Encode Guidelines

- 8.0 Mbps = allows 60 Min. of program
- 6.0 Mbps = allows 90 Min. of program
- 4.5 Mbps = allows 120-130 Min. of program

Some more complete bitrate guidelines:

- <3 Mbps—may not encode without artifacts
- 4.5 Mbps—typical rate for 120-130 Min. of Content
- 6 Mbps—typical rate for 90 Min. of Content
- 8 Mbps—typical rate for 60 Min. of Content
- >6 Mbps—can cause problems—be careful above 6 Mbps!
- >8 Mbps—Not recommended

> **NOTE!**
> Encoding at rates above 6 Mbps should be done with knowledge that some computers that use software MPEG-2 decoders may experience difficulty playing back these files. Encoding above 8 Mbps should only be done if you are certain of what you are doing. Rates above 6 Mbps can be problematic on computer playback. Rates above 8 Mbps can be VERY problematic, especially if you are using AIFF or WAV audio files. (See Chapter 6.)

> **IMPORTANT!**
> The QuickTime MPEG-2 Encoder does **NOT** do NTSC-to-PAL or PAL-to-NTSC Video Standards Conversion. HOWEVER—Compressor 2 now does, so you have this tricky conversion available to you now in software on your own Mac!

131

DVD Studio Pro 4

What Happens during MPEG Encoding

Once the **SAVE** button has been clicked to begin the MPEG Export, the Exporter progress window will open and track the progress of your encode. The thermometer display across the top visually shows the relative progress of the overall encode, both video *and* audio.

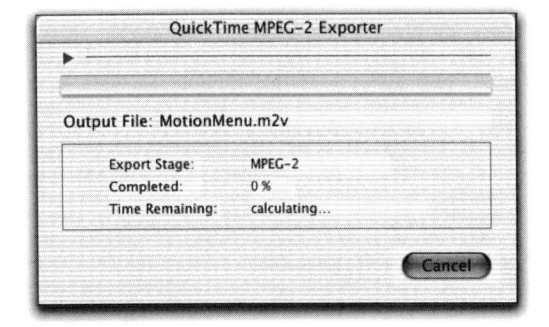

Figure 5-20 — QuickTime MPEG-2 Encoder in action (Preview closed).

The Encode Progress Window

During the encode, the *Progress Window* (see Fig. 5-20) will display the encode parameters on the left side, and the progress data on the right.

"Export Stage" displays the current pass number of a multi-pass export. The "Completed" percentage counter will increment during encoding. "Time Remaining" indicates how long DVD Studio Pro calculates it will take to complete this encode, based on the encode file and the host system parameters. The progress of the encode continues until "Completion"

is at about the 95% point. At that time, the Video will have finished encoding, and the audio will be very quickly extracted from the source Movie.

The progress window initially opens with the preview pane closed, but clicking on the disclosure triangle in the upper left corner will unfold the video preview pane, showing you the encode in progress visually (see Fig. 5-21). If you look closely, however, you may notice that the encode display doesn't move smoothly from frame to frame, but "jumps" in a kind of "jerky" preview. This isn't because your Macintosh isn't fast enough to display all of the frames, but rather because of the nature of MPEG encoding.

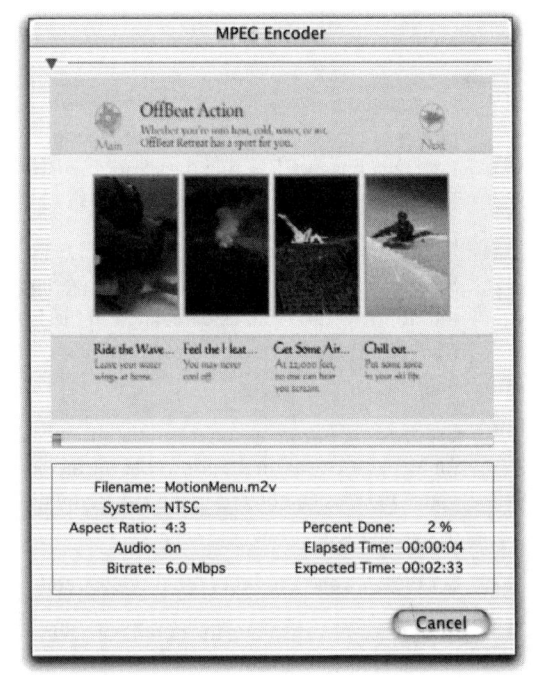

Figure 5-21 — QT MPEG Encoder in action (Preview pane open).

132

Making Video

Face to Face with GOPs—the MPEG "Group Of Pictures"

The jerky frames you are seeing in the preview window (assuming you are not encoding a Final Cut Pro reference movie that has embedded DVD Studio Pro markers that you have placed in Final Cut Pro) is the first frame of each 15-frame GOP, or **Group Of Pictures**.

GOP?

A "GOP" (**G**roup **O**f **P**ictures) is physical evidence of the manner in which MPEG-2 is encoded. Rather than capturing frame-by-frame, like the editable codecs DV or Motion JPEG, MPEG is a codec that encodes the video source in chunks (GOPs), in order to compress the file into a size smaller than is possible with an editable codec. The QuickTime MPEG encoder encodes the program in groups of 15 frames (15 for NTSC—PAL is 12). This makes the finished MPEG-2 file a very compact file, but also makes it difficult to ever edit the file again. Once encoded into MPEG, the individual frames have become the 15 frame GOPs. Because of this phenomenon, MPEG is generally referred to as a "destination codec"—a codec where the assumption is made that the media will not require further editing.

Results of the MPEG Encode

When finished, the encoder will have created an MPEG Stream (a file with an **.m2v** extension) and an AIFF (PCM) Audio file (**.aif**). (See Fig. 5-22.) The AIFF file can be further compressed by encoding it into Dolby Digital using Compressor 2, Apple's newest AC-3 encoding application. If you are running an older version of DVD Studio Pro, you may still have A.Pack installed. DVD Studio Pro has supported DTS audio streams (.cpt) since V 3.02, so you may also consider using that as an alternative. Details of this process can be found in Chapter 6, "Making Audio for DVD."

Figure 5-22 — Files created by MPEG Encoding MPEG (L), AIFF (R).

> **NOTE!**
> You MUST have rendered your movie completely prior to beginning an MPEG encode/export. Unrendered movies will not successfully encode into MPEG.

> **TIP!**
> You can *batch encode* MPEG-2 files from your Final Cut Pro sequences by using the Batch Export function, and selecting the MPEG-2 format as your codec choice. You can also batch using Compressor 2.

> **IMPORTANT!**
> No matter which method you use, each encode should generate an MPEG-2 video stream and a synchronized audio file in some DVD-compatible format—AIFF, AV, or AC3.

QuickTime MPEG-2 Encoder Performance Benchmarks

The QuickTime MPEG-2 encoder's performance is dependent on the CPU speed of the Mac system that

133

it is running on, as well as the codec used to capture the video.

As the data becomes more dense (as would happen with an uncompressed codec), the time required for encoding increases.

The QuickTime MPEG-2 encoder is optimized for the DV codec, which has a built-in advantage. The DV codec actually partially compresses are the video data (at about 5:1) during the recording process, so there are less video data to worry about than in other codecs.

Given DV source footage, Table 5-1 displays the approximate encode times you can expect for a minute of footage:

Table 5-1 Typical MPEG Encoding Times

CPU speed	Encode time
733 MHz	2.0 x runtime
867 MHz	1.5 x runtime
Dual-800 MHz	1.2 x runtime
Dual-1GHz	1.0 x runtime
Dual-1.25 GHz	under 1.0 x runtime
Dual-2.00 GHz	under 1.0 x runtime
Dual-2.50 GHz	under 1.0 x runtime

While you can select other formats to export to from QuickTime, you eventually *MUST* create MPEG-2 or H.264 streams for use with DVD Studio Pro. Installing DVD Studio Pro will not prevent you from using the QuickTime Pro Player to export other types of media, as usual.

And, of course, as we have mentioned, you do NOT have to encode your video content into MPEG-2 before you begin authoring in DVD Studio Pro.

Using QuickTime MPEG-2 Encoding in OS 9

You should have been upgraded by now to OS X, but in case you still have an old Mac around running some OS 9 applications (you diehard), here's how to use QuickTime encoding in OS 9:

> **IMPORTANT!**
> OS 9 ONLY: Be sure you have allocated enough RAM memory to the QT Player—you'll need around 50-60 MBytes or more to do MPEG-2 encoding successfully.

To allocate RAM to the QuickTime Player under OS 9:

- Locate the QuickTime Player application icon.
- Be sure the application is not launched.
- Click it <u>once</u> to select it.
- Use Cmd-I ("Get Info") to open the Get Info window.
- Select "Memory" pop-up
- Allocate at least 50,000K in the "Preferred Size" field.
- Close the Get Info window.

You need QuickTime Pro, not just QuickTime Player

It's important to be sure your QT Player has been upgraded to QT Pro. Check the QuickTime registration screen to be sure it indicates the Pro Player version—this indicates you will be able to use the QuickTime MPEG export function once it has been installed by DVD Studio Pro.

QuickTime Pro has always been a separate upgrade—with DVD Studio Pro, however, the upgrade has been included. In QuickTime 7, the upgrade no longer has a unique visible serial number. (See Figs. 5-23, 5-24, and 5-25.)

Making Video

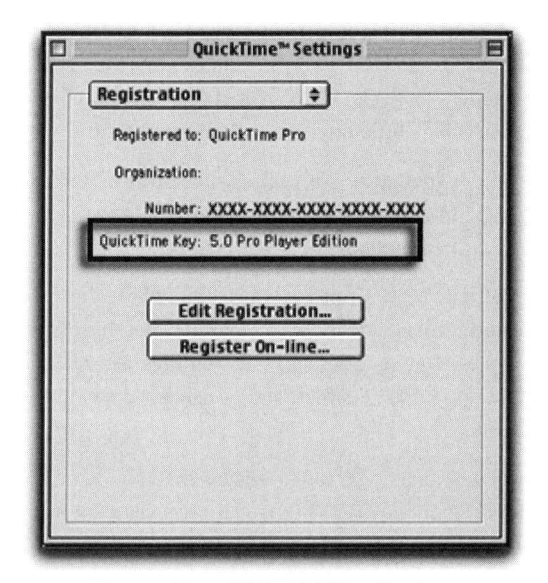

Figure 5-23 — OS 9 QuickTime Settings.

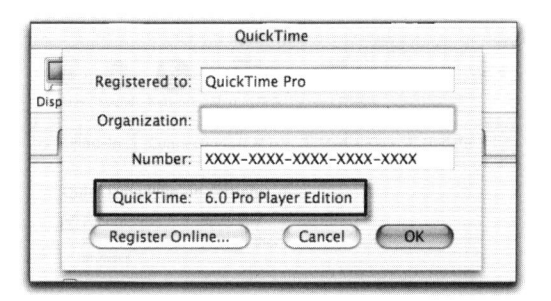

Figure 5-24 — OS X QuickTime 6 Settings.

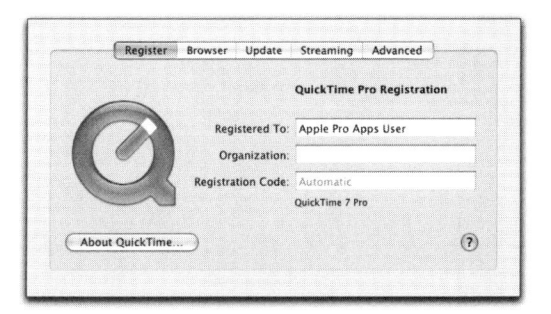

Figure 5-25 — OS X QuickTime 7 Settings.

If you're not sure how to do this, please consult Appendix F, "QuickTime Install/Upgrade Guide," later in this book, or check at **www.apple.com/support**.

Alternative Encoders

There are a wide variety of available software encoders for the Macintosh, and we've rounded them up for you in Appendix B, "Alternative Encoder." Some have unique features found nowhere else, some are faster or slower, but you do have some choices for your DVD encoding needs, and we're happy to present them to you.

Using a DVD Recorder as an Encoder

More recent developments in the DVD world include set-top DVD video recorders that accept video signals directly (like a VCR), but record onto DVD disc instead of a VHS tape. Examples of this kind of DVD recorder can be found from manufacturers Pioneer, Philips, and Panasonic. In some cases, these recorders include a hard drive for recording to prior to making a DVD disc, and even TiVo.

These recorders in many cases will create a fully multiplexed (built) DVD disc, which can be played on standard DVD-Video players. Others of these recorders can be used to record video in a somewhat proprietary format, for playback from the same recorder.

Some of these new recorders use a DVD application format called DVD-VR; a method of directly recording MPEG-2 encoded video onto a DVD disc. In many cases, the video recorded on these set-top DVD recorders can be used for Authoring, provided a method to extract the recorded files exists—and it does! There are now some software solutions avail-

135

able that can extract these encoded MPEG-2 files from the disc and allow them to be used as MPEG-2 assets for DVD authoring.

You can find information about these software tools at the websites of their manufacturers:

- Great Video is offered by Software Architects at **http://www.softarch.com**.

Software Architects claim, on its website [Great Video! 1.1 for Mac can]:

– Read and Convert VR format DVD video recorder files.

– Convert recorded DVD Video Discs in VR format to 21 QuickTime compatible formats!"

- Heuris Extractor is offered by Heuris, Inc. at **http://www.heuris.com**.

Heuris states, on its website:

"[Heuris Xtractor] Rips DVD-VR formatted MPEG streams recorded with Hitachi DZMV100 DVD camera or Panasonic's DMR-E20 line of DVD recorders."

Both applications claim to be able to extract DVD-VR files and repurpose them as DVD authoring assets.

In addition, once a recording has been finalized (if that is needed, as it is on Panasonic DVD-RAM recorders), you can likely extract the VOB files using a marvelous little program called Cinematize 2.

- Cinematize 2 is offered by Miraizon, Inc. at **http://www.miraizon.com/products/products.html**.

Cinematize does not purport to be able to rip protected content, and in fact, it does not. But this can be very handy in extracting your own files recorded on a set-top DVD recorder, and re-using them as DVD assets.

You will want to be sure to review this section of Jim Taylor's online DVD FAQ, as it provides a lot more information about DVD recorders:

[4.3.1] Is it true there are compatibility problems with recordable DVD formats?

You can find the DVD FAQ at this URL: **http://www.dvddemystified.com/dvdfaq.html**.

DVD Studio Pro is very liberal in terms of MPEG-2 assets that it can accept and use effectively.

Making Video

Summary

Although you no longer need to encode video into MPEG before you begin authoring in DVD Studio Pro 2, understanding some of the basics about MPEG encoding will serve you well as a DVD author. All DVDs are based on delivering video using MPEG Compression. HD DVDs allow the use of either MPEG-2 or H.264, as you choose.

DVD's can be created using either MPEG-1 or MPEG-2 video files. MPEG-2 Compression can reduce a 90-minute movie that might require 180 GB (uncompressed) to a file of less than 4 GB that can easily fit into a DVD-5 disc. MPEG-1 Compression can reduce files even further but with some loss of quality.

Running time and encode rate are inversely tied to each other—that is, as the encode rate is reduced, the amount of video that can fit on the disc increases, and vice versa. In general, an encode rate of 8 Mbps will allow a program of about 60 minutes in length, an encode rate of 6 Mbps will permit a program of around 90 minutes, and an encode rate of 4.5 Mbps will permit a program of 120 to 130 minutes running time. Each of these calculations assumes no Subtitles and one Stereo Dolby Digital (AC3) Audio Stream at 192 kbps.

To properly determine the best encode rates for your particular DVD, the process of Bit Budgeting will calculate the available MPEG video encode rates for a given length project with a given number of Audio and Subtitle Streams.

Although hardware encoding cards have existed for years, DVD Studio Pro 2 comes with two software encoding solutions, each based on the QuickTime MPEG-2 encoding engine: Compressor, Apple's new encoding application, and the QuickTime Pro Player, which has a new and improved MPEG encoding engine. Exporting to MPEG through QuickTime is still possible from Final Cut Pro, and now export is also possible from Final Cut Pro 4 through Compressor. In addition, third-party encoders are available from Heuris, BitVice, and Cleaner 6.

DVD Studio Pro 2 includes an onboard MPEG-2 encoder, which is also based on the QuickTime MPEG-2 encoder. This encoder can create MPEG assets in the background, like iDVD, or it can postpone encoding until time for the project's Build phase.

NTSC and PAL video programs require separate encoding in order to accommodate their different frame rates and image sizes.

Note that information in this summary primarily deals with SD DVD, and not the new HD DVD format. Higher MPEG-2 encoding rates are required for HD content, due to the significantly larger number of pixels contained in the frame.

Chapter 6

Preparing Audio Assets

DVD Studio Pro 4

Figure 6-1 — Three Audio Formats—AIFF, Dolby Digital, and WAV.

Goals

By the end of this chapter, you should:

- Understand what **Audio Assets** are, and how they work
- Understand **Uncompressed** vs. **Compressed** audio files
- Understand about PCM, AIFF, WAV, AC3, and DTS formats
- Understand how to create **AIFF** and **WAV** files from QuickTime
- Understand how to create **Audio files** from Final Cut Pro
- Understand how to encode **Dolby** Streams with Compressor 2
- Understand how to encode **Dolby** Streams with A.Pack
- Understand how to use encoded **Audio Streams**
- Understand how to use uncompressed **Audio files**
- Understand why **Dolby Digital** encoding is beneficial
- Know the difference between **Dolby Digital** and **Dolby Surround**
- Understand about **DTS** audio streams
- Understand how **DTS** fits into DVD authoring
- Understand how to encode **DTS** streams from PCM files

What Audio Formats Can I Use with DVD Studio Pro?

Audio formats usable for DVDs, and for authoring in DVD Studio Pro include: **PCM** (*Pulse Code Modulation*), **Dolby Digital** and **Dolby Surround** (both in *AC-3* format), **DTS** (*Digital Theatre Systems*), and **MPEG** Audio (.MPA files). Of these, MPEG audio is the least used, and has never been DVD legal in Region 1 titles (DVDs for use in North America).

My personal opinion is that you will have consistently better luck with your DVD if you use Dolby Digital (AC-3) files, especially if you have long programs (over 60 minutes on the disc). If you are planning to use more than one audio stream in a Track, you will find that Dolby Digital is indispensable.

That said, here's our roundup of permissible audio file formats:

PCM FILES

AIFF FILES (Audio Interchange File Format)

AIFF is the Macintosh form of PCM audio, and, as with all PCM formats, is best used sparingly in DVD authoring, if at all (see Fig. 6-2).

Preparing Audio Assets

Figure 6-2 — AIFF File Icon.

- AIFF is typically used only for Stereo files, because of the extremely high encode rates involved.

- AIFF Stereo files use an encode rate of approx. 1.6 Mbps. Multichannel PCM files take up even MORE bandwidth, and are rarely, if ever, used in DVDs for that reason.

- Nonetheless, AIFF files can be used directly in DVD Studio Pro for authoring.

- AIFF is an uncompressed audio format, and requires far more DVD playback bandwidth than compressed formats like Dolby Digital (AC-3), DTS, or MPEG audio. As a consequence, less bandwidth is available for the encoded video, which might compromise the quality of the finished video project.

- AIFF audio files can also create playback problems in DVD players with marginal performance. Because of the increased bandwidth requirements, software DVD players in inexpensive PCs and laptop computers may occasionally experience playback difficulties.

IMPORTANT!
PCM files (*AIFF, WAV*) are best used for DVD programs less than 60 minutes in length. We *do not* recommend the use of PCM audio if your DVD requires more than one audio stream in a Track.

.WAV FILES

Figure 6-3 — WAV file Icon.

- WAV files are PCM audio files formatted for use on PCs, instead of AIFF as on Macs (see Fig. 6-3).

- The biggest difference between WAV and AIFF is the order in which the bytes of digital data are stored in the file.

- In WAV files, the audio data should be identical to that in the same file in AIFF format.

- WAV files can be easily converted into AIFF files using the QuickTime Pro player. You need the Pro Player to enable the `File > Export...` menubar command.

TIP!
WAV files can be used directly as a source for authoring in DVD Studio Pro, and as a source for audio encoding in Compressor 2 (and A.Pack)—just import them! (This is not a well-known feature!)

Dolby Digital (AC-3)

Dolby Digital is the current Dolby Laboratories name for what we knew earlier as *AC-3* files (see Fig. 6-4). Dolby Digital bitstreams, encoded using a sophisticated bit-reduction algorithm (that's where the AC-3 comes from), comprise the bulk of audio tracks used in many DVDs today. I'll still refer to them here as AC-3 for convenience.

DVD Studio Pro 4

Figure 6-4 — Dolby Digital File Icon.

- **AC-3** Streams are usually created at encode rates from 96 kbps (Mono) to 192 kbps (stereo), 384 Kbps (5.1), and up to 448 Kbps.

- **AC-3** Audio is *not legal for DVD* at rates above 448 Kbps!

- **AC-3** provides advantages by reducing the size of the finished audio file, and by reducing the bandwidth required for playback of audio. In fact, it is possible to put 8 AC-3 Stereo programs in *less* bandwidth than one AIFF stereo audio program.

- **AC-3** allows delivery of all audio track configurations from Mono (1/0, "Center") to 5.1 (3/2 + LFE "L, C, R, Ls, Rs, Sub"). The different track configurations are usually identified by use of the logos displayed in Figures 6-5, 6-6, and 6-7, which are available upon a license request from Dolby Labs. (Browse **www.dolby.com** for information.)

Figure 6-5 — Stereo (2.0) Icon.

Figure 6-6 — Surround (5.1) Icon.

Figure 6-7 — Dolby Digital Logo (requires Dolby license to use).

Dolby Surround (4-ch L, C, R, S) [AC-3-encoded]

Dolby Surround (see Fig. 6-8) *does not* generically mean the same thing as *Dolby Digital*, although it is often misused this way. Let's clarify:

- **Dolby Surround** refers to 2-channel audio files that contain four discreet channels of audio information (Left, Right, Center, mono Surround), electronically matrixed together into a 2-channel delivery format called Lt/Rt (*Left Total/Right Total*). For

142

Preparing Audio Assets

Figure 6-8 — A Dolby Surround format Icon (Left, Right, Surround).

many years, this was the standard of feature film sound.

- **Dolby Surround** tracks on a DVD can be played back through a Dolby Surround decoder (called *Dolby Pro Logic* in consumer equipment), if the bitstreams have been properly tagged as such. Dolby Digital encoders, like Compressor 2 and A.Pack, will allow you to set this property, should you need to do so. (Refer to the section "Encoding Quick Guide," further in this chapter.)

MPEG Audio (".MPA")

Figure 6-9 — A QuickTime MPEG audio file icon.

MPEG Audio is another form of DVD-compliant audio (see Fig. 6-9). Usage in DVD primarily refers to audio compressed according to MPEG-1 Layer 1 or Layer 2. This provides for Stereo audio tracks, usually created at encode rates similar to Dolby Digital (192–384 kbps).

- Despite the similarities in their underlying technical standards, the form of MPEG Audio used in DVD is *not* **MP3**. DVD Studio Pro does provide a method of converting MP3 files from your music library into usable assets, but despite this slight-of-hand, MP3 is not natively legal in the DVD-Video world.

- MPEG Audio does also provide for multichannel sound (again, per the MPEG spec), but this format may be difficult to create.

- MPEG Audio is DVD legal in Region 2 (Europe, primarily), but does not necessarily enjoy compatibility with all DVD players.

- MPEG Audio is *not* DVD legal in Region 1 DVD players, nor generally used, by convention.

- MPEG Audio can potentially cause difficulties in some regions (like Region 1), if players are not provided with MPEG audio playback/decoding. This is especially true in DVD players manufactured for use in Region 1.

> **NOTE!**
> MPEG Audio Streams can be created using Compressor 2's MPEG-1 presets, but its use as an audio format for DVD is not universal, and can be problematic. The QuickTime player can playback MPEG audio files created with Compressor 2.

DTS (supported since DVD Studio Pro 3.02)

The DTS format had its auspicious beginnings at the theatrical release of Jurassic Park (June 1993), and has been a consistent player in digital audio ever since (see Fig. 6-10). DTS formats for DVD exist, and DTS has encoding software available that runs on Mac OS X (information is at **www.dtsonline.com**).

143

Figure 6-10 — The DTS logo.

- **DTS** (Digital Theatre Systems) audio streams are being used today in some commercial DVDs, and since Version 3.02, DVD Studio Pro has supported authoring of DTS Audio streams (.cpt).

- **DTS** is an *optional audio format*, according to the DVD specification, and cannot be the only format of audio present on a DVD in any region. There must be at least a PCM or Dolby Digital Stream.

Audio Formats Upcoming for HD DVD and BD

As the audio formats for both HD DVD and BD solidify in the coming months, we will update the book's website to include this information. Expect to see Dolby Digital +, MLP (*Meridian Lossless Packing*), and DTS-HD as some of the contenders, along with other formats such as AAC (*Advanced Audio Codec*).

What Audio Format Should I Use for My DVD?

Your choices are actually pretty simple:

AC-3 or *PCM* (AIFF from Macs, WAV from PCs), and here's why:

AC-3 solves many issues: If your project requires more than one audio program accompanying a Track, our recommendation would be to stick with Dolby Digital AC-3 files. When multiple audio streams are used in a Track, the limited bandwidth required of an AC3 stream is beneficial.

Both Compressor 2 and the (now obsolete) A.Pack application do a very efficient job of creating encoded Dolby Digital AC-3 files from your uncompressed AIFF or WAV source files. In addition to being smaller files, these streams can preserve precious video encode rate, allowing higher quality MPEG encodes.

TIP!
Use AC-3 compression for DVDs that use more than one audio program in a Track (alternate audio streams).

AIFF files are simple to use and instantly available upon the successful export of your video program from Final Cut Pro, iMovie, and the like, as well as QuickTime exports. They have also created many hardware encoder boards. They also require little or no further processing or down-conversion to be usable as an authoring asset. PCM WAV files can easily be transferred to your Mac via network or hard drive.

If you have a simple DVD that is relatively short (less than 60 minutes), you can probably use AIFF files successfully.

BUT—the tough part of using AIFF files is the bandwidth (playback bitrate) required to deliver the audio. Uncompressed AIFF files use up about 1.6 Mbps of the (nominally) available 9.8 Mbps DVD playback bandwidth. That's a *lot* of bandwidth to give up for just audio!

Since the 9.8-Mbps DVD playback bandwidth must deliver *all* video, audio, and subtitle streams selected,

Preparing Audio Assets

you can see that more than one AIFF file can burn up your precious DVD bandwidth all too quickly. In DVD Studio Pro, using more than two AIFF files will likely choke the Preview mechanism, making it difficult to test your DVD.

> **REMEMBER!**
> AIFF files are best used for DVD programs less than 60 minutes in length. We DO NOT recommend using AIFF when your DVD exceeds 60 minutes in length, *or requires more than one audio program.*

How to Create AIFF Audio Files (in General)

Figure 6-11 — AIFF File Icon.

AIFF audio files are a standard byproduct of video editing, and are easily created by a wide variety of tools (see Fig. 6-11). AIFF audio tracks are integral to Final Cut Pro, Premiere, After Effects, and iMovie, and are a very standard format used by QuickTime.

AIFF is an *uncompressed* format, where the sound data have been digitized, but have not been compressed to reduce its size.

AIFF audio files can also be created as part of an MPEG-2 Export from Final Cut Pro or an MPEG-2 encode in Compressor. The QuickTime Pro Player will also create an uncompressed AIFF sound file when a QuickTime movie is exported to MPEG-2 (QuickTime 6).

> **IMPORTANT!**
> AIFF files for DVD use *must be at the 48 KHz sample rate* or above. Timing and sync problems will be created if the original audio is at 44.1KHz prior to import, or before it is encoded into Dolby Digital form.

How to Create an AIFF File Using Final Cut Pro

One very easy way to create an AIFF audio file for DVD is by exporting a 48K audio soundtrack from Final Cut Pro, using **File > Export > Audio to AIFF(s)** (see Fig. 6-12) or by exporting your video content into a QuickTime movie. In either case, the channels of audio mixed in the Final Cut Pro timeline will become a simple stereo soundtrack, if the settings are correct (see Fig. 6-13).

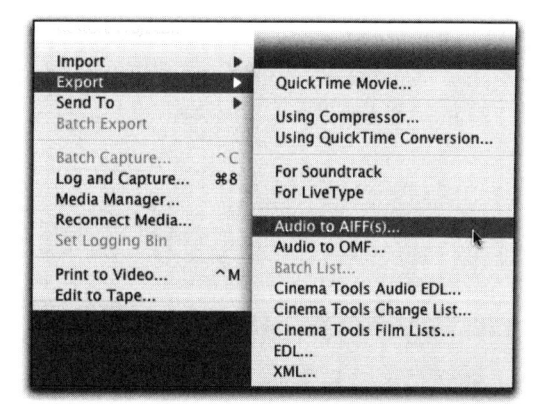

Figure 6-12 — Final Cut Pro Export Audio to AIFF command.

Figure 6-13 — Verifying Final Cut Pro has Stereo Mix selection.

145

DVD Studio Pro 4

If you export only the audio to AIFF using the command `File > Export > Using QuickTime Conversion > Format: AIFF`, be sure to leave "Rate" set to 48KHz, and "Sample Size" set to *at least* 16 bits in the settings dialog (see Fig. 6-14).

If you export to MPEG-2 directly through Compressor 2, an AIFF file and/or AC3 stream can be created automatically as part of the MPEG encode process. (We covered this fairly well back in Chapter 5.)

How to Create an AIFF File Using iMovie

iMovie HD's new "`File > Share`" command provides for an audio-only export, but you might not think so at first glance. You first share the iMovie using the "`QuickTime > Expert Settings`" mode, and then configure the QuickTime export as 48 kHz, Stereo (L/R), 16-bit, just as in exporting from Final Cut Pro. (See Fig. 6-15.)

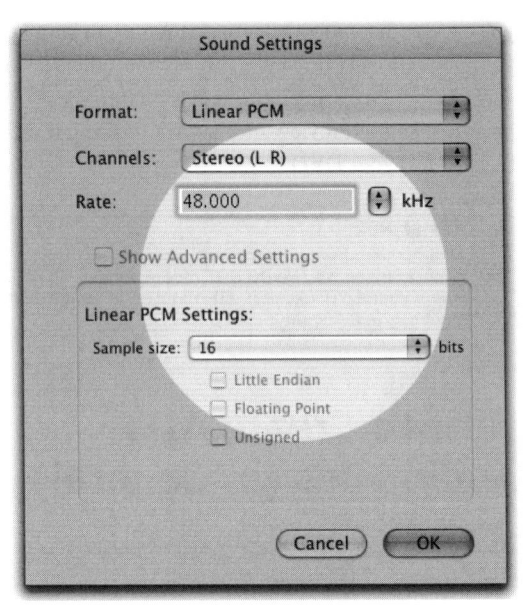

Figure 6-14 — FCP Audio Settings for DVD.

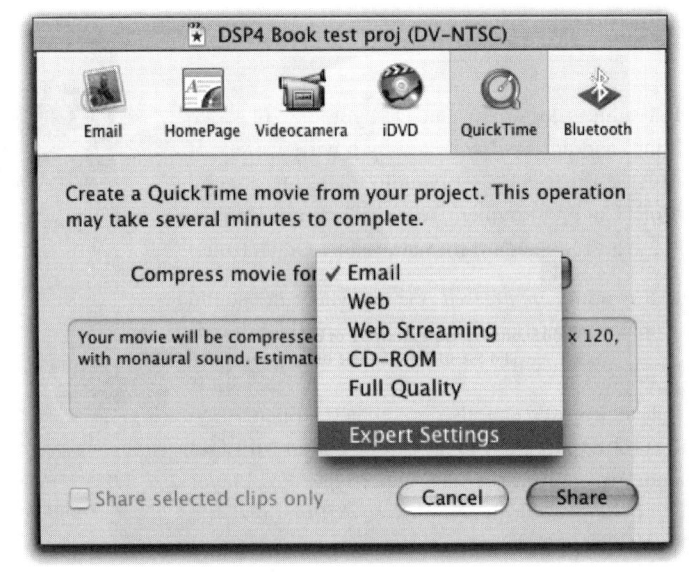

Figure 6-15 — The iMovie "share" command and dialog.

146

Preparing Audio Assets

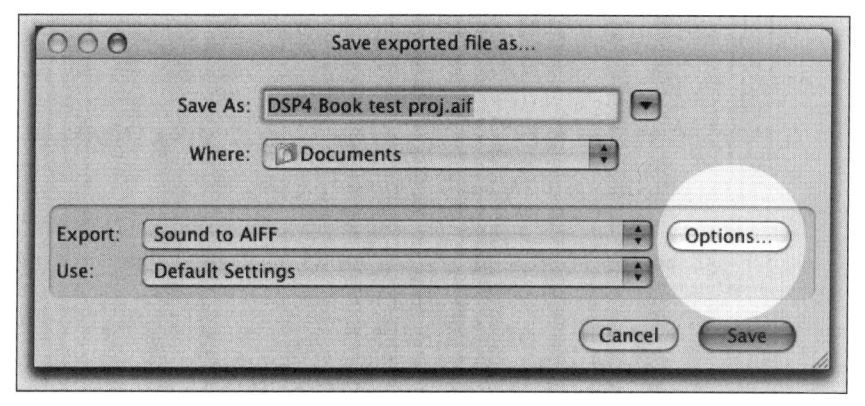

Figure 6-16 — The iMovie HD file export dialog.

Once you have initiated the "Share" (actually an Export), be sure to select the "Sound to AIFF" setting in the *Export:* pop-up, and don't forget to click the "*Options…*" button to open the "Sound Settings" dialog, and confirm the same 16-bit 48 KHz settings as in Figure 6-14.

When the AIFF audio file has been exported, it's saved as a QuickTime AIFF file. It can be used for Authoring in the DVD Studio Pro project by simply dragging and dropping into a project Element. You can also Import the file into the Asset Tab, if you prefer to be formal. You may also further process that file by encoding into Dolby Digital AC3, as this chapter outlines. Exporting from iMovie does not give you the option to encode directly into AC3. (See Fig. 6-16.)

> **TIP!**
> If you are importing sound from a CD library for use in a DVD, be sure to *Sample Rate Convert* the 44.1 KHz file to 48 KHz by Exporting it using the QuickTime Pro player.

> **IMPORTANT!**
> You will have better luck with your DVD if you use Dolby Digital AC3 files, especially if you have long programs (>60 minutes), or are planning to use more than one audio stream in a Track.

If you are looking for instructions on how to use Compressor 2 to create Dolby Digital streams, skip ahead to the section entitled "Encoding AC-3 Audio Streams Using Compressor 2," starting on page 163.

The following section is for those users who may still be using DVD Studio Pro 2 or 3, and would like to learn how use A.Pack.

Encoding AC-3 Audio Streams Using A.Pack

In *A.Pack*, Dolby Digital bitstreams can be created in two different methods—individually, in the *Instant Encoder*, or in multiples, using the *Batch List*.

> **NOTE!**
> Audio for DVD *must* be at the 48 KHz sample rate. Trying to create AC-3 files from 44.1K files will cause unpredictable timing problems later.

About the A.Pack Instant Encoder

Use the A.Pack *Instant Encoder* to encode individual AIFF or WAV files into AC-3 one at a time. All

147

DVD Studio Pro 4

encode settings must be verified by hand, and only one file at a time can be created.

About the A.Pack Batch List

Use the A.Pack *Batch List* to encode multiple files in a single operation. These files may be all the same format, or a mixture of different formats. (See the section on Using the A.Pack Batch List later in this chapter for details on using this interface.)

Launching A.Pack

To launch the A.Pack application, locate the *Application Icon* (see Fig. 6-17) and double-click. (You may also launch A.Pack by double-clicking on the icon of an existing AC-3 audio file, but that will load the selected file into the AC-3 monitor window.)

Figure 6-17 — A.Pack Application Icon and AC3 File Icon.

What Is A.Pack?

A.Pack is an application that was used in DVD Studio Pro 1, 2, and 3 to convert uncompressed audio files (AIFF or WAV) into the Dolby Digital format (also called AC-3). Dolby Digital is a preferred audio format for DVDs, and one of the two mandatory forms of audio for DVDs worldwide. A DVD must have either a PCM or Dolby Digital soundtrack to comply with the DVD Spec except in Region 2, where MPEG audio still is an alternative to AC3.

Starting with DVD Studio Pro 4, Compressor 2 replaces A.Pack's capabilities.

Why Use Dolby Digital?

Compressing PCM audio files into Dolby Digital AC-3 audio streams not only reduces the size of the audio data, but more importantly, it reduces the *bandwidth* (data rate) required to playback the audio. The unused data rate can be used to encode higher-quality video streams.

Dolby Digital also allows the creation of 5.1 Surround Sound files, which would otherwise be very difficult to create using PCM files, and would be troubling to use in a DVD.

A.Pack is certified by Dolby Labs and designed to make easy work of the job of creating high-quality Dolby Digital (AC-3) streams, whether they are stereo or in a surround format. It has both an *Instant Encoder,* for encoding a single file, and a *Batch List* processor for encoding a group of files in a continuous manner (even unattended).

This Section will guide you through the details using A.Pack to create Dolby Digital compressed sound files for your DVD.

A.Pack Quick Encode Guide for Stereo Files

To Quickly create a Dolby Digital file from an AIFF or WAV Master:

- *Launch A.Pack*—unless you have modified the A.Pack preferences, the *Instant Encoder* will open (see Fig. 6-18a).

- In the *Audio Tab*, adjust the Settings for the file desired:

148

Preparing Audio Assets

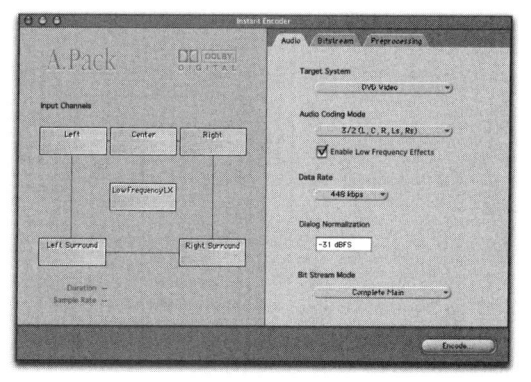

Figure 6-18a — A.Pack Instant Encoder Window.

- Leave *Target System* at **DVD-Video**.
- Set *Audio Coding Mode* to **Stereo 2/0**.
- Set *Data Rate* to **192 Kbps** for Stereo.
- Set *Dialog Normalization* to **–31**; hit **ENTER**.
- Leave *Bitstream Mode* at "**Complete Main**."

- In the *Bitstream Tab*, leave the settings unchanged.
- In the *Preprocessing tab*, turn Compression **Off**.
- In the same tab, turn *RF Overmodulation Protection* **Off**.
- Load the file to be encoded into A.Pack's file grid source tabs—drag files into tabs, or click a channel's tab to open a Dialog box.
- Click *Encode*; Set the Destination Location and Final File Name.

NOTE!
Failure to Load an Input File will cause an error message (see Fig. 6-18b).

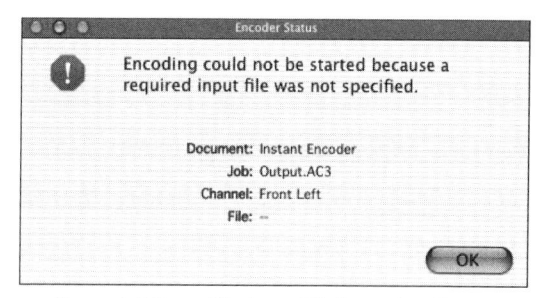

Figure 6-18b — "No Input File" warning dialog.

Preparing to use A.Pack

(Additional detailed information about using A.Pack can be found by consulting Appendix B of the DVD Studio Pro 2 or 3 *Help* file.)

IMPORTANT!
AIFF or WAV (PCM) files for DVD use must be at the **48 KHz** sample rate. If you are importing sound from a *CD library*, be sure to *Export* the captured 44.1 KHz file to 48 KHz using the QuickTime Pro player.

Using the A.Pack Instant Encoder

To create a Dolby Digital encode using the Instant Encoder window, you will need to make settings choices in the three available tabs that organize sets of parameters. Note that the manner of changing settings is common to both the *Instant Encoder* and the *Batch List*.

In the Audio Tab:

- Set the *Target System* (always DVD Video).
- Set the *Audio Coding Mode* (channel configuration).
- Set the *Data Rate* (the Encode Bit Rate).

149

- Set the *Dialog Normalization* Setting.
- Set the *Bit Stream Mode* (a MetaData label).

In the Bitstream Tab:

- If this is a 5.1 encode, set the *Center Downmix level*.
- If this is a 5.1 encode, set the *Surround Downmix level*.
- If this is an Lt/Rt encode, set *Dolby Surround mode*.
- If a Copyright Exists, *check the checkbox*.
- If this is the Original Bitstream, *check the checkbox*.
- If needed, check the *Audio Production Information* box.
- Set the *Peak Mixing Level*, based on mixing room info.
- Set the *Room Type*, based on mixing room information.

In the Preprocessing Tab:

- Set/Verify the *Compression setting*.
- Set/Verify the *General settings*.
- Set/Verify the *Full Bandwidth Channels settings*.
- Set/Verify the *LFE Channel setting*.
- Set/Verify the *Surround Channels settings*.

The File Grid (Input Channels):

- Load the Source files into the appropriate Grid locations: Left into Left, Right into Right, etc. (See Fig. 6-19.)

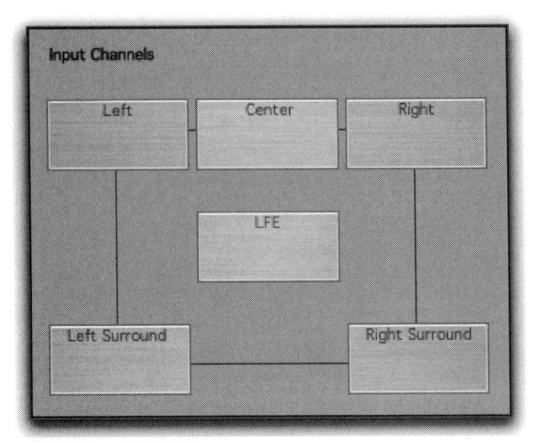

Figure 6-19 — A.Pack Input Grid in Surround 5.1 Mode.

Loading the File Grid

Before encoding can take place, the *File Grid Channels* must be loaded with the source files required for the encode. To do so, click on a Channel in the Grid, and use the File Dialog box to locate the file for that channel, or... (easier) drag and drop a file onto the proper channel in the Grid. Select the appropriate channel from within the file if a dialog is offered. (Mono files will not require choosing a channel from within the file. AIFF files are frequently stereo, and will offer this choice to be made).

> **TIP!**
> Use the *Batch List* (detailed later) when you have a lot of files to convert and wish to do so all at once or unattended. The *Instant Encoder* will require manually loading each file and setting the job parameters by hand.

Settings Preferences in the Audio Tab (in Detail)

Setting the Target System

To properly use A.Pack, you have to identify the system for which the Audio bitstream is being created:

Preparing Audio Assets

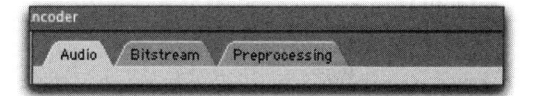

Figure 6-20 — The Audio Tab.

DVD-Video, DVD-Audio, or Generic AC-3. The proper selection is always *DVD-Video* (see Fig. 6-20 and 6-21).

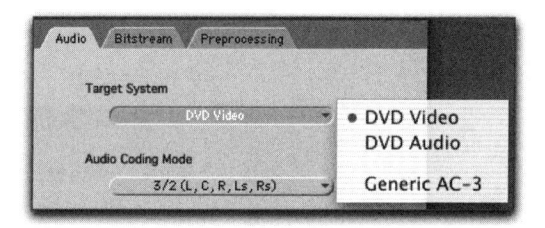

Figure 6-21 — The Target System.

Setting the Audio Coding Mode

- *Audio Coding Mode* defines the channel configuration of the Dolby Digital bitstream you are encoding. It can vary from Mono to 5.1 with or without LFE. NOTE that Dolby denotes the channel configuration as a two digit series, i.e., 3/2, or 2/0, where the first digit indicates the number of Front Channels (1, 2, or 3) and the second digit indicates the number of surround channels (0, 1, or 2) (see Fig. 6-22).

  Most DVD audio programs are usually Stereo (2/0) or 5.1 Surround (3/2 w/LFE).

- Select the desired *Audio Coding Mode* pop-up to indicate the number of channels you are encoding—Standard Stereo is 2/0 (L, R). Some programs you encode may be 5.1 or other channel configurations.

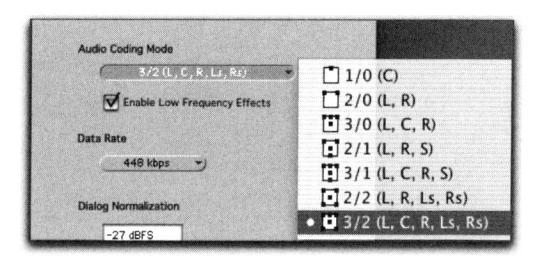

Figure 6-22 — The Audio Coding Mode pop-up.

- If you are using the LFE channel, check the *Enable Low Frequency Effects* checkbox. Leave it off if you are not doing a format that requires it.

- Once you've selected the Audio Coding Mode, you MUST continue on to verify or set the Data Rate.

Setting the Data Rate (the All-Important Encode Rate)

Setting a **Data Rate** for Encoding Audio is analogous to setting the MPEG Encode rate for Video (see Fig. 6-23). The accepted standard rates for AC-3 are 192 kBps for Stereo and 384 kBps for 5.1. Note that the allowable encode rates displayed in the pop-up will vary depending on the Audio Coding Mode (channel configuration) you have selected.

> **WARNING!**
> Under *no* circumstances should a Dolby Digital data rate *ever* exceed 448 Kbps for DVD. You'll probably find Dolby Digital encode rates below 192 kbps unsatisfactory for Stereo Sound files, but possibly workable for Mono sound files. Rates under 96 Kbps are not recommended for use at all.

151

DVD Studio Pro 4

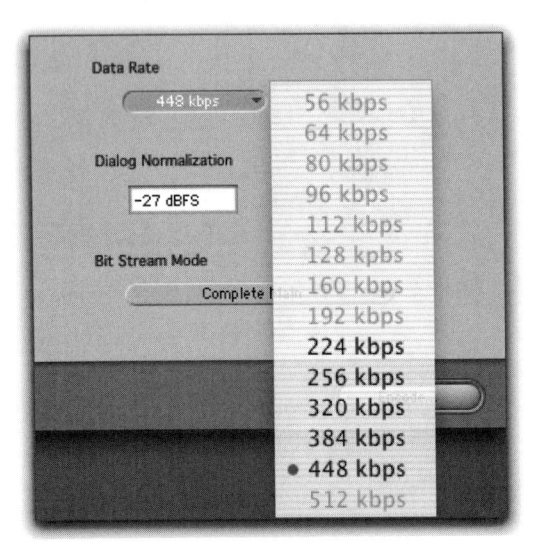

Figure 6-23 — The Data Rate (encode rate) pop-up.

Setting the Dialog Normalization Setting

This is probably THE most misunderstood function in A.Pack! Dialog Normalization is a setting in the Dolby AC-3 spec that allows the Decoder to dynamically adjust the playback level, as needed (see Fig. 6-24). The setting programmed at the time of encode tells the Decoder how much attenuation to add during playback. For typical Stereo files that have already been leveled, this function may be better left disabled.

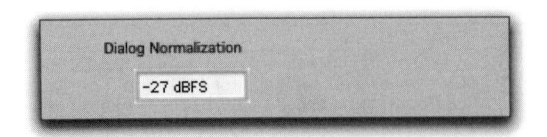

Figure 6-24 — The Dialog Normalization parameter.

- To set the *Dialog Normalization* value, click in the field and adjust the value. Valid values are from –1 to –31. To *disable* the Dialog Normalization function, set it to –31.

- You *MUST* be sure to hit "ENTER" or "RETURN" after entering the value, to close the field. If not, the value may not be set properly. To verify, check the file's MetaData once your encode has been completed. (See the section on Monitoring your Dolby Digital encodes further on in this chapter).

Setting the Bit Stream Mode Parameter

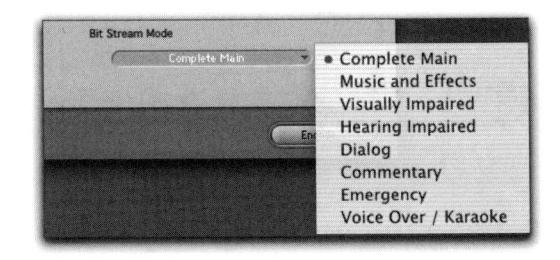

Figure 6-25 — The Bit Stream Mode pop-up.

Dolby "MetaData" (labels or information) may be programmed into the AC-3 file, to define the purpose of this particular Audio bitstream. These labels include *Complete Main, Music and Effects, Commentary, Dialog*, and more. (See Fig. 6-25.)

- To set the Bit Stream mode, click on the *Bit Stream Mode* pop-up, and select the desired label. In many cases, Complete Main is the appropriate setting—it is always the *default* value. In any case, this setting does NOT affect the way in which the Dolby Decoder reproduces the bitstream. It's just a label. (See Fig. 6-26.)

152

Preparing Audio Assets

Settings in the Bitstream Tab (in Detail)

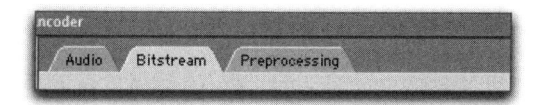

Figure 6-26 — The Bitstream Tab.

If this is a 5.1 encode, set the Center Downmix Level:

The Center channel of a Surround Sound bitstream can be mixed into the Analog Left/Right Stereo Outputs of your DVD player, attenuated per the downmix level programmed in at the time of encoding. Setting this parameter controls the level of the Center channel present in the final Stereo Downmix.

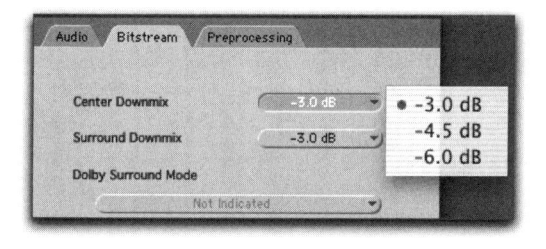

Figure 6-27 — Setting the Center Downmix parameter.

- To set the *Center Downmix* level, click on the Center Downmix pop-up and select the appropriate setting. (See Fig. 6-27.)

In many cases, the default setting of -3.0 dB may be perfectly appropriate. For further information, you may wish to consult the Dolby *Metadata Guide* PDF, available from the Dolby website. We've included it in the Dolby folder on the accompanying DVD.

If this is a 5.1 encode, set the Surround Downmix Level:

This parameter controls the level of the Left and Right Surround channels in the final Stereo Downmix.

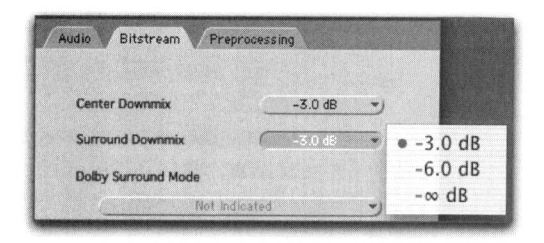

Figure 6-28 — Setting the Surround Downmix.

Surround Sound bitstreams can output their mono or stereo surround information from the Analog Front Left/Right Outputs of your DVD player, if this information is encoded into the bitstream.

- To set the *Surround Downmix* level, click on the Surround Downmix pop-up and select the appropriate setting. (See Fig. 6-28.)

In many cases, the default setting of -3.0 dB may be perfectly appropriate. At this setting, the left and right surround channels are each attenuated 3 dB and sent to the left and right front channels, respectively. At –6 dB, the output routing is the same as –3dB, but the signal is attenuated 6 dB. At the –∞ dB (Infinity) setting, the surround channel(s) are discarded.

If this is an Lt/Rt encode, set Dolby Surround mode:

Dolby Surround Mode is a special two-channel configuration, which actually contains FOUR channels of discrete audio information: Left, Center, Right, and a Mono Surround channel, electronically matrixed into two channels. These files are known by

153

the special code **Lt/Rt** (Left Total/Right Total). If the file is Lt/Rt, setting this property can cause a Dolby Surround Pro Logic decoder to engage in order to properly decode four-channel Dolby Surround information encoded in the file's two channels.

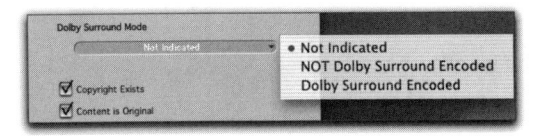

Figure 6-29 — Dolby Surround Mode.

Dolby Surround Mode would only be required if the original sound mix was a four-channel mix containing L, C, R, and mono Surround) that has been processed by a *Dolby Matrix encoder*. If this is true, then set the Dolby Surround Mode property to *Dolby Surround Encoded*. (See Fig. 6-29.)

If the original sound mix was NOT a four-channel mix with mono surround, then Dolby Surround Mode is NOT required, and this property should be set to NOT Indicated, or *NOT Dolby Surround Encoded*. If you are NOT sure about this setting, leave it Not Indicated. Odds are you will be correct. It may also be possible to use this setting to decode files prepared in the Ultra-Stereo format, since that is a similar matrixed L, C, R, S format as Dolby Surround.

If a Copyright Exists, check the designated checkbox:

Figure 6-30 — The Copyright and Content checkboxes.

Dolby MetaData can contain a flag bit to indicate if the audio content of the bitstream is currently under Copyright. (See Fig. 6-30.)

- The Default setting is "on." To turn this flag "off," click on the Copyright Exists checkbox to uncheck it. Click again to check it if needed.

If this is an Original Bitstream, check the checkbox:

Dolby Metadata can contain a flag bit to indicate if the audio content of the bitstream is the Original Dolby Bitstream, or a copy. (See Fig. 6-30.)

- The Default setting is "on," meaning this is the original bitstream created of this content. To denote this is not the original bitstream, click on the *Content is Original* checkbox, to uncheck it. Click again to check it if needed.

If appropriate, check the Audio Production Information box:

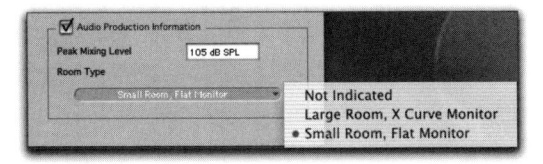

Figure 6-31 — Setting Audio Prod'n Info.

The default of this parameter is OFF, but by turning it ON, it is possible to advise Dolby Digital decoders that the accompanying reference information is valid, and it may be used to adjust playback levels. (See Fig. 6-31.)

Set the Peak Mixing Level, based on mixing info:

This parameter defines the Peak mixing level used during the surround mix, but is only referenced by the decoder if the Audio Production Information is checked on.

Preparing Audio Assets

Set the Room Type, based on mixing room information:

This parameter defines the Type of mixing room or stage used during the surround mix, but is only referenced by the decoder if the Audio Production Information is checked on. The Large room X-curve parameter would be checked if the audio file was created in a traditional theatrical mixing (dubbing) stage, whose audio characteristics follow the Academy Curve—a specific equalization curve that attenuates high frequencies to emulate the playback characteristics of a movie theater.

Settings in the Preprocessing Tab

Setting the Compression Factor

A.Pack can apply different types of Program Compression to assist in leveling out the dynamic range (not always a good thing), depending on the kind of program you are encoding—Film, Music, or Speech. You may find, however, that turning Compression off may be the right solution for material that is already dynamically leveled or doesn't have large dynamic range swings.

- To apply a Program *Compression* type, or to deselect Compression, click on the *Compression* pop-up (as shown in Fig. 6-32) and drag to select your choice.

Setting the RF Overmodulation Protection

- According to Dolby Labs, this parameter is designed to protect against overmodulation

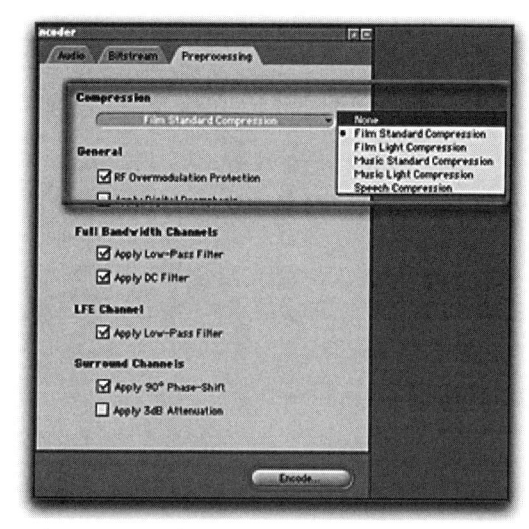

Figure 6-32 — Setting the Compression parameter.

when a decoded Dolby Digital bitstream is RF modulated. When enabled, the Dolby Digital encoder includes pre-emphasis in its calculations for RF Mode compression. The parameter has no effect when decoding using Line Mode compression.

NOTE! Except in rare cases, this parameter should be disabled.

A.Pack can set or disable the RF Overmodulation Protection bit through a simple checkbox. Check if carefully, as it usually defaults to "On."

- The Default is "on" (enabled). (See Fig. 6-33.) To disable the RF Overmodulation Protection bit, deselect the check box for this parameter.

NOTE that in recent versions of A.Pack, this may default to OFF.

DVD Studio Pro 4

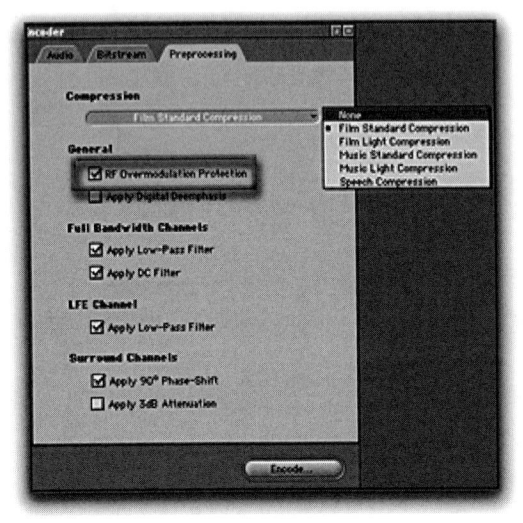

Figure 6-33 — Setting the RF Overmodulation parameter.

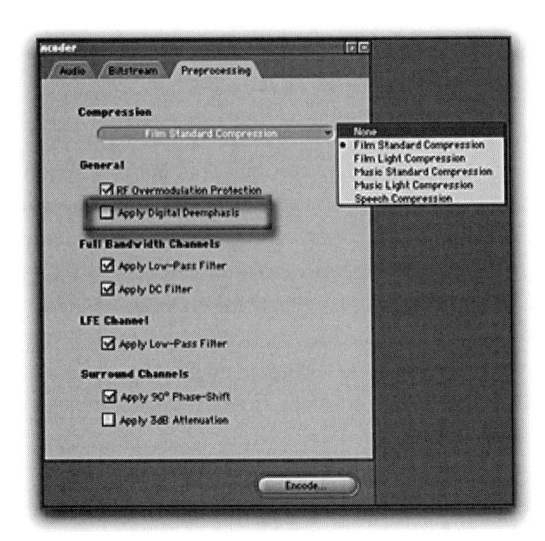

Figure 6-34 — Setting the Digital Deemphasis parameter.

Setting the Digital Deemphasis Parameter

Believe it or not, there is NO official Dolby explanation for this parameter in the Dolby Metadata Guide. (See Fig. 6-34.)

- Since the parameter defaults to off, it's best to leave it that way.

Setting the Low-Pass Filter (for Full Bandwidth Channels)

According to Dolby Labs, this parameter determines whether a lowpass filter is applied to the main input channels of a Dolby Digital encoder prior to encoding (see Fig. 6-35). This filter removes high-frequency signals that are not encoded. At the suitable data rates, this filter operates above 20 kHz. In all cases, it prevents aliasing on decoding and is *normally switched on*. This parameter is not passed to the consumer decoder.

- To deselect *Apply Low-Pass Filter* (which defaults to "*on*"), click on the checkbox to deselect it (see Fig. 6-35).

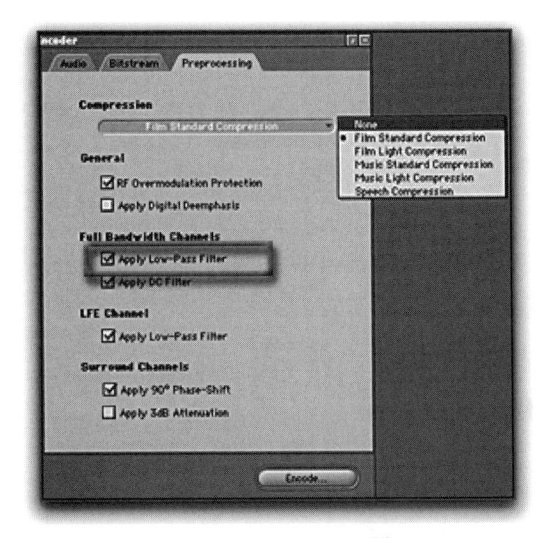

Figure 6-35 — Setting the Low-Pass Filter parameter.

156

Preparing Audio Assets

Setting the DC Filter (for Full BW Channels)

According to Dolby Labs, this parameter determines whether a DC blocking 3 Hz highpass filter is applied to the main input channels of a Dolby Digital encoder prior to encoding. It is used to remove DC offsets in the program audio and *would only be switched off in exceptional circumstances*. This parameter is not carried to the consumer decoder.

- To deselect *Apply DC Filter* (which defaults to *on*), click on the checkbox to deselect it (see Fig. 6-36).

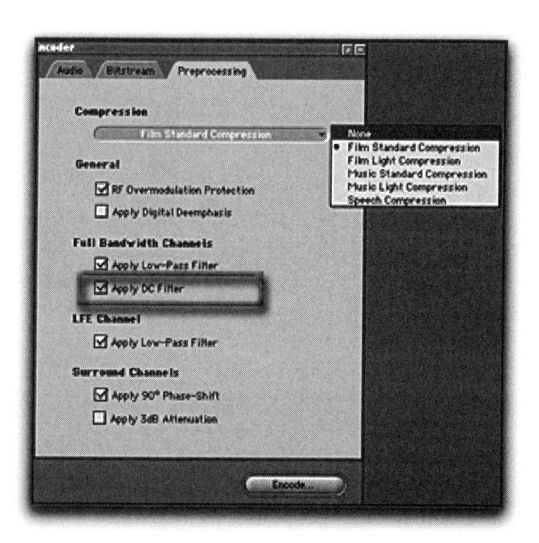

Figure 6-36 — Setting the Full BW Channels DC Filter parameter.

Setting the Low-Pass Filter (for LFE Channel)

According to Dolby Labs, this parameter determines whether a 120 Hz 8th order lowpass filter is applied to the LFE channel input of a Dolby Digital encoder prior to encoding. It is ignored if the LFE channel is disabled. This parameter is not sent to the consumer decoder. The filter removes frequencies above 120 Hz that would cause aliasing when decoded. *This filter should only be switched off if the audio to be encoded is known to have no signal above 120 Hz.*

- To deselect *Apply Low-Pass Filter* (which defaults to "*on*"), click on the checkbox to deselect it (see Fig. 6-37).

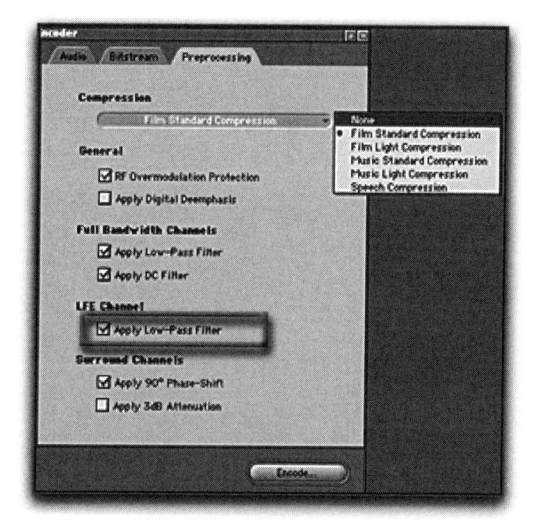

Figure 6-37 — Setting the LFE Low-Pass Filter parameter.

Setting the Surround Phase Shift

According to Dolby Labs, "This parameter causes the Dolby Digital encoder to apply a 90-degree phase shift to the surround channels." This allows a Dolby Digital decoder to create an Lt/Rt downmix simply. For most material, the phase shift has a minimal impact when the Dolby Digital program is decoded to 5.1 channels, but provides an Lt/Rt output that can be Pro Logic decoded to L, C, R, S, if desired. However, for some phase-critical material (such as music), this phase shift is audible when listening in 5.1 channels. Likewise, some material downmixes to

157

a satisfactory Lt/Rt signal without needing this phase shift. It is therefore important to balance the needs of the 5.1 mix and the Lt/Rt downmix for each program. *The default setting is Enabled.*

- To disable the *Apply 90-degree Phase-Shift to Surround Channels* parameter, click on the Apply 90-degree Phase-Shift checkbox (as shown in Fig. 6-38) to uncheck the selection.

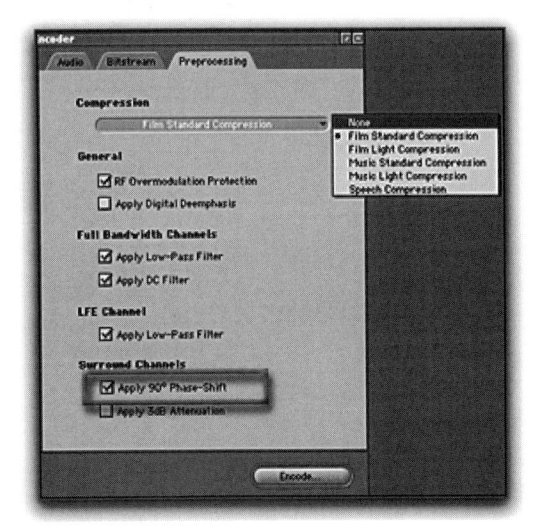

Figure 6-38 — Setting the Surround Phase-Shift parameter.

Setting the 3 dB Attenuation in Surround Channels

According to Dolby Labs, the *Surround 3 dB Attenuation* parameter determines whether the surround channel(s) are attenuated 3 dB before encoding. The attenuation actually takes place inside the Dolby Digital encoder (in this case, A.Pack). It balances the signal levels between theatrical mixing rooms (dubbing stages) and consumer mixing rooms (DVD or TV studios). Consumer mixing rooms are calibrated so that all five main channels are at the same sound pressure level (SPL). For compatibility reasons with older film formats, theatrical mixing rooms calibrate the surround channels 3 dB lower in SPL than the front channels. The consequence is that signal levels on tape are 3 dB louder. Therefore, to convert to a consumer mix from a theatrical calibration, it is necessary to reduce the surround levels by 3 dB by enabling this parameter.

- To apply the Surround Attenuation, click on the *Apply 3dB Attenuation* checkbox, shown in Figure 6-39.

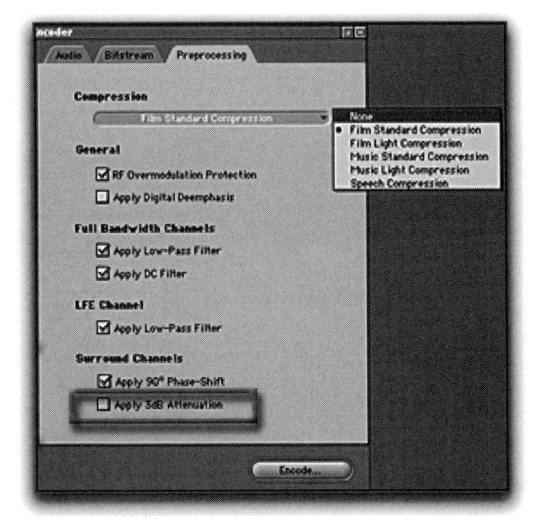

Figure 6-39 — Setting the 3dB Surround Attenuation parameter.

Using the A.Pack Batch List

Use the *Batch List* (see Fig. 6-40) to encode multiple AIFF or WAV files into AC3 in an unattended manner. If you have used Compressor, Cleaner, or Media Cleaner Pro, you are probably familiar with the batch process concept. It saves you time, by freeing you up to continue your creative work while the computer does the AC-3 encoding automatically, unattended.

Preparing Audio Assets

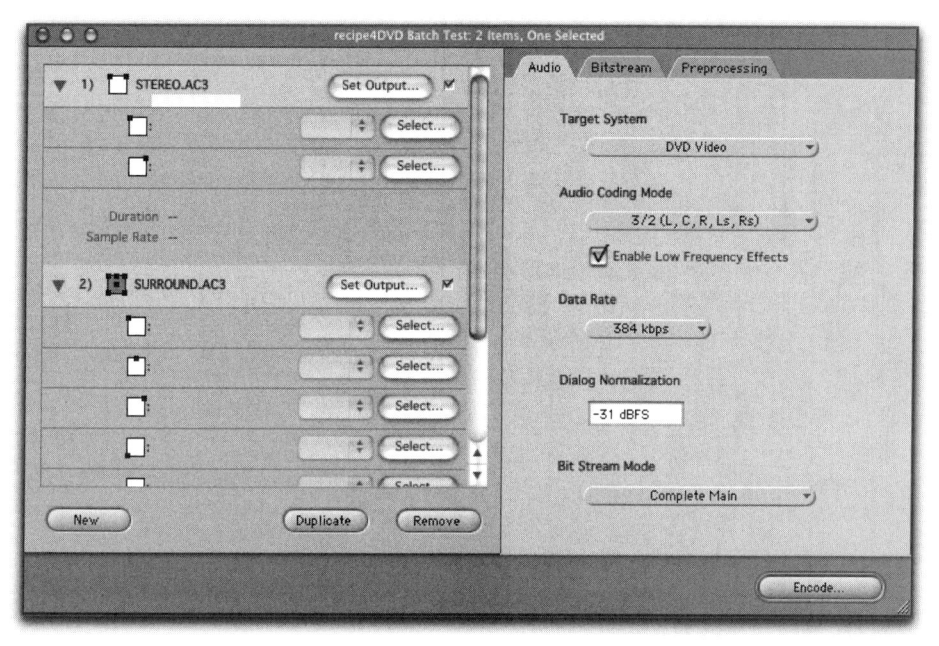

Figure 6-40 — A.Pack Batch List.

To use the A.Pack Batch List process:

- Select **File > New Batch List...** from the menu (Fig. 6-41) or open an existing Batch List with the **Open...** command.

Figure 6-41 — The Batch List command.

- A new Batch List will open with a default "Job" created; (see Fig. 6-42). Each individual Encode will use a separate "Job" in the batch list. Each entry in the Batch List has space for the details to be programmed.

- Modify the *Audio Coding Mode* in the same way you configured the Instant Encoder. In the *Audio Tab* of the Batch List, select the desired configuration from the Audio Coding Mode pop-up.

- *Load the Source File(s)* into the appropriate channel tabs. This can be done by clicking on a "Select" button and then navigating to a file, or by dragging and dropping a file onto a channel box.

- Select the destination and filename of the completed encode using the "*Set Output*" button. The Checkbox will enable or disable this file in the Batch.

- Adjust the encode Job settings in the Settings Tabs, in the same manner as you would adjust the settings in the Instant Encoder.

159

DVD Studio Pro 4

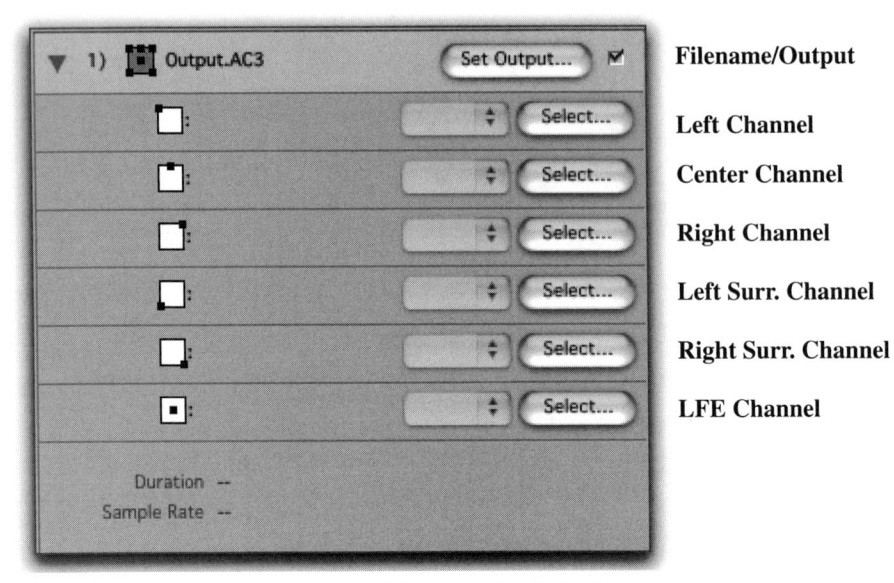

Figure 6-42 — A Batch List default job.

- *Add Jobs* (additional encodes) as needed, using the **Batch Menu > "New Job"** command (see Fig. 6-43), or use the *"New/Duplicate/Remove"* buttons at the bottom to add, duplicate, or remove encode Jobs.

- You can remove jobs from the list using the **Batch > Remove JOB** command, or the *Remove* Button—SELECT the desired Job first, though, before issuing that command (see Fig. 6-44).

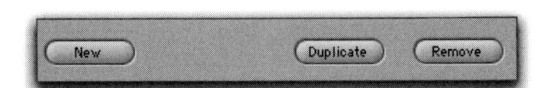

Figure 6-44 — Batch list Job buttons New, Duplicate, Remove.

> **TIP!**
> To encode many similar encode jobs but with different source files, set up the encode settings for the first Job *completely*, then **DUPLICATE** that job, changing the Source File afterwards—this will make for a MUCH faster setup of the batch list.

- You can also adjust the settings of each job by selecting multiple jobs in the Batch List and changing settings in the tabs.

- If you wish to remove a file from the Batch List, use the **"Batch > Unflag for Encoding"** command; to restore it, use the **"Batch > Flag for Encoding"** command. (See Fig. 6-45.)

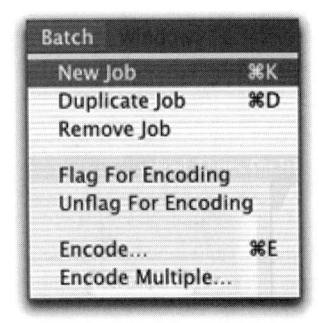

Figure 6-43 — The Batch Menu New Job Command.

Preparing Audio Assets

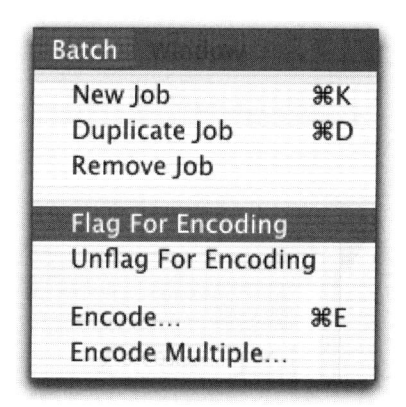

Figure 6-45 — Flag [file] For Encoding sets the Checkbox "on" Unflag [file] For Encoding turns it "off."

- You can also use the *Checkbox* next to the "Set Output" button to accomplish the same thing. (See Fig. 6-46.)

- Once all files are properly set up, *Encode a single file* using the **"Encode..."** Menu option, or *Encode All <u>enabled</u> files in the Batch List* by using the **"Encode Multiple..."** command (see Fig. 6-47).

- When completed, you can import the finished Dolby Digital files into DVD Studio Pro for use in authoring your DVD!

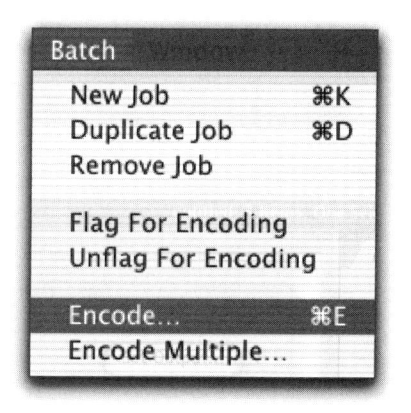

Figure 6-46 — Batch List Job Output controls.

Setting the A.Pack Batch List Encode Job Settings

Review the Encode Job settings listed earlier in the *Instant Encoder Window*—the parameters are identical, and so is the method of setting these parameters for both the Batch List and Instant Encoder.

Monitoring Your Dolby Digital Encodes

To be sure that the encode has been created properly, it is possible to playback an encoded AC-3 file using A.Pack's Monitor Window (Command-2 from the Keyboard). This will give you an audio playback of the file, and also allow you to look at the MetaData that has been created from your encode parameters.

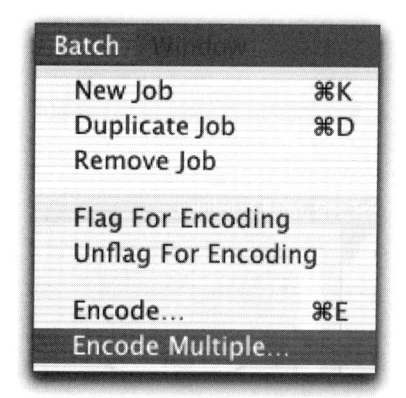

Figure 6-47 — "Encode..." and "Encode Multiple..." commands.

161

DVD Studio Pro 4

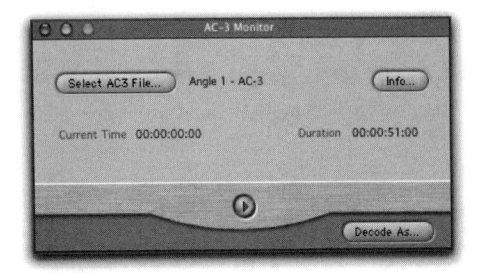

Figure 6-48 — The AC-3 Monitor Window.

- You can load a file into the empty Monitor using the *"Select AC-3 file..."* button, **or**

- Drag an AC-3 file onto the window. (See Fig. 6-48.)

- To view the embedded MetaData, click on the *"Info..."* button. The *Stream Information* panel (see Fig. 6-49) will open for the currently loaded file.

The Log Window

Immediately after encoding, the Log window will appear, confirming the status of the AC3 encode jobs

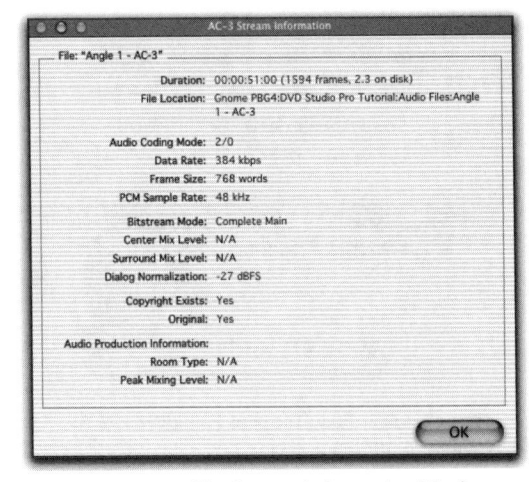

Figure 6-49 — The Stream Information Window.

in that batch. You will find the Log window will accumulate encoder status messages until such time as the "Clear" function is invoked to clear the Log window. Log window messages may be copied and printed or emailed as the need may be.

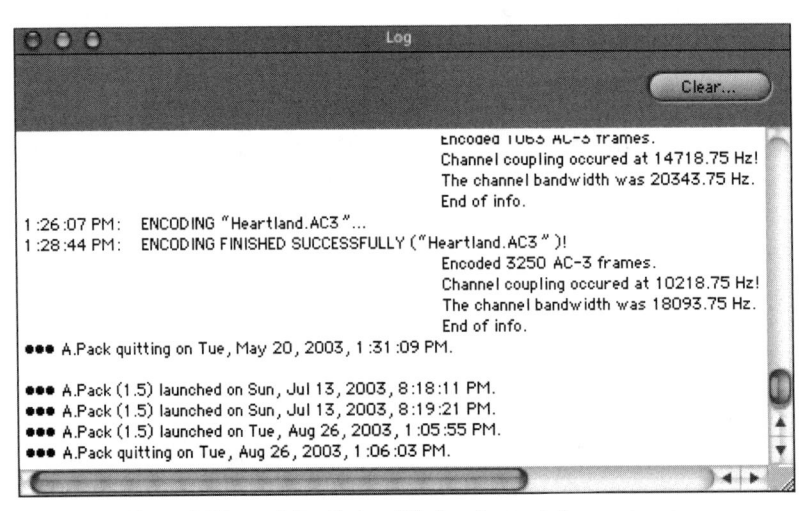

Figure 6-50a — A.Pack's Log Window for encoder messages.

162

Preparing Audio Assets

Encoding AC-3 Audio Streams Using Compressor 2

This section will guide you through the details using Compressor 2 to create Dolby Digital compressed sound files for your DVD.

In Compressor 2, Dolby Digital bitstreams are always created using a Batch List process—there is no "Instant Encoder." Files for conversion are first added to a *Batch List* document, then *Submitted for Encoding*.

The process outlined below is similar to that outlined previously for A.Pack, but procedures and screens are different for Compressor 2.

> **NOTE!**
> Regardless of which application you use for encoding, source audio for DVD *must* be at the 48 KHz sample rate, and not 44.1 KHz. Trying to create AC-3 files from 44.1 K files will cause problems later.

What Is Compressor 2?

Compressor 2 is an application that replaces the AC-3 encoding function that A.Pack provided in DVD Studio Pro 1, 2, and 3. Compressor 2 can convert uncompressed audio files (AIFF or WAV) into the Dolby Digital format (also called AC-3). Dolby Digital is a preferred format for DVDs, and one of the two mandatory forms of audio for DVDs worldwide. A DVD must have at least one PCM or Dolby Digital soundtrack to comply with the DVD Spec.

Why Use Compressor 2 and Dolby Digital?

Compressing PCM audio files into Dolby Digital AC-3 audio streams not only reduces the size of the audio data, but more importantly, it reduces the *bandwidth* (data rate) required to playback the audio. The unused data rate can be used to encode higher-quality video streams. Dolby Digital also allows the creation of 5.1 Surround Sound files, which would otherwise be very difficult to create using PCM files, and would be troubling to use in a DVD.

Compressor 2 is designed to make easy work of the job of creating high-quality Dolby Digital (AC-3) streams, whether they are stereo or in a surround format. It has a *Batch List* processor for encoding a group of files in an continuous manner (even unattended).

About the Compressor 2 Batch List

Use the Compressor 2 *Batch List* to encode multiple files in a single operation. These files may be all the same format, or a mixture of different formats (AIFF or WAV). Regardless of whether you are encoding Stereo or Surround files, you will use the Batch process in Compressor 2. There is no *Instant Encoder Window* in this program.

Launching Compressor 2

To launch the Compressor 2 application, locate the *Application Icon* (see Fig. 6-50b) and double-click. (You may also launch Compressor 2 by double-clicking on the icon of an existing batch file, but that will load the selected batch into Compressor 2's *Batch Window*.)

163

DVD Studio Pro 4

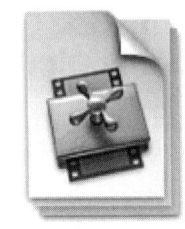

Figure 6-50a, b — Compressor 2 Application Icon (L) and Batch File Icon (R).

Using Compressor 2 for Encoding Stereo Files

Loading Stereo Files

To create a Dolby Digital Stereo encode, you first *load the file(s)* that will be encoded, then *make settings choices* in the Encoder Pane's three parameter tabs, and finally *submit the job*. Note that the manner of changing settings is common to both Stereo files and Surround Sound Groups, but the settings are different for each.

To create a Dolby Digital Stereo file from an AIFF or WAV master, follow these steps:

- Launch Compressor 2; the *Batch Window* opens (Fig. 6-51).

- Load the file to be encoded into Compressor 2's Batch List. You can drag files (Fig. 6-52, left), or click the *Import File* button to open a Dialog box (Fig. 6-52, right). The *Import File* button will open a dialog where you can select a file (not shown). (The *Import Surround Group* button to the left of the *Import File* button is only for Surround Sound Encodes.)

Selecting an Encode Format Preset

- Once the file exists in the Batch Window, Click on *Setting*, and from the pop-up, select the desired *Encoding Format*—in this case, Dolby 2.0 for stereo (see Fig. 6-53a).

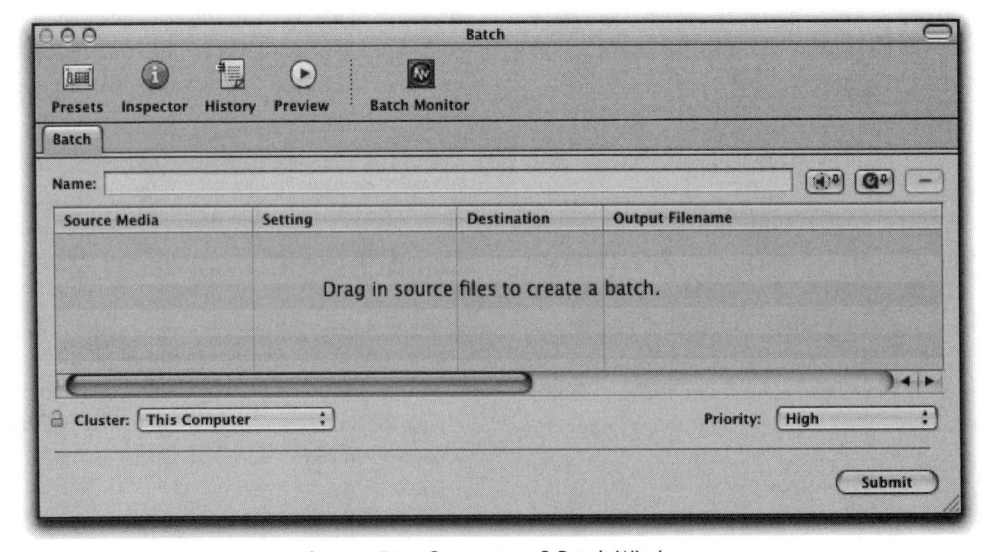

Figure 6-51— Compressor 2 Batch Window.

Preparing Audio Assets

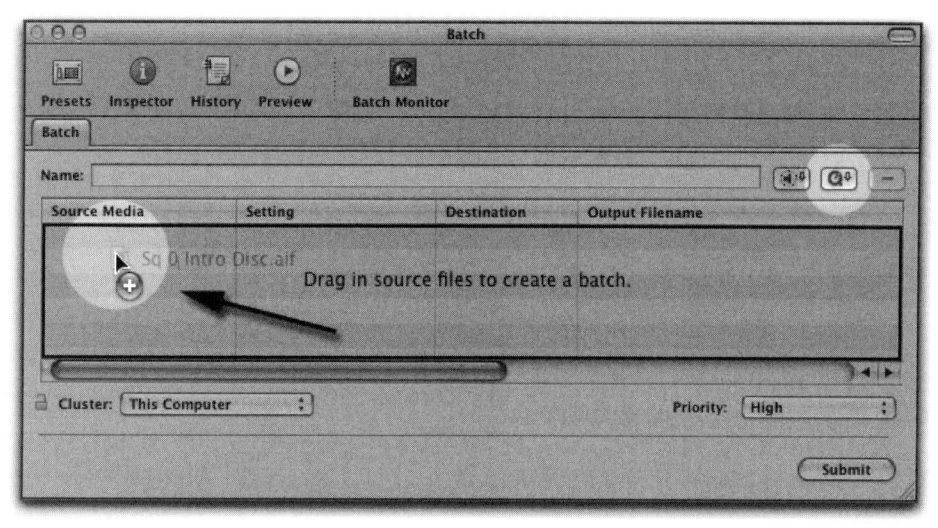

Figure 6-52 — Adding (L) or Importing Files (R) into Compressor 2.

Set Destination and Output Filename

- Once a format is chosen, set the encode's *Destination* location and *Output Filename*; you may use the default settings, if you like them. If not, change them this way:

  – To modify the *Destination*, click that pop-up and select the location where you want the encoded file saved. (See Fig. 6-53b.)

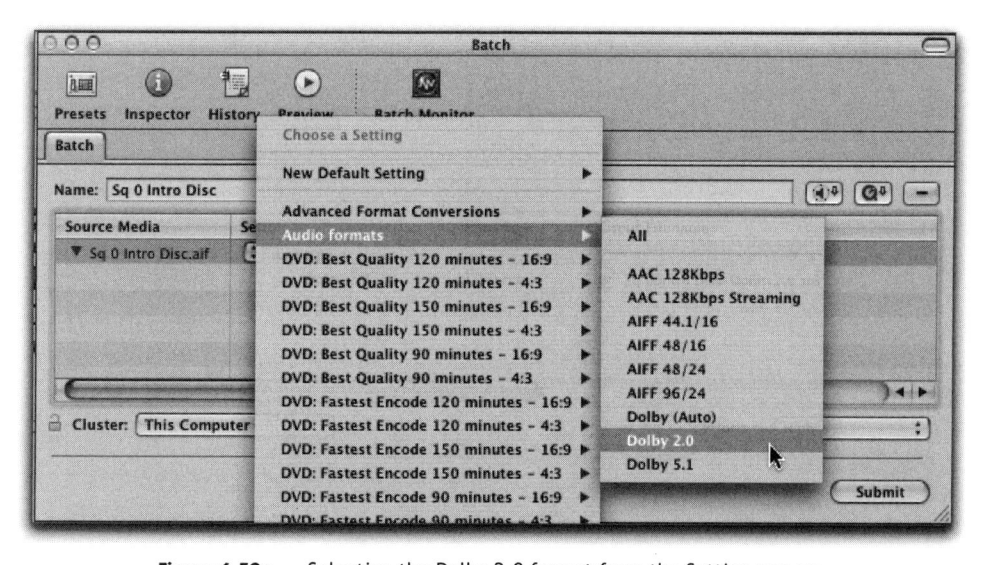

Figure 6-53a — Selecting the Dolby 2.0 format from the Setting pop-up.

165

DVD Studio Pro 4

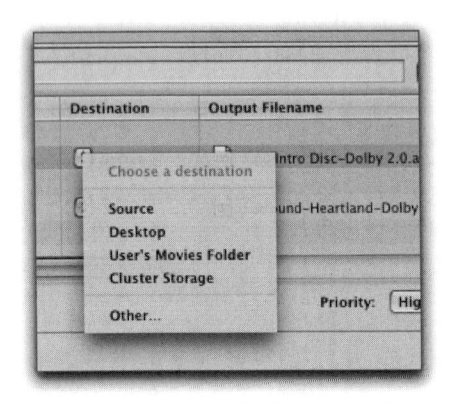

Figure 6-53b — Changing the Destination from the pop-up.

- To modify the *Output Filename*, double-click the name, and modify the name field as you wish.

Using the Inspector for a Closer Look

Once the file settings have been adjusted, you may want to verify the Dolby Digital 2.0 encode settings—to do so, click on the Dolby Digital setting (as in Fig. 6-54, #1), then click the Inspector button (Fig. 6-54, #2). If you use a preset with known values, you don't need to perform this step every time.

Verifying the Encode Preset Settings

Once the Inspector has opened, select the *Encoder Pane* if it isn't already selected (see Fig. 6-55). In the Encoder Pane, check the following settings:

- In the *Audio Tab*, check these Settings at least once:
  - *Verify Target System* at **DVD-Video**.
  - *Verify Audio Coding Mode* to **2/0 (L/R)**.
  - *Verify sample Rate* is **48 kHz.**
  - *Verify Data Rate* is **192 Kbps.**
  - *Verify Bitstream Mode* is **"Complete Main."**
  - *Verify Dialog Normalization* is **–27**; some people prefer the -31 setting for a louder file (read the *Dolby Metadata Guide* in

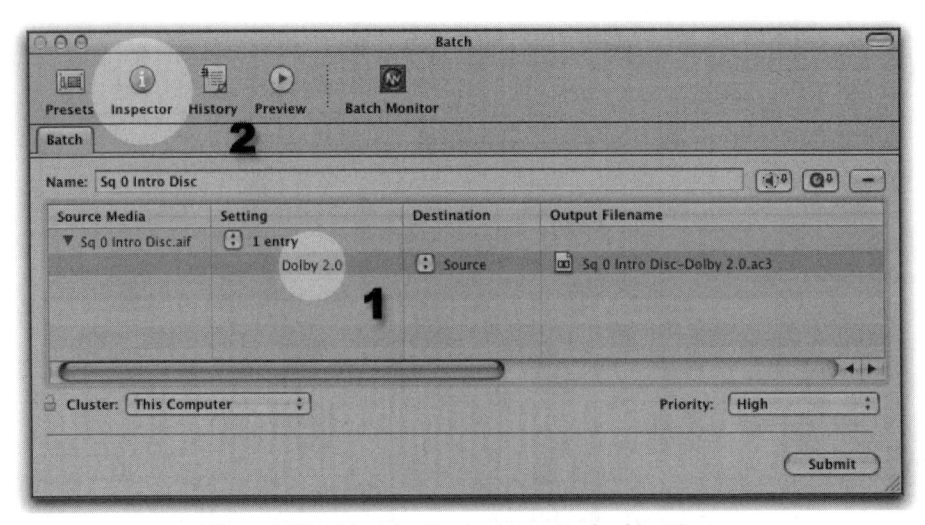

Figure 6-54 — Opening the Inspector to check settings.

Preparing Audio Assets

the Dolby folder of the accompanying DVD).

- In the *Bitstream Tab*, leave the settings unchanged.

- In the *Preprocessing tab*, turn *Compression* Off.

- In the same tab, verify *RF Overmodulation Protection* is **Off**.

- If you have changed anything, click "*Save As...*" to save the settings in a new Preset. Enter the new Preset name in the dialog box that appears.

Submitting the Job for Encoding

When you have finished verifying your settings, Click back on the Batch Window, and click the *Submit* Button. The encode job will be submitted to the *Batch Monitor* for processing, where you can follow its progress.

> **NOTE!**
> Be sure you have set the Destination and Output Filename!

As simple as that seems, that's all there is to creating a Dolby Digital Stereo encode in Compressor 2.

Read the *Batch Monitor* Help file for information about how to use it.

Using Compressor 2 for Surround Sound Encoding

To create a Dolby Digital Surround encode, you first *load the files* that make up a Surround Sound Group, then *make settings choices* in the Encoder Pane's three parameter tabs, and finally *submit the job*. Note

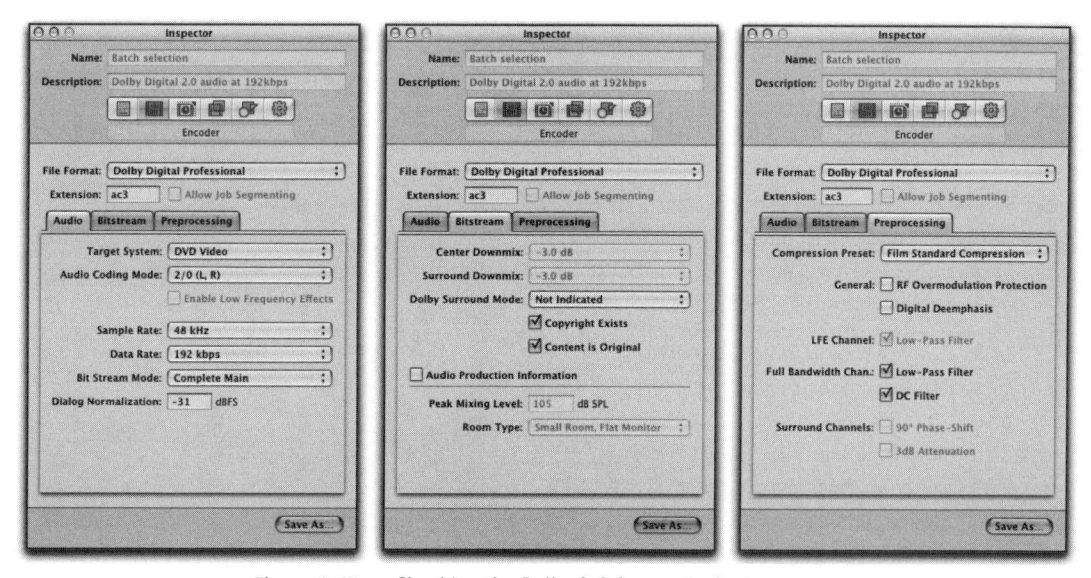

Figure 6-55 — Checking the Dolby 2.0 format in the Inspector.

167

DVD Studio Pro 4

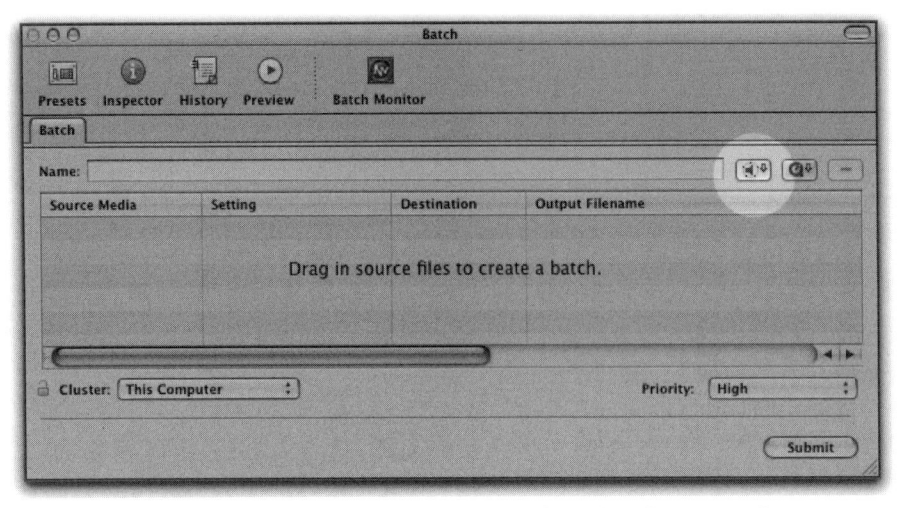

Figure 6-56a — Importing a Surround Sound Group into Compressor 2.

that the manner of changing settings is common to both Surround Sound Groups and Stereo files—the settings are different for each.

Making Surround Sound encodes using Compressor requires knowledge of a few more settings, so pay close attention to the Encode Pane screen shots—they appear similar, but are different.

Loading a Surround Sound Group

- To load the files that are to be encoded as a surround Sound Group, click the *Import Surround Sound Group* button (see Fig. 6-56a) to open a file input grid (see Fig. 6-56b).

- In the *File Input Grid*, drag the individual files required into their proper channel boxes, or click on a channel box to open a dialog box, then select the proper file. It isn't clear from the documentation if Stereo AIFF files containing L and R, or Ls and Rs will be properly placed if dragged into mono channel boxes.

- Once the correct channels have been populated, click on OK to close the file grid. An entry for the *Surround Sound Group* will now have been made in the Batch List. Note

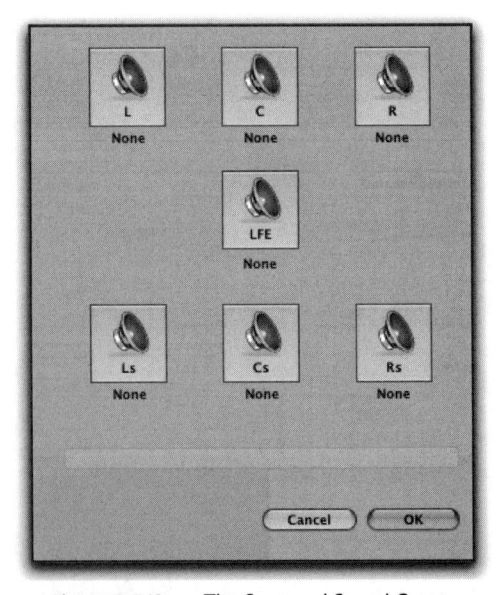

Figure 6-56b — The Surround Sound Group file input grid.

Preparing Audio Assets

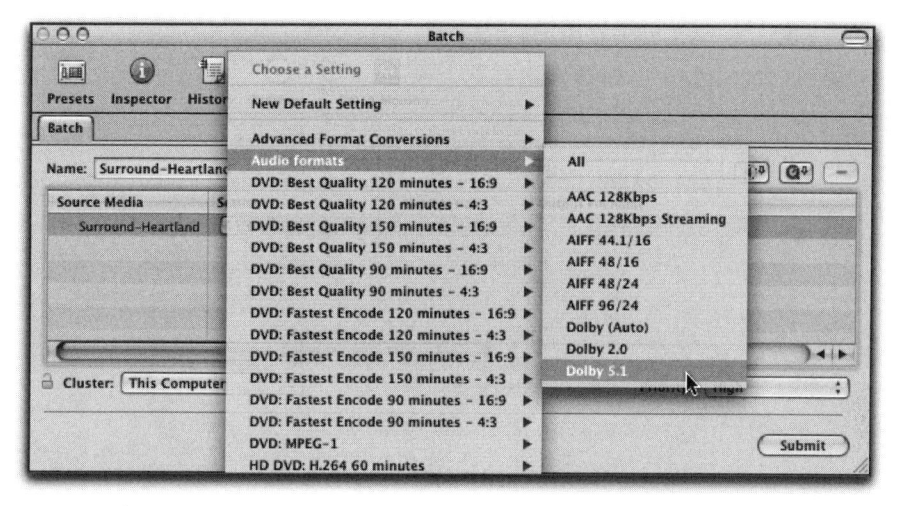

Figure 6-57a — Selecting the Dolby 5.1 format from the Setting pop-up.

that you may not need to fill all of the channels if you aren't making a 6.1 stream.

- Click on *Setting*; from the pop-up, select the desired *Encoding Format*—in this case, Dolby 5.1 for Surround (see Fig. 6-57a).

- Set the encode's *Destination* Location and *Output Filename*; you may leave the default settings, if they're OK for you.

    – To modify the *Destination*, click the pop-up and select the location where you want the encoded file saved (see Fig. 6-57b).

    – To modify the *Output Filename*, click the name, and modify it as you wish.

Once the file settings have been adjusted, you may want to verify the Dolby Digital 5.1 encode settings—to do so, click on the *Dolby 5.1* setting (Fig. 6-58, #1), then click the Inspector button (Fig. 6-58, #2). If you use a preset with known values, you won't need to perform this step every time.

Once the Inspector is visible, select the *Encoder Pane* if it isn't already selected, using the second icon

Figure 6-57b — Changing the Destination from the pop-up.

from left at the top (see Fig. 6-59). Check the Encoder Pane settings in the following tabs, as listed:

In the Audio Tab:

- Set the *Target System* (always DVD Video).
- Set the *Audio Coding Mode* (channel configuration).

169

DVD Studio Pro 4

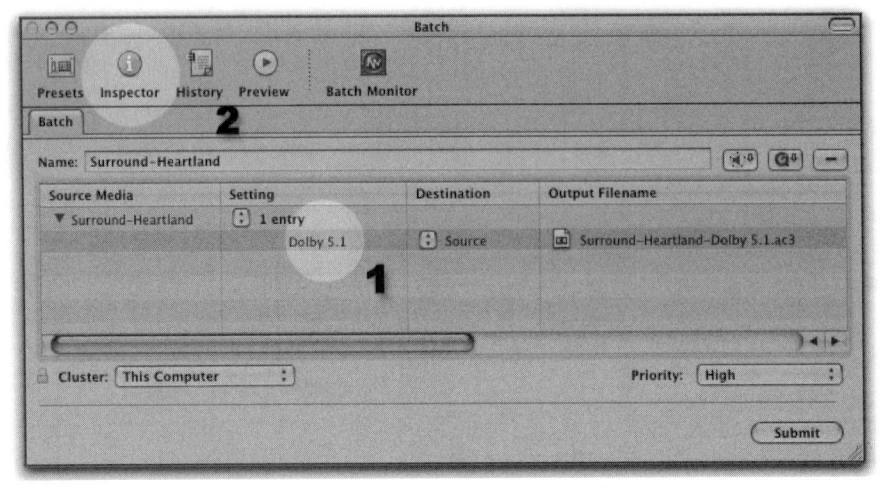

Figure 6-58 — Opening the Inspector to check settings.

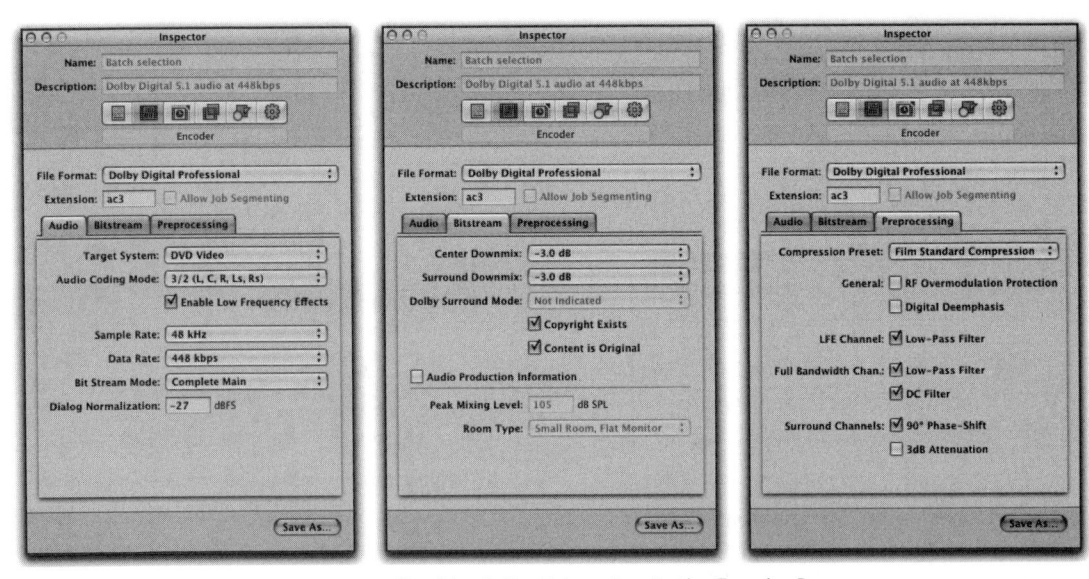

Figure 6-59 — Checking Dolby 5.1 settings in the Encoder Pane.

- Set the *Data Rate* (the Encode Bit Rate).
- Set the *Dialog Normalization Setting*.
- Set the *Bit Stream Mode* (a MetaData label).

In the Bitstream Tab:

- If this is a 5.1 encode, set the *Center Downmix level*.
- If this is a 5.1 encode, set the *Surround Downmix level*.

Preparing Audio Assets

- If this is an Lt/Rt encode, set *Dolby Surround* mode.
- If a Copyright Exists, *check the checkbox*.
- If this is the Original Bitstream, *check the checkbox*.
- If needed, check the *Audio Production Information* box.
- Set the *Peak Mixing Level*, based on mixing room info.
- Set the *Room Type*, based on mixing room information.

In the Preprocessing Tab:

- Set/Verify the *Compression setting*.
- Set/Verify the *General settings*.
- Set/Verify the *Full Bandwidth Channels settings*.
- Set/Verify the *LFE Channel setting*.
- Set/Verify the *Surround Channels settings*.

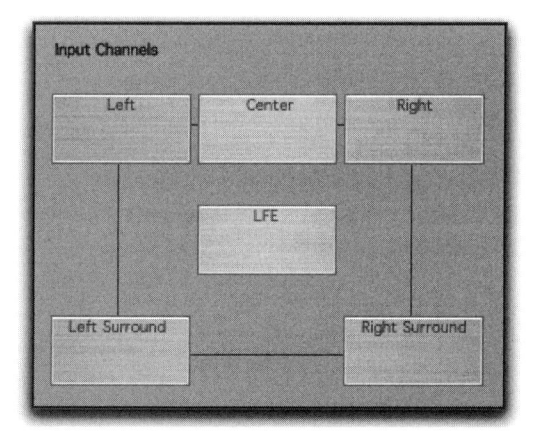

Figure 6-60 — Compressor 2 Input Grid in Surround 5.1 Mode.

Loading the File Grid

Before encoding can take place, the *File Grid Channels* must be loaded with the source files required for the encode. To do so, either:

- Click on a Channel in the Grid, and use the File Dialog box to locate the file for that channel, **or...**
- Drag and drop a file onto the proper channel in the Grid. Select the appropriate channel from within the file if a dialog is offered. (Mono files will not require choosing a channel from within the file. AIFF files are frequently stereo, and may offer this choice to be made—refer back to Fig. 6-56b.)

Dolby Encoding Preferences Outlined in Detail

Dolby Preferences in the Audio Tab

Figure 6-61 — The Audio Tab.

Setting the Target System

To properly use Compressor 2, you have to identify the system for which the Audio bitstream is being created: DVD-Video, DVD-Audio, or Generic AC-3. The proper selection for DVD is always *DVD-Video* (see Figs. 6-61 and 6-62).

171

DVD Studio Pro 4

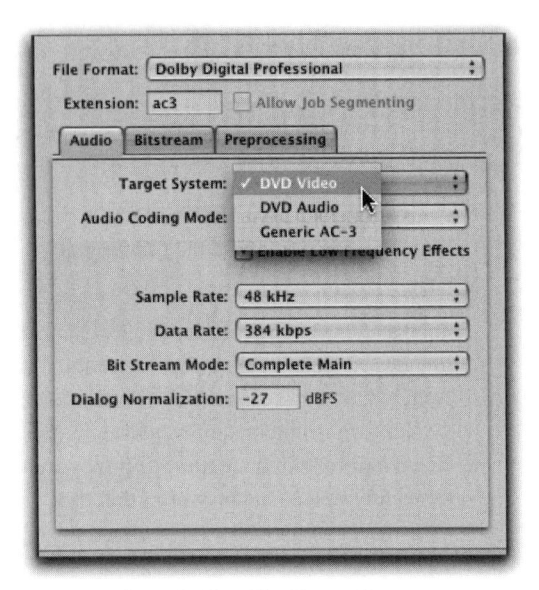

Figure 6-62 — The Target System.

Setting the Audio Coding Mode

- *Audio Coding Mode* defines the channel configuration of the Dolby Digital bitstream you are encoding. It can vary from Mono to 5.1 with or without LFE. NOTE that Dolby denotes the channel configuration as a two digit series, i.e., 3/2, or 2/0, where the first digit indicates the number of Front Channels (1, 2, or 3) and the second digit indicates the number of surround channels (0, 1, or 2) (see Fig. 6-63 top image).

Note that most DVD-Video soundtracks are either Stereo (2/0) or 5.1 Surround (3/2 w/LFE).

- Select the desired *Audio Coding Mode* pop-up to indicate the number of channels you are encoding—Standard Stereo is 2/0 (L, R). Some programs you encode may be 5.1 or other channel configurations (3/2 + LFE in Dolby-speak).

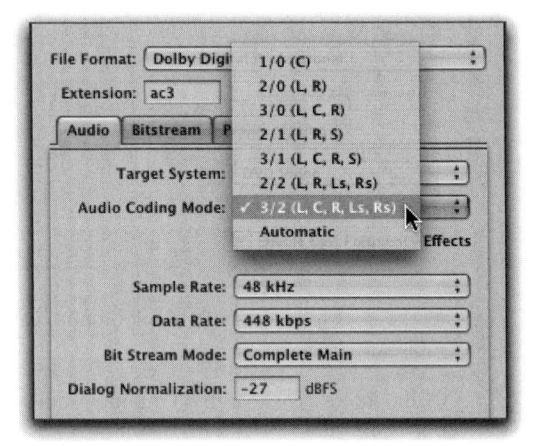

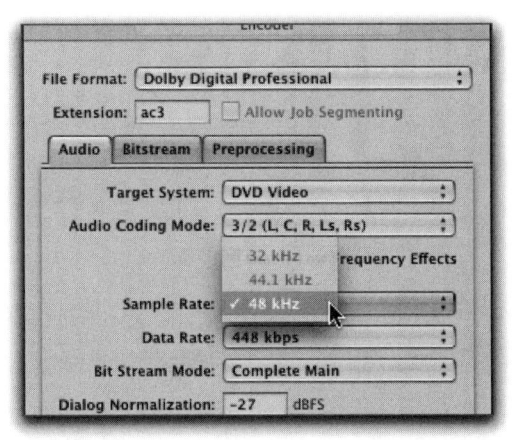

Figure 6-63 — The Audio Coding Mode pop-up.

- If you are using the LFE channel, check the *Enable Low Frequency Effects* checkbox. Leave it off if you are not doing a format that requires it.

- Once you've selected the Audio Coding Mode, you MUST verify or set the Data Rate.

172

Preparing Audio Assets

Verify the Sample Rate of the Source File(s)

Verify the *Sample Rate* for your source files. Note that the standard is 48 kHz for DVD, and most AC-3 encoders prefer or insist on this sample rate. The allowable sample rates displayed in the pop-up (see Fig. 6-63, bottom image) will vary depending on encode source selected. In all cases, match this setting properly with the actual sample rate of the source files. *THIS IS NOT THE DATA RATE—THAT'S THE NEXT SETTING.*

Setting the Data Rate (the All-Important Encode Rate)

Setting a *Data Rate* for Dolby encoding audio is analogous to setting the MPEG Encode rate for Video. (See Fig. 6-64.) The accepted standard rates for AC-3 are 192 kbps for Stereo, and 384 kbps for 5.1. Note that the allowable encode rates enabled in the pop-up will vary depending on the Audio Coding Mode (channel configuration) you have selected.

> **WARNING!**
> Under no circumstances should a Dolby Digital data rate *ever* exceed 448 Kbps for DVD. You'll probably find Dolby Digital encode rates below 192 kbps unsatisfactory for Stereo Sound files, but possibly workable for Mono sound files. Rates below 96 Kbps are not recommended for use at all.

Setting the Dialog Normalization Setting

This is probably THE most misunderstood function in Dolby compression! Dialog Normalization is a setting in the Dolby AC-3 spec that allows the Decoder to *dynamically* adjust the playback level. The setting programmed at the time of encode tells the Decoder how much attenuation to add during playback. For

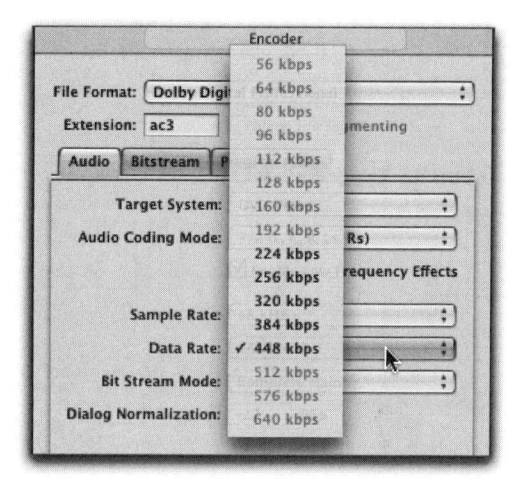

Figure 6-64 — The Data Rate (encode rate) pop-up.

Figure 6-65 — The Dialog Normalization parameter.

typical Stereo files that have already been leveled, this function may be better left disabled.

- To set the *Dialog Normalization* value, click in the field and type in a new value. Valid values are from –1 to –31. To *disable* this function, set it to –31 (see Fig. 6-65).

- You *MUST* be sure to hit "ENTER" or "RETURN" after entering the value, to close the field. If not, the value may not be

173

DVD Studio Pro 4

set properly. To verify, check the file's MetaData once your encode has been completed.

Setting the Bit Stream Mode Parameter

Figure 6-67 — The Bitstream Tab.

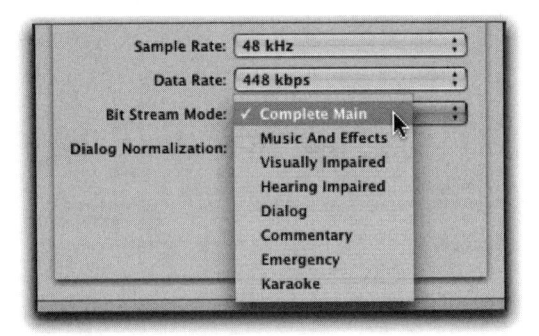

Figure 6-66 — The Bit Stream Mode pop-up.

Dolby "MetaData" (labels or information) may be programmed into the AC-3 file, to define the purpose of this particular Audio bitstream. These labels include *Complete Main, Music and Effects, Commentary, Dialog*, and others (as shown in Fig. 6-66).

- To set the Bit Stream Mode label, click on the *Bit Stream Mode* pop-up, and select the desired choice. In many cases, *Complete Main* is the appropriate setting—it is always the *default* value. In any case, this setting does *not* affect the way in which the Dolby Decoder reproduces the bitstream. It's just a label.

Settings in the Bitstream Tab (in Detail)

If this is a 5.1 encode, set the Center Downmix level:

The Center channel of a Surround Sound bitstream can be mixed into the Analog Left/Right Stereo

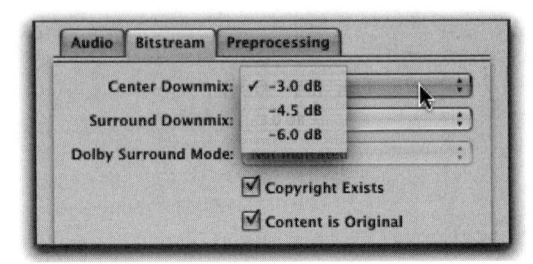

Figure 6-68 — Setting the Center Downmix parameter.

Outputs of your DVD player, attenuated per the downmix level programmed in at the time of encoding. Setting this parameter controls the level of the Center channel present in the final Stereo Downmix. (See Fig. 6-67.)

- To set the *Center Downmix* level, click on the Center Downmix pop-up and select the appropriate setting (see Fig. 6-68).

In many cases, the default setting of –3.0 dB may be perfectly appropriate. For further information, consult the Dolby *Metadata Guide* PDF, available from **www.Dolby.com**. We've included it in the Dolby folder on the accompanying DVD.

If this is a 5.1 encode, set the Surround Downmix level:

This parameter controls the level of the Left and Right Surround channels in the final Stereo Downmix.

Surround Sound bitstreams can output their mono or stereo surround information from the Analog Front

Preparing Audio Assets

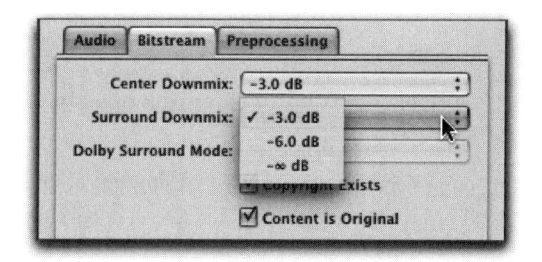

Figure 6-69 — Setting the Surround Downmix.

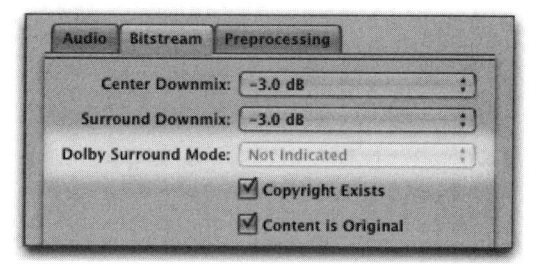

Figure 6-70 — Dolby Surround Mode.

Left/Right Outputs of your DVD player, if this information is encoded into the bitstream.

- To set the *Surround Downmix* level, click on the Surround Downmix pop-up and select the appropriate setting (as in Fig 6-69).

In many cases, the default setting of –3.0 dB may be perfectly appropriate. At this setting, the left and right surround channels are each attenuated 3 dB and sent to the left and right front channels, respectively. At -6 dB the output routing is the same as -3dB, but the signal is attenuated 6 dB. At the –∞ dB (Infinity) setting, the surround channel(s) are discarded.

If this is an Lt/Rt encode, set Dolby Surround mode:

Dolby Surround Mode is a special two-channel configuration, which actually contains FOUR channels of discrete audio information: Left, Center, Right, and a Mono Surround channel, electronically matrixed into two channels. These files are known by the special code **Lt/Rt** (Left Total/Right Total). If the file is Lt/Rt, setting this property can cause a Dolby Surround Pro Logic decoder to engage in order to properly decode four-channel Dolby Surround information encoded in the file's two channels.

Dolby Surround Mode would only be required if the original sound mix was a four-channel mix containing L, C, R, and a mono Surround that has been processed by a *Dolby Matrix encoder*. (See Fig. 6-70.)

- If this is true, then set the Dolby Surround Mode property to *Dolby Surround Encoded*.

- If the original sound mix was NOT a four-channel mix with mono surround, then Dolby Surround Mode is NOT required, and this property should be set to *NOT Indicated*, or *NOT Dolby Surround Encoded*.

- If you are *not* sure about this setting, leave it *Not Indicated*. Odds are you will be correct. It may also be possible to use this setting to decode files prepared in the Ultra-Stereo format, since that is a similar matrixed L, C, R, S format as Dolby Surround.

If a Copyright Exists, check the designated checkbox:

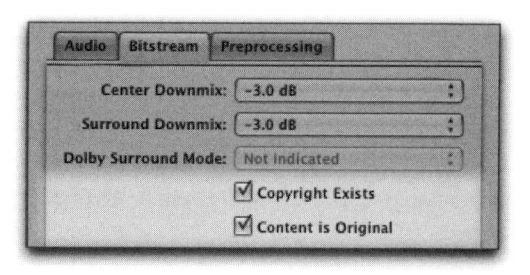

Figure 6-71 — The Copyright and Content checkboxes.

Dolby MetaData can contain a flag bit to indicate if the audio content of the bitstream is currently under Copyright. The Default is "on."

175

- To turn this flag "off," click on the *Copyright Exists* checkbox to uncheck it. Click again to check it if needed (see Fig. 6-71).

If this is an Original Bitstream, check the checkbox.

Dolby MetaData can contain a flag bit to indicate if the audio content of the bitstream is the Original Dolby Bitstream, or a copy (see Figure 6-71).

- The Default setting is "on," meaning this is the original bitstream created of this content. To denote this is not the original bitstream, click on the *Content is Original* checkbox, to uncheck it. Click again to check it if needed (see Figure 6-71).

If appropriate, check the Audio Production Information box.

Figure 6-72a — Setting Audio Prod'n Info.

The default of this parameter is OFF, but by turning it ON, it is possible to advise Dolby Digital decoders that the accompanying reference information is valid, and it may be used to adjust playback levels (see Fig. 6-72a).

Set the Peak Mixing Level, based on mixing information.

This parameter defines the Peak mixing level used during the surround mix, but is only referenced by the decoder if the Audio Production Information is checked on.

Set the Room Type, based on mixing room information.

This parameter defines the Type of mixing room or stage used during the surround mix, but is only referenced by the decoder if the Audio Production Information is checked on. The Large room X-curve parameter would be checked if the audio file was created in a traditional theatrical mixing (dubbing) stage, whose audio characteristics follow the Academy Curve—a specific equalization curve that attenuates high frequencies to emulate the playback characteristics of a movie theater.

Settings in the Preprocessing Tab (in Detail)

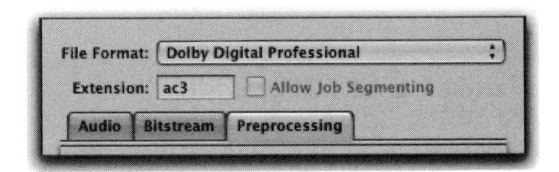

Figure 6-72b — The Preprocessing Tab.

Setting the Compression Factor

Dolby encoders can apply different types of Program Compression to assist in leveling out the dynamic range (not always a good thing), depending on the kind of program you are encoding—Film, Music, or Speech. You may find, however, that turning Compression off may be the right solution for material that is already dynamically leveled, or doesn't have large dynamic range swings.

Preparing Audio Assets

- To apply a Program *Compression Preset*, or to deselect Compression, click on the Compression Preset pop-up (as shown in Fig. 6-73), and drag to select your choice. The default is Film standard Compression, and that may not be proper for your particular content, especially if it is music, and already levels off nicely in a mix.

- Verify the setting is not "*on*" (not enabled—see Fig. 6-74).

- To disable the *RF Overmodulation Protection* bit, deselect the check box for this parameter (as shown in Fig. 6-74).

    NOTE that in Compressor 2, this should default to "off." Some earlier versions of A.Pack defaulted to "on."

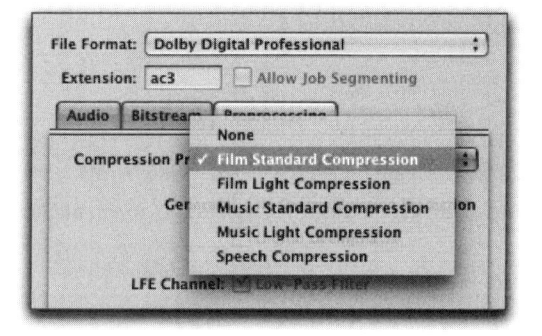

Figure 6-73 — Setting the program Compression parameter.

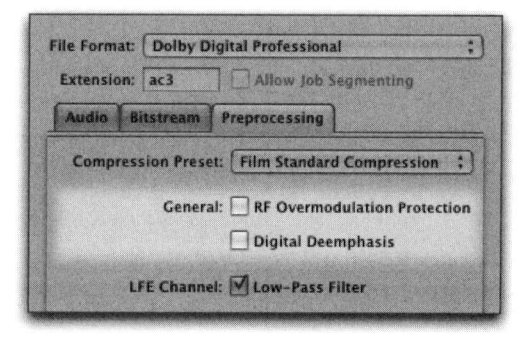

Figure 6-74 — Setting the RF Overmodulation parameter.

Setting the RF Overmodulation Protection

- According to Dolby Labs, this parameter is designed to protect against overmodulation when a decoded Dolby Digital bitstream is RF modulated. When enabled, the Dolby Digital encoder includes pre-emphasis in its calculations for RF Mode compression. The parameter has no effect when decoding using Line Mode compression.

> **NOTE!**
> Except in rare cases, this parameter should be disabled.

Compressor 2 can set or disable the RF Overmodulation Protection bit through a simple checkbox. Check it carefully, as it usually defaults to "on"—"off" is the preferred state.

Setting the Digital Deemphasis Parameter

Believe it or not, there is NO official Dolby explanation for this parameter in the Dolby Metadata Guide. (See Fig. 6-75.)

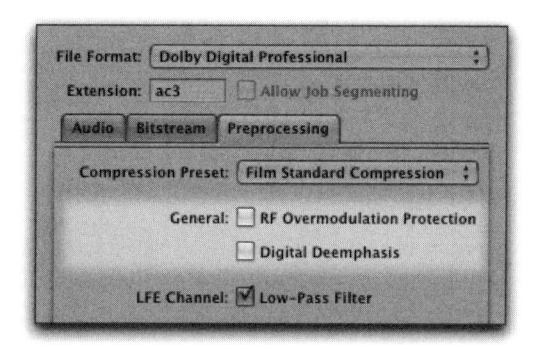

Figure 6-75 — Setting the Digital Deemphasis parameter.

177

DVD Studio Pro 4

- Since the parameter defaults to "off," it's best to leave it that way.

Setting the Low-Pass Filter (for LFE Channel)

According to Dolby Labs, this parameter determines whether a 120 Hz 8th order low-pass filter is applied to the LFE channel input of a Dolby Digital encoder prior to encoding. It is ignored if the LFE channel is disabled. This parameter is not sent to the consumer decoder. The filter removes frequencies above 120 Hz that would cause aliasing when decoded. *This filter should only be switched off if the audio to be encoded is known to have no signal above 120 Hz.*

- To deselect *LFE Low-Pass Filter* (which defaults to "on"), click on the checkbox to deselect it (see Fig. 6-76).

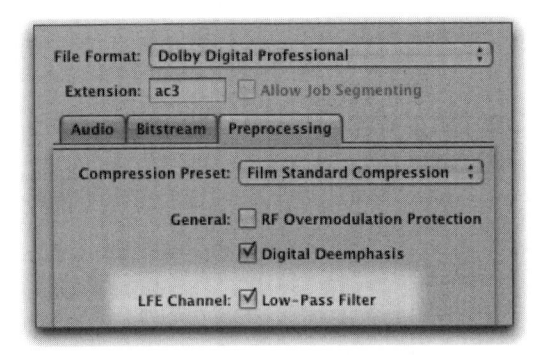

Figure 6-76 — Setting the LFE Low-Pass Filter parameter.

Setting the Low-Pass Filter (for Full Bandwidth Channels)

According to Dolby Labs, this parameter determines whether a low-pass filter is applied to the main input channels of a Dolby Digital encoder prior to encoding. This filter removes high-frequency signals that are not encoded. At the suitable data rates, this filter operates above 20 kHz. In all cases, it prevents aliasing on decoding and *is normally switched on*. This parameter is not passed to the consumer decoder.

- To deselect *full Bandwidth Channel Low-Pass Filter* (which defaults to "on"), click on the checkbox to deselect it (see Fig. 6-77).

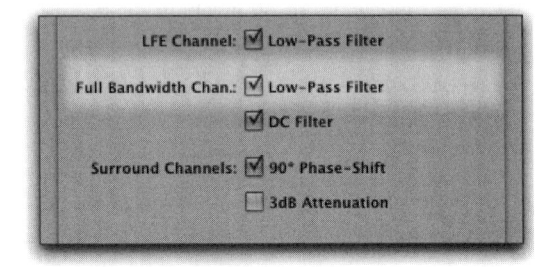

Figure 6-77 — Setting the Low-Pass Filter parameter.

Setting the DC Filter (for Full Bandwidth Channels)

According to Dolby Labs, this parameter determines whether a DC blocking 3 Hz high-pass filter is applied to the main input channels of a Dolby Digital encoder prior to encoding. It is used to remove DC offsets in the program audio and *would only be switched off in exceptional circumstances*. This parameter is not carried to the consumer decoder.

- To deselect *Apply DC Filter* (which defaults to *on*), click on the checkbox to deselect it (see Fig. 6-78).

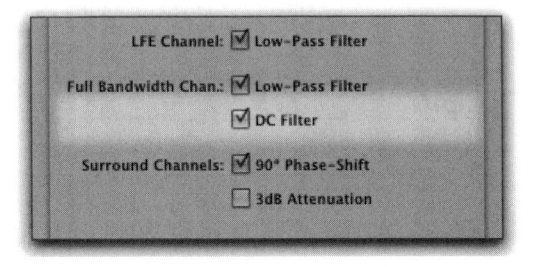

Figure 6-78 — Setting the Full BW Channels DC Filter parameter.

Preparing Audio Assets

Setting the Surround Phase Shift

According to Dolby Labs, "This parameter causes the Dolby Digital encoder to apply a 90-degree phase shift to the surround channels." This allows a Dolby Digital decoder create an Lt/Rt downmix simply. For most material, the phase shift has a minimal impact when the Dolby Digital program is decoded to 5.1 channels, but provides an Lt/Rt output that can be Pro Logic decoded to L, C, R, S, if desired. However, for some phase-critical material (such as music), this phase shift is audible when listening in 5.1 channels. Likewise, some material downmixes to a satisfactory Lt/Rt signal without needing this phase shift. It is therefore important to balance the needs of the 5.1 mix and the Lt/Rt downmix for each program. *The default setting is Enabled.*

- To disable the *Apply 90-degree Phase-Shift to Surround Channels* parameter, click on the Apply 90-degree Phase-Shift checkbox (shown in Fig. 6-79), to uncheck the selection.

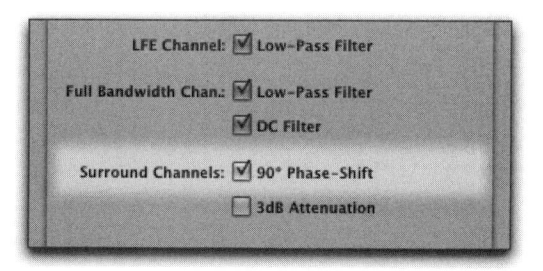

Figure 6-79 — Setting the Surround Phase-Shift parameter.

Setting the 3 dB Attenuation in Surround Channels

According to Dolby Labs, the *Surround 3 dB Attenuation* parameter determines whether the surround channel(s) are attenuated 3 dB before encoding. The attenuation actually takes place inside the Dolby Digital encoder (in this case, Compressor 2). It balances the signal levels between theatrical mixing rooms (dubbing stages) and consumer mixing rooms (DVD or TV studios). Consumer mixing rooms are calibrated so that all five main channels are at the same sound pressure level (SPL). For compatibility reasons with older film formats, theatrical mixing rooms calibrate the surround channels 3 dB lower in SPL than the front channels. The consequence is that signal levels on tape are 3 dB louder. Therefore, to convert to a consumer mix from a theatrical calibration, it is necessary to reduce the surround levels by 3 dB by enabling this parameter.

- To apply the Surround Attenuation, click on the *Apply 3 dB Attenuation* checkbox, shown in Fig. 6-80.

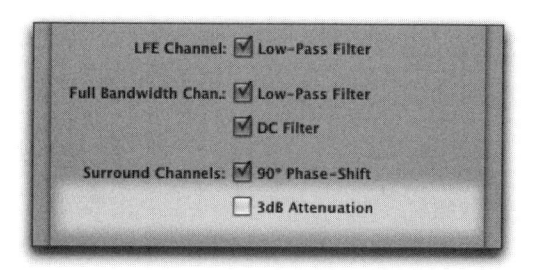

Figure 6-80 — Setting the 3 dB Surround Attenuation parameter.

Encoding DTS Streams for DVD

To encode DTS (Digital Theatre Systems) audio streams for use with DVD Studio Pro 3 or 4, you will need a DTS encoder—there is a software version for OS X (see Fig. 6-81). This encoder allows you to pack and encode PCM files into proper DTS streams (.*cpt*) for use with DVD Studio Pro (see the DTS folder in the accompanying DVD for more information).

179

DVD Studio Pro 4

Figure 6-81 – The DTS Standalone Encoder logo.

A Simple Overview of the DTS Process

Encoding for DTS is a process similar to Dolby Digital encoding, but it requires an additional pre-processing step, called *packing*. The DTS software encoder installation will provide you with the components required to perform all of the steps required for DTS—packing, encoding, and encrypting (see Fig. 6-82).

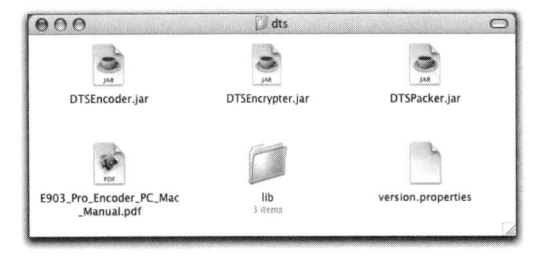

Figure 6-82 – DTS Software Encoder Components, as Installed.

Preparing Files for Packing

Standard AIFF files can be used to make a Stereo DTS encode, and pairs of AIFF files can be used to supply the channels required for a DTS Surround Encode. Surround pairs are traditionally prepared in this manner:

L/R	Left Front with Right Front,
Ls/Rs	Left Surround with Right Surround, and
C/Sub	Center with Subwoofer (LFE).

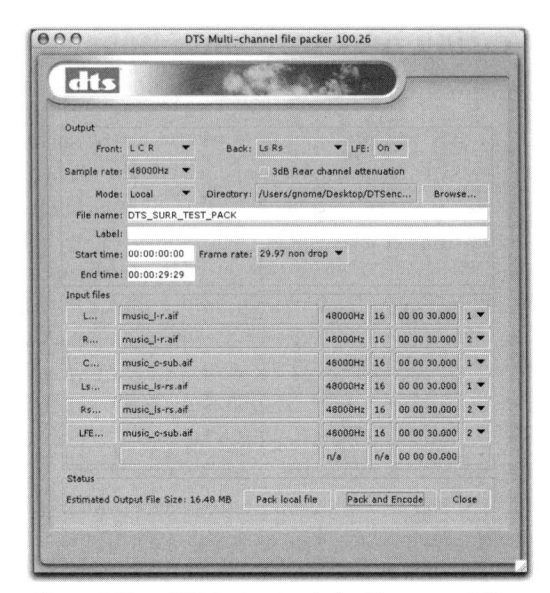

Figure 6-83 – DTS Packer loaded with surround files.

Once prepared, they can be loaded easily into the DTS Packer (see Fig. 6-83).

Encoding a DTS Packed File into a DTS Bitstream

Once the file has been successfully packed to prepare it for encoding (see Fig. 6-84) it can be used as source for the *DTS Encoder*, to create the actual DTS bitstream (see Fig. 6-85).

Figure 6-84 – DTS Packer successful.

Note that in Figure 6-85, the encoder has been set to encode the *compact (.cpt)* bitstream format (the *Output Type*: setting)—this is essential in order to use these streams in DVD Studio Pro 3 and 4.

180

Preparing Audio Assets

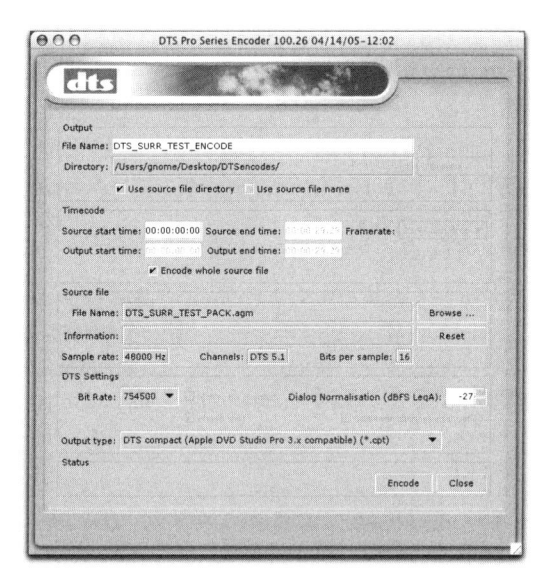

Figure 6-85 — DTS Encoder Loaded with Packed File.

Figure 6-86 — DTS Encoder successful!

For more information, visit the DTS website (**www.dtsonline.com**). You may also review the DTS information we have provided on the DVD.

Many thanks to my friend Scott Esterson of DTS ("the biggest man in showbiz") who was kind enough to provide a copy of the DTS OS X Software encoder for our review.

When the encoding process is finished (see Fig. 6-86), the files are ready to be used in your DVD Studio Pro project. Have fun!

DVD Studio Pro 4

Summary

DVD Video discs may contain up to eight audio programs in each of the Track presentations created. Each of these audio programs may include anywhere from Mono to full 5.1 Surround Sound channels, depending on the method of creation of the audio program. The audio source for these programs can be PCM files created on either Macintosh (AIFF files) or on a PC (WAV files), but may also be of other permissible formats, as dictated by the DVD Video Spec.

While DVDs can use these PCM audio files directly after being exported from NLEs (*nonlinear video edit systems*) or captured and edited in DAWs (*Digital Audio Workstations*), a better utilization of the DVD's audio bandwidth can be accomplished by using the Dolby Digital Audio encoding format. Often improperly called AC3, Dolby Digital encoding provides a lossy encoding format that should create a fairly accurate version of the original audio program while using a far lower data rate. The recovered data rate may be channeled into the MPEG video encode to make better looking video for the DVD.

To facilitate creating Stereo and Multichannel Dolby Digital bitstreams, the DVD Studio Pro 2 application suite comes with its own Dolby Digital (AC3) encoding software: A.Pack.

A.Pack is a very powerful Dolby Digital-compliant encoder for preparing AC-3 bitstreams from AIFF or WAV source files. To create Stereo, you can use a single AIFF file, but to create full 5.1 Surround, you will need six discreet Mono sound files, one for each channel.

The Dolby Digital (AC3) bitstreams provide DVD-compliant sources of Audio Tracks at a fraction of their PCM bandwidth requirements.

In addition, Dolby Digital (AC3) has become a worldwide DVD standard for audio in the years since the format was first approved for DVD Region 1. Today, DVDs from every region can play back Dolby Digital (AC3) soundtracks.

Chapter 7

Authoring Tracks

DVD Studio Pro 4

Figure 7-1 — The Track Editor.

Goals

By the end of this chapter, you should:

- Understand what **Track** Elements are, and how to create them
- Understand what a **Clip** is
- Understand what a **Bitstream** is
- Be able to Author **Video Streams** into a Track Element
- Be able to Author **Audio Streams** into a Track Element
- Be able to Author **Subtitle Streams** into a Track Element
- Be able to Author **Video Angles** into a Track Element
- Be able to use the **Track Editor**
- Be able to use the **Track Inspector**

Authoring Tracks

This chapter covers how to create Tracks, and how to enhance them with additional Audio programs and even Subtitles. (See Fig, 7-1.) You will also find information on using Multiple Angles and Mixed Angles in this chapter. We'll briefly mention Markers and Stories in this chapter as well, but Chapter 8 is entirely devoted to those two elements.

What Is a Track?

Tracks are, without doubt, probably the most basic element used in DVD—in fact, a proper DVD really cannot be created without at least one Track (or Slideshow). Tracks are the *containers* used for assembling a DVD presentation that uses video files, audio files, and perhaps even subtitles that may appear with that video content.

A Track is used to organize video and audio bitstreams and, in many cases, it is also used to display subtitles, watermarks, or "buttons on video" as well.

184

Authoring Tracks

If you have seen the "White Rabbit" feature in "The Matrix" DVD, or a New Line Cinema "Infinifilm" DVD, you've already seen "buttons on video."

Since the advent of DVD Studio Pro 2, the method of Authoring Tracks is very different from the earlier version, with some interesting new and improved functionality. There is also a cool feature, called "mixed angles," used to create multiple-angle tracks using only small video segments instead of full-length video programs.

At a bare minimum, you'll need some video content or a still graphic image to create a track. The DVD Spec has no provision for playing an Audio-only presentation, so if this is your goal, it's normal to create audio-driven presentations like songbooks using a Stillframe graphic and a sound file in a Menu Element or Slideshow (see Chapter 9, "Menus and Buttons," or Chapter 10, "Slideshows"). You may also convert a Slideshow into a Track, if it suits your purposes.

Track Structure and Limits

In planning our content for the DVD, we need to remember that a Track can only organize content within these limits:

- Up to 9 Video streams, as main or alternate presentations (angles)
- Up to 8 Audio streams, for alternate languages, or commentaries (may be Stereo or Surround)
- Up 32 Subtitle streams for subtitles, graphics, or buttons on video
- Up to 99 Chapter Markers, for navigation and Story markers;
- Up to 255 cell markers, including the 99 Chapters, above; for things like marking Button Highlights and setting the Dual-Layer Breakpoint
- Up to 98 Stories, or fewer if more tracks or slideshows are created.

In your entire DVD, you can only create a maximum of 99 Tracks, Stories, and Slideshows, *combined*. This means, if you already have planned for six slideshows, you will only be able to create 93 (99 – 6) Tracks and/or Stories, maximum. This is limited by the DVD Spec, and cannot be changed. But you will find that in most cases, you will have more than enough "containers" to finish your DVD in fine style.

How Many Tracks Do I Need for My DVD?

That's a *great* question—the answer depends on what you wish to do with your DVD, and how you are going to organize your content. As an example, let's say you are trying to make a simple home video DVD, and you have edited your content into a nice 60-minute movie. The goal would be a relaxing, uninterrupted one-hour presentation. The rules of the DVD "Spec" indicate that *seamless* playback of your entire movie can only happen if it is authored into a Track in its entirety. That's right, you will have to put the whole 60-minute file in one track. But what if you want to navigate around and skip forward and back through your vacation highlights, just like in commercial DVDs? That's easy—you can define Markers in the Track to do just that—in fact, you can organize your Movie into "Chapters," just like in Hollywood DVDs by using the Markers capability of the Track Editor (see Chapter 8, "Markers and Stories").

I'm Not Sure How to Structure My DVD (Yet)

No problem—DVDs are amazingly fluid little buggers, and can be created in hundreds of different ways. Let's say you are creating an Instructional DVD instead, and your content is actually five Segments, each one several minutes long. Since each Segment is really a self-contained lesson, you *could* build five Tracks, each one containing one of the five Segments. The beauty of this approach is that you can have lots of navigation possibilities when you create multiple tracks, as each Track can be programmed with an "End Action" (Jump When Finished).

Playing Video Clips Individually

Figure 7-2 shows our training content authored into five individual Tracks, and the return arrow paths indicate each movie is designed to return back to the Menu when finished. This is a typical Menu-and-Track navigation arrangement in DVD.

This kind of arrangement is great for multiple pieces of content, but not good if the goal is *sequential playback* of all Tracks in order. For that, we'd need to rearrange the structure. Let's experiment with that...

Playing Video Clips Sequentially

Figure 7-3 shows our five content clips assembled into one Track element, arranged for *seamless* play-through. A Track will always play its content completely from beginning to end—unless interrupted—before it will navigate to its End Jump.

The solid arrows indicate that this Track will play through and, because all of the clips are in one Track, the playback is *seamless*. When finished, the Track will return to the Menu via the *End Jump*. The dashed lines indicate possible navigation paths to Chapter Markers, if we define those Markers in the Track and link them to Menu Buttons. (Marker 1 is always defined at the beginning of each Track.) That takes care of the seamless play-through, but what if we wanted each clip in our content to be accessible as an individual element as well? Hmmm... Figure 7-3 does have seamless playback, but does not give us the individual playback we had in Figure 7-2—are we at an impasse? Maybe not!

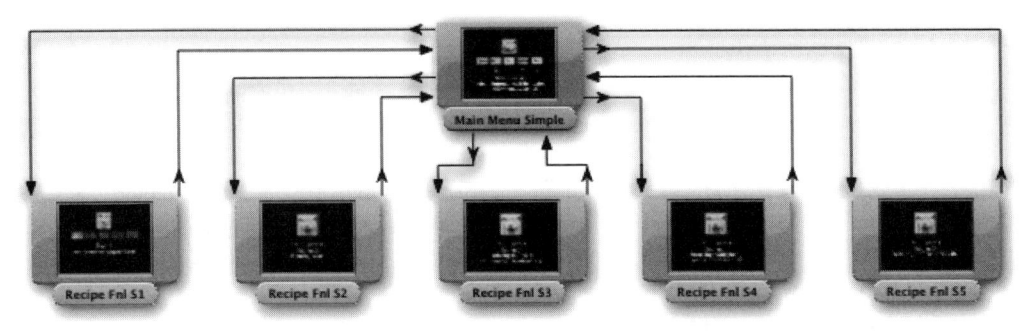

Figure 7-2 — Five Video Clips arranged as individual Tracks.

Authoring Tracks

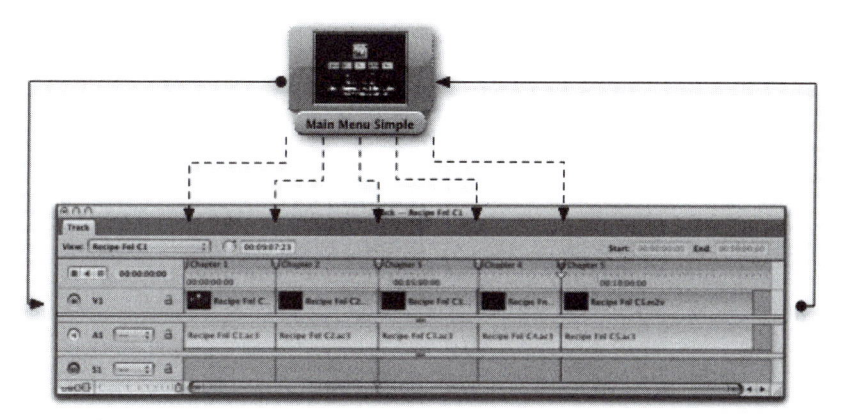

Figure 7-3 — A Typical DVD Movie—one Track with Chapter Markers.

Playing Video Clips Sequentially and Individually

We can combine the functionality of our two previous versions by using DVD Studio Pro *Stories*. The Track still has the five clips in sequence, but we've defined each clip as a separate element, a *Story*—the video between two Markers. A Story can be played from a Menu Button, just like a Track. This arrangement is the best of both worlds, as we can choose to play all of the clips, or just a single one. (See Fig. 7-4a.)

Could we have done this differently? Sure! That's the beauty of DVD Studio Pro—there's always more than one way to do practically anything. To get continuous playback, individual Tracks can be linked

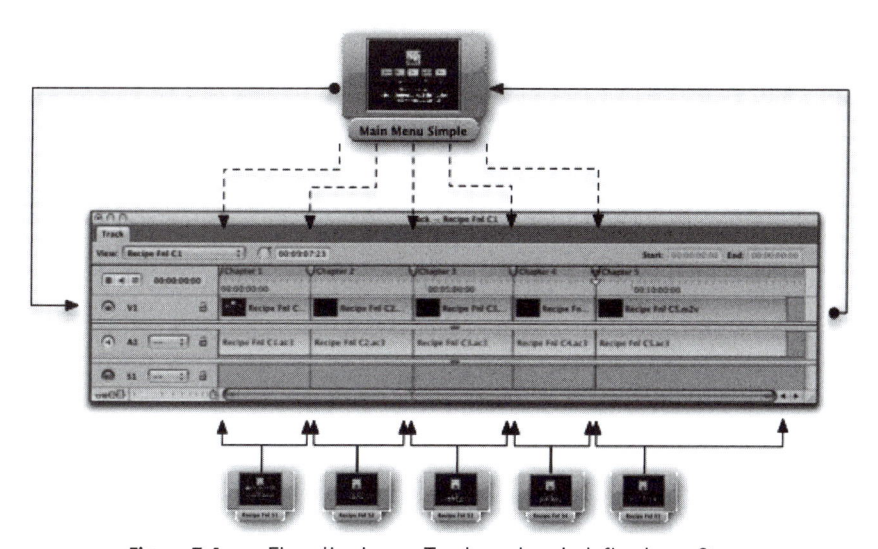

Figure 7-4a — Five clips in one Track, and each defined as a Story.

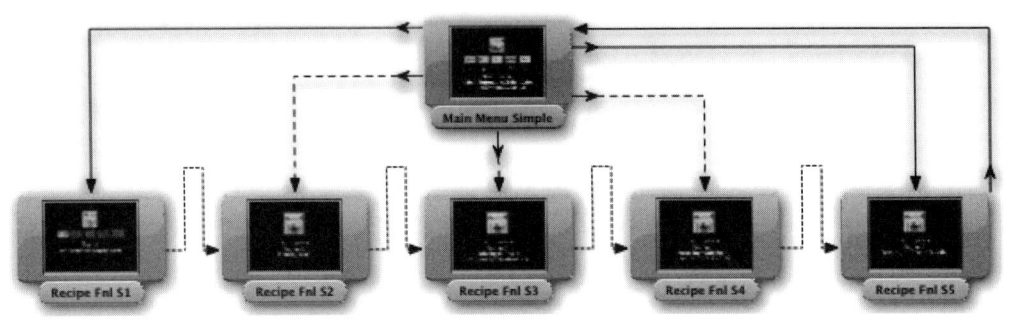

Figure 7-4b — Five Tracks in a row, linked sequentially.

together using their End Jumps, with each Track connected to the next one in sequence. This kind of arrangement is good, but not great—if the goal is *seamless* playback of Tracks in this kind of structure, we need to be aware of some special circumstances.

When Can a "Nonseamless" Transition Look "Seamless"?

This really isn't a trick question; the answer is "when you can't tell the difference"! This points out one very interesting aspect of DVD players. Each player includes a video frame buffer, to reassemble and "hold" the video frames being decoded from the MPEG bitstreams on the DVD. The interesting trick here is that the last frame played into the frame buffer will be held *infinitely* on screen until the next incoming frame of video replaces it. This means, the last black frame of a video segment will be frozen on screen without a glitch, until the next incoming frame of your video replaces it. Now, what if that next frame also happens to be, let's say *black*, as in the start of a fade up from black; what do you think the DVD viewers will see? Correct! They will think they are experiencing a seamless transition from the previous segment into this new segment. In fact, they have been fooled—tricked! If we used this technique in Figure 7-4b, the tracks would appear to be playing seamlessly, even though they really aren't! Of course, this presumes your content clips DO fade up and down from black. If they do not, this technique does not work.

Tracks and Chapters and Stories

So we have seen that you can build this same project using five Tracks, or using one Track and dividing it into five Stories! How will you know which way is best? By reading this book thoroughly, and lots of experimentation! We'll cover lots of options in authoring within this book, and the DVD accompanying this book will provide you with some real-world projects you can use to investigate these options!

> **BIG TIP!**
> *How* you prepare your video while editing can have a serious impact on how effectively you can display it in your finished DVD.

If your goal is to have seamless playback, either edit your content as a continuous movie with fades in and

Authoring Tracks

out of segments, or export it in segments that fade up from black, and fade down to black when finished. Always leave a little black in between segments—this gives you the best flexibility.

Working with Tracks

Working with tracks in DVD Studio Pro is very straightforward. There are two tools you will need to create, edit, and fine-tune the Track to your needs.

The Track Editor

The *Track Tab* in DVD Studio Pro contains the *Track Editor* timeline interface. This gives you access to the bitstream areas for nine videos, eight audios, and 32 subtitles. You will use the *Track Editor* to assemble the contents of your Tracks.

The Track Inspector

The *Track Inspector* gives you access to important parameters of each track you create, including the display mode (aspect ratio), Closed Caption information, Macrovision settings, and newly expanded User Operations support in detail (UOPS are Remote Control functions).

You Need Separate Video and Audio for DVD Authoring

Your Video and Audio files may be sourced from a NLE, where they have been edited together as a synchronized program and then exported, or they may have been captured and/or encoded from an existing videotape master. Either way, during Authoring, we'll refer to video and audio as *elementary* streams—that's because the DVD Video Spec requires that the video and audio components of a Track exist as separate files. Encoding from Final Cut Pro 4 or through Compressor or the QuickTime™ encoder will separate your movie into these elementary streams. Audio may be used as AIFF, or further compressed into Dolby Digital (see Chapter 6, "Making Audio").

Later on, in the Build process (Chapter 14), we will recombine these separate, elementary files into the DVD you expect to see.

Creating a Track— (Track Editor Basics)

What Do I Need to Create a Track?

You will find that every new DVD project you create will already have one empty Track element created and waiting for you to assemble assets into. When you need more, you can easily create them, either empty or by creating them using assets, as outlined below.

Generally, each Track combines one video clip with one or more associated audio programs. You can make a silent video-only Track, but in most DVDs, audio accompanies the video. Compatibility is also helped by the inclusion of an audio track, even if it is silent! You can also make a Track by using one or more Still Images, or by converting a Slideshow into a Track (both features began in DVD Studio Pro 2). (See Chapter 10, "Authoring Slideshows" for info about this last feature.)

Creating a Track without an Asset

DVD Studio Pro has so many new ways to streamline your authoring that it may take you a while to

DVD Studio Pro 4

use them all. You might want to create an empty Track first, without using movies or MPEG files, then load it later on, once the Assets have been delivered or created.

To create an empty Track manually, you can:

- Select **Project > Add to Project > Track** in the menubar (see Fig. 7-5),

- Click the "*Add Track*" tool from the Menu Editor's Toolbar (if it's not there, you can Customize the Toolbar [see Fig. 7-6] to add it).

- Use Keyboard Shortcut **Cmd-Control-T**. (See Fig. 7-6.)

- Control-Click in the Outline Tab and Select **"Add > Track"** (see Fig. 7-7).

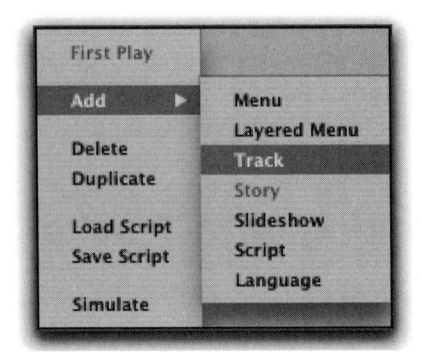

Figure 7-7 — The Control-Click "Add Track" dialog.

Adding a Video Asset to an Empty Track

Once a new empty Track Element exists, a video Asset can be assigned by selecting the desired video clip from the Assets Tab, or a video file can be authored in from the Palette or Finder Window.

To add a video asset, click on it to select the file, then do one of these:

- Drag it *onto* a Track in the Outline Tab's Tracks Folder (see Fig. 7-8), **or**

- Drag it *into* the V1 stream of the Track Editor (see Fig. 7-9).

NOTE!
The first video asset can ONLY be dragged into the V1 stream area (if the Track is empty). V2–V9 are *not* valid destinations until V1 has been authored. Remember, the asset file does not have to be imported into the DVD Studio Pro Asset Tab before use.

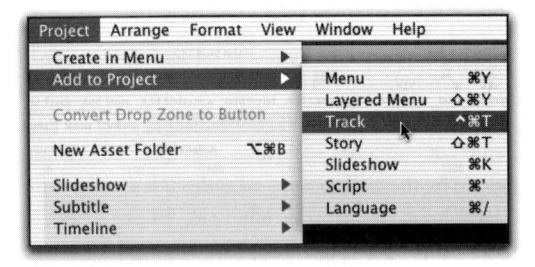

Figure 7-5 — Adding a Track from the Menu.

Figure 7-6 — The Add Track tool in the Toolbar, with Keyboard equivalent.

190

Authoring Tracks

Immediately upon releasing the mouse, the Video Asset will appear in the V1 stream area of the selected Track (see Fig. 7-10). Note that no audio stream was added during this operation!

Using a DVD Studio Pro feature, both Audio and Video can be added to the Track editor at the same time (see Fig. 7-11).

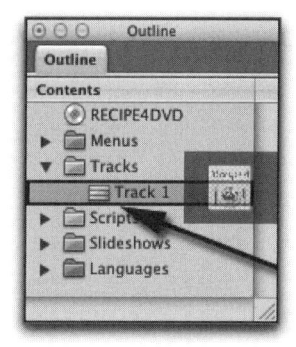

Figure 7-8 — Drag a video asset onto the Track to add it.

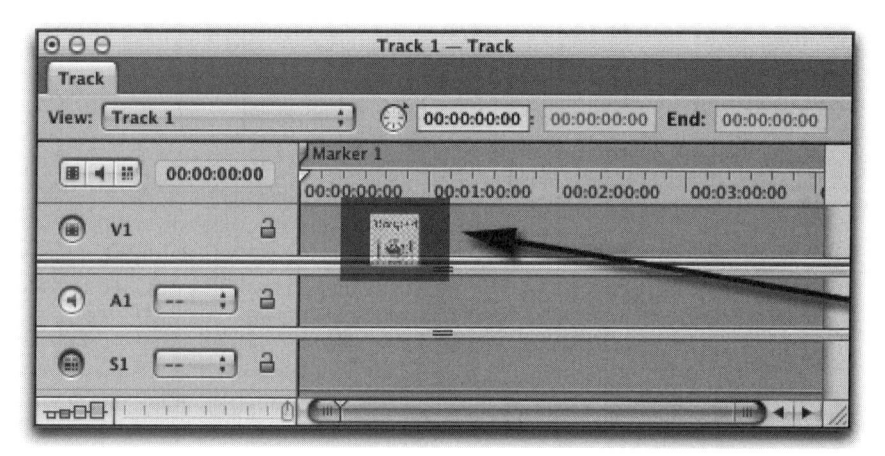

Figure 7-9 — Drag a Video Stream into V1 to add it to the track.

Figure 7-10 — Adding only Video results in this Track.

191

DVD Studio Pro 4

Figure 7-11 — Adding Video and Audio results in this Track.

Find Matching Audio for a Video Asset Automatically

DVD Studio Pro seeks to minimize the steps required to create your DVD whenever possible—one example of this is "*Find matching audio when dragging*" in the *Track Preferences* pane (see Fig. 7-12).

When enabled (checked), the "*Find matching audio when dragging*" preference tells DVD Studio Pro to look for a matching audio file whenever you drag an MPEG video asset into your DVD. If it can find an audio file with a matching root filename (the part to the left of '.m2v') in the *same folder* as the video, it will author them together as an *asset pair*, saving you the trouble of having to author the audio separately. If an audio filename that exactly matches the root filename of the video file being dragged *cannot* be found, no audio will be linked to that video. If you are using a non-MPEG movie file, video and audio are added together automatically.

This "Find matching audio" feature is a great timesaver, especially if you are creating lots of Tracks directly from files in the Palette or the Macintosh Finder.

TIP!
You can override the current setting of the "Find matching audio" preference by holding the Command key while dragging (⌘-**Drag**).

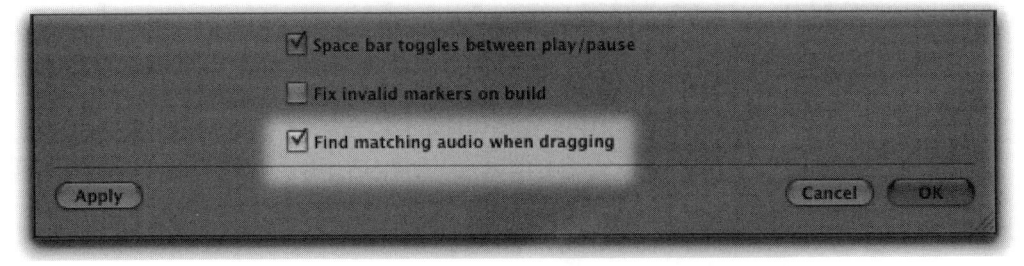

Figure 7-12 — Track Preferences Pane—note bottom checkbox.

192

Authoring Tracks

Adding Audio Streams by Hand

Of course, the *Find matching audio preference* does not prevent you from adding additional or alternate audio streams into your Track, up to the eight audio stream limitation of the DVD Video Spec.

To add an audio stream by hand:

- Click on an audio file in the Palette, Finder, or Asset Tab.

- Drag the file directly into an Audio Stream area and release the cursor. Note that this will not necessarily time-align the audio file, so you may have to tweak it into sync.

Adding AC-3 Audio Streams

Dolby Digital AC3 streams are different from AIFF—they can have an embedded timecode value. You can use this when adding an AC3 stream to a Track.

To add an AC3 stream using its timecode:

- *Option-Drag* the desired AC3 file from the Asset Tab to an Audio stream area.

> **NOTE!**
> The AC3 file must be read (parsed) in order for the timecode to be determined; therefore, it must be available from the Assets Tab. Drag it there first before you attempt to use it for authoring.

When you hold down the Option key while dragging an AC3 file, DVD Studio Pro is forced to determine if the timecode in the AC3 file falls within the region contained in the V1 video stream.

- If yes, the AC3 stream will be synchronized with the V1 Video stream.

- If no, the AC3 stream will be placed at the beginning of the stream.

> **NOTE!**
> This technique does not work if there is more than one Video clip in V1, or if the audio stream area where you are dragging already has a clip.

You can add up to a total of eight Audio streams in each Track. Each Audio stream may be mono, stereo, or contain a surround sound program. It may be PCM, Dolby digital, or DTS. How you prepare these tracks is explained in Chapter 6.

Some Assets to Practice With

In the DVD accompanying this book, you will find the Recipe4DVD Assets we used in this example shown in Figure 7-13. Video Clips and their matching AC3 audio streams can be found in the "Video+Audio" folder. You can practice making Tracks by using these files, or your own. As a bonus, if you assemble the Tracks from the assets provided, it will teach you more about the DVD authoring process!

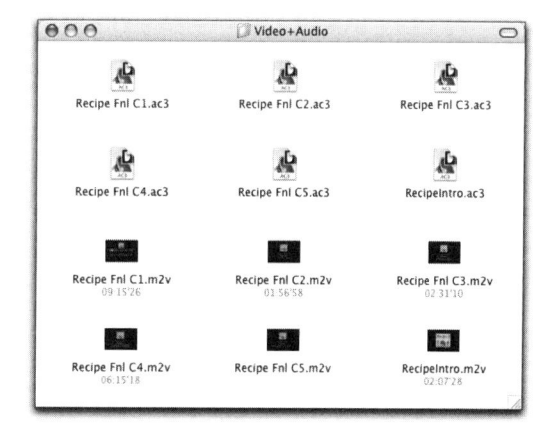

Figure 7-13 — Recipe4DVD Video+Audio assets folder.

What about Markers?

When dragging an asset that includes Markers already encoded into it, these Markers will automatically appear in the Track Editor timeline. If DVD

DVD Studio Pro 4

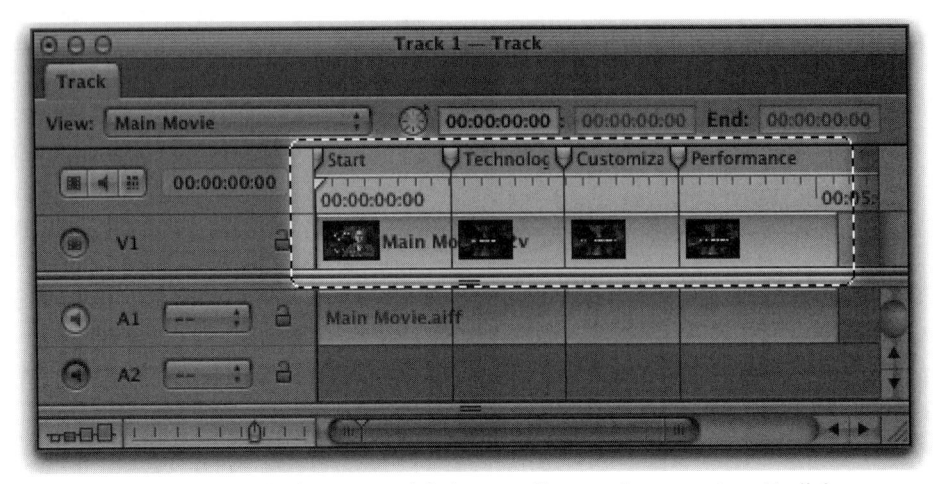

Figure 7-14a — Markers created during encoding can show up automatically!

Studio Pro has not already parsed the file, it may take a few seconds for it to read the information in the file and create the Markers. The old DVD Studio Pro 2/3 Tutorial *Main Movie* has Markers. If you still have this asset, you can import this movie and experiment with it during Chapter 8 (see Fig. 7-14a).

In the event that embedded Markers do not automatically appear (although they should), Control Click on the Video Asset in the Track and select "Add Embedded Markers" in the pop-up. You can also import a List of Markers from this same pop-up (see Fig. 7-14b).

Can I Still Use the Track Inspector to Set the Video Asset?

If you're asking this question, you are probably upgrading from DVD Studio Pro 1, where this was a common function. You'll find this is NOT possible in the current versions DVD Studio Pro, as the *Track Editor* takes its place. (We will cover the *Track Inspector* later in this chapter).

Figure 7-14b — Adding Embedded Markers or a List by hand.

Creating a Track with an Asset Automatically

To create a Track using either an encoded MPEG video stream or a QuickTime movie file, you can

194

Authoring Tracks

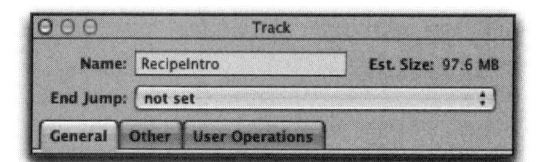

Figure 7-15 — The Track Inspector top area.

select a file from the Palette, the Asset Tab, or even the Finder Window and do one of the following:

- Drag it onto the Disc Name in the Outline Tab (Fig. 7-16).

- Drag it onto the Tracks Folder in the Outline Tab (Fig. 7-17).

- Drag it <u>into</u> the Tracks List area in the Outline Tab (Fig. 7-18).

- Drag it onto <u>any</u> empty area in the Outline Tab (Fig. 7-19).

Figures 7-16 through 7-20 show what these operations look like.

Figure 7-16 — Dragging an Asset <u>onto</u> the Disc Name.

Figure 7-17 — Dragging an Asset <u>onto</u> the Track Folder.

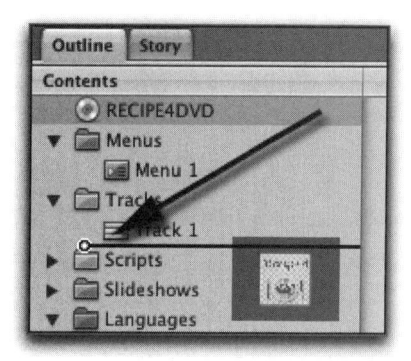

Figure 7-18 — Dragging an Asset <u>into</u> the Track Folder list.

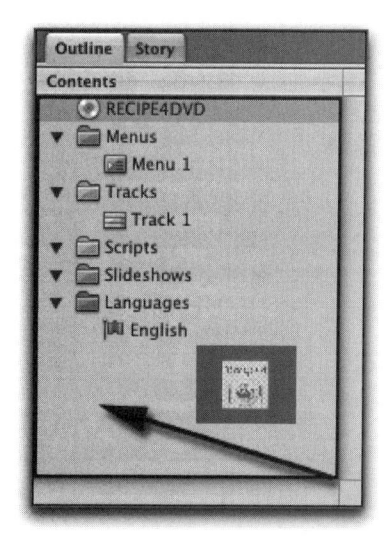

Figure 7-19 — Dragging an Asset <u>into</u> Outline Tab— Note selection highlight!

195

DVD Studio Pro 4

When you release the mouse, the Track will appear in the Track Folder, *and it will have the name of the asset file you used!*

Figure 7-20 — New Named Track created by Dragging Asset file.

A hint for you: When you drag into the *Outline Tab*, be sure to look for the dark selection highlight around the Tab's perimeter—this mean's you're in the right place to drop the asset, (refer to Fig. 7-19).

> **NOTE!**
> Dragging a video Asset to an existing Track name in the Outline Tab will add or append the selected asset into the named Track's V1 stream. This may or may NOT be what you are trying to accomplish! All assets that share the same stream must be of the same kind, and same aspect ratio. In DVD Studio Pro 2, 3, and 4, you can have more than one Clip in a Track, unlike in DVD Studio Pro 1 (see the section on "Adding Additional Clips" to a Track below).

> **POWER USER TIP!**
> Dragging Multiple Video files to one of the locations above will create multiple Tracks at one time. You can select multiple files at one time in the Finder Window, Palette Tab, and the Asset Tab. Audio will drag with them if you have that preference set. *You can Shift-Click*, or *Click-Drag* the Thumbnails in the Palette to select a group of clips at one time.

Adding Additional Clips to a Track

Adding one or more additional clips *into* the Track Editor will append the clip(s) to the end of the current contents of the Track.

To add a clip to an existing Track with video, do one of the following:

- Drag one or more clips <u>onto</u> the Track in the Outline Tab, **or**

- Drag one or more clips <u>into</u> the V1 Stream Area in the Track.

Audio will drag with them if that preference is set (see Fig. 7-21).

> **NOTE!**
> If you Drag multiple clips into the Outline Tab, or onto the Tracks Folder—you'll get Multiple Tracks instead!

You can add Markers and define Stories in a Track with multiple clips (see Chapter 8).

Duplicating Clips in a Track

To duplicate a clip in a video stream, do one of the following:

- *Option-Click* the desired stream, and drag a copy to an empty part of the Stream area, **or**

- *Control-Click* the desired stream, and select "*Duplicate Media Clip*" in the pop-up.

To duplicate a clip in a different video stream, do one of the following:

- *Option-Click* the desired stream, and drag it to a new Stream area, **or**

- *Shift-Option-Click* the desired stream, and drag it to a new Stream area; the start time will be matched to the start of the original.

Authoring Tracks

Figure 7-21a — Multiple Clips appended in one Track's Timeline.

> **NOTE!**
> When duplicating clips, remember that the more video you add into your project, the more disc space you use. Whenever possible, use commands or Scripting in order to re-use assets without taking up additional disc space.

You Can Author with Movie Files as Well as MPEG Streams

Since the advent of DVD Studio Pro 2, you have no longer needed to import (or encode) an Asset before you use it, and, in most cases, just dragging the file you want to use into your project will create the Track! This is 21st century *"drag-and-drop,"* and very time-saving.

In DVD Studio Pro, the onboard MPEG encoder will encode your movies into MPEG-2 files in the background while you author OR WHEN YOU BUILD (this preference has been moved from the *General Preferences* to the *Encoding Preferences Pane* in DVD Studio Pro 4). (You can read more about encoding MPEG-2 in Chapter 5, if you skipped it.)

Using a QuickTime movie (.mov) that contains one video and one audio program, you can directly author a Track by dragging and dropping this file into any location that will create a movie. You may also continue to use traditional elementary MPEG video Streams and their associated audio streams.

Creating a Track Using the Menu Editor

To create a Track using the Menu Editor, do the following:

- Click on the desired file from the Palette, Finder, or Asset Tab;
- Drag the file directly onto the Menu Editor Screen and hold it.

Holding a dragged file on the Menu Editor screen will cause the display of one of the context-sensitive *Drop Palettes* (see Fig. 7-21b), which give you intelligent choices of what to do with this movie. In the case of a video asset, you will probably want to select *Create Button and Track*, which will create the Track automatically and connect it to the Button it makes in the Menu (see Fig. 7-21c). We'll cover a *lot* more on Drop Palettes in Chapter 9.

197

DVD Studio Pro 4

Figure 7-21a — Creating a Track in the Menu Editor by Dragging assets.

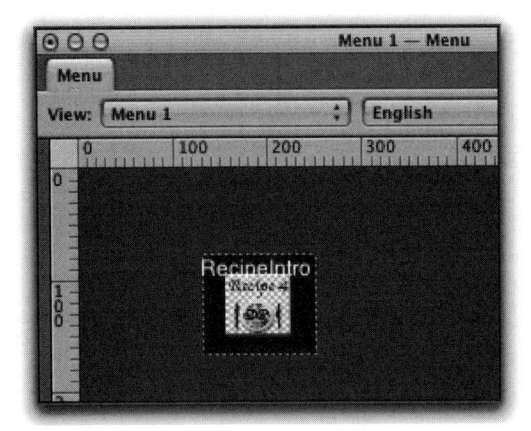

Figure 7-21b — The Track's Button created in the Menu Editor.

> **NOTE!**
> Dragging an MPEG stream or movie file into the *Assets Tab* will NOT automatically create a Track Element, but it WILL add that file into the available authoring assets. To use that asset, you will need to author that asset into an existing Track, or create a new Track for it.

You Don't Need a Menu to Make a DVD...

If your DVD doesn't require a menu, there's nothing to stop you from authoring your DVD project by creating a simple Video track, and then building and burning the disc. Just be sure you set that Track as the *First Play* action (see Chapter 11, "Making Connections").

Using the Track Editor

The Track Editor is included in the default Extended or Advanced Configurations. If the Track Editor is not visible in its Quadrant, click on the *Track Tab*. If you are in the Basic configuration, you can reveal the bottom quadrant, or switch to Extended or Advanced view.

To select which Track to edit:

- *Double-Click* the Track's name in the *Outline Tab* Track list, **or**
- Select the desired Track from the *View*: pop-up in the Track Tab.

If you're new to DVD Studio Pro, the first thing you might notice about the *Track Editor* is the *timeline*. DVD Studio Pro 1 did not use a timeline while authoring tracks, so if you haven't see it before, it might take a second for you to get used to. If you are familiar with nonlinear video editing, you are probably an old hand when it comes to timelines.

About the Track Tab Controls

Along the top of the Track Editor (Fig. 7-22), you will find these tools, starting from left:

- The *View*: pop-up (to select which Track to edit in the Track Editor),
- The *Playhead location field* (shows/sets Playhead current timecode), and

198

Authoring Tracks

Figure 7-22 — The Track Editor's Top tools.

- The *Clip Start*: and *End*: time displays (for currently selected clip). These fields may also be used to edit Start and End times.

Immediately below these controls are:

- The three "Show Multiple" Stream selectors, which control the display of the Stream Areas for Video (Fig. 7-23a), Audio (Fig. 7-23b), and Subtitles (Fig. 7-3c).

- The "current Cursor location" display, which always shows the current position of the Cursor within the Timeline (to the right of the three stream selectors in Fig 7-23a).

- The Playhead (Inverted Yellow Triangle); shows the current position of the movie while playing (right under the Chapter 1 marker in Fig 7-23a).

If you haven't changed your Marker labeling preferences, the default Marker called Chapter 1 should also be at the left end of the timeline. Chapter 1 (or whatever you may choose to rename it to) is a very important Marker. It is required in order for the DVD player to know where this Track begins, and it cannot be removed from the Track. The gray area directly above the timeline numbers is where Markers are created and displayed (see Chapter 8 for more about creating Markers).

Stream Areas

Immediately beneath the top tools we have been looking at are the Stream areas where you can organize the Video, Audio, and Subtitles that make up this particular Track. Each of your up to 99 tracks may

Figure 7-23a — The Show Multiple Video Streams tool.

Figure 7-23b — The Show Multiple Audio Streams tool.

Figure 7-23c — The Show Multiple Subtitle Streams tool.

199

have a completely different configuration or combination of these items, with one exception—there must *always* be a picture of some kind within a Track, or that Track will be illegal in your DVD, and your project will likely not *Build*.

Organizing the Stream Display

The Stream areas are very flexible to display and can be rearranged quite easily. The divisions between the Video, Audio, and Subtitle Stream areas are movable, and can be easily rearranged. Streams may be hidden or revealed by clicking on the horizontal marks in the middle of the stream dividers (as in Fig. 7-24), and dragging up or down to reveal more or fewer streams, as needed. Depending on the current size of the Track tab, any number of Video, Audio, and Subtitle Streams may be displayed. When it comes to displaying Tracks and Streams, having a large display or a dual-monitor setup can really come in handy—more screen real estate can mean a much larger Track tab—a benefit in DVD Studio Pro, as it is in Final Cut Pro.

You will find having a larger timeline area comes in handy for editing not only DVD Stream data, but also when creating Subtitle events (see Chapter 12, "Creating Subtitles").

Customizing the Stream Area Display

Each of the Stream areas has its own visibility tools.

- The *Viewer Stream Select* turns on visibility of each Stream in the Viewer (see Fig. 7-25). This affects the Simulator as well as the Viewer playback. Note that Video Stream V1 must *always* remain on.

Selecting a Video Stream other than V1 results in the playback of that angle (Video Stream) during simulation and preview. If this is a "mixed angle" Track, display will alternate between the selected Stream (where video exists in that Stream) and Stream 1 (where video does not exist in the selected Stream). Audio and Subtitle Streams are optional during Simulation.

- The *Stream Lock* (the padlock icon) locks the contents of a selected Stream, preventing editing (see Fig. 7-26). There is a Stream lock for each Video, Audio, and Subtitle stream.

Figure 7-24 — Adjusting stream visibility and arrangement.

Authoring Tracks

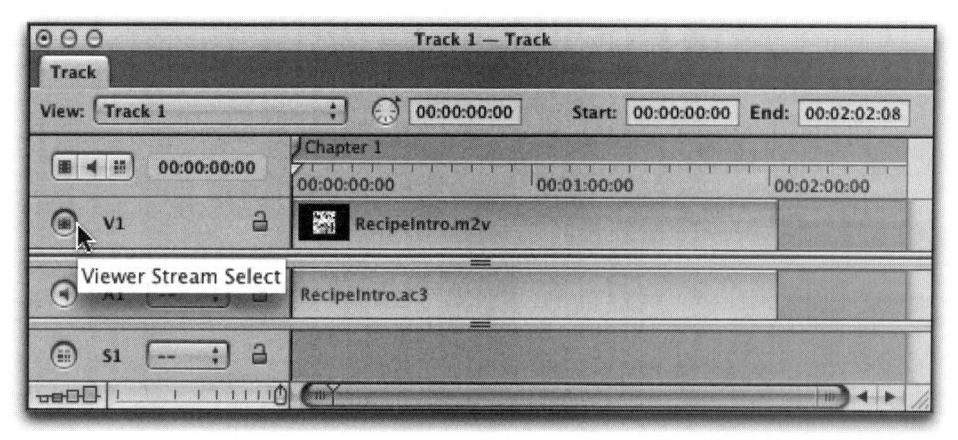

Figure 7-25 — Stream Select Icon—on means Stream is visible in Viewer.

Figure 7-26 — Stream Lock tool (hash marks means Stream is locked).

- Press Shift-F4 to lock all streams simultaneously, or select **Project > Timeline > Lock All Streams** in the Menubar.
- *Audio Stream Language Select* pop-ups (Fig. 7-27). Selects a label for the Language associated with that Stream. This label is read by DVD players and used to identify the language contained within that Audio stream to the DVD viewer.

IMPORTANT!
It is always advisable to set this property correctly during authoring. If not set, the DVD player will usually report an "unspecified" language, and, more importantly, certain automatic language selection functions may not perform properly.

NOTE!
There is a Preference for automatically selecting a Language tag in the *Track Preferences*.

201

DVD Studio Pro 4

Figure 7-27 — Stream Language Select tool (Audio, Subtitles only).

Figure 7-28 — Using the Language Selection pop-up for a Subtitle Stream.

More Track Display Selection Tools

At the bottom of the Track Editor are three display selection tools that closely resemble similar tools in Final Cut Pro:

- *Stream Height Select*—Timeline Bottom, extreme left (Fig. 7-29). There are four bars of varying height, each of which selects a Stream Height level. There are four levels of Stream Height selectable, which control the height of all streams visible in the Track Editor. These can be seen in Figures 7-30a, 7-30b, 7-31a, and 7-31b.

- The *Subtitle Stream Language Select* pop-up shown in Figure 7-28 selects a label for the Language associated with that Stream.

This label is read by DVD players and used to identify the language contained within that Subtitle stream to the DVD viewer.

You will find more information about Subtitles in Chapter 12.

Figure 7-29 — Stream Height Tool.

Authoring Tracks

Figure 7-30a — Maximum Stream Height.

Figure 7-30b — Large Stream Height.

Figure 7-31a — Medium Stream Height (Default).

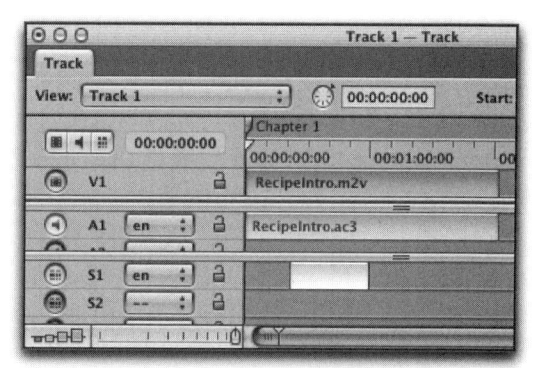

Figure 7-31b — Small Stream Height.

- *Timeline Zoom Tool*—Timeline Bottom left (see Fig. 7-32).

The Timeline Zoom determines how much of the current timeline is visible within the Track Editor. The Zoom level is selectable in a smooth sliding scale using this tool. The degree of zoom can also be selected using ⌘ = and ⌘ –.

- *Timeline Scroll Tool*—the bar to the right of the Zoom (see Fig. 7-33).

This Tool can be dragged to resize the Zoom of the Timeline and to locate the visible position of the Timeline. Dragging either end allows scaling of the display, while dragging from the middle allows scrolling of the displayed portion of the timeline. The little inverted triangle shows the current position of the Playhead.

Figure 7-32 — Timeline Zoom Tool (also can use ⌘ = and ⌘ -).

> **TIP!**
> Hold shift while scrolling to keep the Playhead visible!

How the Timeline Works—Zero-Based versus Asset-Based TC

The Timeline in the Track Editor defaults to a measurement of *elapsed* time—basically, the Track's running time beginning at what is usually called "Hour Zero"—SMPTE value 00:00:00:00.

Although MPEG files may contain timecode values other than Hour Zero embedded during encoding, the default time base is called "Zero-Based Timecode." In this mode, the duration of the MPEG video asset is counted in Hours, Minutes, Seconds, and Frames elapsed from Hour Zero at the Start of Track (Chapter 1). The location of any marker within the MPEG file can thus be defined by its distance from the beginning of the stream.

As an alternative, SMPTE timecode embedded within the MPEG file can be displayed instead, by using the *Asset-Based Timecode* mode, which can be selected by *Control-Clicking* on the Timeline ruler (see Fig. 7-33). Be sure to click on the Timeline ruler, and not *within* the Stream areas or on a Stream (Clip), because clicking in those locations will bring up an entirely different set of options.

Trimming Assets in the Timeline

Beginning with DVD Studio Pro 2, you have the ability to trim the head and tail (the beginning point, and the end point, or duration) of any stream in the Track Editor. Place the cursor at the beginning or end of any stream, and you will see the trim cursor reveal itself.

To Trim Streams using the cursor:

- Click on the stream's beginning with the Trim Cursor (see Fig. 7-34). Drag to the right from there to trim the *start point*.

Remember that the V1 Stream must always begin at Chapter 1, so after any trim you make, V1 will always shuffle down so its first frame is aligned with Chapter 1.

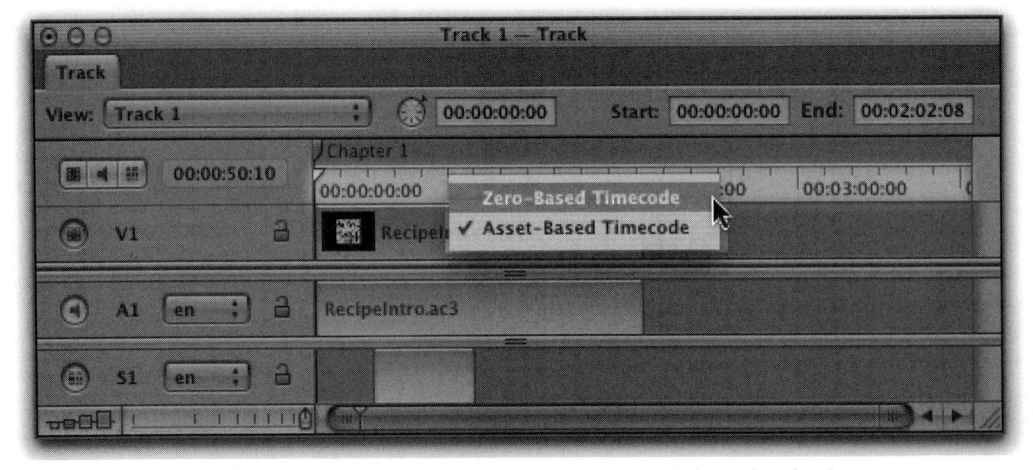

Figure 7-33 — Control-Click on the timeline for Timecode mode selection.

Authoring Tracks

Figure 7-34 — Trimming the Video Stream beginning (head).

To trim a Stream's Tail, or its duration (see Fig. 7-35), do one of the following:

- Click the Stream's end with the trim cursor and drag left to trim back the ending point to shorten the Stream, **or**

- Click in the End Time field of the selected Stream, and you can edit the Stream's numerical duration.

Figure 7-35 — Trimming the Video Stream duration (tail).

In each case, the current cursor position will be shown in the cursor time field, and in the Start or End field as well.

> **NOTE!**
> If you are trimming an encoded MPEG-2 asset, trims can only occur on *GOP boundaries* (I-frame headers). This is due to the DVD Spec, and cannot be changed. This means the trim points are limited to the locations of the GOP sequence as encoded, not *Compression marker* locations (forced I-frames).

You will find you can also trim Audio and Subtitle streams this way.

Why Trim an Asset Stream?

Many times encoded MPEG files will include material not needed in the final encode—color bars and tones at the beginning, or perhaps an extra-long roll-out or tail of black.

To facilitate removing these extraneous items, it is nice to have trimming tools within the DVD Authoring program. If they were not available, the only recourse would be to re-encode, or to try to work around the defect by defining Markers and then using Stories or scripted playback to "play around" the undesired segments. Trimming is far easier.

Trimming Streams Using the Asset Inspector

Sometimes, you will require more precise trimming than is convenient using the cursor. A stream's duration or starting timecode may be trimmed using the numerical "*Clip Start Trim*" and "*Duration*" fields present in the Inspector. Be sure to make the Inspector visible before you try adjusting these durations.

To trim the start time numerically:

205

DVD Studio Pro 4

- *Click* on the Video or Audio stream to be trimmed, then
- *Click* on the up/down arrows next to the Inspector's Clip Start Time field; **or**
- *Click* in the Clip Start time field to place the cursor where needed, or *double-click* on a pair of numbers to select them, or type *Cmd-A* (⌘-**A**) to *Select All* in the desired field. When you have selected the desired numbers, delete and/or retype numbers as needed (see Fig. 7-36, left).

To trim the duration numerically, use one of the following methods:

- *Click* on the Video or Audio stream to be trimmed, then
- *Click* on the up/down arrows next to the Inspector's Duration field, **or**
- *Click* in the field to place the cursor where needed, or *double-click* on a pair of numbers to select them, or type *Cmd-A* (⌘-**A**) to *Select All* in the desired field. When you have selected the desired numbers, delete and/or retype numbers as needed (see Fig. 7-36, right).

Track Inspectors

The Inspectors play an important role in DVD Studio Pro by defining certain aspects of an Element's functionality, once it's been created.

- The *Track Inspector* top area contains a *Name* field, and a pop-up to select or modify this Track's *End Jump* property.
- The *Track Inspector* has three Tabs, to organize the information available about this element: *General, Other,* and *User Operations* (see Fig. 7-37).

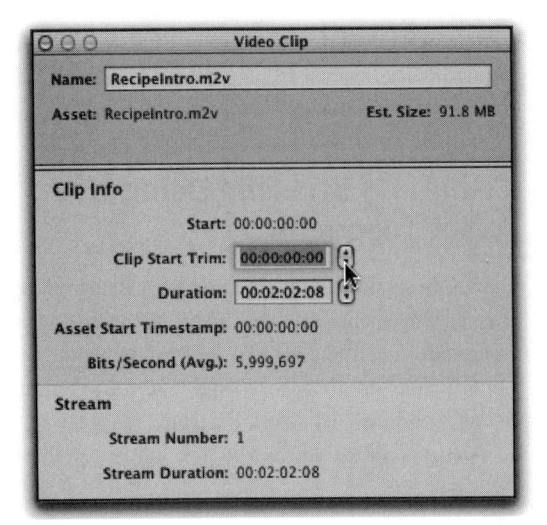

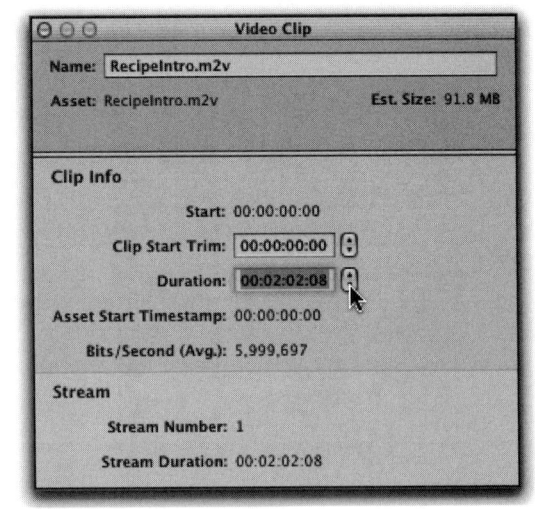

Figure 7-36 — Trimming Start Time (L) and Duration (R) numerically.

Authoring Tracks

The Track Inspector General Tab

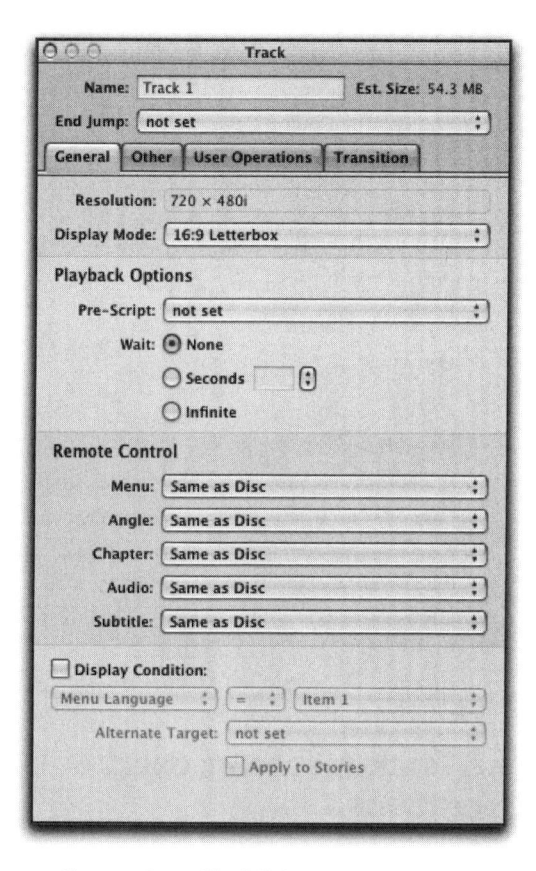

Figure 7-37 – The DVDSP4 Track Inspector's General Tab.

- *Resolution*: Reports on the video resolution of the asset contained in this Track.

- *Display Mode*: Controls the MPEG video playback aspect ratio—4 × 3 is normal. If this Track is using 16 × 9 assets, you can choose how they will appear on a 4 × 3 display—as Pan & Scan, Letterbox, or both.

- *Playback Options*:

- *Pre-script*: Allows selection of a script to be executed prior to playback. The Script runs each time this Track is navigated to.

- **Wait: Includes three options:**
  - *None* processes the End Jump immediately, no delay.
  - *Seconds* allows for a timed pause after playback. When set to a specific length, the DVD player will pause for that length before processing the *End Jump* action.
  - *Infinite* will cause the DVD player to display this Track's last frame indefinitely. The DVD player will usually respond to a *PLAY* command to continue past this point. This may vary by player, so check carefully when you use this feature.

- *Remote Control*: Allows individual override of the Global Disc Remote Control properties. Use this carefully, and evaluate its proper operation before you replicate thousands of discs using this feature. It's best to Build this disc, and Burn it to check your project on a set-top DVD player.

  Same as Disc means to follow the Remote Control properties set in the Disc Properties *General* and *Advanced Tabs*.

- *Display Condition*: Under certain specified conditions, the display of this Track may be changed, and a different Track substituted. To specify such a condition, check this checkbox to activate the Display Condition dialog and set the conditions for display of an alternate target.

- An *Alternate Target* element may also be defined.

207

DVD Studio Pro 4

- Check the *Apply to Stories* checkbox if you wish to apply the Display Condition settings to Stories based on this Track.

The Track Inspector Other Tab

The Track Inspector *Other* Tab includes the following options (see Fig. 7-38).

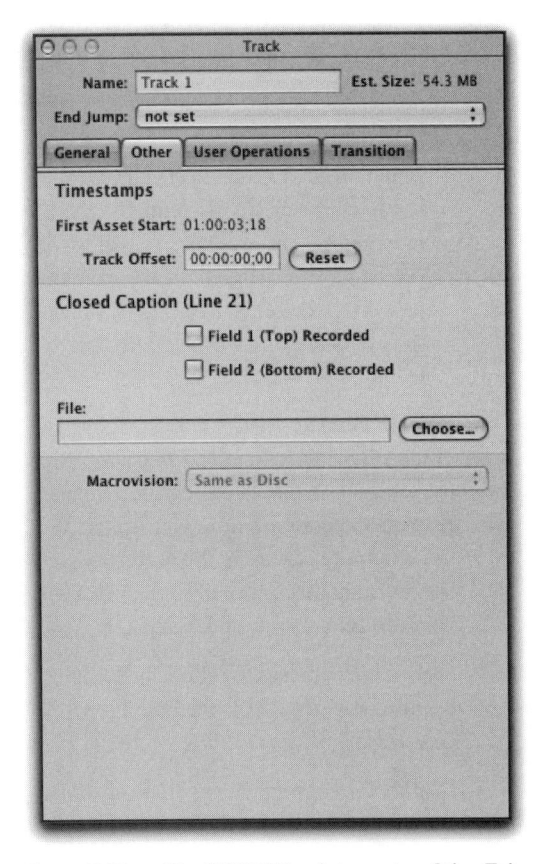

Figure 7-38 — The DVDSP4 Track Inspector *Other* Tab.

- *Timestamps*:

    Displays the timestamp of the First Video Asset in V1; *Track Offset* allows modification of the Timecode reference used by the "Asset-Based Timecode" feature of the Track Timeline. If a timecode value is inserted in this field, and the asset in V1 has Zero Hour timecode, you can program a SMPTE time for the timeline of this Track.

- *Closed Caption (Line 21)*:

    Select a checkbox for which Field to use (generally Field 1) and click *Choose* to open a dialog to find the Caption file for this track. This must be a valid .CC file (see Chapter 17, "Advanced Authoring" for more on Closed Captioning).

- *Macrovision:*

    Allows override of the global Macrovision setting of the DVD Disc; however, Macrovision cannot be selected unless the Disc Master Macrovision property has been enabled. If Macrovision has been *enabled* in the Disc Inspector, then this property will allow it to be disabled for this track, or for the Type (1, 2, or 3) to be changed. (See Chapter 17 for more on Macrovision.)

The Track Inspector User Operations Tab

Have you ever wanted to lock certain functions of the DVD remote control, but weren't sure how to do it? *User Operations* are the answer (see Fig. 7-39). *UOPs*, as they are called, are the remote control functions used by the viewer to control playback and navigate through Menus while watching the DVD. The DVD author (that's you!) has the ability to lock (or leave unlocked) these functions using these controls. Since DVD Studio Pro 2, you have had complete control over the UOPs on each Track, and each Menu. (Slideshows do not give UOP control.)

Authoring Tracks

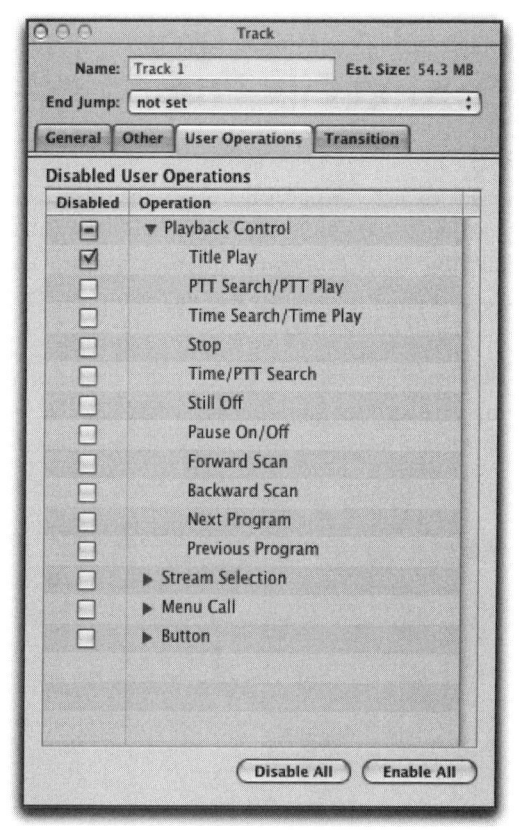

Figure 7-39a — The DVDSP4 Track Inspector User Operations Tab.

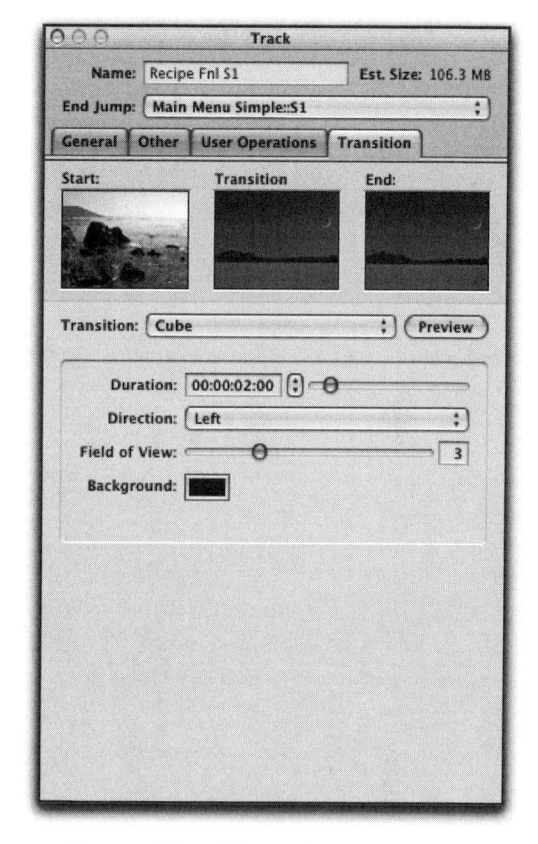

Figure 7-39b — The DVD Studio Pro 4 Track Transition Inspector.

> **NOTE!**
> Menu UOPs already have a few controls locked for you, so you may find not much needs to be tweaked there.

The Transition Inspector

New since DVD Studio Pro 3, Transitions are a great new way to add creative options to a DVD during Authoring, without generating thousands of dollars of new Video compositing charges for your client.

The Transition Inspector is straightforward, offering a *Transition*: pop-up to allow selection of the desired Transition type, and a *Preview* button right next door, which will preview the chosen Transition in the center panel of the three-pane display at the top (see Fig. 7-39b).

The three-pane image display shows the following:

- *Start*: The last frame of this Track, OR, a default seacoast view to hold the "A" effect position.

- *Transition*: When the Preview button is clicked, the current settings of the chosen Transition are previewed here, in real time (isn't QuickTime wonderful?). Yes, it's

209

small, but it's in real time. Absent any content in this track, or with an unspecified "End," Preview will use the default "day-to-night" stills to show you how the transition will look.

- *End*: The frame that the Transition will end on, or a default nighttime shot to hold the "B" effect position.

The bottom pane will change with each Transition, as it displays the specific Options available for that Transition.

The Clip Inspector— Viewing Clip Properties

Each Clip in your Track is a unique entity, and its properties can be investigated in the Inspector. Clicking on any Video or Audio clip within a Track will bring that Clip into the Inspector and provide information about the clip and the relevant stream (see Figs. 7-40 and 7-41).

Note that the information shown in this Inspector is different than the information that will be displayed if you click on a Video or Audio *Asset* in the *Asset Tab*.

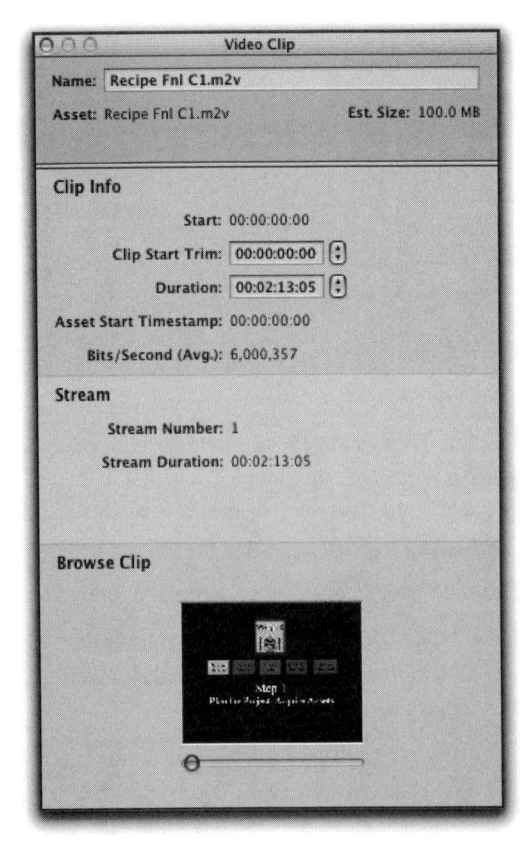

Figure 7-40 — Video Clip Inspector.

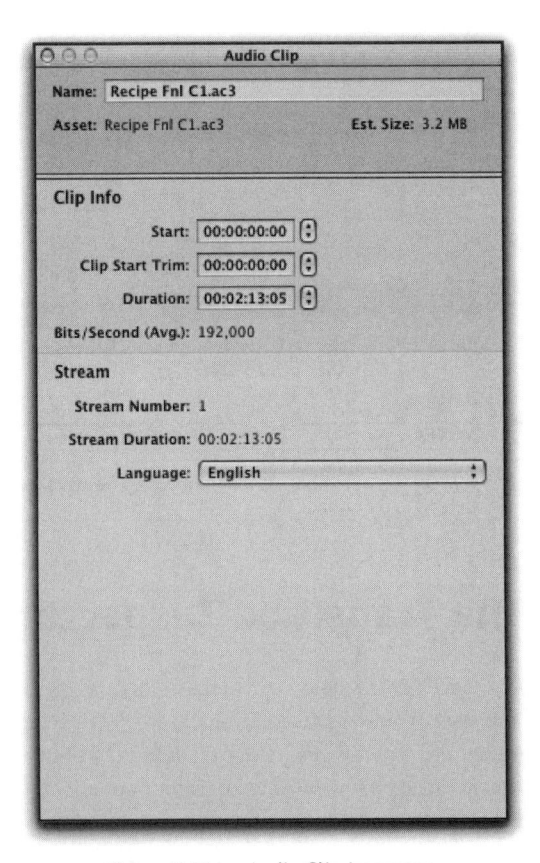

Figure 7-41 — Audio Clip Inspector.

Authoring Tracks

Even though they exist inside of Tracks, selecting a Subtitle event in any Subtitle Stream (for whatever reason, Apple also calls those *Clips*, too) will bring up the *Subtitle Inspector*. We'll cover the Subtitle Inspector and all things Subtitle-related in Chapter 12.

The Clip Inspector shows information about both the clip and the stream. In the top area, you find:

- *Name*: Enter an alternate name for the clip, which applies to the clip in this track's timeline only—it does not affect the asset's name.
- *Asset*: Shows the clip's actual filename.
- *Est. Size*: Shows how much disc space this clip requires.

Clip Info

- *Start*: Shows the clip's start time in the stream (V1 is always zero!)
- *Clip Start Trim*: Enter an amount of time by which to offset the clip's beginning. This does not affect the clip's overall length or its position in the timeline. Choose the part of the asset that the clip should use. This setting is only valid if the clip's length has been trimmed so that you are not using the entire asset.
- *Duration*: Shows the clip's length. You can enter a new length to trim the end of the clip.
- *Asset Start Timestamp*: Displays the timecode of the asset's first frame.
- *Bits/Second (Avg.)*: Shows the clip's bitrate information.

To use the start and end values to position and trim a clip:

- In the Track Editor, select the clip you want to trim. Its start and end values appear in the Track Editor's *Start*: and *End* boxes.
- Enter a new *Start Time*. The beginning of the clip moves to that time. This *moves* the entire clip, but does not *trim* it.
- Enter a new *End Time*. The end of the clip is *trimmed* to that time.

You can use the Clip Inspector to trim a clip:

- Select the clip you want to trim. The clip will appear in the Clip Inspector.
- Enter a new *Clip Start Offset* to move the clip's start time.

The clip's duration or its start time on the timeline will not be changed—only the part of the asset used by the clip is affected.

- Enter a new *Duration* value to trim the clip's end.

Clip Conflicts

There are a number of rules to keep in mind when you trim and position a clip.

Video Clips

Dragging clips to the V1 stream is different from dragging audio clips. Gaps are not allowed between clips in the V1 stream, and there must be a clip at the first marker (the Track beginning point).

The following situations cause video clips already in the V1 stream to move. This can cause problems if the video clips have associated audio or markers, since they do not move with the video.

- *If you drag a video clip to the start of a V1 stream that already has a clip assigned*, the new clip becomes the first clip and the

211

original one (and any that follow it) shuffle down the timeline.

- *If you drag a video clip to the point where two existing video clips meet in a V1 stream*, the new clip is inserted between them.

- *If you drag a video clip to the stream, and the "Find matching audio when dragging" preference setting is enabled*, the audio in the A1 stream is trimmed or moved to accommodate the new video clip, even if that clip does not have matching audio. This maintains sync between the track's existing video and audio clips.

Stream information and settings

- *Stream Number*: Shows the Stream Number for this clip.

- *Stream Duration*: Shows the Duration of this stream.

- *Language*: Applies to Audio Clips and Subtitle* Clips, but not Video Clips) and shows the current language assigned to this Audio Stream. You can change languages by using the *Language*: pop-up, and selecting a new choice from the menu.

> **NOTE!**
> The Language property also applies to Subtitle Clips, but is controlled in the Subtitle Inspector (see Chapter 12).

Browse Clip

For Video clips, the *Browse Clip* function provides a scroll bar that can be dragged to scan through the video content of the clip.

To Delete a Clip from within a Stream:

- Click on the Clip to select it, and hit the *Delete* Key, **or**...

- *Control-Click* the Clip and select *Delete Media Clip* from the pop-up.

About DVD Data Rates

The DVD format is designed to deliver a magnificent digital picture, along with pristine digital sound and graphical subtitle overlays, all within a delivery bandwidth that is around 9.8 Megabits per second... How fast is that? Well, if you have a DSL line at home, your DSL line typically maxes out at 1.5 Megabits per second, or roughly one sixth of that data rate. (A T-1 Network Connection also runs at 1.5 Mbps.)

DVD uses the 9.8 Mbps of bandwidth to play the user's selected choices of Video, Audio, and Subtitle streams. Now let's be clear—the DVD can only play one video, audio, and subtitle at a time, but you can choose one Video you want to see, one Audio to hear, and one Subtitle to watch from the authored choices.

Because of the way DVD shares the bandwidth among the streams, we need to be careful of our Video and Audio encoding rate choices. A good general rule of thumb is NOT to use uncompressed audio (i.e., AIFF or WAV sound files) if the running time of your project is more than 60 minutes, or if you plan on using more than ONE Audio Program. Under these circumstances, you really should compress your sound files into Dolby Digital AC-3 files using A.Pack.

A few notes on video encoding rates too, while we're here—in order to put 120 minutes of video on your DVD, you'll need to be encoding your video at around 4.5 Mbps. If your encode rate is higher, say around 6 Mbps, you will only be able to put around 90 minutes of video on the DVD. Above 6 Mbps, and you will possibly run into difficulty playing back your DVD on PCs that use software DVD/MPEG decoder and on certain PC laptops as well. At 8 Mbps, you'll only be able to store 60 minutes of Video in that DVD, less if you wind up using AIFF

Authoring Tracks

Audio files. A good rule is to stay between 4.5 Mbps and 6 Mbps for most of your work, even if you only have 60 minutes of video or less—if you NEVER will need to play back this DVD on a computer (fat chance of that!) you may encode your video at 8 Mbps—but you should be CAREFUL of ever encoding the video above the 6 Mbps rate.

If you are going to create a multi-angle disc, you will need to be aware of the limitation on encoding rate, and some other issues as well, such as Closed GOP and matching the GOP structure in your encodes. If you are using the Bit Budget spreadsheet included with this book's DVD, you'll have no problem determining the proper encode rate.

About Multi-Angle Tracks (and Mixed-Angle Tracks, Too)

Once you've created a basic video track, there are two ways of converting it into a multi-angle track in DVD Studio Pro—Multi-Angle and Mixed-Angles.

To create a multi-angle Track, each angle needs to have the same duration. The assets must also be angle-compatible with each other, meaning the GOP structure must match, and encode rates must be within tolerance for the number of angles being authored.

Encoding all angle content from Final Cut Pro or Compressor 2 is a good way to ensure this. Be especially careful not to create mismatched Chapter Markers that do not fit in the regular GOP sequence (15 frames NTSC, 12 frames PAL). Doing so will prevent that video from being used as angle content.

To create a Multi-Angle Track:

- Select the desired Track in the Track Editor (see Fig. 7-42).
- Author the V1 stream with the basic video.
- Locate a compatible asset in the Palette, Finder, or Asset Tab.
- Drag the Angle asset into an empty Stream area (Stream 2–9).
- Align the first frame to the first frame of the Stream.
- Be sure the stream durations match.

Figure 7-42 — A Multiple Angle Track.

213

About Mixed-Angle Tracks

Mixed-Angle Tracks allow you to use an Angle asset that does not have to match the duration. You can use smaller clips, or even trim clips to focus on a specific section. The assets must still be compatible with each other, meaning the GOP structure must match, and encode rates must be within tolerance for the number of angles being authored.

To create a Mixed-Angle Track:

- Author the V1 stream as needed. If you are adding multiple clips or still images, place them all and make any necessary adjustments before adding anything to streams V2 through V9.
- Create a marker where the first mixed-angle segment should begin.
- Add the second video asset to the track's second video stream (V2) at the marker.
- Continue placing video assets to the next available streams until they are all in place and lined up with the marker.
- Add a marker at the end of the angle clips.
- If you are adding an additional mixed-angle section to the track, add another marker and repeat the previous steps.

To create a mixed-angle Track, as in multiple-angle tracks, you must have properly prepared the Video during encoding, with matching GOP lengths. (See Fig. 7-43.)

Previewing a Track in the Viewer

Once a Track Element has been created (even if it is only a single video angle), the Viewer can be used to verify its functionality.

To check a Track Element in the Viewer:

- *Double-click on the Element in the Outline Tab* to make the Track and Viewer Tabs visible with that Element loaded, then

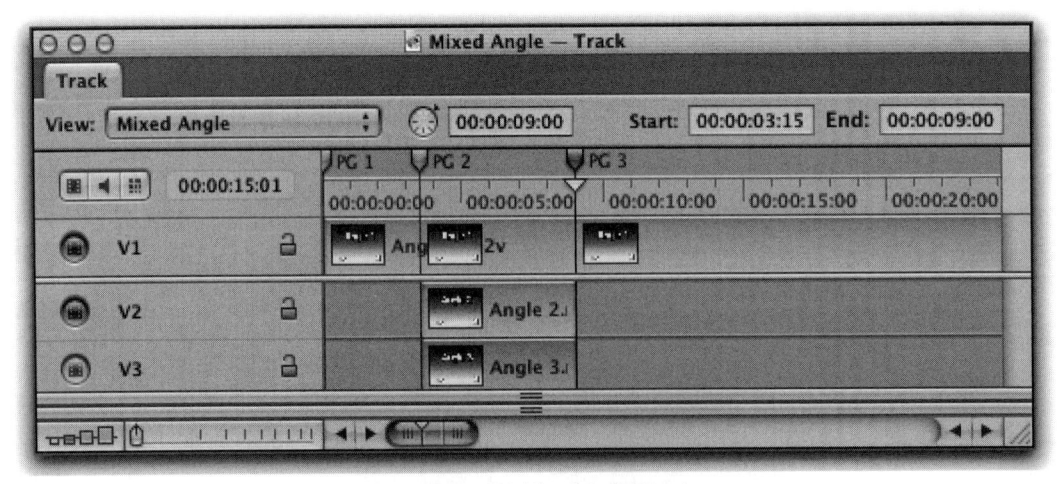

Figure 7-43 — A Mixed-Angle Track.

Authoring Tracks

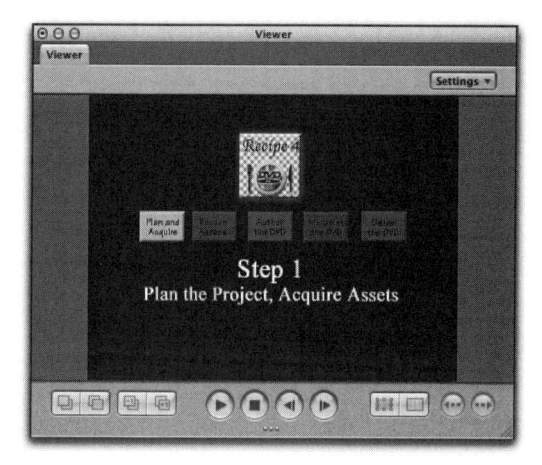

Figure 7-44 — Checking a Track in the Viewer.

- Use the playback control buttons in the Viewer window (see Fig. 7-44).

Controlling the Viewer

While the Viewer is loaded with a Track, you can exercise limited playback control with the button tools at the bottom of the screen.

The Viewer Controls include (see Fig. 7-45).

Play/Pause   Stop   Step Back   Step Fwd

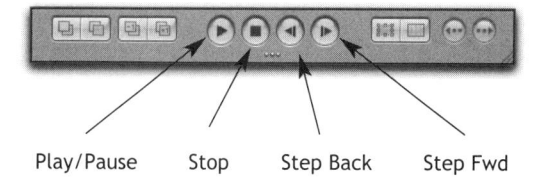

Figure 7-45 — Viewer Controls.

Keyboard Commands for Viewer Controls

- *Play/Pause* — Spacebar or "L" key
- *Stop* — "K" key
- *Step Back* — Left Arrow
- *Step Forward* — Right Arrow

The Step Forward and Step Back buttons default to moving the playhead one frame at a time in the chosen direction.

Adding other keys will modify these capabilities:

- *Shift-Arrow* will move the playhead one second forward or back.
- *Option-Arrow* will move the playhead one (GOP) at a time (usually 15 frames in NTSC, 12 frames in PAL).
- *Control-Arrow* will move the playhead one Marker at a time.

You can also click a Skip Button and continue to hold down the mouse button to repeatedly step the playhead.

If you *PAUSE* playback, the Playhead remains where it stops; if you *STOP* playback, the Playhead snaps back to the position where playback began.

Configuring Streams for Preview

You can select the stream combination you wish to Preview by using the Viewer Stream Select buttons, located along the left side of the Track Editor (see Fig. 7-46). Each stream has a button that can be highlighted. You can select one each from the available Video Streams, Audio Streams, and Subtitle Streams that you have authored into that Track. The Selected Stream combination will play for the duration of the preview.

If you wish to experiment with switching Angles, Audio, or Subtitles, you will need to use the Simulator, as outlined in Chapter 14.

Previewing Angles

To prepare to view an Angle in the Viewer:

215

DVD Studio Pro 4

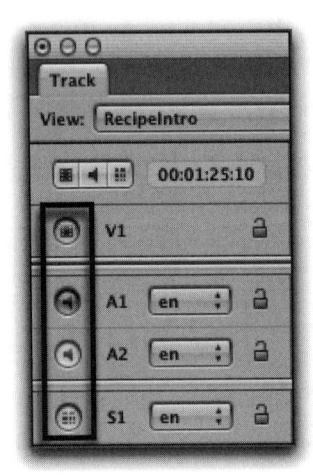

Figure 7-46 — Track Viewer Stream Select buttons.

- Click the Viewer Stream Select button in the Track Header to select the desired angle. The selected Angle will now be readied for playback in Preview. (See Fig. 7-47.)

Figure 7-47 — An Angle Stream Select Button, selecting V3.

How Angle Playback Works in Viewer

During Preview, the selected Angle will play when content is available in that Stream, but the Viewer will default down to a lower-numbered Angle (usually Angle 1) when content does NOT exist in the selected stream. This is exactly the way the DVD will play, so it gives a good approximation of how the Angle content will look when the DVD is finished.

How the Playhead Works

The Playhead (inverted yellow triangle) is the visual indicator of where you are in the Timeline, whether playing or standing still. When playing, the playhead moves along the timeline, indicating relative position within the Track. In addition, the Cursor Time indicator will update, indicating the timecode position. As a visual aid, there is a tiny playhead Icon that appears in the Timeline Scrollbar, to show where the playhead is with reference to the currently visible portion of the timeline. If it is not visible within the Scrollbar, the playhead is either before or after the visible section of the timeline.

To Set the playhead location:

- Drag the Playhead (or click the Cursor) to the desired location on the Timeline, **or…**

- Enter a timecode value in the *playhead timecode* field, adjacent to the *View* pop-up.

When setting the playhead, do so *only* in the time-coded portion of the Timeline—clicking *above* the Timeline will create a Marker!

Once the playhead position has been established, playback will begin from that point if "*Play*" is selected in the Viewer. When playback stops, the playhead will return to that location.

Authoring Tracks

To Set the Playhead Using Keyboard Shortcuts:

Keyboard shortcuts can be used to move the playhead:

- *Arrow keys*: Move by one frame either way.
- *Shift–Arrow keys*: Move by one second either way.
- *Option–Arrow keys*: Move by one GOP either way.
- *Control–Arrow keys*: Move to the next or previous marker.
- *Command–Arrow keys*: Move to the start or end of the selected clip.
- *Up and Down Arrow keys*: Move to the next edge of a clip or marker.
- *Home and End keys*: Move the playhead to the start or end of the timeline.

The Viewer Is for Quick Checks Only...

The Viewer is designed for quick convenience viewing, and does NOT have many of the features available in the *Simulator*, including chapter skips, Menu and Title buttons, and even slow or fast play modes. It's just a quick reference for checking stream playback, reviewing alternate audio tracks, and verifying Subtitles.

For more on the Simulator, see Chapter 14.

How to Make Your Tracks Do More

Adding Markers and Stories can unlock additional capabilities of the DVD medium:

Markers can define an entry point into a Track, and *Stories* can define both an entry and an exit point, or multiple entry points to be played in sequence.

The next chapter (Chapter 8) will detail how to use these important DVD functions.

Summary

Tracks are one of the most important DVD building blocks.

In DVD Studio Pro, Tracks are the containers that are used to collect and organize Video, Audio, and Subtitle Streams. Tracks can build presentations that can contain multiple video programs (otherwise known as "angles"), multiple audio soundtrack bitstreams, and multiple Subtitle Streams.

In DVD Studio Pro, Tracks are assembled in a new Track Editor timeline, which provides a video-editing-like environment in which to align Video, Audio, and Subtitle Streams. This environment also provides a timeline workspace in which to create the various Markers that delineate special locations within the bitstreams.

Tracks are also the containers where Stories can be created, each of which can further divide the video content into easily played chunks, or Subpicture Overlay file-clips, if you will.

While retaining the ability to create Multiple-Angle presentations where all video angles are the same duration, DVD Studio Pro has the ability to use small pieces of MPEG encodes to create mixed-length Angle Tracks.

Tracks are also the containers where Subtitles are created and proofed.

Tracks are generally made from video content that has been encoded into MPEG-2 or MPEG-1, but in DVD Studio Pro, it is also possible to use unencoded QuickTime movie files as the video content, allowing DVD Studio Pro to do the encoding once the project is ready to be built, or while working in the background.

Chapter 8

Creating Markers and Stories

DVD Studio Pro 4

Goals

By the end of this chapter, you should:

- Understand Track **Markers**
- Know **how to create Markers**
- Know **how to define their type**
- Understand **the differences** between Markers
- Understand **Stories**, and how they work
- Know **how to create Stories**
- Know **how to create Story Markers**
- Know **how to modify Story End Jumps**
- Know **how to use Stories** in DVD Studio Pro Authoring

About Markers

What Is a Marker?

Sometimes the beginning of a track isn't the only place we would like to jump to. Markers can define selected locations within the video track, to denote navigation points of interest, beginnings and endings of segments, or scenes within a film.

In DVD Studio Pro 4, you can have up to 255 Markers to your track, of which 99 may be defined as *Chapters*. You can use these Markers for a number of different purposes:

- Markers can define Chapters, like Tab stops on a typewriter.
- Markers can define starting points for Buttons over Video.
- Markers can define Story segment entry and exit points.
- Markers can define the Dual-Layer Break Point.
- Markers can define locations where Mixed-angle segments begin.
- Markers can define a location for a DVD@ccess event to occur.
- Markers can define locations to be accessed by Scripts and Buttons.

Markers as Chapters

Up to 99 Markers can be designated to define Chapters at selected locations within the video track. Chapters denote navigable locations of interest, like scene changes within a film. Most movies on DVD have Chapter Menus to access the Chapter select buttons directly, and you can use this feature the same way. Chapters give you the ability to locate to a specific location in a Track, but it is presumed you will play from that point to the Track's end. The most significant feature of a Chapter is that it can be navigated to from a Menu Button, Script, or other Connection. Chapters may also have Menu and End Jump attributes, but these should be used with great care. They can interrupt the playback of a Track if not programmed correctly.

Markers for Buttons over Video

Active Buttons may be placed over Video streams to give them menu-like interactivity. Markers created for this purpose are called Interactive Markers and define the beginning and ending location of Button Highlights that occur over MPEG video streams.

Creating Markers and Stories

Marker for a Dual-Layer Breakpoint

A special-purpose Marker, the Dual-Layer breakpoint, defines the location within an MPEG stream where a DVD-9 may safely switch layers during playback. Setting this location correctly can dramatically minimize the interruption of the flow of the movie.

Markers for DVD@ccess Functions

DVD@ccess ("DVD Access") is a web/file-enablement tool that is a part of DVD Studio Pro. DVD@ccess events may be added to Markers and will trigger when the Marker is reached during normal playback. See Chapter 17, "Advanced Authoring" for more information on DVD@ccess.

Markers to Define Story Segments

Stories use Markers to define Entry and Exit points for Story segments. Stories themselves define segments within existing Tracks that can be played as unique, individual programs. For example, a 60-minute training video Track in your DVD could be defined as six 10-minute long stories, each of which could be played as a unique individual element. Defining them as Stories would permit access to each segment individually, while preserving the ability to play through the entire 60-minute video Track seamlessly.

Creating Markers in Tracks

Markers in a Track can be created in three ways:

1. You can manually place the markers along the track's timeline.

2. You can place the markers during the editing process, if you are editing your video using Final Cut Pro 3.0.2 or later, Final Cut Express, or iMovie 3.0 or later. These markers are automatically imported into DVD Studio Pro when you import the video asset. They may also be embedded into an MPEG stream created from that video asset if it is encoded using Compressor 2 (or using the QuickTime™ MPEG encoder from earlier versions of DVD Studio Pro and Final Cut Pro).

3. You can import a text file ("Marker List") that contains a list of markers.

Regardless of the method you choose, you can still edit the Marker's position and characteristics as needed.

About Marker Locations

Creating a marker within a Track is easily accomplished using the *Track Editor Tab*. Some small limitations do exist:

- The Track Editor can only be used to set Markers in specific "legal" locations in the encoded MPEG file (called "GOP headers"), which are determined by the specific encoder and encoding scheme. Since MPEG encoders usually create GOPs of 12–15 frames, your Marker is not necessarily going to be Frame-accurate, if set within the Track Editor.

> **TIP!**
> If you need to define Marker locations that are frame-accurate, and not necessarily located at the beginning of a GOP, create those Markers in *Final Cut Pro 3.02* or later, *Final Cut Pro Express, iMovie 3.0* or later, or during encoding in Compressor.

DVD Studio Pro 4

You can define, modify, or verify Marker placement in the Track Editor, and set their characteristics using the Marker Inspector. Stories and Story Markers also have their own Inspectors, as you will see later.

Chapter 1 is always defined for you when a new Track is created. You cannot delete or move this Marker. You may add up to 254 additional Markers, of which 98 may be additional Chapters.

Marker Icons are visible in the Marker area directly above the Timeline in the *Track Editor* Tab (see Fig. 8-1).

To Add a Marker in the Track Editor:

- Click in the *Marker Area* above the Timeline at the location where you want to add a Marker. A Marker will be created at that point (see Fig. 8-1), **or...**

- Navigate the Playhead to the location you would like the Marker, then type the "M" key. (This is exactly like Final Cut Pro.) When using the "M" key, a *Chapter* Marker is created by default, colored purple.

Remember that the Playhead may be moved in controlled amounts using Keyboard Shortcuts (see Chapter 7).

Displaying Marker Info quickly

Figure 8-1 shows a "Marker tip" for Chapter 2, which reveals the Marker's name, timecode position, and attributes (this one's a *Chapter*). To cause the Marker tip to appear, hover the cursor over a Marker for a few seconds.

Markers position themselves automatically

Since Markers created in MPEG Streams in the Track Editor must line up with the GOP headers encoded into that MPEG stream, the position of a new Marker will be adjusted, or "snapped" to the GOP specified in the Track Preferences Pane. *This restriction applies even if you are setting Markers into an un-encoded Movie (.mov).* (See Fig. 8-2.)

> **NOTE!**
> Regardless of the method you choose to create Markers, you will want to verify the "Marker Snapping" behavior in the Track Preferences Pane. The "Snap to:" default is "Nearest GOP."

Figure 8-2 — Marker "snapping" preferences.

Figure 8-1 — Markers in their native habitat—the *Track Editor*.

Creating Markers and Stories

Defining Markers in Tracks with Multiple Clips

Tracks containing more than one Video Clip offer special opportunities to create Markers and to control the Playhead.

Figure 8-3a — The Clip pop-up menu, when no marker exists at Clip End.

To add a Marker at the end of a Clip where no Marker exists:

To add a Marker at the end of a Clip where no Marker exists, follow these steps:

- *Control-Click* on that Clip to display a pop-up menu (see Fig. 8-3a).

- Select *Add Marker to Clip End*, or *Add Chapter to Clip End*. *Add Chapter* will add a Marker that is flagged as a Chapter. *Add Marker* adds an unspecified Marker Type (see the *Marker Inspector*). **Or...**

- *Command-Arrow* to navigate the Clip end, and type "M" to add a Chapter.

To position the Playhead on a Marker:

- *Control-Click* on a Marker and select *Set Playhead Here* from the pop-up menu.

> **REMEMBER!**
> The keyboard can also be used to position the Playhead.

To move an existing Marker:

- Drag the marker to a new position. As you move it, the timecode of the marker's position appears in the Track Editor's *cursor timecode display field*. **Or...**

- Enter a new time in either the zero-based or asset-based timecode field of the *Marker Inspector General Tab* (see Fig. 8-5a). You can also nudge the up/down arrows to move the Marker forward and backward in the timeline. **Or...**

- Drag the slider underneath the video thumbnail in the *Marker Inspector*.

As you move a Marker, the thumbnail image in the video stream will change to reflect the picture at the new location.

To delete a marker:

- *Control-click on the Marker*, then choose *Delete Marker* in the pop-up (see Fig. 8-3b), **or...**

- Select the desired Marker, then press the *Delete key*.

To delete all markers:

- Select **Edit > Delete All Markers** in the menu bar, **or...**

223

- *Control-click* in the Track Editor *Marker Area*, then select *Delete All Markers* from the pop-up (see Fig. 8-3c).

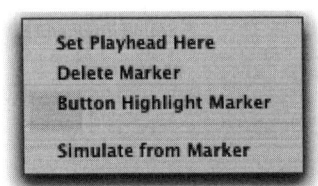

Figure 8-3b — Control-Click on the Marker to get this pop-up.

Figure 8-3c — Control-Click in the Marker Area to see this pop-up.

Importing Markers

In many cases, you will be using video assets created in Final Cut Pro 3.02 or later, iMovie 2.0 or later, Final Cut Express, or encoded in Compressor, any or all of which may have embedded Markers that can be read by and used in DVD Studio Pro. For more information on how to best use these assets, see Chapter 9, "Authoring Menus" for examples of how to import assets and connect Markers automatically.

To Import a Clip with Embedded Markers:

- Drag the Clip directly into V1 in a Track.
- If there are no Markers in the timeline, embedded Markers will appear shortly after the clip is placed into the stream area.

To Import Embedded Markers from a Clip if they don't appear:

- Control-Click on that Clip to display the pop-up menu (see Fig. 8-3a).
- Select *Add Embedded Markers*.

Markers that are embedded in the Clip (from Final Cut Pro 3.02 +, Final Cut Express, iMovie 3.0+, or Compressor) will be added to the timeline if they weren't previously automatically created on import.

Importing Markers from a Text File

A simple text file may be created with timecode points to define Markers, and imported into a Track in DVD Studio Pro.

The following simple rules apply:

- A valid Marker entry must begin with a timecode value, in 00:00:00:00 format (HH:MM:SS:FF).
- A line that does not begin with a timecode value will be ignored, and may be used to embed a comment in the text file.
- After the timecode value, a name may be included for the Marker. Use a Comma, Space, or Tab to delimit the name from the timecode.
- It is not necessary to list Markers in timecode order.

> **NOTE!**
> This file MUST be saved as ASCII plain text, and not in a Word processing format, like *.doc* or *.rtf*. Plain text! We have provided a suitable text file for you to experiment with in the DVD.

Creating Markers and Stories

About Marker Types

There are four types of markers that can be defined. The type of Marker is indicated by its color, or colors, if more than one type has been assigned to that Marker. Markers with more than one type have a split Marker symbol to display the appropriate colors.

Marker Types and their respective colors:

- *Chapter Markers*—purple
- *Button highlight Markers*—orange
- *Cell Markers*—green
- *Dual-layer Break Marker*—black dot in the middle

A Marker can be more than one of these types, or just one. A Marker's type is defined by selections made in the Marker Inspector. It's also possible to set a marker as a button highlight type by Control-clicking on it and choosing *Button Highlight Marker* from the pop-up menu.

The distance between markers has significance, as well. For example, the length of a Story Segment is defined by the distance between the Chapter Marker defined as the entry Point and the next Chapter Marker. Likewise, the length of time a Button Highlight stays "on" is determined by the distance between a Button Highlight Marker and the following marker used as the breakpoint to terminate that highlight.

Chapter Markers (Purple)

All markers start out as Chapter markers by default. While a Track is playing, the *Skip Next* and *Skip Previous* buttons on the DVD remote control can be used to skip forward and backward between Chapter Markers. Chapter Markers are also used to define Track segments for Stories.

> **NOTE!**
> Only Chapter markers can be used to form connections with Menu Buttons and Scripts.

The *Skip Next* and *Skip Previous* buttons can act differently on different DVD players. See Chapter 17 for more information on this troubling phenomenon, and steps to prevent it.

> **IMPORTANT!**
> Subtitles cannot cross chapter markers. When you build your project, any subtitle clips that reach a chapter marker are clipped at that point. Be sure to check this when using Subtitles.

Button Highlight Markers (Orange)

Button Highlight Markers are used to activate the highlights of "buttons over video." These buttons are actually configured as part of a subtitle clip. The buttons appear once you reach the marker and disappear when you reach the next marker (unless it has also been configured as a button highlight marker). This feature allows you to provide choices to viewers while they watch the track part of the title. (See Chapter 12, "Creating Subtitles" for more information on using buttons over video.)

Dual-Layer Break Markers (the Black Dot)

A layer-break Marker is used to identify the point at which the DVD player may switch between Layer 0 and Layer 1. Properly set, this layer break is invisible. Try to set this breakpoint at a natural scene change, or in Black (black is best). You might try leaving it on Auto to see where DVD Studio Pro sets it first.

225

Verification of this may be tricky, because building a disc image or volume to your hard drive will *not* create a layer break. Only when the DVD is manufactured in replication will the layer break become evident (or, hopefully, not).

These days, you may also Burn a Double-Layer DVD+R or –R disc, to test yourself on a set-top DVD Player. Please be aware that this test may not be conclusive, as double-layer recordable discs, while convenient, may not accurately reflect how your DVD-9 will play once replicated. DO NOT TRUST THE TEST RESULTS YOU GET FROM THESE DISCS.

Cell Markers (Green)

Cell markers are used for things like DVD@ccess points, or to terminate a button highlight that was switched on by an earlier Button Highlight Marker. Cell Markers *cannot* be navigated to by Menus, nor can they be used to define Story Segments. Both of those functions require Chapter Markers.

About the Marker Inspector

Once the Marker has been established, a Marker Icon will appear in the area over the timeline (see Fig. 8-1). Activating the Inspector and clicking on the Marker Icon will display the *Marker Inspector*.

Top Area

The Marker Top Area includes the Name Field and the End Jump action property (see Fig. 8-4).

- *Name:* Displays the name for the marker. Edit the name here. Each marker created will have a default name associated with it, which can be set using the *Marker Prefix*

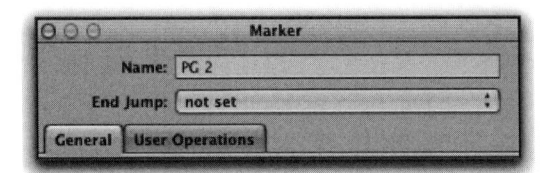

Figure 8-4 — The Track Marker Top Area.

(root) Name selection in the *Track Preferences Pane*. Each marker in sequence will be named and sequentially numbered in the order it is created—Marker 1, Marker 2, Marker 3, and so on.

More about Marker Names

> **NOTE!**
> Marker names may also be tied to the timecode location of the Marker, if the "Timecode Based" preference is checked. Further, if Auto update is checked, the name of the Marker will change if it is moved to a different location within the Track! Be careful using this feature!

- *End Jump:* Selects the project element to jump to (if any) when the current marker finishes playing (this occurs at the end of this marker's segment). While this parameter may be assigned a value in more complex scripts and projects, it is typically left at not set.

If this parameter is set, it will prevent the seamless playback of this Track in its entirety.

Marker Inspector General Tab

Marker Functions
(see Figs. 8-5a and 8-5b):

- *Thumbnail:* The thumbnail is the video frame where the marker is located. You can drag the slider beneath the thumbnail image to move the marker to a new position.

Creating Markers and Stories

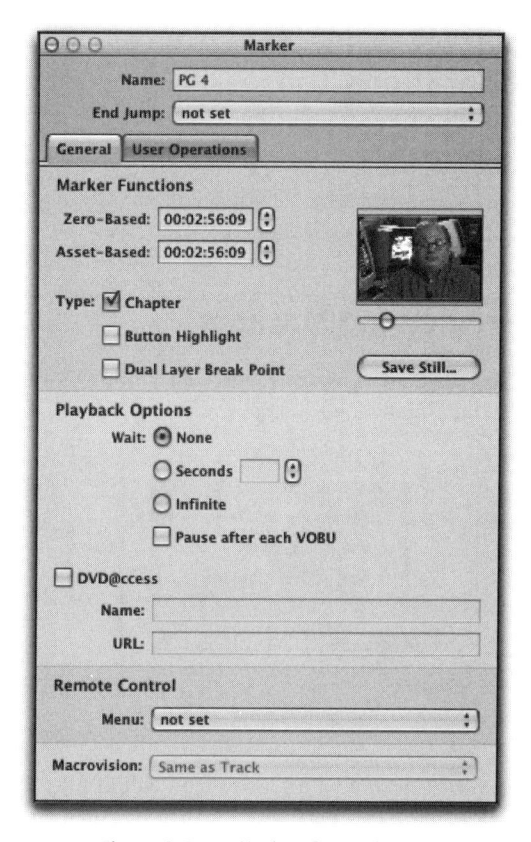

Figure 8-5a — Marker General Tab.

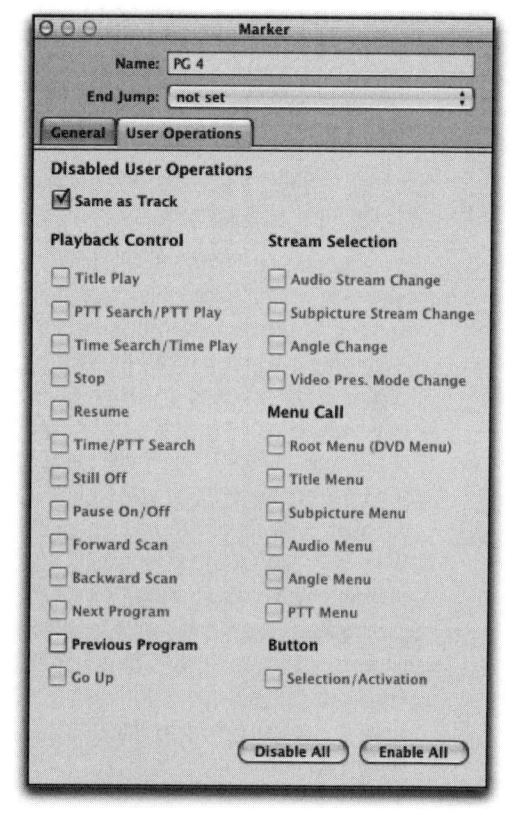

Figure 8-5b — Marker UOP Tab.

- *Save Still:* Click to save the thumbnail image as a video resolution TIFF file. You can import this file for use as a menu background or shape asset. This is useful as a way to create a still image from a frame of a video asset.

- *Zero-Based:* Shows the marker's timecode based on the timeline's zero-based mode (based on the timeline's first frame being 00:00:00:00). You can reposition the marker by entering a new value or using the up/down arrows. The new position must correspond to a GOP boundary, since the arrows jump one GOP at a time.

- *Asset-Based:* Shows the marker's timecode based on the timeline's asset-based mode (based on the timecode of the timeline's first clip or a value entered in the Track Inspector's Other tab).

- *Type:* Select the function to assign to the marker: Chapter, Button Highlight, or Dual Layer Breakpoint. You can select one or all functions.

Playback Options:

- *Wait:* Use this control to set how long the DVD player displays the last frame of the

227

marker's video before processing the End Jump setting.

- *None*: Immediately jumps to the End Jump setting.

- *Seconds*: Shows the marker's last frame for the number of seconds you enter.

- *Pause after each VOBU*: If selected, is used to pause the track's playback and wait for the viewer to start it again. According to Apple's DVD Studio Pro manual: "The length of a VOBU varies depending on whether it is a still image or full-motion video. When you place a still image in the video stream, it lasts for one VOBU regardless of its duration. With video, a VOBU can be from 0.4 to 1 second long. For this reason, "Pause after each VOBU" is generally only used on markers at still boundaries. You should avoid using it on markers with video since it will cause the playback to stop about once a second.

- *Infinite*: Shows the marker's last frame indefinitely.

- *DVD@ccess:* Allows a URL to add web-like functionality to your project if it is played on a computer. See Chapter 17 for more information on *DVD@ccess*.

- *Remote Control:* Selects a target to jump to if the viewer presses the Menu button on the DVD remote control. See Chapter 17 for more information on *Remote Control Settings*.

- *Macrovision:* Choose the Macrovision setting to apply to this marker's video. This is Analog Copy Protection. See Chapter 17 for more information on Macrovision Settings.

Marker Inspector User Operations Tab

Leaving "Same as Track" checked will make this Marker follow the Track's User Operations as set in the *Track Inspector*. See Chapter 17 for more on programming User Operations.

Previewing Markers

Previewing from Markers is (technically) not possible, but Simulation is possible.

Figure 8-6 — Control-Click on the Marker to get this pop-up.

Control-Clicking on a Marker icon will display the pop-up menu (see Fig. 8-6), from which a Simulation may be started.

> **TIP!**
> As a workaround, locate the Playhead to the desired Marker, then Control-Click in a Stream area and select *Play* from there.

Some Important Marker Do's and Don'ts

- The Start Marker is created automatically when the Track is created and needs no additional definition. It may be renamed, but not deleted.

Creating Markers and Stories

- The End of Track does not officially appear in the Track Editor. It's technically a Marker, but it cannot be moved, nor renamed.

- Markers should not be placed closer together than 1 second.

- Markers should not be placed earlier than 00:00:01:00 in the time code stream, or unpredictable behavior can result.

- Markers should not be placed closer than 1 second to the end of the track.

- Subtitles cannot continue underneath a Chapter Marker unbroken. See Chapter 12, "Creating Subtitles."

Interactive Markers (Buttons over Video)

DVD Studio Pro allows you to add buttons on top of moving video in Motion Menus, and you can do the same thing over moving video inside of a Track.

The *Button Highlight Markers* allow you to author a Button with functioning highlights to provide interactivity on top of the moving video in the Track. Since they will require a Subtitle Event to build on, see Chapter 12 for more information.

About Stories

Stories are built by creating or selecting Markers within video Tracks to define the beginnings (and possibly the endings) of Story segments and then defining which of those segments constitutes the Story, and in what order they will play. Finally, an End Jump destination is defined that will execute when all Story segments have completed playback.

You may also set a Story PreScript to execute before Story playback begins, but Pre-scripts in DVD Studio Pro have been a little unpredictable in the past—test them carefully before using them or use in-line Scripts instead. (See Chapter 13, "Basic Scripting for DVD Studio Pro.")

In some cases, Stories are simply used to add functional playlists by segmenting the video content; in other cases you can use a Track with Stories to divide a long, continuous encode into bite-sized segments that have a defined beginning and ending point.

Contrast the Story function with that of a Chapter Marker, which defines a beginning point but assumes that playback will continue until the entire Track has been played through. The ability to predetermine an exit point that occurs *before* the *Track End Marker* is the most significant feature of a Story. Stories give you a great deal of flexibility in the way you can segment and reuse your video content, without using it more than once in the project.

How Do Stories Work?

A Story is created by making a playlist of multiple Markers, each of which defines an *Entry Point* for a story segment. The Story Segment plays until it encounters a Marker *not* listed in the Story Markers list. At that point, the Story Segment completes playback, and one of two things happens:

- If another Story Marker exists in the Story, navigation is transferred to that Story Marker, and the Story continues to play; **or**

- If no additional Story Markers exist, the Story is completed, and navigation is directed to the Story's *End Jump* location.

There is no requirement that Story Markers be in chronological order, making it possible to re-order the flow of segments as you see fit.

> **NOTE!**
> If no End Jump is programmed into the Story, behavior is unpredictable, ranging from STOP to an unexpected Jump to an unspecified Marker. The specific behavior you experience depends on which DVD player you are using.

> **NOTE!**
> You cannot add a story to a specific Track unless that Track is currently selected in the Outline tab. If you "*Add a Story*" without a specific Track selected, it will add the Story to the first Track listed in the Tracks folder.

Creating Stories

To make a Story, you will need the following:

- An authored video Track

- *Chapter Markers* created within that Track to define the Story Segment beginning and ending points; you will assemble these Markers to create the Story (which is, basically, a Playlist)

- A *Destination* for the Story's *End Jump*, when it is finished playing

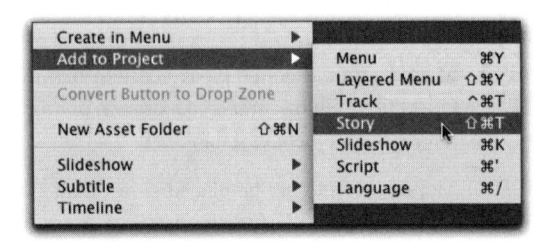

Figure 8-7 — Add Story Menu command.

Figure 8-8 — Add Story Toolbar Icon.

> **REMEMBER!**
> There is a 99 Element limit in the Disc for Tracks, Stories, and Slideshows combined. Since each DVD has at least one Track, you may not really define more than 98 Stories, but each Story may have up to a maximum of 99 Story segments defined within it.

To Create a New Story:

To create a new Story, use one of the following methods:

- Select the Track to receive the new Story, then

- Choose **Project > Add to Project > Story**, in the menubar (see Fig. 8-7), **or**

- Type *Command-Shift-T* (⌘-**Shift-T**), **or**

- Click the "*Add Story*" Icon in the toolbar, if visible (see Fig. 8-8), or

- *Control-Click* on a Track in the Outline Tab, and select *Add Story* from the pop-up (see Fig. 8-9).

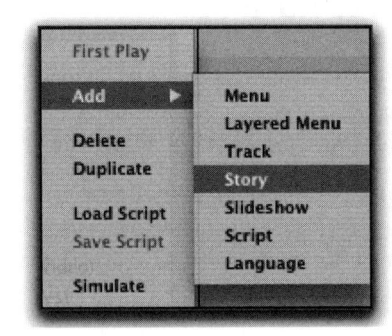

Figure 8-9 — Control-Click on a Track to Add a Story.

Once the Story is defined, you can use the *Story Editor* (Fig. 8-10), the *Story Inspector* (Figs. 8-13 and 8-14), and the *Story Marker Inspector* (Fig. 8-15) to fine-tune the characteristics of the Story.

Creating Markers and Stories

The Story Editor Tab

To create a Story, you will define a list of *Story Markers* within the *Story Editor* Tab—the order and location of the Story Markers defines how the story plays out. You can have many different Stories defined using different combinations of the same Markers.

To view the Story Editor, double-click on the Story name you wish to edit from within the Outline Tab. The *Story Editor* Tab will appear. If it remains hidden, use the Window menu to make it visible (type ⌘-**8**).

Using the Story Editor to Define the Story Segments

The left side of the Story Editor shows the existing Chapter Markers in the Current Track, while the right side of the Story Editor is where you will create a list of Story Markers, chosen from the available Chapters on the left side.

Along the top of the Story Editor is the *View*: pop-up, shown close-up in Figure 8-11a, where you can select which story is displayed in the Story Editor. The field adjacent to the *View*: pop-up displays the name of the currently selected Story (as shown in Fig. 8-10).

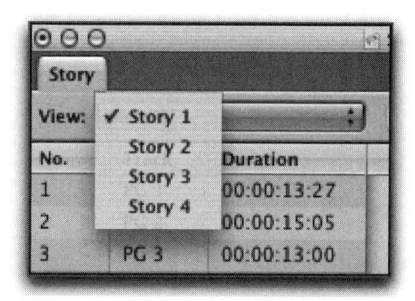

Figure 8-11a — The *View*: pop-up in action.

To Add a Story Marker

To add a Story Marker to the current Story, follow these steps:

- Drag any Track Marker from the left side of the Story Editor to the right side to create a

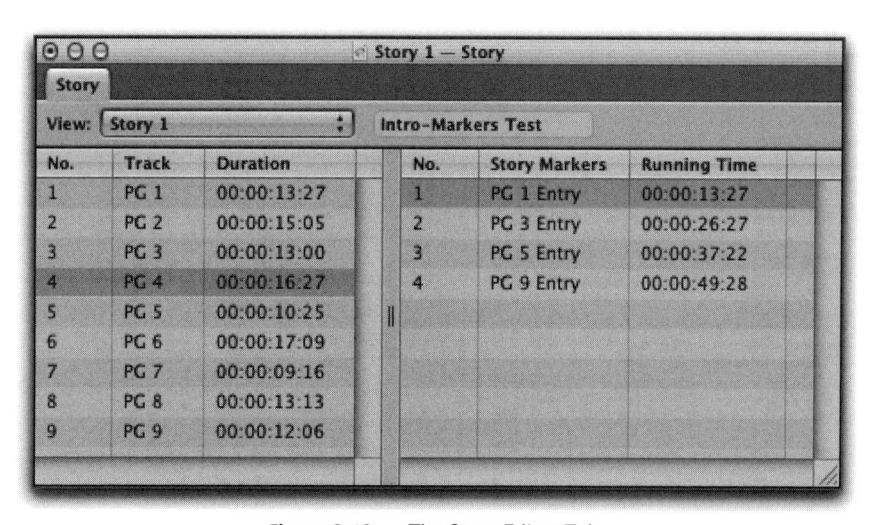

Figure 8-10 — The Story Editor Tab.

231

DVD Studio Pro 4

Story Marker (see Fig. 8-11b). The Story Markers will accumulate in the order in which they are created, but you may reorder the Story Markers as you see fit at any time. Just click on them and drag them up and down in the list order.

- Continue defining Story Markers until you have created the number of segments you need for this particular Story.

To Move an Existing Story Marker

To Move an Existing Story Marker, do this:

- Click and drag a Story Marker up or down in the Entry Point list (the right side of the Story Tab). The black line highlight will show where the moved Story Marker will appear when you release the cursor.

To Delete a Story Marker

To delete a Story Marker, do this:

- Click on the desired *Story Marker*, then hit the "Delete" key. The Story Marker will be removed from the list. Remaining Story Markers will "ripple up" to fill in the gap created by the deletion.

To Reassign the Track Marker Assigned to a Story Marker

To reassign the Track Marker assigned to a Story Marker, do one of the following steps:

- *Control-Click* on the Story Marker, and select a new Track Marker from the pop-up, **or**

- Select the desired Story Marker, open the Story Marker Inspector (see Fig. 8-15), and select a new Track Marker from the *Track Marker:* pop-up.

To Complete the Story Definition

To finish up, define the *End Jump* action for this Story in the Top Area.

Figure 8-11b — Dragging a Chapter to create a Story Marker.

232

Creating Markers and Stories

> **IMPORTANT!**
> Always set the *End Jump* action in the Story Inspector.

Defining the Story by Defining the Marker Order

The order in which the Story Markers appear determines the order in which the Story segments will play. Remember that each segment will play from the named Chapter Marker at its *Entry Point* to the next Chapter Marker it finds in its path that is not named in the Story Markers list.

About the Story Inspector

Top Area

The Story Inspector Top Area contains the *Name* field, and the End Jump Action. The default *End Jump* is "*Same as Track*" (as in Fig. 8-12), but you will probably want to redefine this for each Story.

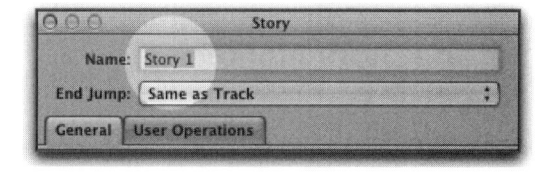

Figure 8-12 — Renaming the Story in the Inspector.

To Rename the New Story:

You can rename the Story's default Name ("Story 1") from the Outline Tab or from within the Story Inspector by doing one of the following steps:

- In the *Outline Tab*, click on the Story name, then click again to open the Name Field for editing. Edit or replace the name as needed, **or**

- In the *Outline Tab*, click on the Story name, then open the Inspector.

- In the *Story Inspector* Top Area, click in the *Name*: field, then edit or replace the name as needed (see the highlighted area in Fig. 8-12).

To Redefine the Story's End Jump Action:

- Click on the *End Jump* property in the Story Inspector, and select a new destination from the pop-up.

> **NOTE!**
> Do NOT confuse this with defining an End Jump for a *Story Marker*, using the *Story Marker Inspector*. Defining an End Jump for any Story Marker other than the last one in the story will *interrupt* the flow of the Story, and is probably NOT what you are looking to do.

General Tab

The Story Inspector's General Tab, as shown in Figure 8-13, contains certain controls that can be used to shape the Story's behavior, as detailed below:

- **Stream Options:**
  - All Audio and Subtitle streams in the scrolling list default to being enabled, but you can deselect the checkboxes to disable any Audio or Subtitle streams you do not want to be available when this story plays.

- **Playback Options:**
  - *Pre-Script*: The script you choose in this pop-up menu will run as soon as the story is selected. (See Chapter 13 for information on creating and using pre-scripts.)
  - *Wait*: Use this control to set how long the DVD player displays the last frame of the

233

DVD Studio Pro 4

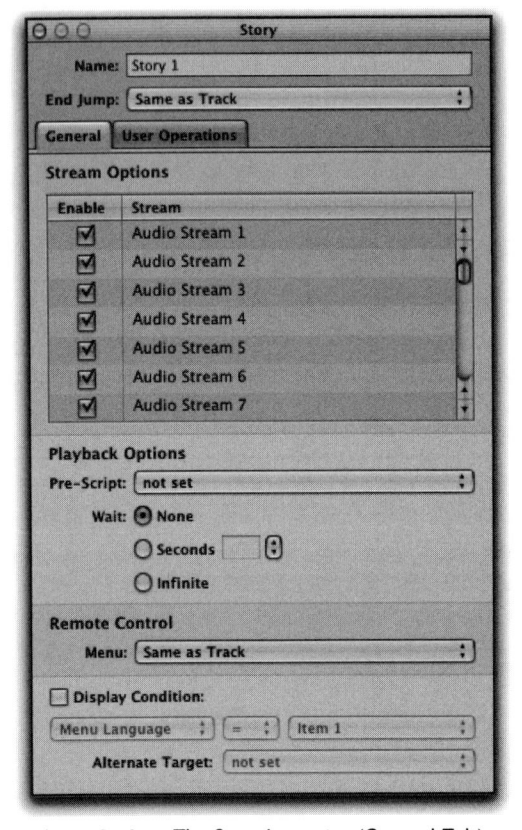

Figure 8-13 — The Story Inspector (General Tab).

story's video before exercising the End Jump setting.

- *None*: Immediately processes the End Jump action.

- *Seconds*: Displays the story's last frame for the number of seconds you enter.

- *Infinite*: Displays the story's last frame indefinitely.

• *Remote Control*: This pop-up menu allows you to set the destination element when the viewer uses the Menu Button of the DVD remote control while this Story is playing. (See Chapter 17, for more information on remote control settings.)

• *Display Condition*: Select this checkbox to set a Display Condition to control whether this story should play or not, and to define what should play if not this story. (See Chapter 17 for more information on using a Display Condition.)

The User Operations Tab

As in other Inspectors, UOPs (*User Operations*) control the function of the DVD remote Control. Certain functions may be selectively disabled using the Inspector's UOP Tab (see Fig. 8-14).

You will note that in more recent versions of DVD Studio Pro, the UOPs are grouped in logical arrangements. All of the UOPs within a group may be disabled by selecting the Group Name. Individual UOPs may be disabled by clicking on the disclosure triangle to open the Group, then disabling the desired UOPs.

You can find more information on User Operations in Chapter 17.

Setting Story Marker Properties

With the Inspector visible, selecting an Entry Point from the Story Marker list will cause the *Story Marker* Inspector to appear (see Fig. 8-15), displaying the settings for that Story Marker. Do *not* confuse this with *Track Marker* Properties, which we previously covered earlier in this chapter. Track Marker properties are *not* associated with Story Markers!

- **Top Area:**
  - *Name*: Enter the name for the story entry. This does not affect the marker's name,

234

Creating Markers and Stories

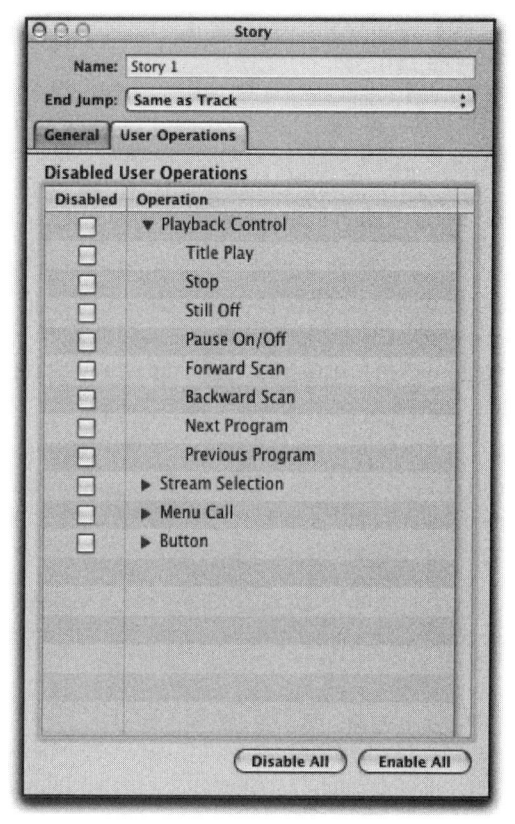

Figure 8-14 — The Story Inspector UOP Tab.

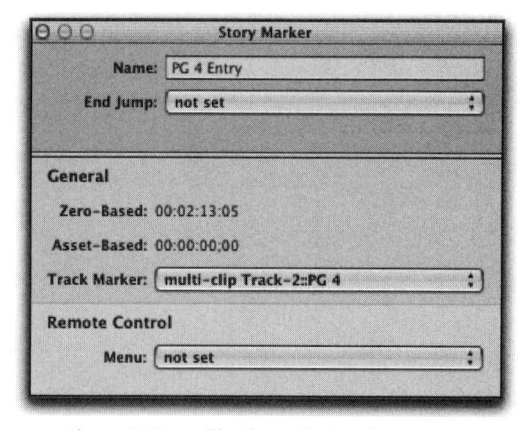

Figure 8-15 — The Story Marker Inspector.

which remains visible in the Track Marker pop-up.

- *End Jump*: Choose the project element to jump to when the entry finishes playing (this occurs just before the next entry is reached). This is typically left at "not set." In most cases, if you leave it at "not set," each story entry is automatically connected to the next entry. This setting can be used by specialized Scripts that dynamically control which parts of the Track play.

• **General:**

- *Zero-Based*: Shows the story entry's timecode based on the timeline's zero-based mode (assumes the timeline's first frame is 00:00:00:00).

- *Asset-Based*: Shows the marker's timecode based on the timeline's asset-based mode (based on the timecode of the timeline's first asset).

- *Track Marker*: Choose a marker from the source list to assign to this entry.

• **Remote Control:**

- *Menu*: Use this pop-up to choose the place in the project to jump to if the viewer uses the remote control *Menu* button. See Chapter 17 for more information on programming using the DVD Remote Control.

235

DVD Studio Pro 4

Summary

Markers

Markers are special locations within an MPEG stream that can perform one of several important functions—they may define navigation destination points, turn button highlights on and off, activate DVD@ccess functions, and define the layer break on a dual-layer DVD.

Up to 255 Markers may exist in any one Track, but of those, only 99 may be defined as Chapters. This is an important limitation to keep track of, as only Chapter Markers may define Story Markers.

Markers may be defined during authoring, but only at an existing location known as a GOP Header. This means a Marker may be off by as many as 11 to 14 frames from the precise location desired. To solve this, Markers may be defined into QuickTime movies prior to encoding if you are creating content using FCP 3.02 or later, iMovie 3, or encoding with Compressor 1 or 2.

In all cases, Markers define entry points into an MPEG Stream, but unless they are used in a Story, they do NOT define an exit point.

Stories

Stories are collections (playlists) of up to 99 segments, each of which is defined by a beginning and ending Chapter Marker within the Track in which the Story is defined. Stories may *only* be defined within the Track that contains the video content you wish the Story to play.

Once a new Story has been defined, the Story Editor is used to create the playlist of up to 99 Story Segments and to define the End Jump Action as well. It also enables Audio and Subtitle streams.

The Story plays the Segment list in order, beginning at the first Story Marker. Each segment plays until it encounters a Marker that is NOT named in the Story Marker playlist.

Once all segments have been played to completion, the Story's End Jump action is processed.

While Markers allow you to jump *into* a Track, you typically play to the Track's End Marker. Using a Story allows you to Jump *into* a Track and exit the Track at some defined location *prior* to the ending.

Chapter 9

Authoring Menus

DVD Studio Pro 4

Goals

By the end of this chapter, you should be able to:

- Understand the role of the **Menu**
- Understand the role of the **Menu Properties**
- Create **Still** Menus, **Audio** Menus, **Motion** Menus
- Author a still image file into a **Still Menu**
- Author a layered PSD file into a **Still Menu** Item
- Author an Audio stream in a **Still Menu**
- Set Menu parameters to make a **Motion Menu** loop
- Set Menu parameters to make an **Audio Menu** loop
- Create a composite **Still Menu Picture** using PSD layers
- Author an MPEG stream into a **Motion Menu** Item
- Author an Audio stream into a **Motion Menu** Item
- Set the Loop parameter to make a **Motion Menu repeat**
- Understand how the **Button Container** works
- Understand what **Buttons and Highlights** are
- Understand how to set basic Highlights into a **Menu**
- Understand how to **map Highlights using a Subpicture Overlay**
- Understand how to make **Switched Photoshop Layer Menus**

About DVD Menus

If Tracks are the meat and potatoes of your DVD projects, then Menus without a doubt are their heart and soul.

One of the most powerful features in DVD, the interactivity between the user and the DVD itself, takes place in Menus.

Menus can provide an almost infinite number of functions within your DVD—from a simple jump to a video Track selection to initiating a carefully scripted set of actions and behaviors. Menus can also provide a welcoming or familiar "environment" or "experience" for the DVD viewer, coloring both their expectation and their enjoyment of the DVD. In fact, a quick perusal of the available Hollywood features on DVD will reveal a number of DVDs whose menu environments not only reflect the feature film, but in many cases carry the theme, look, and feel all the way through. In these cases, the DVD menus have become an extension (and many times an amplification) of the original feature film.

But your DVD doesn't have to be a Hollywood feature (or even a *Bollywood* feature) to be a candidate for a great set of menus. Practically *any* DVD can benefit from a set of well-designed menus. But to understand how to make menus, you first need to understand what they are, what they do, and how they do it.

Authoring Menus

About DVD Studio Pro's Menu Editor

As we have hinted at in earlier chapters, one of the most significant features of DVD Studio Pro 2, 3, and 4 lies in the Menu editing capabilities of the new version. These versions have integrated some *serious* menu creation capabilities, including a very powerful video compositing engine to make motion menus.

Menu Creative Functions Are Built-in

In earlier DVD authoring applications, heavy reliance was made on outside applications like Adobe Photoshop or After Effects to create the background still images or motion graphics. Menu assets needed to be built as complete images before they could be imported as assets and used in a DVD. As a consequence, it was very cumbersome to make any changes to a menu once it had been imported and used. As you might expect, the menu would need to be modified in its original application, then re-exported for use in the DVD. *Not* intuitive at all...

In DVD Studio Pro 4, the Menu Editor has not forsaken its original functions. You can still use the Menu editor to assemble graphics images and overlay files that have been built to the proper graphics specifications (see Chapter 15, "Graphics Issues for DVD Images").

The DVD Studio Pro Menu Editor is a very powerful compositing tool, capable of assembling background pictures from graphics or video, as well as adding stylized text and graphics to the background picture being built. This combination of tools allows for a level of menu control and creativity that was just not possible in DVD Studio Pro 1.

What the New Menu Editor Can Do

The Menu editor allows you to create a menu from scratch using any of the following items:

- Background Pictures (made from Stills, PSD files, or Motion assets)
- Shapes (tiny pieces of graphics images)
- Patches (graphics with built-in motion effects)
- Drop Zones (graphics placeholders that can contain still images or moving video and may also supply special effects)

You may add many of these items and composite these into the final menu. In addition, the Menu editor no longer requires MPEG video files, but can composite QuickTime background movies and movies in Drop Zones into a finished MPEG Motion Menu. Making incremental changes is as simple as then touching up the Menu inside of DVD Studio Pro 2 and rebuilding the project.

Round-Trip Links to Outside Editors

You may use external applications in conjunction with DSP due to the ability to easily link DSP Assets to the editing applications that created them. As an example, a Motion Menu created in Apple *Motion*, can be reopened and tweaked in Motion, then saved and used immediately in DSP with the updates. This "round-trip" editing allows similar linking to Final Cut Pro for video editing, Adobe After Effects for motion compositions, and Adobe Photoshop for still or layered menus.

DVD Menu Basics

DVD *Menus* are the "traffic cops" of DVD navigation—they are the points at which the DVD user can make choices that actively direct or modify the course of the DVD's playback.

These choices are made in the Menus by selecting and activating one of the Menu's up to 36 Buttons, which represent the available destination Elements, or "targets" in that DVD menu. Once a button is Activated, the *connection* (*command*) programmed into that Button's "Target" property passes the DVD navigation over to the selected destination Element. You can view the Button's Target in the Button Inspector or in the Connections Tab.

DVD Menus can also use their Buttons to change certain DVD playback parameters—Audio Stream selection, Angle Stream selection, and Subtitle Stream selection (and Subtitle visibility). If the buttons themselves don't actively change these parameters (by using the Streams properties in the Inspector before jumping to the desired Track), these attributes can be changed in a specially created Menu. These menus are generically referred to as "Interactive Menus"—they can be known as Chapter Menus, Language Menus, Angle Menus, Audio Menus, Setup Menus, etc. When using Interactive Menus, in many cases, the buttons are linked to special Scripts that set or modify values held in one or more of the General Parameters or System Parameters (memories). You can read more about SPRMs and GPRMs in Chapter 13, "Basic Scripting for DVD Studio Pro 4."

Menu elements in DVD Studio Pro can also be used to simply display images—either graphics or video, although video is usually displayed using a Track element. One advantage of Menus over Slideshows for the display of a sequence of still images is the ability to add buttons over the image in menus, allowing for greater interactive control of the DVD.

DVD Studio Pro 2 Slideshows don't allow buttons to be created on top of the images, so you are limited to timing the images using the pause function, or using the Next and Previous keys on the DVD remote control to move forward or backward through the Slideshow.

Understanding Menus in General

A DVD Menu is like a sandwich, in a way. The three layers of the menu include:

- The *Background* picture, which may be a still image or a movie;
- An optional *Subpicture Overlay* (which contains pixels that determine the menu highlights that will be displayed), and
- The *Buttons* themselves, each of which define a "hot zone" within the full SD menu image of 720 × 480 (NTSC) or 720 × 576 (PAL) or the larger HD DVD menu sizes.

The *Background* picture is used to convey the menu information to the DVD user; this may be in the form of text or graphics icons, or both—the goal of a good menu is to clearly communicate what functions are possible using the buttons on that particular menu.

The *Subpicture Overlay* can be used to define the menu highlights that are visible within the Button hot zone when the Button is in the Selected or Activated state.

Buttons are created to define the "hot zones" on the Menu image, each of which specifies the area where highlights may be displayed for that button. Buttons also generally contain a command to navigate to a subsequent Element. This Element then performs the function described for that button in the background picture.

Authoring Menus

Understanding Button "States" in Menus

To understand the function of the Menu better, it's helpful to appreciate how the buttons actually signal their different "states" of readiness, sometimes called their "modes."

Buttons may have one of three distinct states:

- *Normal*—Normal state is the absence of either the Selected or Activated states described below. In this state, the buttons are usually transparent and do not display a highlight—instead, they display the text information or graphical icon which informs the user as to this button's function in this menu.

- *Selected*—On each DVD menu, there is always ONE button that is in the Selected state. This is required by the DVD specification. As the DVD user navigates through the menu, the currently "selected" button should display a noticeably different appearance, indicating that this is the button where the cursor currently resides. "Selecting" a button first is a requirement for activating it. (Navigation in the menu is controlled by the DVD Remote Control, or by the computer cursor [mouse] on a DVD-equipped Computer).

- *Activated*—Activated state exists when the OK or Enter key is pressed *while* a button is selected. In the Activated state, the button should change to a different color or appearance from the "selected" state highlight, to identify to the DVD user that his or her activation command has been acknowledged. Of course, the activated state only exists for a very brief period of time before the button command is processed, and the DVD navigates to a new Element.

Buttons can jump to a Track or Slideshow for playback, a further Menu for a more refined selection, or to a Script, to perform the indicated function the button represents (like selecting an Angle, Audio, or Subtitle Stream for playback).

In all cases, the *Color Settings* control the visible colors for each of the states. In many cases, Normal is left transparent, leaving the Background picture to show the "normal state" graphics for all of the buttons. Selection and Activation modes are generally programmed to provide different colors, to avoid confusing the DVD viewer.

If a Subpicture Overlay file is NOT used to map the highlights, the entire button may take on a highlight color and transparency to indicate Selection or Activation.

Types of Menus

In DVD Studio Pro, different types of Menus can be built using different kinds of Assets. How the menus will appear to the end user is then a function of what type of assets are used to assemble them.

Only *Standard* menus use the "Simple Menu Highlighting" technique described in the previous paragraphs. But another kind of Menu can provide a more graphically rich menu—*Layered Menus*. Let's start with the simple ones, created using the *Standard Method*. Later, we'll look at *Layered Menus*, which use Photoshop .PSD files to determine the highlights.

What Constitutes a "Standard Method" Menu?

A Standard Method Menu is the three-piece sandwich we described earlier, created by assembling:

- A *Background Picture* of some kind (a still or a video clip),
- A *Subpicture Overlay* file (optional, used to define the menu's highlights), and
- The *Buttons* themselves.

DVD Studio Pro calls this method of assembling a menu with a Subpicture Overlay and Buttons the *Standard Method*, using what I often refer to as Simple Menu Highlighting. This method is, in fact, the method used on many of the commercial DVDs released by Hollywood studios. (I'll bet you will notice it more often now when you are watching DVDs.)

The *Background picture* may be a still or a video clip, but it must be the proper size and frame rate for the video standard used in the project (720 × 480 for NTSC, or 720 × 576 for PAL). Menus may also use the newer HD DVD sizes 1280 × 720, 1440 × 1080, and 1920 × 1080.

The *Subpicture Overlay file*, although optional, is a very important component of the menu. The pixels in this file are used to define the specific highlights desired for each button. Its size is also defined by the video standard of the DVD project, just as with the Background Picture above.

The *Buttons* are actually a set of numerical coordinates, which define the rectangular "hot zone" location within which that button has control over the Menu screen. Once a particular button is *Selected* or *Activated*, the Menu Highlight for that *button* state is displayed. To accomplish this, the DVD player activates the appropriate pixels contained within that area of the Subpicture Overlay file. The colors that get displayed on-screen for those pixels are usually not the colors of the pixels in the Subpicture Overlay file, but rather are a combination of that pixel's color, and the *Color Settings* for that *Menu* (or *Button*). The "key color" of the pixels in the Subpicture Overlay may be determined by either Grayscale values, or by using the colors Blue, Red, Black, and White, as we will see in just a bit. If a Subpicture Overlay file has not been specified, the entirety of the button shape can be highlighted, but this technique may not be effective for every menu.

Stillframe Menus

Standard Menus that contain a still image as the Background picture are generically called *Stillframe Menus*. In DVD Studio Pro, practically any graphic format can be used—many types of flat graphic image files and even a layered .PSD can be used to create the Background picture (see Chapter 15, "Graphics issues for DVD Images").

Buttons can then be created over the background picture, and Connections made using the Button Inspector, either by Control-Clicking the Button or by using the Connections Tab.

The simplest stillframe menus don't even *use* a Subpicture Overlay file to create the Menu Highlights. Instead, the Color Settings act on the shape of the button hot zone itself, causing it to take on the highlight colors as programmed. This is a pretty rudimentary method of highlighting, but it can be effective in certain simple menu designs—in fact, you will find that the gorgeous motion menus in the *Gladiator* DVD actually use this very simple technique, and quite successfully, at that!

Stillframe Menus with Audio

Ordinary Stillframe Menus (not layered menus) created using any graphics file (even .PSD layers) may contain audio streams. In that case, they may be referred to as Audio Menus. The audio file adds an extra level of production value to the menu, but not as much as a video clip in the background picture. Menus containing an audio file add a *time* compo-

Authoring Menus

nent to the menu, meaning it's no longer just an MPEG still image looping infinitely. Every Audio Menu will have a running duration and, more importantly, it will require the *loop* property to be set so that the menu will continue to display even after one full playback of the audio file.

Motion Menus

Motion Menus contain Video streams or movie files as the background picture. Motion Menu images have traditionally been created in an external application, like Final Cut Pro, Adobe After Effects, discreet Combustion, Avid, Media100—even the QuickTime player—any program that could do video compositing. In DVD Studio Pro 2, this is no longer required. Using the new Menu editor, it is now possible to create a Motion menu by compositing the desired movie, graphics, and text elements inside of DVD Studio Pro, without resorting to an external application.

Motion Menus can use a Subpicture Overlay to display Highlights, or they may display highlights without using a Subpicture Overlay. They may also contain Audio programs.

Adding a Subpicture Overlay

Adding a Subpicture Overlay file to the menu allows you to define the highlights of the menu with much greater precision, since the pixels contained within the Overlay file are used to define the shape of the menu highlights displayed. So while you don't *need* a Subpicture Overlay, your highlights will potentially look *much* nicer if you do use one. There are three different systems for defining highlights: Simple Colors, Advanced Colors Chroma, and Advanced Colors Grayscale.

About the Subpicture Overlay File

The Subpicture Overlay file is a separate flat file, or a single layer of a Photoshop PSD file. The Subpicture Overlay file may be created using the *Simple* Color scheme, using just Black pixels to indicate highlights, or by either of the *Advanced* Color methods. The *Chroma* scheme defines highlights by using the four-color DVD standard (Black, Red, Blue, and White) while the *Grayscale* scheme uses grayscale values of 100% (Black), 66% (Dark Gray), 33% (Light Gray), and 0% (White) to define the highlight shapes. Regardless of whether you use the Chroma scheme or the Grayscale scheme, it is typical to avoid using the white color controller, because it maps to the entire menu screen and is difficult to use effectively.

About Color Settings

One of the more confusing aspects of menu highlights is that the colors or grayscales specified in the Subpicture Overlay file are not usually the colors that will be seen in the finished DVD menu. This is because the DVD Menu Editor allows you to *map* or *assign* the colors or grayscale values in the Subpicture Overlay file to any of a number of different colors in the finished menu. If you are familiar with the concept of chroma-keying in video editing, you already know how one color may be used to "key in" or "matte in" a different color or image. It is essentially the same in DVD Menu highlighting: The Color Settings define (or map) which color will be displayed in the finished DVD menu for each of the colors or grayscale values in the Subpicture Overlay file. The Color Settings can also define the opacity of this color, so you have lots of possibilities for your menus. To allow for variations in Menu design, there are three different sets of Color Settings, and Buttons may be assigned to any one of the three Color Sets.

243

DVD Studio Pro 4

Layered Menus

Layered Menus are very different from Standard Menus. Layered Menus use a special Photoshop .PSD file that contains layers within it to define the unique graphics for each Button, in each of the different Button Highlight states—Normal, Selected, and Activated. This is a very special kind of highlighting that is only possible in a few DVD Authoring programs, and DVD Studio Pro is one of these.

To create a Layered Menu, the special Photoshop file is authored into a *Layered Menu* Element instead of a Standard Menu, and the Background Picture is defined from within the available layers. Once the buttons have been created, the *Layered Button Inspectors* are used to define which layers will display for each state of each button. In many cases, the graphic image for the *Normal* state is provided in the background picture of the Menu element and not necessarily by a unique layer for each button. While the *Selected* state should always have a unique graphic for that display, it is possible to bypass the *Activated* state and not to create a special graphic for it.

Pros and Cons of Layered Menus

One positive attribute of Layered Menus is that since each button can display a full 24-bit graphic image, rich graphical images can be created, and problems traditionally associated with the 8-bit Subpicture Overlay file can easily be avoided.

The Layered Menu does have some disadvantages, though, that should give one pause before jumping in whole hog: Layered Menus accomplish their magic by building multiple menus (invisibly), one for each possible combination of button states. For example, in a four-button menu, you would have nine unique images if you programmed a layer for each button in both Selected and Activated states. As you navigate around on a Layered Menu, you are actually navigating to one of these other invisible menus, each of which presents a graphic picture of some valid combination of button highlights. This navigation is *slow*. So slow, in fact, that Layered Menus can be noticeably slower to navigate on some DVD players.

Common Factors

What you have hopefully noticed in this discussion is that ALL menus contain a Background Picture of some kind and Buttons to activate the desired functions. Most Menus usually use a Subpicture Overlay to create the all-important Menu Highlights, but some do not.

About the Menu Editor

The *Menu Editor* interface exists in the *Menu Tab*. In addition to the Menu editing area (shown with rulers enabled in Fig. 9-1), it contains the *View:* pop-up, used to select which Menu is currently being edited (at top left), and the *Menu language select* pop-up (to its right). DSP can be used to create Menus in multiple languages.

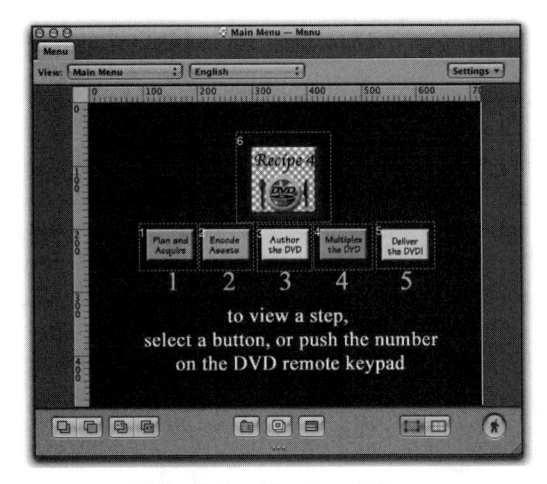

Figure 9-1 — The Menu Editor.

Authoring Menus

Menu Editor Pop-Up Menus

The top area of the Menu Editor contains (from the left):

- *View pop-ups*: When the Menu Editor Tab is visible, you can select which Menu to edit by using the leftmost pop-up menu, instead of double-clicking it in the Outline Tab. The right pop-up allows you to choose a language to configure for this menu (assuming additional Menu languages have been created in the Outline tab). *See Menu Languages*, later, for more information.

Menu Editor Settings: Pop-Up

At the top right is a comprehensive *Settings:* pop-up (see Fig. 9-2).

Figure 9-2 — The Settings Pop-up revealed.

The *Settings:* pop-up contains the following functions that control how the Menu Editor displays the Menu currently being edited:

- *Auto Assign Buttons Now* and *Auto Assign Buttons Continuously*: DVD Studio Pro can be set to automatically assign button navigation settings to the current menu configuration while authoring.

- *Display Background, Display Overlay,* and *Display Composite*: The Menu editor can be set to display either the menu background or the menu overlay by itself or both as a composite image. The "Q" key is a shortcut for this function.

- *Title Safe Area* and *Action Safe Area*: Discussed in more depth in Chapter 15, the Title Safe and Action Safe areas can be overlaid on the Menu editor display, to determine if your menu contents are all placed within the "safe display areas." All text and titles (and buttons) should be within Title safe, and all other menu graphics should be within the Action safe area. Keyboard shortcut *Command-Shift-E* (⌘-**Shift-E**) can toggle the Title safe area on and off (the DSP manual *still* erroneously states this as Command-E), and *Command-Option-E* (⌘-**Option-E**) can toggle the action safe area on and off.

- *Square Pixels* and *Rectangular Pixels*: Menu images may be viewed as square pixels or displayed in the rectangular video pixel aspect ratio (as either 4:3 or 16:9, depending on menu size). Use the keyboard shortcut P to toggle the display type.

- *Show Single Field*: Allows for smoother playback of field-based video sources, but may compromise playback of frame-based video.

Menu Editor Bottom Tools

Along the bottom of the Menu editor is an additional set of Menu tools, which can control button func-

245

DVD Studio Pro 4

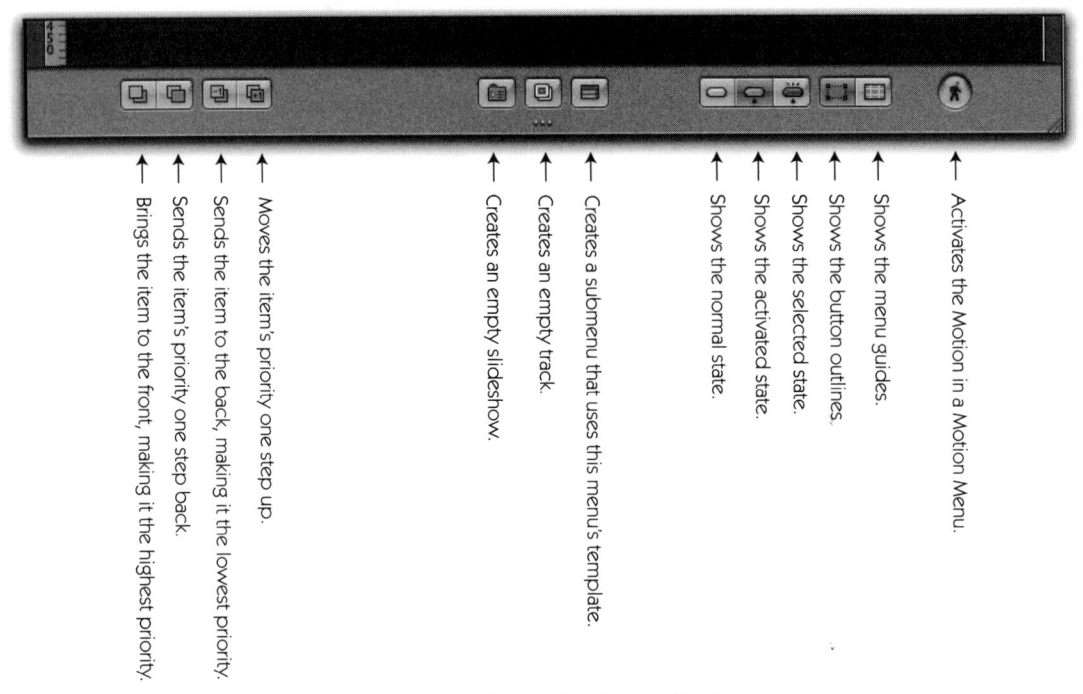

Figure 9-3 — The Menu editor Bottom Tool pane.

Labels (left to right):
- Brings the item to the front, making it the highest priority.
- Sends the item's priority one step back.
- Sends the item to the back, making it the lowest priority.
- Moves the item's priority one step up.
- Creates an empty slideshow.
- Creates an empty track.
- Creates a submenu that uses this menu's template.
- Shows the normal state.
- Shows the activated state.
- Shows the selected state.
- Shows the button outlines.
- Shows the menu guides.
- Activates the Motion in a Motion Menu.

tions, add additional elements to the current Menu, set the current button highlight state, and specify if a motion menu is displayed "in motion" or "at rest" (Fig. 9-3).

- *Arrange controls*: These buttons allow you to change the priority of the selected item, letting you control which items have a higher priority by bringing them forward, or by sending lower-priority items to the back.

- *Add Submenu, Add Slideshow*, and *Add Track buttons*: These buttons allow you to create a new element in your project and add a button to the current menu that connects to them. You can also press *Command-Option-Y* (⌘-**Option-Y**) to create a submenu, *Command-Option-K* (⌘-**Option-K**) to create a slideshow, and Command-Option-T (⌘-**Option-T**) to create a new track. The new button added to the menu uses the menu's default button style.

- *Button state selections*: These buttons allow you to see the selected button in any of the three states (*Normal, Selected*, or *Activated*). You can also use keyboard shortcut key W to step through these options.

- *Button Outlines button*: This feature can display or hide the button *outlines*, which show each button's hot zone and the button name. Turning off the button outlines for buttons without an asset or a shape can make them disappear in the Menu.

- *Guides button*: You can show or hide the menu guidelines created by dragging from the rulers into the Menu Editor. This does not affect the dynamic guides that appear

246

Authoring Menus

when you drag items in the menu. Keyboard shortcut *Command-semicolon* (⌘**-;**) can be used to show or hide these guides.

- *Motion button*: Applies to standard menus only. Use this button to start or stop playback of video and audio assets assigned to the menu. This is useful for motion menus and still menus that include audio, providing a preview of how the motion elements appear. You can also press the *Space bar or Command-J* (⌘**-J**) to turn the motion on and off.

> **NOTE!**
> This button does not work the same way as the similar button in iDVD. In DVD Studio Pro, the Motion button only affects the menu *preview*, and will not affect the motion of the finished menu.

The Bottom Tool pane may also be closed, if it is no longer needed. To close it, click on the three dots in the center bottom of the editor. (See Fig. 9-4.)

Getting Started with Menus

Adding a Standard Menu

Each DVD Studio Pro project contains one empty standard Menu to begin with, by default. Additional Menus are easily added.

To add a Standard Menu, use one of the following methods:

- Select **Project > Add to Project > Menu** in the menubar, **or**
- Click *Add Menu* in the toolbar (if visible), **or**
- Type *Command-Y* (⌘**-Y**) on the keyboard, **or**
- *Control-Click* in the Outline Tab, then select **Add > Menu** in the pop-up.

Adding a Layered Menu

Additional layered Menus are just as easily added as standard menus.

To add a Layered Menu, do one of the following:

- Select **Project > Add to Project > Layered Menu**, **or**
- Click *Add Layered Menu* in the toolbar (if visible), **or**
- Type **Command-Shift-Y** (⌘**-Shift-Y**) on the keyboard, **or**
- *Control-Click* in the Outline Tab, select **Add > Layered Menu** in the pop-up.

Remember that layered menus will require a properly configured Photoshop file with layers to create not only the menu background picture, but also the button highlights for the *normal, selected,* and *activated* states for each button.

Figure 9-4 — Menu editor Bottom Tool pane closed.

247

DVD Studio Pro 4

Adding Submenus

Submenus may be added to a current Menu by clicking on the *"Add Submenu"* button in the bottom Menu Tools (see Fig. 9-3). This will add a menu of the same type and configuration as the current menu. It will also add a button to link to that submenu.

Menu Names

By default, the first Menu (one is always created in a new DVD Studio Pro project) is named "Menu 1." When subsequent new menus are created, they are given a default name of "Menu X" (where X equals the next number in sequence, starting with "Menu 2").

Renaming Menus

To rename a menu, do one of the following sets of steps:

- Display the *Outline* Tab if not already opened;
- Open the *Menus* folder if not already opened (click the Triangle);
- Click on the desired Menu in the outline tab to select it, then
- Click again on the Menu name to select the name field, then
- Type a new name or modify the name as desired, **or**
- Display the Outline Tab if not already opened;
- Open the Menus folder if not already opened (click the Triangle);
- Click on the desired Menu in the outline tab to select it, then
- Show the *Menu Inspector*;
- Click in the *Name* field of the Menu Inspector to select it, then
- Type a new name or modify the name as desired.

Editing a Menu

Before a Menu can be edited, it must be opened in the Menu Editor.

To open a Menu into the Menu editor:

- Display the *Outline* Tab if not already opened;
- Open the *Menus* folder if not already opened (click the Triangle);
- Click the desired Menu in the Outline Tab if the Menu Tab is opened, **or**
- Double-click the desired Menu in the Outline Tab to force the Menu Tab to display, **or**
- If the Menu editor Tab is already displayed, select the desired menu from the *View:* pop-up at the top.

> **NOTE!**
> The speed with which the display will refresh and change is VERY dependent on the system CPU speed—expect slow changes if your system has less than the 733-MHz CPU speed recommended for DVD Studio Pro.

Once the desired Menu has been opened into the Menu editor, you can proceed to create or edit the menu. To modify the Menu layout at any time, you only need to reopen the desired menu in the Menu editor tab.

Authoring Menus

Adding Assets to Menus

Menu assets can be added by several means, including:

- Dragging an Asset to the Menu Editor from the Palette, or Finder,
- Specifying a Menu asset in the Menu Inspector (if the Asset has already been imported into the Asset Tab), or
- Using a preset Template or Style from the Palette.

Since DVD Studio Pro 2, the Menu editor has had very special tools that can be used with Drag-and-Drop editing. Let's take a look at *Drop Palettes* and how they work in DVD Studio Pro. You will find a complete summary of how Drop Palettes work at the end of this chapter.

Drop Palettes for Drag-and-Drop Menu Editing

DVD Studio Pro has a remarkable Menu Editor with a very interesting twist: *Drop Palettes*. These unique context-sensitive tools pop up in the Menu Editor each time you drag something into the Menu Editor Tab and pause while holding it.

Which *Drop Palette* you get depends on what Asset or project Element you drag, what kind of Menu you are currently editing (Standard or Layered), and where you pause the cursor (e.g., on the empty area, on a *Button*, or on a *Drop Zone*). Different Assets and Elements will be able to do different things in different places, so you will want to take a moment to carefully read the options presented in the Drop Palette.

Of special note is the behavior of DVD Studio Pro when dragging a Video Asset—if you have selected the "*Find Matching Audio when dragging*" preference in the Track Preferences, these drag-and-drop operations will also drag a matching audio file if one can be found in the same folder as the Video Asset. Of course, you can also select a Video Asset with Audio from the Asset Tab and drag them to the Menu.

How the Drop Palettes Work

The Drop Palette will not appear unless you pause and hold the cursor in place for a few seconds. This *Drop Palette delay* time can be set in the Menu Preferences Pane.

> **HINT!**
> If the Drop Palette you expect to see doesn't appear, you are probably not paused in the proper location in the Menu.

Moving the cursor will cause the current Drop Palette to disappear and, when you again pause, a new one will appear. Remember, this function is *context sensitive*, so changing your pause point from the *Empty Area* to a *Button* or *Drop Zone* will change the Drop Palette that appears.

In each case, when a Drop Palette appears, a *default* option appears first in the list, colored orange; other options, if any exist, are listed below the default; displayed options may also show secondary operations in smaller type. Some Drop Palettes don't have more than one option, but most do.

- The *default option* will be applied to the operation if you don't hold the cursor long enough for the Drop Palette to appear.

When dragging multiple Assets or Elements to the Menu and invoking an option that creates multiple Buttons, the resulting Buttons will be aligned in tight proximity and overlapping—just as if you had used the *Duplicate* command multiple times. This can cause problems later on.

- You will want to separate the Buttons to avoid *overlap*, which will cause problems in the proper selection of the Menu Buttons.
- Drop Palette actions that create new Buttons will use the *default Button Style* for that Menu. Buttons may create thumbnail images, if the default style is set to do so.

Adding a Menu Template from the Palette

DVD Studio Pro contains 40 prepared Menu templates for your authoring convenience, and any one of them may be selected and added to a new menu using the Palette.

To apply a Menu template:

- Open the desired Menu Element into the Menu editor,
- Click the desired Template to select it in the Palette, then
- Click the "Apply" button in the Palette, **or**
- Drag the desired Template to the Menu Editor and hold it
- Select "Apply to Menu" or "Apply to Menu, Add All Buttons" from the Drop Palette.

About the Menu Inspector

The Menu Inspector is used in conjunction with the Menu Editor, to display and allow adjustment of the Menu's properties. The Menu Inspector has a Tabbed display, like many of the other Inspectors we have looked at. Each of the Tabs displays information concerning a specific part of the Menu's functions.

The Menu Inspector Top Area

The Top Area contains the Menu name field. You will notice that the top area for a Standard Menu includes the Background file name, but the Top area for a Layered Menu does not, as can be seen in Figures 9-5a and 9-5b. There are also other differences between the two types of Menu Inspectors, as we will see.

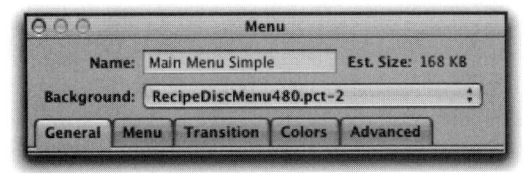

Figure 9-5a — Standard Menu Inspector Top area.

Figure 9-5b — Layered Menu Inspector Top area.

The Menu Inspector General Tab

The Standard Menu Inspector General Tab contains vital data regarding the function of the basic Menu. The display will appear the same when either a flat image file (TIFF, PICT) or layered .PSD is used for the Background picture (see Fig. 9-6a), but when a layered image file is used as the menu Background, the layers in the .PSD are displayed in the "Background Layers" areas of the Menu Tab (see Fig. 9-6b). Depending on how this menu is created,

Authoring Menus

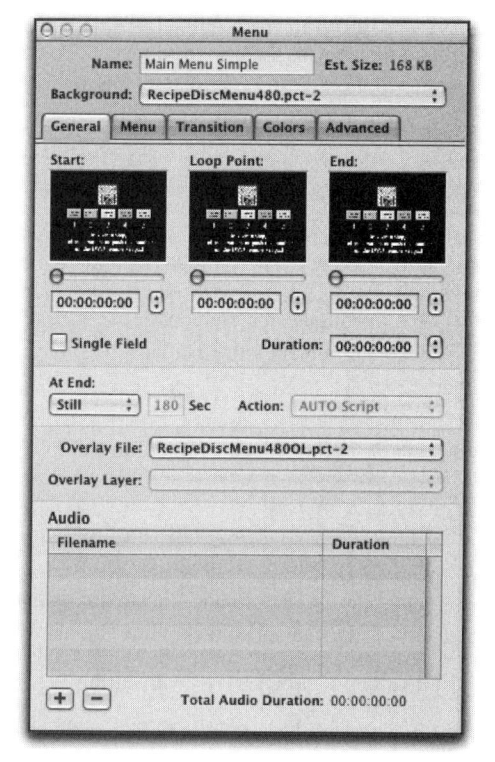

Figure 9-6a — Std Menu with TIFF.

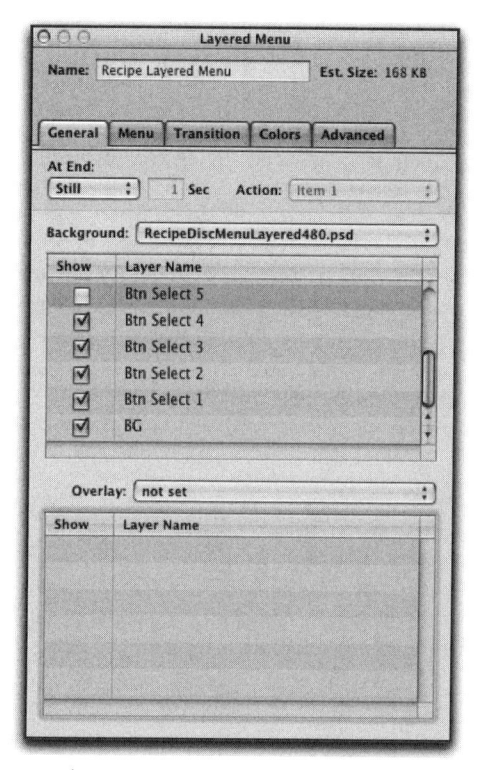

Figure 9-6b — Std Menu with PSD file.

Background Layers may default to being visible (checked) or not, and unwanted layers may be deselected (unchecked) to be turned off or made inactive on this menu.

Notice that the Standard Menu Inspector is organized in sections, with Motion Menu loop settings at the top, Timeout Action properties and General settings in the middle, and the Audio Selection area at the bottom (see Fig. 9-6a for a Standard Menu with Flat File). When a layered PSD file is used in a Standard, nonlayered Menu, the Background Picture is assembled by selecting PSD Layers to be composited for display as the finished Background Picture (see Fig. 9-6b).

Standard Menu Inspector General Tab Settings

Motion Menu settings

These settings apply whenever the menu has a motion background or has had audio assigned to the menu.

- *Start*: Sets the first frame for the motion background. Drag the slider or adjust the timecode field to set this value.

- *Loop Point*: When the motion background is looped, this value defines the point to which it will return. By default, this value is equal to the Start timecode value. This is also

251

where the button highlights will appear. Button highlights cannot move, so they are best kept suppressed until any opening transition has completed. Drag the slider or adjust the timecode field to set this value.

- *End*: Defines the end frame for the motion background. When the last frame of the Background is reached, the *At End* setting becomes active. Drag the slider or adjust the timecode field to set this value.

- *Duration*: Automatically calculated to reflect the time between the Start and End points. Adjusting Duration only affects the End setting.

- *Single Field:* Optimizes playback when using field-based video sources.

- *At End*: Sets the motion menu action to occur when the Background End point is reached during playback.

  Choose one of these settings:

  − *Still*: Freezes the video's last frame once the video asset finishes playing. If there is an audio file assigned, it will also play once and then stop.

  − *Loop*: Cycles the motion background or audio file continuously.

  − *Timeout*: Using the Secs (delay) and Action settings, this setting configures the menu to automatically jump to the element specified by the Action setting if a menu button has not been selected for the specified amount of time.

Menu timeout and jump action

If a menu timeout duration is set, the navigation passes to the element selected in the Action property after the specified amount of time. If the menu has a video asset assigned as the background, the timeout's countdown does not start until the video finishes playing. You'd generally use a timeout setting with titles played at a sales kiosk, where you want to have something playing on-screen as much as possible.

- *Secs*: Enter the amount of time, in seconds, to wait for a Menu button to be pressed.

- *Action*: Choose the element to navigate to once the timeout value is reached.

General Settings:

- *Overlay*: Choose the overlay file for this menu.

- *Audio*: Choose an audio file for this menu, or create a playlist of multiple audio files by using the "+" key to add files to the list, and the "−" key to delete them from the list.

PSD Background and Overlay layer settings:

- *Background Layers*: This field is active if you assign a layered PSD file as the menu background. Layers in the file are listed with checkboxes to select which ones are used for the background picture. More than one layer may be selected to create the Background.

- *Overlay*: This field is active whenever you assign a layered .PSD file as the menu subpicture overlay. One layer can be selected as the active overlay layer from the pop-up menu.

Layered Menu Inspector General Tab Settings

When a Layered Menu Element is created and used, the Inspector's *General Tab* display is different, as shown in Figures 9-6c and 9-6d.

Authoring Menus

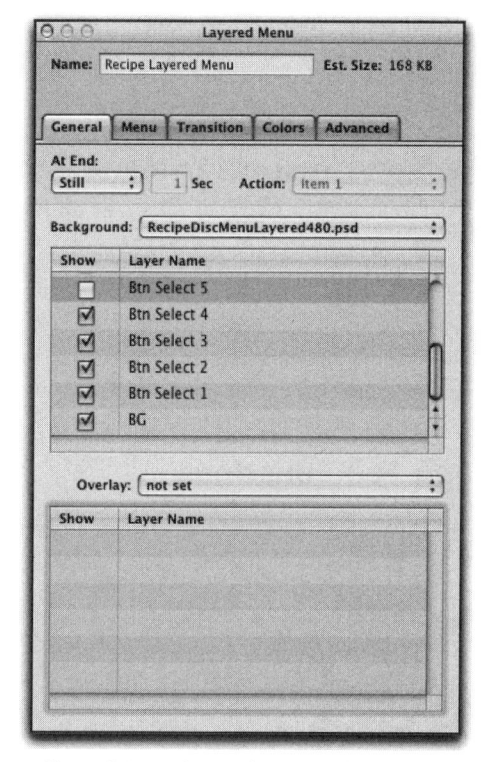

Figure 9-6c — Layered Menu with PSD used as source file only.

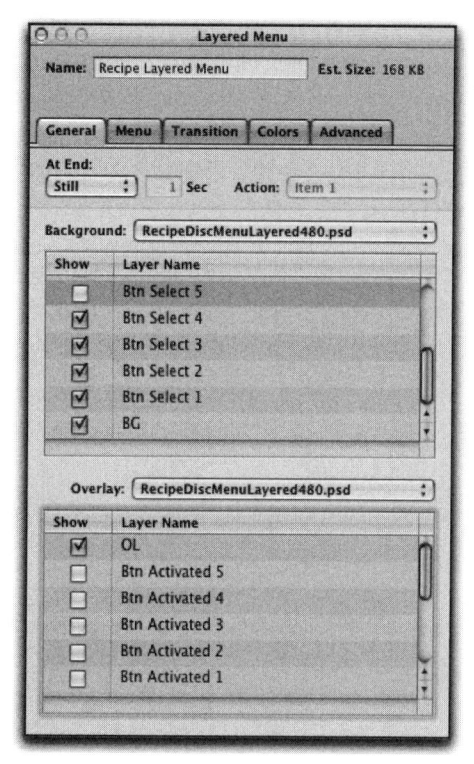

Figure 9-6d — Layered Menu with PSD used as source and overlay.

The Layered Menu Inspector is still organized in sections, but these are laid out differently. Only the Infinite loop ("still") and timeout action properties are at the top, with the Background layer selection area in the middle, and the Overlay Layer selection area at the bottom.

One noticeable difference is the lack of an Audio file property. Layered Menus do *not* use audio files, due to the nature of their operation. Interestingly, however, Layered Menus may use *both* Photoshop Layers *and* a Subpicture Overlay file at *the same time* to create complex menu highlights.

> **NOTE!**
> While multiple layers may be selected to form the Background picture that will be displayed for this Menu, only ONE layer may be selected as an Overlay layer. Of course, the Layered Buttons can select individual layers to create the button selection states.

Layered Menu Inspector Settings

- *At End*: Settings that govern the action of the Menu at the end of its playback.

    Select from one of these settings:

    – *Still*: Loops the video asset infinitely.

    – *Timeout*: Configures the Menu so that if no button is activated within the specified

253

amount of time ("Secs"), the menu automatically jumps to the element specified in the Action property.

Background and Overlay Layer Settings (for PSDs)

- *Background Layers*: This field is active if you assign a layered PSD file as the menu background. Layers in the file are listed with checkboxes to select which ones are used for the background picture. More than one layer may be selected to create the Background.

- *Overlay*: This field is active whenever you assign a layered .PSD file as the menu subpicture overlay. One layer can be selected as the active overlay layer from the list of layers displayed.

The Menu Inspector Menu Tab

The Menu Inspector Menu Tab contains properties concerning the function of the basic Menu. As shown in Figures 9-7a and 9-7b, the Menu Tab functions are nearly identical in a Layered menu, except for the Menu Shadow function, which only exists on a Standard Menu.

Menu Tab Functions

- *Default Button*: Can be used to set a button to be the default selection choice each time

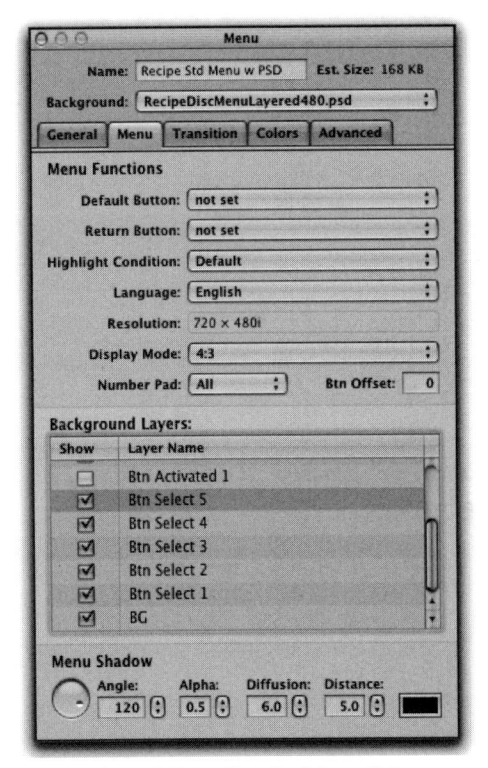

Figure 9-7a — Standard Menu Tab.

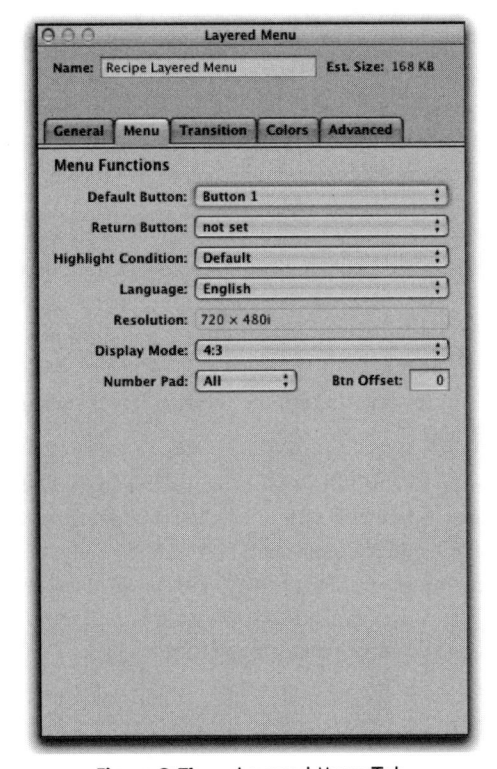

Figure 9-7b — Layered Menu Tab.

254

Authoring Menus

this menu is displayed. Can be overridden by scripting a specific button as a destination for a Jump command, or using the Highlight Condition below.

- *Return Button*: Setting a destination here will activate the Return Button function on this menu; ordinarily off by default.

- *Highlight Condition*: When set to Default, the Default Button property (above) is used. Otherwise, when set to Audio, Subtitle, or Angle, the number of the selected stream type will determine the Menu Button highlighted.

- *Language*: Select the Language tag for this Menu, if not English.

- *Resolution*: Reports the image size for this Menu.

- *Display Mode*: Sets the display aspect ratio for this menu (4 × 3, 16 × 9 Pan-Scan, 16 × 9 Letterbox, 16 × 9 Pan-Scan, and Letterbox).

- *Number Pad*: Choose *All, None*, or a button number from the pop-up menu to define which buttons on this menu may be selected directly by the DVD remote control numeric keypad. If a specific button number is selected, all buttons equal to or less than that number are accessible via the numeric keypad—buttons greater than this value are not accessible from the keypad, but may still be clicked on.

- *Button Offset*: Used with Number Pad above, this property determines the value that will be subtracted from a direct numerical entry on this Menu, to determine which button to activate. As an example, if you have a DVD with multiple chapter Menus, the Chapters may be numbered 21–26, for example, but the buttons will be numbered 1–6 in the Menu. A button offset of 20 will subtract 20 from any direct entry, and correctly select Button 1 on that menu. There is a maximum button offset of 255 on any menu.

Background Layers

This scrollable field allows for the selection of specific Photoshop layers that will be visible in the Menu's Background Picture. Layers with the Show checkbox checked (see Fig. 9-7a) will be visible. In a Standard Menu, this does not control which layers are seen for Button states, as that is only a feature of Layered Menus.

Drop Shadow Settings

- *Menu Shadow*: These settings configure the drop shadow settings for this menu. These settings can be used to provide shadow for text elements, shapes, and drop zones.

The Menu Inspector Transition Tab

The Menu Inspector Transition Tab contains properties that control which Transition is used globally on this Menu. (See Figs. 9-7c and 9-7d.)

The Menu Inspector Color Settings Tab

The Menu Inspector Color Settings Tab contains properties that control which method is used to determine the Menu highlights. These may be set for Menus that use a Subpicture Overlay file (even Layered Menus, despite the fact that this seems at odds with the Switched Photoshop Layer highlights),

DVD Studio Pro 4

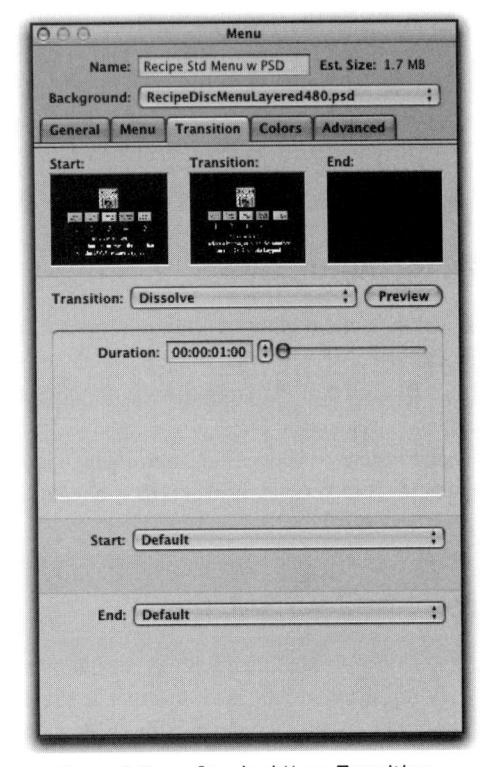

Figure 9-7c — Standard Menu Transitions.

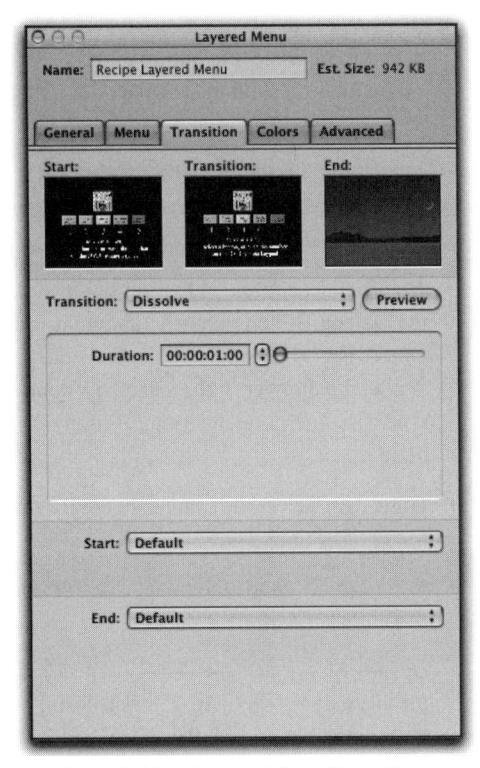

Figure 9-7d — Layered Menu Transitions.

or for menus lacking a Subpicture Overlay file. As shown in Figures 9-8a and 9-8b, the Color Settings Tab functions may use a Simple Overlay Colors (just Black) or Advanced Overlay Colors (see Simple versus Advanced Overlay Colors, below).

Simple Overlay Colors versus Advanced Overlay Colors

There are three different selection schemes for the Overlay Colors: Simple Colors, Advanced Colors (Chroma), and Advanced Colors (Grayscale).

Simple Overlay Colors

The simplest scheme, Simple Overlay Colors, uses black pixels in the Subpicture Overlay file to map the highlights. As seen in Figure 9-8b, there are only three controls—a highlight color and transparency can be set for Normal, Selected, and Activated states.

To set a highlight color:

- Click the Color pop-up and select one of the 16 colors for a highlight.
- Slide the opacity setting to the desired value (0–15), or enter a value in the field to the right of the slider.

256

Authoring Menus

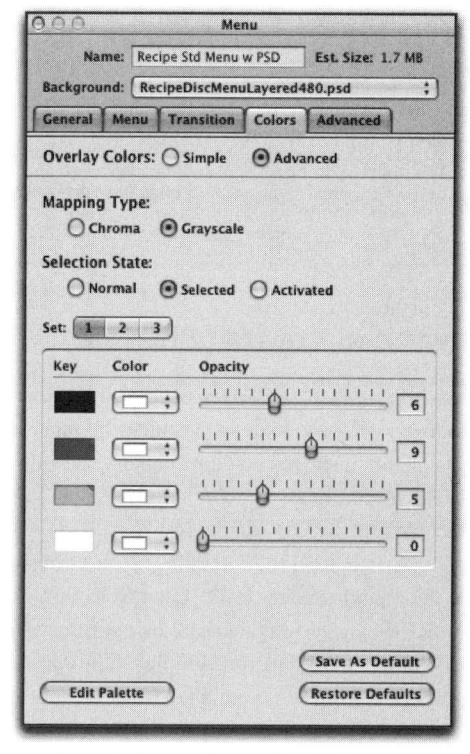

Figure 9-8a — Advanced Color Settings.

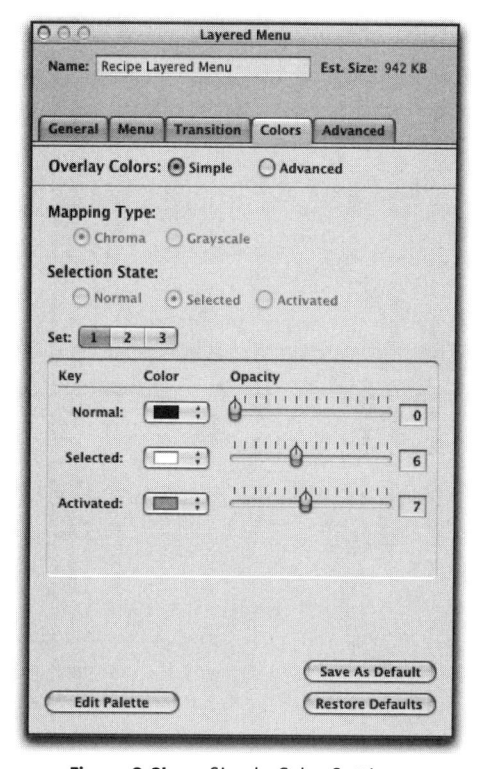

Figure 9-8b — Simple Color Settings.

To edit the colors available in the highlight color palette:

- Click on the edit Palette button, to display the Colors Palette (see Fig. 9-9a).

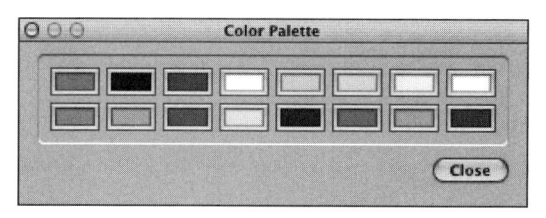

Figure 9-9a — The Color Palette–click a color to open the Color Dialog.

To edit a color in the highlight color palette:

- Click on the desired color, to display the Colors dialog (see Fig. 9-9b).

- Modify the color as desired, then close the Colors dialog.

Advanced Colors (Grayscale)

The *Advanced Colors Grayscale* settings allow the use of more than one level of Gray to map the Menu highlights. For each button state, this setting can map up to 4 grayscale values (Black—100%, Dark Gray—66%, light Gray—33%, and White—0%) to a corresponding color and opacity level. Each

257

DVD Studio Pro 4

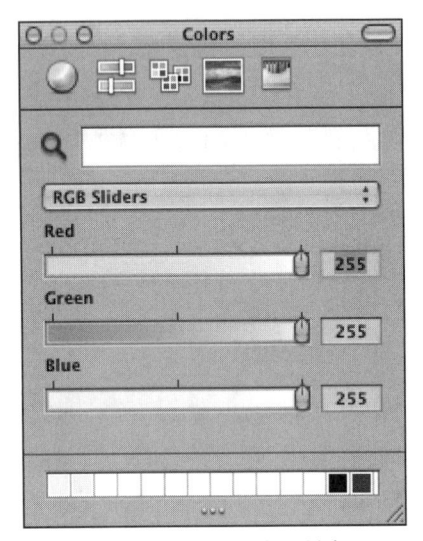

Figure 9-9b — The Colors Dialog.

Selection State (Normal, Selected, Activated) can map these four values, allowing for complex multiple-color highlights or for pseudo-antialiased values.

To set Highlight Colors using Grayscale Values:

- Select the Desired Menu into the Menu editor.
- Display the Menu Inspector.
- Select Mapping Type: Grayscale.
- Select the desired Selection State: Normal, Selected, Activated.

For each Key Grayscale value:

- Select the desired highlight Color from the pop-up.
- Select the desired opacity level.
- Repeat until all Button States have been programmed.

If there is more than one Color Set in use, repeat the above procedure for each of the Color Sets in use in this Menu.

> **NOTE!**
> The Color Settings of the Menu Inspector and the Button Inspector interact—changes to one will be echoed in the other. In other words, Color Settings are global for an entire menu, and not definable button-by-button.

Advanced Colors (Chroma)

The *Advanced Colors Chroma* settings allow the use of four key colors to map the Menu highlight colors. For each button state, this setting can map one of four standard DVD highlight color values (Black—RGB 0,0,0; White—RGB 255,255,255; Red—RGB 255,0,0, and Blue—RGB 0,0,255) to a corresponding color and opacity level. Each Selection State (Normal, Selected, Activated) can map these four color values, allowing for complex multiple-color highlights or for pseudo-antialiased values.

To set Highlight Colors using Chroma Values:

- Select the Desired Menu into the Menu editor.
- Display the Menu Inspector.
- Select Mapping Type: Chroma.
- Select the desired Selection State: Normal, Selected, Activated.
- For each Key color: Black, White, Red, Blue.
- Select the desired highlight Color from the pop-up.
- Select the desired opacity level.
- Repeat until all Button States have been programmed.

Authoring Menus

If there is more than one Color Set in use, repeat the above procedure for each of the Color Sets in use in this Menu.

> **NOTE!**
> As in the previous method, the Color Settings of the Menu Inspector and the Button Inspector interact—changes to one will be echoed in the other.

The Menu Inspector Advanced Tab

The Menu Inspector Advanced Tab contains controls that determine the disabled DVD remote control functions for this Menu, as well as Pre-script, DVD@ccess, and Display Condition controls. User Operation Functions that are checked are *disabled*, while those unchecked remain *enabled*. (See Figs. 9-10a and 9-10b.)

We've covered User Operations before in the Track and Story Inspectors. With the exception of a few UOPs that are already disabled for you, Menu UOPs are locked or unlocked in the same manner as for Tracks and stories. Remember: Checked UOPs are *disabled*; unchecked ones are left *enabled*.

Playback options:

- *Pre-Script*: A script named here will execute before this Menu displays. A Pre-script can determine an alternate Menu to jump to under certain conditions, or may set certain playback conditions prior to this Menus

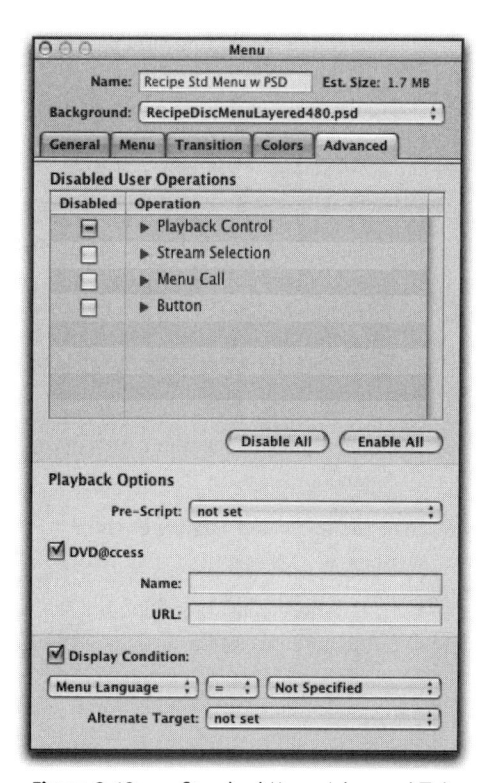

Figure 9-10a — Standard Menu Advanced Tab.

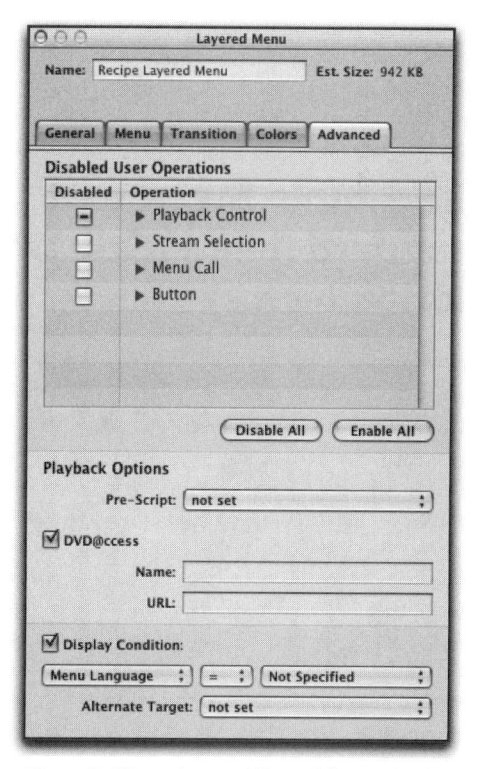

Figure 9-10b — Layered Menu Advanced Tab.

display. See Chapter 13 for information on creating and using pre-scripts.

DVD@CCESS settings:

- *DVD@CCESS*: Checking this box allows you to add a DVD@CCESS link, which can provide a URL-based web or file link when this DVD is played in a Computer. (The name field is only for local reference.) See Chapter 17 for more information on DVD@CCESS and how to use it.

Display Condition settings:

- *Display Condition*: Enables the Display Condition settings, which can control this menu's display, or can define what alternate element should be displayed. See "Display Condition" in Chapter 17 for more information.

The Simple Steps to a Standard Menu (Still or Motion Menu)

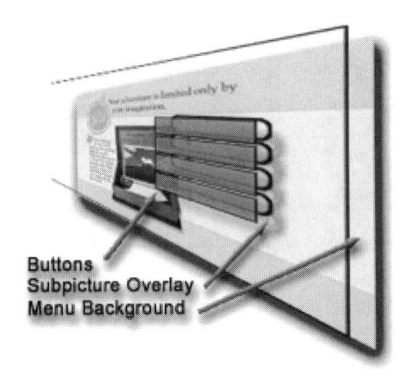

Figure 9-11a — The basics of PICT or MOTION Menu Item.

Here are the **basic steps** to build a typical DVDSP PICT FILE Menu:

- Create an empty Untitled Menu element.
  - Or drag-and-drop the graphic asset onto the Menu Folder.
- Author in the required **Background Picture**.
  - Still image file is 720 × 480 NTSC or 720 × 576 PAL (72 DPI).
  - If you use a video clip, this becomes a loopable motion menu.
- Add sound if desired.
  - This is only possible in a Standard Menu, not Layered.
- Invoke the Menu Editor to create and edit **Buttons**.
  - There is a maximum of 36 buttons per Menu—PERIOD!
  - If multiple aspect ratios are used, this number is apportioned.
- **Create** the required **Buttons**; add commands as needed.
  - Buttons must be square or rectangular in shape.
  - Each button can contain only one connection.
- Generate the **Menu Highlight color settings**.
  - Highlights can use a graphic file as a Subpicture Overlay.y
  - A layer of a PSD file may also be used as an Overlay.

Authoring Menus

- You can also create a menu without a Highlight file.
- **Simulate** the Menu, and verify its operation.

The Simple Steps to a Layered Menu

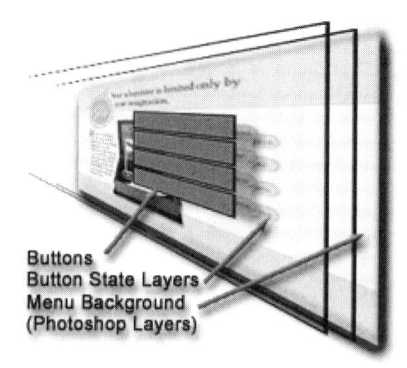

Figure 9-11b — This is a Layered Menu Item.

Here are the **basic steps** to build a Layered Photoshop DVDSP Menu:

- Create an empty menu **Item**; add the Menu PSD asset.
  - Don't drag-and-drop the PSD asset onto the Menu Folder.
- Author the required **Background Picture**.
  - A proper file: 720 × 480 NTSC or 720 × 576 PAL (72 DPI).
  - PSD files require selecting Layers to display "always."
- Invoke the Menu Editor to create and edit **Buttons**.
  - There is a maximum of 36 buttons per Menu—PERIOD!

- **Create** the required **buttons**; add commands as needed.
  - Buttons must be square or rectangular in shape.
  - Each button can contain one JUMP, plus add'l instructions.
- Generate the **Menu Highlight layer scheme**.
  - PSD Menus use layers to switch the graphics for each state.
- **Preview** the Menu, and verify its operation.
  - PSD menus will operate more slowly than Highlighted menus.
- Audio cannot be added to a layered menu.
- Video cannot be used with PSD layered button highlights.

Adding a Background Picture to a Standard Menu

There are two different types of still pictures that can be added to a Standard Menu: a flat image file (like TIFF, PICT, BMP, etc.) or a layered Photoshop .PSD file.

NOTE that we are NOT discussing using this PSD as a source for a *Layered Menu*, but instead we're taking advantage of a new feature in DVD Studio Pro: the ability to use a layered .PSD file as the source for a Standard Menu background picture.

To add a flat background picture to a Standard Menu:

- Import the picture to the Asset bin, then
- Assign it to the Menu using the Menu Inspector, **or**

261

DVD Studio Pro 4

- Drag the Background Picture to the Menu Editor, then
- Select "Set Background" from the Drop Palette.

To add a layered PSD background picture to a Standard Menu:

- Import the picture to the Asset bin, then
- Assign it to the Menu using the Menu Inspector, then
- Select the Layers to be visible in the Background Picture, **or**
- Drag the Background Picture PSD file to the Menu Editor, then
- Select "Set Background—all layers on" from the Drop Palette.

If you prefer, you can set the Background picture with all layers off, and only enable those you wish to see using the Menu Inspector.

Once a Menu Background has been established, a *Subpicture Overlay* file can be selected, and buttons created on the Menu to activate functions or navigation.

Using the Menu Editor to Create Buttons

Once you have created a Menu graphic, you need to create some buttons to make it work. If you are going to be using a Subpicture Overlay file, it is better to assign it before you begin to create buttons. The placement of the Subpicture Overlay will help locate the buttons.

Building Buttons in a menu is actually a pretty easy task; it's getting them to highlight that seems intimidating to some folks. No worry there, though—we'll cover both Simple Menu Highlighting and Switched Photoshop Layer highlighting later in this chapter.

Creating a Button

Unlike DVD Studio Pro 1, there is no "Add Button" tool. Instead, the arrow cursor can be used to draw buttons directly on the Menu editor screen, whenever the Menu Editor is displayed.

To add a Button to the Menu:

- Click with the cursor on the Menu editor and drag to describe a square or rectangular shape; the Button will appear;
- Release the cursor when you have created the button (Fig. 9-12).

The dashed lines show the button size, and the resize handles appear.

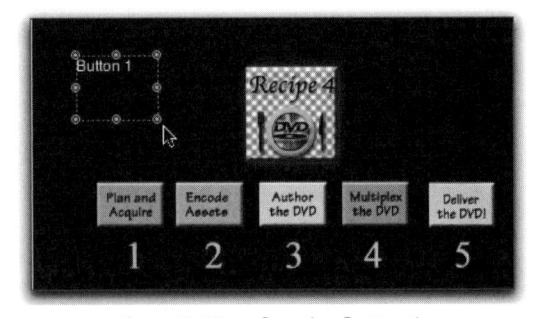

Figure 9-12 — Creating Button 1.

Moving and Aligning a Button

Any menu button may be moved by clicking within it with the arrow cursor and dragging it to a new location. As you drag the button, the dynamic guides will indicate when it is in alignment with other items on the Menu, or if it is centered in the Menu itself. The button will attempt to snap to aligned positions as it

Authoring Menus

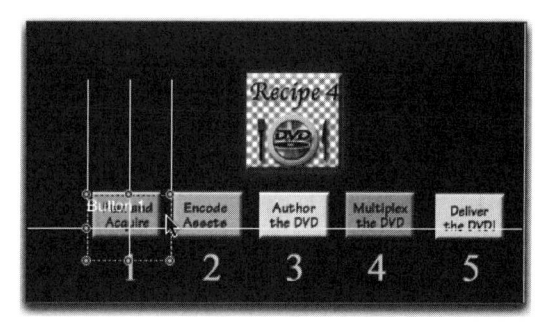

Figure 9-13 — Auto Alignment Guides in action.

is moved, but the snapping is a very light hold, and the button can easily be moved.

Some custom functions are available with modifier keys:

- To temporarily disable the dynamic guides and snapping, hold the *Command* key, then drag.

- To constrain the movement of a button to only vertical or horizontal, hold the *Shift* key while moving the button.

- To clone a button easily, hold *Option*, then click and drag.

- To move the button by nudging it with arrow keys, click the button to select, then use the arrow keys to move it by pixels. *Option-arrow* will move it by 10 pixels; *Shift-Option-arrow* will move it by 20 pixels.

- To resize a button, click on an edge or corner, and adjust the size as needed—see the next section for full details on this.

- To resize a button while retaining its center point, click an edge or corner and start dragging, then press the Option key.

- To resize a button while maintaining a square aspect ratio, hold *Shift* while clicking and dragging on an edge.

Resizing a Button

Once you get the Button placed, you might need to resize it. That's easy to do—when the Menu cursor is placed on an edge or corner of a selected button, the cursor becomes a *resize* tool (showing the double-headed arrow icon).

Clicking on a corner provides a diagonal resize tool...dragging a corner resizes the button in two directions simultaneously...

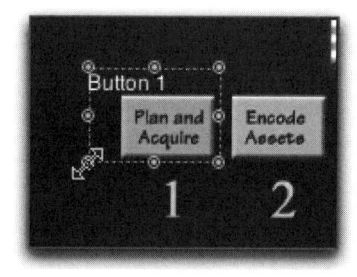

Figure 9-14 — Diagonal resize tool.

Hovering over the top or bottom edge will provide an Up-Down resize tool...dragging resizes the button vertically...

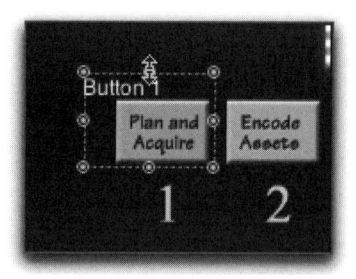

Figure 9-15 — Vertical resize tool.

while hovering over the left or right edge provides Left-Right sizing. Dragging will resize the button horizontally.

263

DVD Studio Pro 4

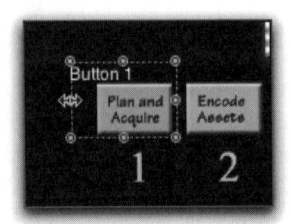

Figure 9-16 — Horizontal resize tool.

Click ON the button to select the button, then move your cursor into place at a corner or edge, click, drag, and resize the button at will. Adjust the shape of the button to fit the graphic in the Background Layer.

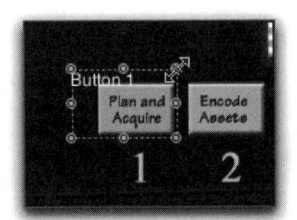

Figure 9-17 — The Diagonal resize tool in action.

Adding More Menu Buttons by Drawing Them

You can add additional menu buttons by drawing additional buttons wherever they are needed—you can draw a button any time the crosshair cursor is visible—just click and drag to create the proper rectangular button size you need. Remember, *36 buttons maximum!*

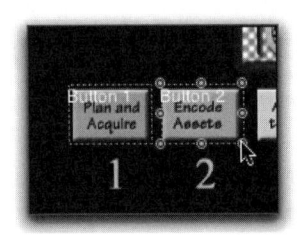

Figure 9-18 — Drawing a new button.

> **NOTE!**
> DVD Menu Buttons must be square or rectangular.

Adding More Menu Buttons by Duplicating Them

Figure 9-19 — Duplicating Buttons.

You can continue drawing button rectangles to add buttons, or you can use the *Duplicate* Command to "clone" them. Select the button (or buttons) you wish to copy first, and then issue the `Edit > Duplicate` command. Remember that each button you clone will contain all instructions, commands, and properties authored into it so far. It is possible to duplicate more than one button at a time.

When you duplicate a button, the clone will appear slightly down and right from the original... like this:

Figure 9-20 — Duplicated Buttons appear offset from original.

264

Authoring Menus

Naming and Renaming Menu Buttons

Note that all buttons are by default named "Button X", where X is a numerical sequence that begins at 1. Naming of the button is accomplished in the Name field of the Button Property Inspector.

To rename a button:

- Click on the button to select it,
- Click in the Name field of the Button Inspector;
- Modify or replace the Button Name text in the Name field.

Double-click to select the entire Name field for easy replacement, then just type a new name or modify the name as needed.

Figure 9-21 — The "Name" field in the Button Inspector top area.

Creating Simple Menu Highlights without Subpicture Overlays

Once the names are in place, you can set up a HIGH-LIGHT scheme—some way of colorizing the buttons to display useful status information to the DVD Viewer.

In their most rudimentary form, menu highlights may be created by setting the appropriate colors for the *Selected* and *Activated* button states, and letting the buttons themselves change color to indicate the highlight modes. By way of an example, let's look at a button from the Recipe 4 DVD tutorial menu:

In the Normal mode, the button looks innocent enough:

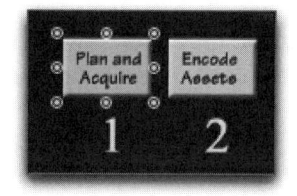

Figure 9-22 — Button 1 as it appears when not selected.

Setting the Button's Highlights Using "Overlay Colors: Simple"

"Simple Overlay Colors" is the least complicated highlight scheme in DVD Studio Pro. To make the difference in the highlight states visible, choose "Simple Overlay Colors" and assign the Apple Shape "Simple Button" to each button.

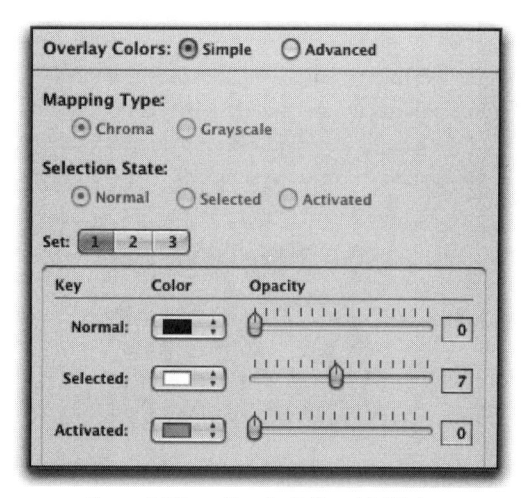

Figure 9-23 — Simple Yellow highlight.

265

DVD Studio Pro 4

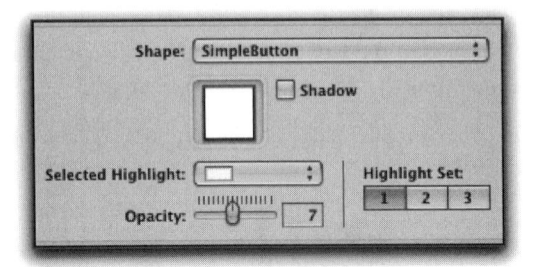

Figure 9-24 — Simple Button Shape

Assigning the Simple Button shape will provide the Button with an overlay to activate the Simple Highlighting. When a button is *Selected* by the DVD user, the Selected color will be visible. When *Activated*, the highlight will disappear. That's why it's called Simple highlighting!

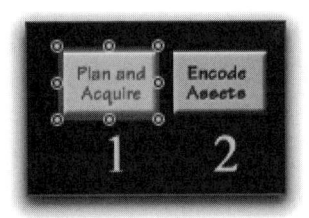

Figure 9-25 — Action button in Selected Highlight Mode (yellow).

When the Button has been *Activated* by the DVD remote control, the Activated Set 1 color will be visible for the split-second just prior to execution of the Button's *connection* (command).

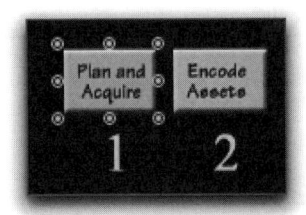

Figure 9-26 — Button 1 in Activated state (no highlight).

Help! I Can't See the Selected and Activated Highlights

To see the highlights when they have been selected, commands in the Menus settings pop-up allow selection and display of the overlay layer. Be sure it is turned on, and the button has been clicked on and selected.

Figure 9-27 — Displaying the Subpicture Overlay.

You can also control whether or not the names and outlines of the buttons are displayed using the Show Button Outline and Names command in the View menu.

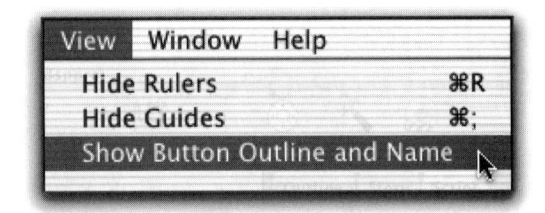

Figure 9-28 — "Show Button Outline and Name" Menu command.

Authoring Menus

Using Color Set 2 and 3

Three different sets of highlight colors can be used, which allow great flexibility in creating complex highlighting schemes. Each Button can be directed to use the Set 1, Set 2, or Set 3 colors in the Button Property inspector. In this way, you can accommodate a menu screen that has buttons requiring different highlights in order to be visible over an inconsistent background picture (imagine a sunset shot on the ocean—blue water below the horizon line, orange-red sunshine above).

To select a Highlight set, first click on the button to be adjusted, then click on the "Highlight Set" property pop-up to select the desired Highlight Set. That button will now follow those Color Settings.

> **TIP!**
> Buttons all default into Highlight Set 1 until changed.

Figure 9-29 — Button "Highlight Set" Property (at right).

Creating Menu Highlights with Subpicture Overlays

In a slightly more advanced form, menu highlights may be created by setting colors for the Selected and Activated state, and additionally using a Subpicture Overlay file that shapes the actual highlights.

Defining the "Subpicture Overlay"

We've heard a lot about the Subpicture Overlay file, but what the heck is it, exactly? Excellent question!

Every DVD player is capable of displaying a "subpicture"—a bitmapped layer of pixels that can be superimposed *on top of* the MPEG video normally running in the background of each DVD Presentation (Track, Slideshow, or Menu).

You can see these pixels quite handily when Subtitles have been created in a Track—but what you may *not* know is that that same Subpicture mechanism also drives the Highlight pixels in DVD Menus.

The goal in creating a good DVD Menu is to create a clear, understandable menu that guides the DVD viewer through the available buttons with clear and unambiguous highlights.

Rather than using the crude ultra-simple Menu highlights that only light up the button rectangles with the Selected and Activated colors, adding an Overlay file can allow for some very creative and artistic highlights, which can assist in creating clear and unambiguous menus!

How Does the Subpicture Overlay Work?

The Overlay file is the middle layer in the Menu sandwich, right in the middle between the bottom layer of the menu sandwich (the Background Picture) and the top layer of the menu (the Buttons).

The Subpicture Overlay file may be a flat image file (TIFF, PICT, etc.), or *one* layer of a Photoshop file. By adding the Overlay file to the menu in between those two other layers, we can specify the shape and color of the Menu Highlights. Review Figure 9-30a

DVD Studio Pro 4

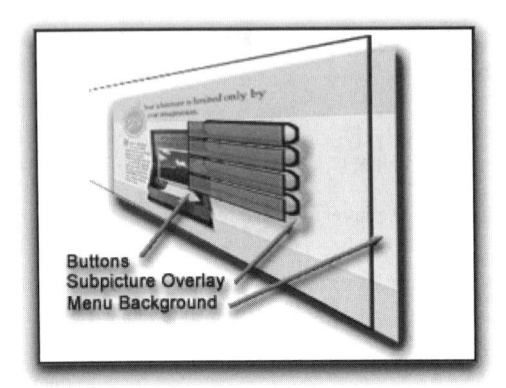

Figure 9-30a — Menu layers in exploded view.

to see how the Menu "layers" work together to create a finished Menu.

This may help visualize the Menu and Overlay relationship more easily. We'll begin with the Recipe4DVD menu with a TIFF file as Background.

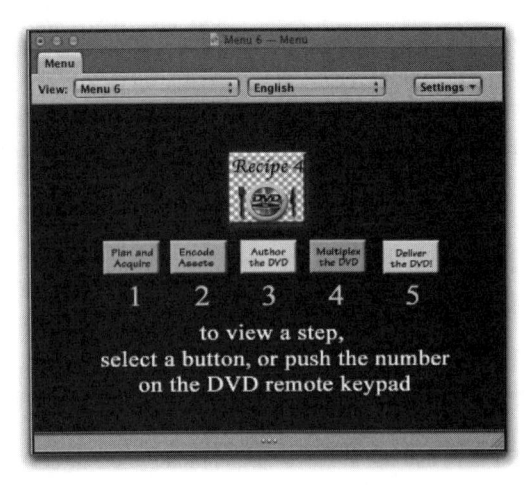

Figure 9-30b — The Background Picture of the Recipe4DVD Menu.

Assembling a Menu with a Subpicture Overlay

If we were to add buttons to the Recipe4DVD menu, and program a simple yellow color for the *Selected* Highlight, then the Button highlights would appear like this:

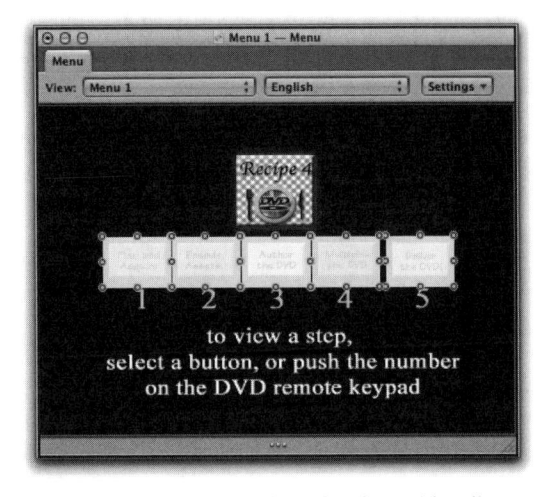

Figure 9-31 — Menu without Overlay, with yellow Selected highlight @ 100%.

Obviously, this is not a pleasant menu to look at—the Menu Button Highlights are obscuring the artwork in the background picture.

Instead, let's add a Subpicture Overlay file to shape the Highlights:

268

Authoring Menus

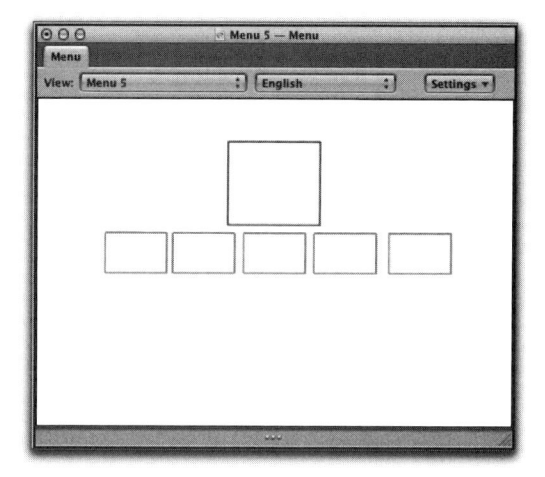

Figure 9-32 — Menu Subpicture Overlay—the boxes outline the buttons in the Menu.

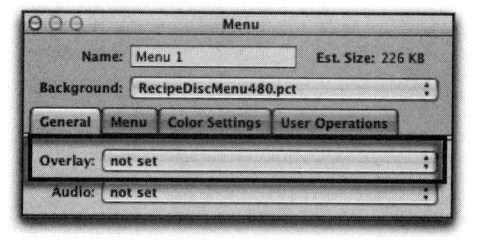

Figure 9-33a — Menu Subpicture Overlay: NOT set.

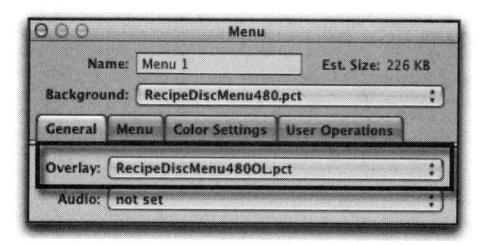

Figure 9-33b — Menu Subpicture Overlay: file set.

Notice that only small rectangles are contained in the overlay file—these rectangles will be used by the Menu to display the highlights in colors and opacities specified by the Menu *Color Settings*, once we specify this file as the Subpicture Overlay file to be used by the Menu.

Of course, the rectangles could be any other shape that would make sense as a Button highlight. We're not restricted as to highlight shapes, *as long as the entire highlight will fit inside of the button itself.*

Specifying the Subpicture Overlay File in the Inspector

To include an Overlay file with the Menu, it's a simple task to specify the filename into the Menu's *Overlay*: property once the file is imported into the Asset bin. Once it's there, click on the *Overlay*: popup and select the desired Overlay file.

It's not a requirement that the Background or Subpicture Overlay file be imported into the Asset bin first before authoring it using the Inspector. Drag-and-Drop works well throughout the Menu editor, and this function is no exception.

Specifying the Background Picture by Dragging

As an alternative, one can drag the Background picture from the Palette or Finder directly to the Menu editor, and select "Set Background" from the Drop Palette that appears (see Fig. 9-34).

Specifying the Subpicture Overlay File by Dragging

A Subpicture Overlay file can also be dragged from the Palette or Finder directly to the Menu editor, and "Set Overlay" selected from the Drop Palette that appears.

DVD Studio Pro 4

Figure 9-34 — Dragging the Background picture to set it.

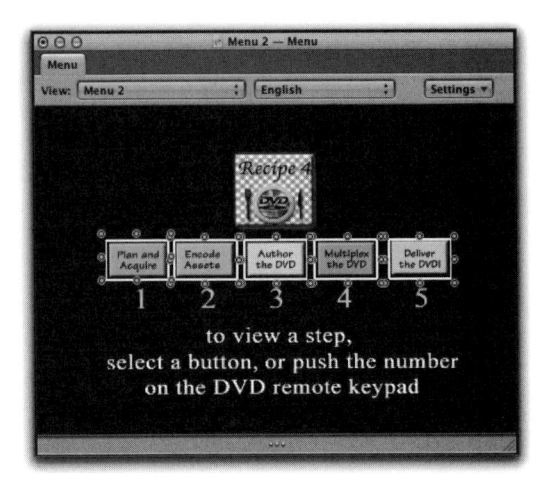

Figure 9-36 — Menu Highlights mapped using the Subpicture Overlay file.

When the Buttons display the yellow Selected state highlights, they are actually displaying the red rectangles in the Subpicture Overlay file according to the rules created in the Color Settings. In this case, we used Advanced Chroma mode to map Red (255) as the Yellow highlight color at 100% opacity. The yellow boxes shown in Figure 9-34 are the result of the interaction of the Color Settings and the Subpicture Overlay file.

Summary of Simple Menu Subpicture Highlights

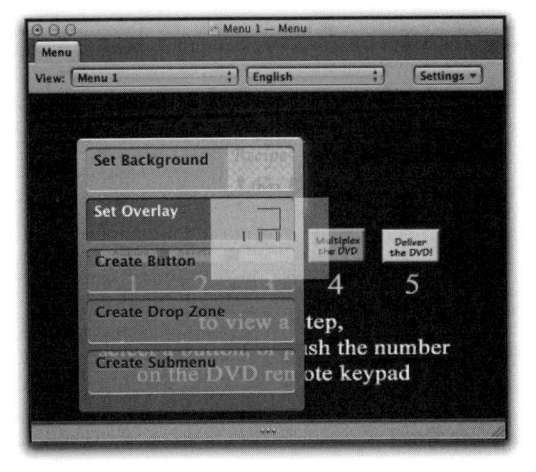

Figure 9-35 — Dragging the Subpicture Overlay file and setting it.

Setting the Selected Highlight color to 100% yellow, let's see what the Motion Menu will wind up looking like (see Fig. 9-36):

So, to summarize how it works, the Button Highlight Colors programmed into the Menu Properties are mapped to the color or grayscale pixels in the Subpicture Overlay file, and the end result is the finished Menu file, with the shaped Highlights.

Authoring Menus

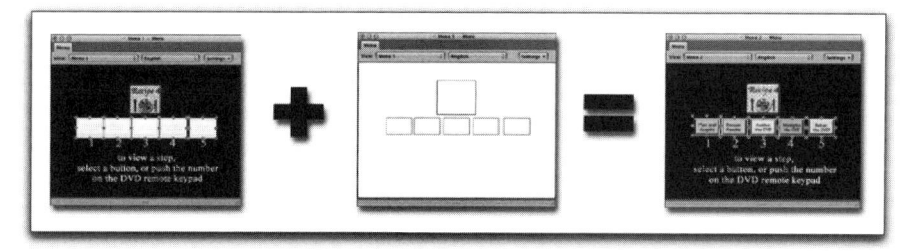

Figure 9-37 — Menu with buttons + Subpicture Overlay = Shaped Highlights.

Creating Layered Menu Highlights with Photoshop Layers

Notice that in the Display properties section, there are also three pop-ups that deal with the Normal, Selected, and Activated STATE of the buttons. These pop-ups control the Photoshop layers that will be switched on and made visible for each state of each menu button. The difference can plainly be seen (and hopefully better understood) in the following example.

Displaying the Normal State Using the Menu Background

Layers selected in the Layered Menu General Tab Background field are used to create the menu background picture. Generally, this picture displays the "Normal State" graphics for the Buttons.

The art contained in the *RecipeMenuLayered480.psd* file includes all layers required to create both the foundation menu picture and all of the Selected and Activated Button State graphics. The "BG" layer is the lowest layer in the Photoshop file and contains the darkened button art, along with the Recipe4DVD tablecloth logo. (See Fig. 9-38.)

Figure 9-38 — The Recipe4DVD menu Background Layer.

Displaying the Normal State Using Layers

A *Normal State* display can also be set by selecting one or more layers using the Normal State pop-up from the Layered Button Layers properties. Regardless of the method used to display the graphic, Button 1 should look like Figure 9-39 in Normal State.

Figure 9-39 — Button 1 in Normal State.

271

DVD Studio Pro 4

Displaying the Selected State Using Layers

In the *Selected State*, specific layers created to show the Selected State graphics can be turned on using the tools in the Layered Button Inspector's *Layers* Tab (see Fig. 9-40):

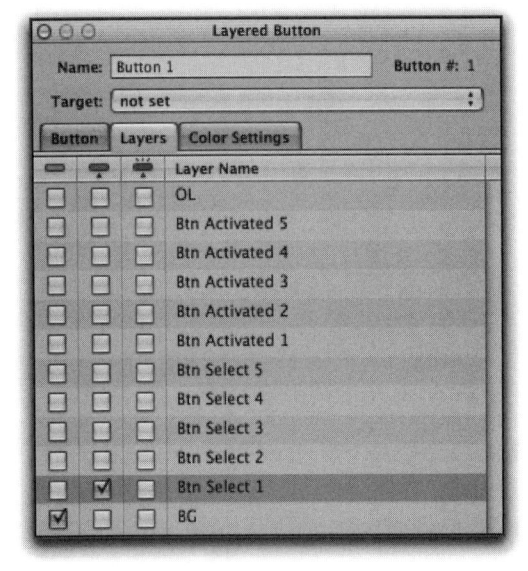

Figure 9-40 — Setting the "Selected State" layers to be displayed.

When shown in its Selected state, Button 1 will now look like this (see Fig. 9-41):

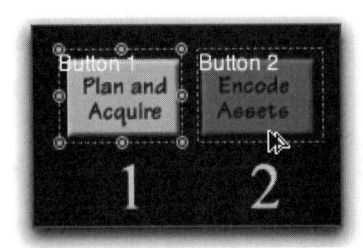

Figure 9-41 — Button 1 with Selected State layer displayed.

It looks this way because the instructions programmed into the *Selected State layer* property (set in Fig. 9-40) will turn ON the layer(s) specified for the *Selected State*. It works the same way for all Buttons and Button States.

Displaying the Activated State Using Layers

In the *Activated State*, a Photoshop layer (or layers) created to show a different specific Activated State graphic would be turned on, using the Button's *Activated State* layer property (see Fig. 9-42):

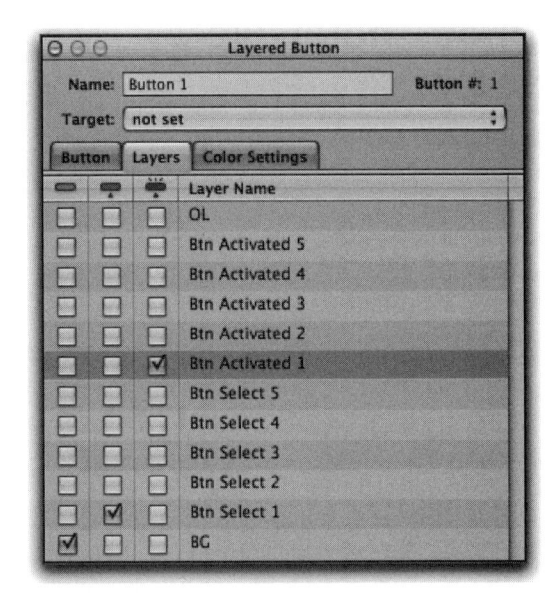

Figure 9-42 — Setting the "Activated State" layers to be displayed.

Accordingly, Button 1 will look like this in the Activated State (see Fig. 9-43):

272

Authoring Menus

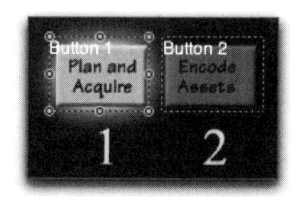

Figure 9-43 — Button 1 with Activated State layer displayed.

Notice that the button layers have been organized into groups, according to the Button State they have been created for—both Selected and Activated States have customized graphics. You will find it easier to assign the Button State layers in your PSD file if you organize the layers in this fashion while editing the file in Photoshop.

Assigning Highlight States Using Layers

To make these button State changes, the Menu artwork must contain all the layers necessary to be switched. If you a look at the *RecipeMenuLayered480.psd* file from the recipe4DVD demo folder, you can see the layers that have been built into the menu file. These layers can be assigned to the appropriate Buttons and their states in the *Layered Button Inspector Layers Tab* (see Fig. 9-44).

Previewing Button Highlight States

The tools provided at the Menu Bottom give you easy access to the button states. Three tools exist, one each for setting: Normal, Selected, and Activated state (see Fig. 9-45). Select a button, then click on the tool for the desired button State.

Figure 9-45 — Button Tools in Normal, Selected, Activated states.

Examples of Stillframe DVD Menus

A Main Menu should clearly lead the user to the destination they seek—in this case, the graphical buttons (static, not motion) display a representative scene, or a glimpse of the destination Menu (in the case of the Chapter Menu).

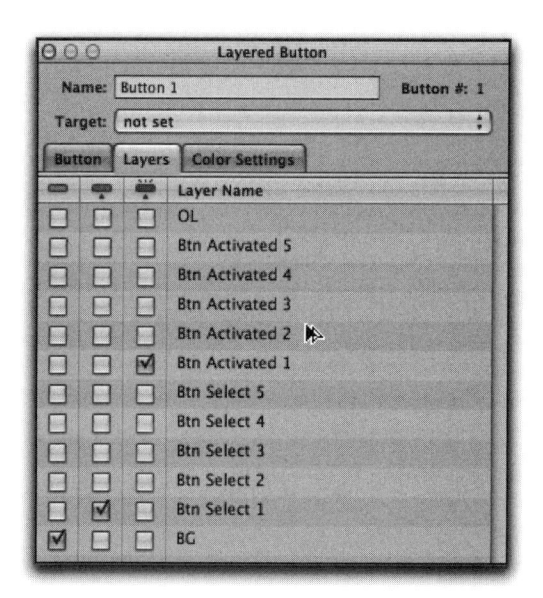

Figure 9-44 — The Layered Button Inspector "Layers" Tab.

273

DVD Studio Pro 4

Figure 9-46 — A typical Main Menu*
Buttons: "Play Feature," "Chapter Menu," "Extras,"
and "Previews."

Figure 9-47 — A typical Chapter Menu.*

Programming the buttons in numerical order, and displaying those numbers in the graphics, will hint to users that the numbers on their remote control can be used to make numerical selections from the DVD menu. This is a function of the DVD Player and SHOULD operate this way on every DVD player—of course, exceptions exist.

---

* Menu images from "A Dog's Tale." Used by permission of Program Power Entertainment.

Making Connections to Buttons

While the vast majority of connections will most likely be made in the Connections Tab (see Chapter 11), it is quite easy to make connections to Buttons while in the Menu Editor.

Making Connections Using the Button Inspector

One way of making connections is to define the Button's Target using the Button Inspector's *Target* property. Located in the top area of each Button Inspector, the *Target* property pop-up can easily create a connection to an Element in the DVD, as long as that element already exists.

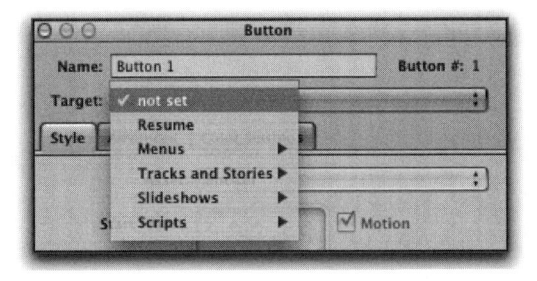

Figure 9-48 — The Button Inspector Target pop-up.

To make a connection using the Button Inspector:

- Display the Button Inspector if it is not already showing;

- Click on (Select) the Menu Button you wish to connect;

- Click on the *Target:* pop-up, and select the project Element to connect to; you can only select from existing Elements.

274

Authoring Menus

Similarly, the Target pop-up can be invoked from the Button itself.

To make a connection directly from the Button:

- Display the desired Menu if it is not showing already, then
- Control-Click on the desired Menu Button, then
- Select the desired project Element to connect to from the *Target:* pop-up menu.

In addition to these direct methods of making connections, the DVD Studio Pro Menu editor contains an extensive array of drag-and-drop tools. These are usable to make connections while performing certain drag-and-drop functions in the Menu editor. You will find information on using the drag-and-drop and image compositing features of the menu editor just ahead in this chapter.

Working with Templates

DVD Studio Pro 4 ships with a large collection of professionally designed Menu templates. Templates are fully built menu designs, including buttons, text, and graphics and, in many cases, graphics Drop Zones are already prepared. Templates are easily customized for your DVD by editing the text, or adding your own, making the proper button connections, and dropping your own movies or graphic elements into the already prepared Drop Zones. In many cases, the menu background contains a Motion element and an audio soundtrack.

To add a template to the current menu, use one of the following methods:

- Select the desired Template in the Palette, then
- Click the "*Apply*" button, lower right of Palette, **or**
- Drag the Template to the Menu Editor, then
- Select "*Apply to Menu*" or "*Apply to Menu, Add All buttons*" in the Drop Palette.

To create a Submenu using a Template, do the following:

- Select the desired Template in the Palette, then
- Drag the Template to the desired button in the Menu Editor, then
- Select "Create Submenu and Apply Template" in the Drop Palette.

The selected template will be used to create a new submenu, which will be connected to the Button automatically.

> **NOTE!**
> If the connection is subsequently broken, the button may erroneously retain the original connection name. Check the Inspector to be certain of the current connection.

Summaries of Dragging Assets and Elements into a Menu

The following summaries will show which options commands will appear in response to the various possible dragging actions, but you should refer to the

275

DVD Studio Pro 4

Apple DVD Studio Pro 4 Manual sections on Menu creation (for Standard Menus starting on Page 285, or for Layered Menus beginning on page 337) for complete descriptions of the many specific actions that take place in response to the Drop Palettes' specific options.

Dragging Assets and Elements into a Standard Menu

While building *Standard* Menus, you are likely to drag Still images or Video Assets into the Editor Tab, as well as project Elements. Different options exist for *Layered* Menus, which we will also cover a little later. The following sections are a summary of your Drop Palette options under different circumstances: dragging to the Empty Area, to a Button, to a Drop Zone, or to various other destinations.

Dragging Assets to a Standard Menu's Empty Area

Different Assets respond differently when dropped in the Empty Area.

Dragging a Video Asset to the Empty Area

The Drop Palette will display:

- **Set Background (Default Action)**: Assigns this Asset as the Menu Background
- **Create Button and Track:** Creates a Track with this Asset, Creates a new Button, links them
- **Create Button:** Creates a New Button with this Asset, but no Track, and no link
- **Create Drop Zone:** Uses this Asset to create a Video Drop Zone in the Background Image

Figure 9-49 — Dragging a Video Asset to the Empty Area.

- **Create Chapter Index or Create Button and Chapter Index:** Create a Chapter Menu, with button links based on Chapter Markers

See the DVD Studio Pro 4 Manual for more specifics.

Dragging Multiple Video Assets to the Empty Area

Figure 9-50 — Dragging Multiple Video Assets to the Empty Area.

The Drop Palette will display:

276

Authoring Menus

- **Create Buttons and Tracks (Connect to Tracks) [Default Action]:** Creates Track and Button for each Asset; links Buttons to Tracks

- **Create Button:** Creates a Button for each Asset but no Track, no link

Dragging a Still Asset to the Empty Area

Figure 9-51 — Dragging a Still Asset to the Empty Area.

The Drop Palette will display:

- **Set Background (Default Action):** Sets the image as the Menu's Background

- **Set Overlay:** Assigns the Asset as the Menu's Subpicture Overlay file

- **Create Button:** Creates a Button with this image as thumbnail

- **Create Drop Zone:** Creates a new Drop Zone using the Image

- **Create Submenu:** Creates a new Submenu using this image as the Button Thumbnail

Dragging Multiple Stills (or a Folder) to the Empty Area

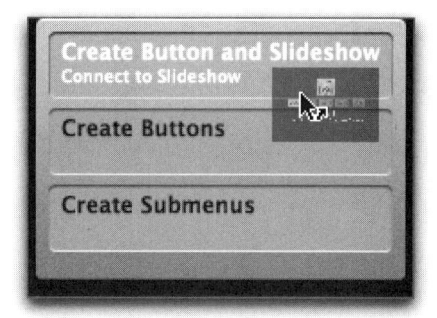

Figure 9-52 — Dragging Multiple Stills (or a Folder) to the Empty Area.

The Drop Palette will display:

- **Create Button and Slideshow (Default Action):** Creates a Slideshow using the images, assigns first image as button

- **Create Buttons:** Creates a Button for each image, with Thumbnail, but no links

- **Create Submenus:** Creates new Submenus using the Images, and linked Buttons with thumbnails

Dragging a Layered Still Image (.PSD) to the Empty Area

The Drop Palette will display (Fig. 9-53):

- **Set Background—All Layers Visible (Default Action):** Assigns the layered file as Menu background, with all layers visible

- **Set Background—No Layers Visible:** Assigns the layered file as Menu background, with no layers visible

- **Set Overlay:** Assigns the layered file as Menu Overlay, with no layers visible

277

DVD Studio Pro 4

Figure 9-53 — Dragging a Layered Still Image to the Empty Area.

Figure 9-54 — Dragging an Audio Asset to the Empty Area.

Dragging a Video Asset with Audio to the Empty Area

- **Create Button:** Makes a New button, assigns the layered file's visible layers as thumbnail

- **Create Standard Submenu:** Makes a new Button, using the visible layers as thumbnail, links to new Submenu using all layers as Background image

- **Create Layered Submenu:** Makes a new Button, using the visible layers as thumbnail, links to new Layered Submenu using all layers as Background image

Figure 9-55a and b — Dragging a Video Asset with Audio to the Empty Area.

Dragging an Audio Asset to the Empty Area

The Drop Palette will display:

- **Set Audio (Default Action):** Assigns this audio asset as the menu audio stream

The Drop Palette will display:

- **Set Background and Audio (Default Action):** Sets Video Asset as Background, Audio asset as audio stream

- **Create Button and Track:** Creates New button with Asset as thumbnail, linked to New Track using Asset

278

Authoring Menus

- **Create Button and Chapter Index:** Creates New button with Asset as Thumbnail, linked to New Chapter Menu created using the chaptered Video Asset

Dragging Assets to a Standard Menu Button

Different Assets will offer different options when connected to a Standard Menu Button.

Dragging a Video Asset to a Standard Menu Button

Figure 9-56 — Dragging a Video Asset to a Button.

The Drop Palette will display:

- **Set Asset (Default Action):** Assigns Asset as the Thumbnail
- Set Asset and Create Track: Assigns Asset as the Thumbnail, linked to new Track using Asset
- **Create Track:** Creates New Track using Asset, links to Track, but no thumbnail
- **Set Asset and Create Chapter Index:** Assigns Asset as the thumbnail, linked to new Chapter Menu using Chaptered Asset
- **Create Chapter Index (uses Chaptered Asset):** Creates new Chapter Menu using Chaptered Asset, links button to Menu

Dragging a Still Image to a Standard Menu Button

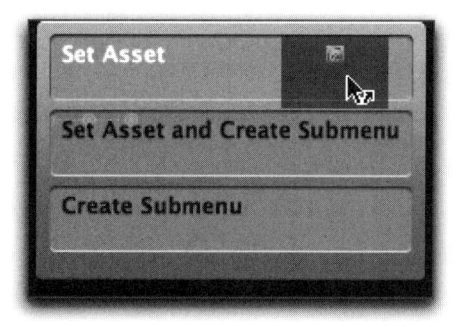

Figure 9-57 — Dragging a Still Image to a Button.

The Drop Palette will display:

- **Set Asset (Default Action):** Assigns image as Thumbnail
- **Set Asset and Create Submenu:** Assigns image as Thumbnail, linked to new Submenu using image as Background
- **Create Submenu:** Creates link to a new Submenu using this image as Background

DVD Studio Pro 4

Dragging Multiple Stills (or a Folder) to a Standard Menu Button

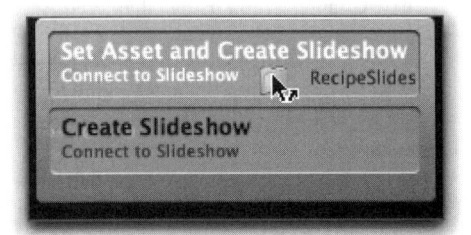

Figure 9-58 — Dragging Multiple Stills (or a Folder) to a Button.

The Drop Palette will display:

- **Set Asset and Create Slideshow (Default Action):** Assigns first still as Thumbnail, Creates new linked Slideshow using images
- **Create Slideshow:** Creates new unlinked Slideshow using all images

Dragging a Video Asset with Audio to a Standard Menu Button

Figure 9-59 — Dragging a Video Asset with Audio to a Button.

The Drop Palette will display:

- **Set Asset and Create Track (Default Action):** Assigns Asset to Thumbnail, Creates new linked track using Assets.
- **Create Track:** Creates new Track using Assets
- **Set Asset and Create Chapter Index:** Assigns video as Thumbnail, creates new linked Chapter Menu using Chaptered Asset
- **Create Chapter Index:** Creates new linked Chapter Menu using Chaptered Asset

Dragging Assets to a Standard Menu's Drop Zone

Interestingly, Assets respond quite similarly in Drop Zones.

Dragging a Video Asset to a Drop Zone

Figure 9-60 — Dragging a Video Asset to a Drop Zone.

The Drop Palette will display:

- **Set Asset (Default Action:** Assigns video asset as the moving image in the drop zone
- **Create Button and Track:** Creates a new Button and a new linked Track

Authoring Menus

Dragging a Still Image to a Drop Zone

Figure 9-61 — Dragging a Still Image to a Drop Zone.

The Drop Palette will display:

- **Set Asset (Default Action):** Assigns this image to the Drop Zone; no other action
- **Create Button:** Creates new Button using this image; no other action; ignores Drop Zone

Dragging Project Elements to Standard Menus

Dragging project elements from the Outline tab to the Empty Area or a button on a standard menu. Different Elements do different things.

Dragging a Track to the Empty Area

The Drop Palette will display:

- **Create Button (Default Action):** Creates a new Button, linked to the Track
- **Create Button and Chapter Index:** Creates a new Button, and a new Chapter Menu using Chaptered Asset

Figure 9-62 — Dragging a Track to the Empty Area.

Dragging Multiple Tracks to the Empty Area

Figure 9-63 — Dragging Multiple Tracks to the Empty Area.

The following option will appear in the Drop Palette:

- **Create Buttons (Default Action):** Creates a new linked button for each Element; Track Asset is used as Thumbnail

Dragging a Story to the Empty Area

Figure 9-64 — Dragging a Story to the Empty Area.

The following option will appear in the Drop Palette:

- **Create Button (Default Action):** Creates a new linked button for the Story; first story clip is used as Thumbnail

281

DVD Studio Pro 4

Dragging Multiple Stories to the Empty Area

Figure 9-65 — Dragging Multiple Stories to the Empty Area.

The following option will appear in the Drop Palette:

- **Create Buttons (Default Action):** Creates a new linked button for each Story; first story clip is used as Thumbnail

Dragging a Slideshow to the Empty Area

Figure 9-66 — Dragging a Slideshow to the Empty Area.

The Drop Palette will display:

- **Create Button (Default Action):** Creates a new linked button for Slideshow; First Slide is used as Thumbnail
- **Create Button and Chapter Index:** Creates a new button linked to new Chapter Menu; first Slide is used as Thumbnail

Dragging Multiple Slideshows to the Empty Area

Figure 9-67 — Dragging Multiple Slideshows to the Empty Area.

The Drop Palette will display:

- **Create Buttons (Default Action):** Creates a new linked button for each Slideshow; first Slide is used as Thumbnail

Dragging a Menu to the Empty Area

Figure 9-68 — Dragging a Menu to the Empty Area.

The Drop Palette will display:

- **Create Button (Default Action):** Creates a new linked button for the Menu; Menu image is used as Thumbnail

Dragging a Script to the Empty Area

The Drop Palette will display:

- **Create Button (Default Action):** Creates a new Button, linked to the Script

Authoring Menus

Figure 9-69 — Dragging a Script to the Empty Area.

Dragging Project Elements to a Standard Menu Button

The following summaries show the options available when dragging to a Button on a Standard Menu.

Dragging a Track to a Button

Figure 9-70 — Dragging a Track to a Button.

The Drop Palette will display:

- **Set Asset (Default Action):** Assigns the Track Asset as the Thumbnail, links Button to the Track
- **Connect to Track:** Links Button to Track
- **Set Asset and Create Chapter Index:** Assigns Track Asset as Thumbnail; creates new Chapter Menu using Chaptered Asset

- **Create Chapter Index:** Creates new Chapter Menu using the Track's Chaptered Asset

Dragging a Story to a Button

Figure 9-71 — Dragging a Story to a Button.

The Drop Palette will display:

- **Set Asset (Default Action):** Assigns Story's first clip as Thumbnail, links Button to Story
- **Connect to Story:** Connects Button to Story; no other action

Dragging a Slideshow to a Button

Figure 9-72 — Dragging a Slideshow to a Button.

283

DVD Studio Pro 4

The Drop Palette will display:

- **Set Asset (Default Action):** Assigns first Slide as Thumbnail; links Slideshow to Button
- **Connect to Slideshow:** Links Button to Slideshow; no other action
- **Create Chapter Index:** Creates new Chapter Menu(s) using Slideshow Slides; lots more…

Dragging a Menu to a Button

Figure 9-73 — Dragging a Menu to a Button.

The Drop Palette will display:

- **Set Asset (Default Action):** Assigns Menu's image to Button Thumbnail; links Menu to Button
- **Connect to Menu:** Links Menu to Button; no other action

Dragging a Script to a Button

Figure 9-74 — Dragging a Script to a Button.

The Drop Palette will display:

- **Connect to Script (Default Action)**

Dragging Templates and Styles to Standard Menus

The following section lists the choices in the Drop Palette that appear when you drag templates and styles from the Palette to the Empty Area or a button on a standard menu.

Dragging a Shape to the Empty Area

Figure 9-75 — Dragging a Shape to the Empty Area.

The Drop Palette will display:

- **Create Button (Default Action):** Creates a new Button using the Shape
- **Create Drop Zone:** Creates a new Drop Zone using the Shape

Dragging a Shape to a Button or Drop Zone

The Drop Palette will display:

- **Set Shape (Default Action):** Assigns the Shape to the Button or Drop Zone

284

Authoring Menus

Figure 9-76a and b — Dragging a Shape to a Button (L) or Drop Zone (R).

- **Create Button:** Creates a new Button using the Shape; Drop Zone is ignored

Dragging a Template to the Empty Area

Figure 9-77 — Dragging a Template to the Empty Area.

The Drop Palette will display:

- **Apply to Menu (Default Action):** Applies Template to the Menu
- **Apply to Menu—Add All Buttons:** Applies Template to the Menu adding all Buttons
- **Create Submenu:** Creates a new Submenu Button linked to a new Submenu using the Template

Dragging a Template to a Button

Figure 9-78 — Dragging a Template to a Button.

The Drop Palette will display:

- **Create Submenu and Apply Template (Default Action):** Creates a Submenu using this Template linked to this Button

Dragging a Button Style to the Empty Area

Figure 9-79 — Dragging a Button Style to the Empty Area.

285

DVD Studio Pro 4

The Drop Palette will display:

- **Create Button (Default Action):** Create a new Button using the Style;
- **Create Button—Set Default Button Style:** Create a new Button using the Style; set the Style as the Default Style
- **Set Default Button Style:** Sets the Default Button style; no other action

Dragging a Button Style to a Button

Figure 9-80 — Dragging a Button Style to a Button.

The Drop Palette will display:

- **Apply to Button (Default Action):** Apply the Button style; no other action

Dragging a Text Style to the Empty Area

Figure 9-81 — Dragging a Text Style to the Empty Area.

The Drop Palette will display:

- **Create Text Object (Default Action):** Creates a new text object using the dragged Style; no other action
- **Create Text Object—Set Default Text Style:** Creates a new text object using the dragged Style; sets the default text Style
- **Set Default Text Style:** Sets the default text Style; no other action

Dragging a Text Style to a Text Object

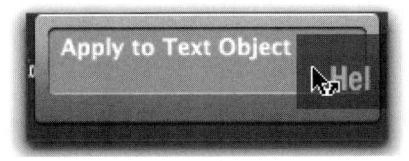

Figure 9-82 — Dragging a Text Style to a Text Object.

The Drop Palette will display:

- **Apply to Text Object (Default Action):** Applies the Style to the text object; no other action

Dragging a Drop Zone Style to the Empty Area

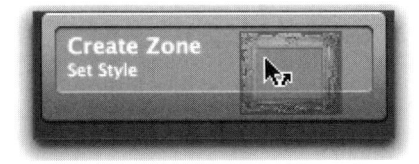

Figure 9-83 — Dragging a Drop Zone Style to the Empty Area.

The Drop Palette will display:

- **Create Drop Zone (Default Action):** Creates a new Drop Zone from the Drop Zone Style

Authoring Menus

Dragging a Drop Zone Style to a Drop Zone

Figure 9-84 — Dragging a Drop Zone Style to a Drop Zone.

The Drop Palette will display:

- **Apply to Drop Zone (Default Action):** Applies the Drop Zone style to the Drop Zone

Dragging a Layout Style to the Menu Editor

Figure 9-85 — Dragging a Layout Style to the Menu Editor.

The Drop Palette will display:

- **Apply to Menu (Default Action):** Applies the Layout Style to the Menu

- **Apply to Menu—Add All Buttons:** Applies the Layout Style to the Menu, adding all Buttons

Dragging Assets to Layered Menus

The following section lists the choices in the Drop Palette that appear when you drag assets to the Empty Area or button in a layered menu.

Dragging Assets to a Layered Menu Button

In many cases, there are fewer options for Layered Menu Buttons than standard Menu Buttons.

Dragging a Video Asset to a Layered Menu's Empty Area

Figure 9-86 — Dragging a Video Asset to the Empty Area.

The Drop Palette will display:

- **Create Button and Track (Default Action):** Creates a new Button and Track, links Button to Track

- **Create Button and Chapter Index:** Creates a new Button linked to a new Chapter Menu using Chaptered Asset

287

DVD Studio Pro 4

Dragging a Video Asset with Audio to a Layered Menu's Empty Area

Figure 9-87 — Dragging a Video Asset with Audio to the Empty Area.

The Drop Palette will display:

- **Create Button and Track (Default Action):** Creates a new Button and Track with Audio, links Button to Track

- **Create Button and Chapter Index:** Creates a new Button linked to a new Chapter Menu using Chaptered Asset

Dragging a Single-Layer Still Image to a Layered Menu's Empty Area

Figure 9-88 — Dragging a Single-Layer Still Image to the Empty Area.

The Drop Palette will display:

- **Set Overlay (Default Action):** Assigns the image as the Menu's Subpicture Overlay file

- **Create Standard Submenu:** Creates a new Button linked to a new Submenu using the image as Background

Dragging a Layered Still Image (.PSD) to a Layered Menu's Empty Area

Figure 9-89 — Dragging a Layered Still Image to the Empty Area.

The Drop Palette will display:

- **Set Background—All Layers Visible (Default Action):** Assigns the image as the Menu Background—All Layers turned on

- **Set Background—No Layers Visible:** Assigns the image as the Menu Background—No Layers turned on

- **Set Overlay:** Assigns the image as the Menu Subpicture Overlay file

- **Create Standard Submenu:** Creates a new Button linked to a new Standard Submenu using Chaptered Asset

- **Create Layered Submenu:** Creates a new Button linked to a new Layered Submenu using Chaptered Asset

Authoring Menus

Dragging Multiple Stills (or a Folder) to a Layered Menu's Empty Area

Figure 9-90 — Dragging Multiple Stills (or a Folder) to the Empty Area.

The Drop Palette will display:

- **Create Button and Slideshow (Default Action):** Creates a new Button linked to a new Slideshow using the images

Dragging a Track to a Layered Menu's Empty Area

Figure 9-91 — Dragging a Track to the Empty Area.

The Drop Palette will display:

- **Create Button (Default Action):** Creates a new Button linked to the dragged Track
- **Create Button and Chapter Index:** Creates a new Button linked to a new Chapter Menu using the Chaptered Asset

Dragging a Story to a Layered Menu's Empty Area

Figure 9-92a — Dragging a Story to the Empty Area.

The Drop Palette will display:

- **Create Button (Default Action):** Creates a new Button linked to the Story

If a story link doesn't exist, an error message will appear:

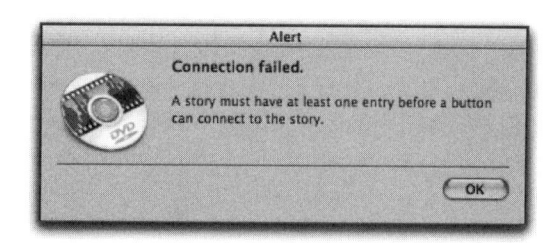

Figure 9-92b — Story Link "failure to link" message.

Dragging a Slideshow to a Layered Menu's Empty Area

Figure 9-93 — Dragging a Slideshow to the Empty Area.

289

DVD Studio Pro 4

The Drop Palette will display:

- **Create Button (Default Action):** Creates a new Button linked to the Slideshow; first Slide is Thumbnail
- **Create Button and Chapter Index:** Creates a new Button linked to a new Chapter Menu using Chaptered Asset

Dragging a Menu to a Layered Menu's Empty Area

Figure 9-94 — Dragging a Menu to the Empty Area.

The Drop Palette will display:

- **Create Button (Default Action):** Creates a new Button linked to the Menu; assigns Menu image as Thumbnail

Dragging a Script to a Layered Menu's Empty Area

Figure 9-95 — Dragging a Script to the Empty Area.

The Drop Palette will display:

- **Create Button (Default Action):** Creates a new Button; connects Script to Button

Dragging Assets to a Layered Menu Button

The following summaries show Drop Palette actions when dragging to a Layered Menu Button.

Dragging a Video Asset to a Layered Menu Button

Figure 9-96 — Dragging a Video Asset to a Button.

The Drop Palette will display:

- **Create Track (Default Action):** Creates Track using asset; links Track to Button
- **Create Chapter Index:** Creates new Chapter Menu using Track's Chaptered Asset

Dragging a Video Asset with Audio to a Layered Menu's Button

Figure 9-97 — Dragging a Video Asset with Audio to a Button.

290

Authoring Menus

The Drop Palette will display:

- **Create Track (Default Action):** Creates a new Track using Assets; links to Button

- **Create Chapter Index:** Creates a new Button linked to new Chapter Menu using Chaptered Asset

Dragging a Single-Layer Still Image to a Layered Menu's Button

Figure 9-98 — Dragging a Single-Layer Still Image to a Button.

The Drop Palette will display:

- **Create Standard Submenu (Default Action):** Creates a new Standard Menu using the Image as Background; links to Button

Dragging a Layered Still Image to a Layered Menu's Button

Figure 9-99 — Dragging a Layered Still Image to a Button.

The Drop Palette will display:

- **Create Standard Submenu (Default Action):** Creates a new Standard Submenu using the Image as Background; links to Button

- **Create Layered Submenu:** Creates a new Layered Submenu using the Image as Background; links to Button

Dragging Multiple Stills (or a Folder) to a Layered Menu's Button

Figure 9-100 — Dragging Multiple Stills (or a Folder) to a Button.

The Drop Palette will display:

- **Create Slideshow (Default Action):** Creates new Slideshow linked to Button

Dragging Project Elements to a Layered Menu Button

The following section lists the Drop Palette options that appear when you drag project elements to a Button in a layered menu.

Dragging a Track to a Layered Menu Button

The Drop Palette will display:

- **Connect to Track (Default Action):** Links Track to Button

291

DVD Studio Pro 4

Figure 9-101 — Dragging a Track to a Button.

- **Create Chapter Index:** Creates linked Chapter Menu using Track's Chaptered Asset

Dragging a Story to a Layered Menu Button

Figure 9-102 — Dragging a Story to a Button.

The Drop Palette will display:

- **Connect to Story (Default Action):** Links Button to Story

Dragging a Slideshow to a Layered Menu Button

Figure 9-103 — Dragging a Slideshow to a Button.

The Drop Palette will display:

- **Connect to Slideshow (Default Action):** Links Slideshow to Button
- **Create Chapter Index:** Links Button to new Chapter Menu created using Slideshow Asset

Dragging a Menu to a Layered Menu Button

Figure 9-104 — Dragging a Menu to a Button.

The Drop Palette will display:

- **Connect to Menu (Default Action):** Links Button to Menu; no other action

Dragging a Script to a Layered Menu Button

Figure 9-105 — Dragging a Script to a Button.

The Drop Palette will display:

- **Connect to Script (Default Action):** Links Script to Button; no other action

Authoring Menus

Summary

Menus are the lifeblood of many DVDs. Menus give the DVD user an opportunity to make selections that can control the navigation flow and the presentation options of the DVD. Menu choices are made by selecting and activating from the Buttons, which indicate the available choices of that particular Menu. Buttons use Highlights to indicate their State—Normal, Selected, or Activated. When Activated, a button typically transfers navigation control to the element programmed as the Target for that Button.

DVD Studio Pro provides a Menu editor that allows for the creation of DVD menus from scratch, which include graphics, text, motion assets, and effects (drop zones and patches).

Menus may be created in DVD Studio Pro as Standard Menus, using ordinary graphics and standard menu highlighting, or as Layered Menus, which use specially prepared Photoshop files that include graphics layers to represent the background image, as well as the selected and activated states for each button of that menu. It may also include normal state graphics for the buttons, or the normal button image may be a part of the background picture.

There may be a maximum of 36 Buttons on each Menu. In DVD Studio Pro, there is a limit of 10,000 Menu elements, but in practical terms, most DVDs do not use anywhere near that number of Menus. Buttons define the hot zones on a Menu (a rectilinear shape) within which the Menu Highlights may be Activated and displayed. Buttons also contain linkages to other elements within the current DVD to continue the navigation, or to modify the current DVD presentation.

Button Highlights inform the DVD user which Menu Button is currently selected, and can also display the activated highlight, to confirm that a Button has been executed.

Chapter 10

Authoring Slideshows

DVD Studio Pro 4

Goals

By the end of this Chapter you, should:

- Understand what a **Slideshow** is
- Understand how to create a **Slideshow**
- Understand what a **Slide** is
- Understand how to create a **Slide**
- Understand what graphics formats can be used for **Slides**
- Understand what **audio formats** can be used in a Slideshow
- Understand how to **add audio** to an individual slide
- Understand how to **add overall audio** to the Slideshow
- Understand the **Slideshow Inspector**
- Understand the **Slide Inspector**
- Understand the **Pause** function
- Understand how to Convert a Slideshow into a **Track**

Slideshow Basics

Slideshows are linear sequences of graphic images, which may or may not have audio accompanying them individually, or as an overall soundtrack underneath all the slides. A Slideshow can display a sequence of up to 99 images, with each image timed to pause for a certain length of time, or images may require user interaction in order to advance the slideshow; in fact, you can combine both of these techniques, if you wish—any way you want to build it, DVD Studio Pro can accommodate you.

Creative Issues

How you choose to use a slideshow in your DVD is totally up to you (see Fig. 10-1)!

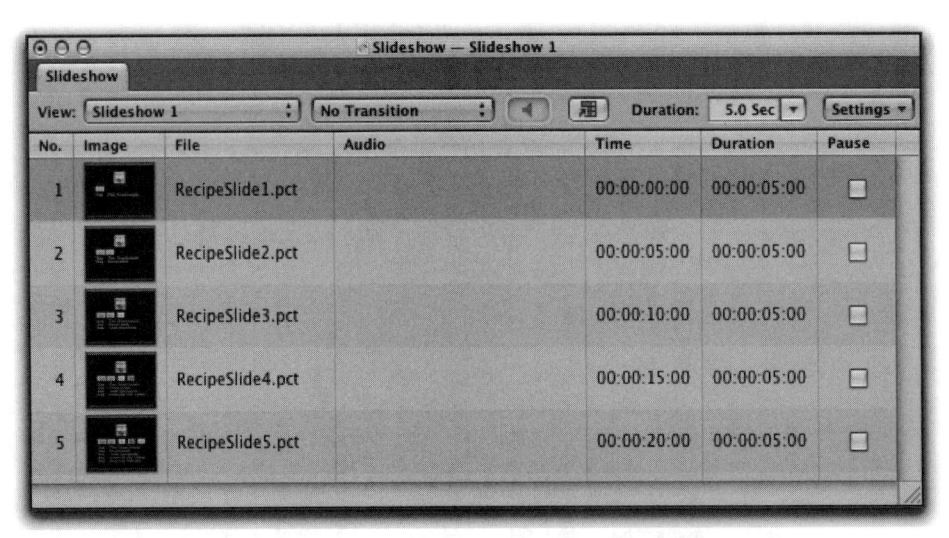

Figure 10-1 — The Slideshow Editor Tab in action.

Authoring Slideshows

Slideshows can serve a multitude of creative purposes, including:

- Biography cards for talent (company execs, seminar speakers, etc.)
- Outlining the plot of a book, or the plot of a movie's story.
- Creating a musical jukebox or songbook (good for kids' DVDs!)
- Showing an album of graphics or photos relevant to the DVD content.
- Explanation of Human Resources, Job requirements, or forms.

May other possibilities exist, of course; the previous are some common uses.

> **NOTE!**
> Unlike DVD Studio Pro 1, DVD Studio Pro 2, 3, and 4 cannot use video as elements in a slideshow. Assemble the video in a Track instead.

Graphics Formats Usable for Slides

Slideshow graphical images can be of practically any format:

- PICT files
- BMP files
- JPEG files
- QTI (*QuickTime Image*) files
- TGA (*Targa*) format files
- TIFF (*Tagged Interface File Format*) format files
- Photoshop layered .PSD files using the 8-bit RGB mode

> **NOTE!**
> Unlike menus, Slideshows do not allow for .PSD layer selection to create a composite image. The image displayed in a slide will be that of the layers that were visible when the file was last saved.

You will find that the same rules of graphics creation for DVD Menu images apply to graphics creation for Slideshows, including *Action Safe* and *Title Safe* zones, NTSC Color Saturation, and control of the RGB Values of Black (16, 16, 16) and White (235, 235, 235). See Chapter 15, "Graphics Issues for DVD Images" to review these issues. Of course some of these rules are now changing with the advent of widescreen video formats and High Definition Monitors.

Slide Sizes and Aspect Ratios

Slides may be sized for either SD projects (720 × 480 NTSC, 720 × 576 PAL) or the various image resolutions usable for HD DVD projects (**720 × 480, 720 × 576, 1280 × 720, and 1920 × 1080**).

For SD DVD:

Slides for SD DVD are easiest to use when they are prepared at the standard sizes for NTSC or PAL graphic images listed above. While images prepared in nonstandard sizes will be scaled by DVD Studio Pro upon import, it's better to begin at the proper size. There is nothing to prevent you from setting a 4 × 3 image into a 16 × 9 matte and using that for a widescreen slide, should you need to blend into an otherwise 16 × 9 SD slideshow.

Slides for HD may be created in any of the sizes acceptable for HD video, but may also be created in nonstandard sizes, as can those for SD. Nonstandard size slides will be scaled to fit the selected project Aspect Ratio.

(We've included in the book DVD a folder of Photoshop images in various sizes and aspect ratios that can help you experiment with display screen res-

297

DVD Studio Pro 4

olutions—look for the folder named "Grids and Blocks.")

Once a slide image file has been imported into the Asset Tab, it will be converted into an MPEG image when you build the project. Scaling will automatically occur if you import an image that is NOT an exact match with an NTSC or PAL video resolution—whether SD or HD. When an image scales, the Background Color that has been set in the General Preferences pane (under "Slideshows and Tracks") will fill in the gaps in the image size.

> **WARNING!**
> *Image Scaling* can cause some unusual image distortions, to say the least. It is better to prepare your slide images using the size techniques in Chapter 15, and import standard sizes rather than relying on DVD Studio Pro to scale your images.

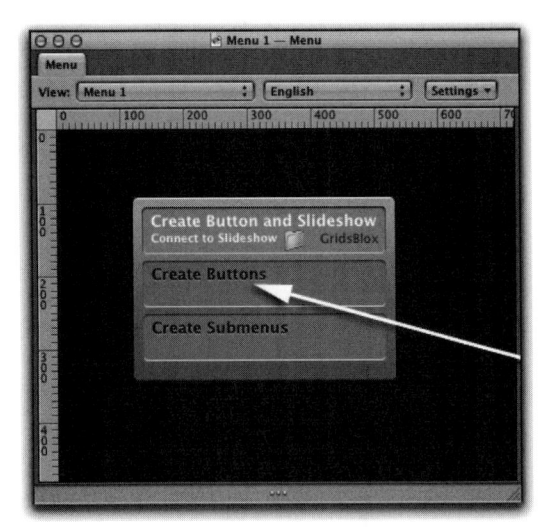

Figure 10-2a — Drag a Folder of Graphics to a Menu to Make a Slideshow.

Creating a Slideshow

When you create a new DVD Studio Pro project, there is no default Slideshow created. You may create a slideshow easily with the Menu Editor, if you are using the Basic Configuration.

To Create a Slideshow in the Menu Editor

- Dragging images (or a Folder of them) to a Menu screen will create a Slideshow automatically (see Fig. 10-2a).

This is a "down and dirty" quick way to get a slideshow, but you may prefer to build a Slideshow in the Slideshow Editor, using the more traditional methods. As a first method, you can create an empty Slideshow, and then add media to it. You will find this is easier to do in the Extended or Advanced Configuration, where you have access to the Outline Tab.

To Create an Empty Slideshow:

- Select in Menu: **Project > Add to Project > Slideshow**.
- Click on the *Add Slideshow* button in the Toolbar (if visible) (see Fig. 10-2b).

Figure 10-2b — The Add Slideshow Tool (Cmd-K).

Authoring Slideshows

- Type *Command-K* (⌘-**K**) on the keyboard.
- Click on the *Add Slideshow Button* at the bottom of the Menu Editor Window (if it is visible).

Once you have created a new Slideshow item, it will appear inside of the Slideshows folder in the Outline Tab (see Fig. 10-2c). Slideshows are numbered in the sequential order in which they are created, but can be easily renamed or resequenced at any time (see the section titled *The Slideshow Inspector*, later in this chapter).

Figure 10-2c — A Slideshow in the Outline Tab.

Figure 10-3 — Hold the cursor over the Slideshow to check its contents.

> **TIP!**
> At any time, pausing the cursor over any Slideshow in the Outline Tab and pausing there will display the "tool tip," which indicates its current population. In this case, Slideshow 1 is empty (see Fig. 10-3).

To add assets into the Slideshow, double-click on the Slideshow entry in the Outline Tab to open the Slideshow Editor in its Tab. You will also want access to the Asset Tab, the Palette, and/or the Finder Window, depending on where your slide images are located. Also, open the Inspector. You will use both the Slideshow Editor and the Inspector while creating Slideshows.

Meet the Slideshow Editor

The Slideshow Editor (Fig. 10-4) contains many powerful features that are now consolidated into a simple and concise control panel.

Along the top of the Slideshow Editor are the controls for:

- *View: pop-up*–selects which slideshow to edit.
- *Transition*:—allows selection of a Transition effect (Transitions were added in DVD Studio Pro 3).
- *Overall Audio well*—a drop bin for an overall soundtrack.
- *Convert to Track tool*—converts this slideshow into a Track—useful if you wish to add multiple audio tracks or other Track-specific features (Chapter-based Cell commands, and similar).
- *Duration*:—a pop-up/text field to enter slide duration.
- *Settings: pop-up*—this consists of four options:
  - *Fit to Audio*—spaces all slides evenly over the soundtrack.

299

DVD Studio Pro 4

Figure 10-4 — The Slideshow Editor when empty.

- *Fit to Slides*—trims audio file to fit total of slide durations.
- *Loop Audio*—loops the overall audio file if needed.

The above function only if an overall audio file has been added to the slideshow.

- *Manual Advance*—click to set "Pause" on all slides.

Sets the Pause attribute on all slides—be aware that "Pause" will interrupt the overall audio playback.

The *Slideshow Editor* occupies the remainder of the window, and it is here you will build the sequence (event list) of slides to be shown. The Editor uses multiple columns to display the attributes of each slide in the Slideshow:

- *No.*—the event number in the Slideshow sequence
- *Image*—a thumbnail miniature of the slide graphic
- *File*—the filename of the image file
- *Audio*—the name of any individual audio file used to accompany this individual slide (if any)
- *Time*—the beginning time of this slide in the Slideshow
- *Duration*—the duration of this slide—need we say more?
- *Pause*—a checkbox that sets an Infinite Pause for this slide

Selecting Slideshow Assets

The Slideshow Editor no longer lists the assets to be used in Authoring (and has not since DVD Studio Pro 1)—you will have to locate the proper graphics files from the Asset Tab (if you have already imported them into your project), or else use the Palette or the Finder.

Regardless of where your source file or asset is located, dragging an image file from one of those locations into the Slideshow Editor will create a slide (an event). The slides will be listed in the Slideshow

Authoring Slideshows

Editor in the order in which they are created, which will also be the order in which they are displayed in the finished Slideshow, unless you change this order. As you would expect, changing the order of events is easy—as simple as dragging a slide from one position to another in the event list. You may also add a slide into the middle of an existing slide list; just drag it and place it somewhere other than at the end. Additionally, you may drag an entire folder of images directly into the Slideshow Editor, or onto the Slideshow Element in the Outline Tab (but my experience is that they may not wind up in the original order!).

Adding Slides to the Slideshow

You can use your own slides for this if you wish, but we've provided a folder full on the DVD for you to use to practice with ("Recipe Slides").

- In the Folder called *Recipe Slides* in the ***Recipe4DVDsource* folder on the DVD, you will find slides named RecipeSlide1 thru RecipeSlide5.

- Click the asset called *RecipeSlide1* and drag it into the slideshow event list. The ghost image of the asset file will accompany your cursor into the Slideshow Editor. (See Fig. 10-5.)

- Once you release the mouse, the slide event will appear in the Slideshow Editor. (See Fig. 10-6.)

To Delete Slides from a Slideshow:

- Click to Select a slide or Shift-Click to select multiple slides, then

- Hit *Delete* to remove the selected slides, or...

- Choose **Edit > Delete** from the Menu.

Need More Slides for the Slideshow?

Additional slides can be added to the Slideshow (up to the limit of 99 slides) at any time—even after you begin editing the order and adding sound.

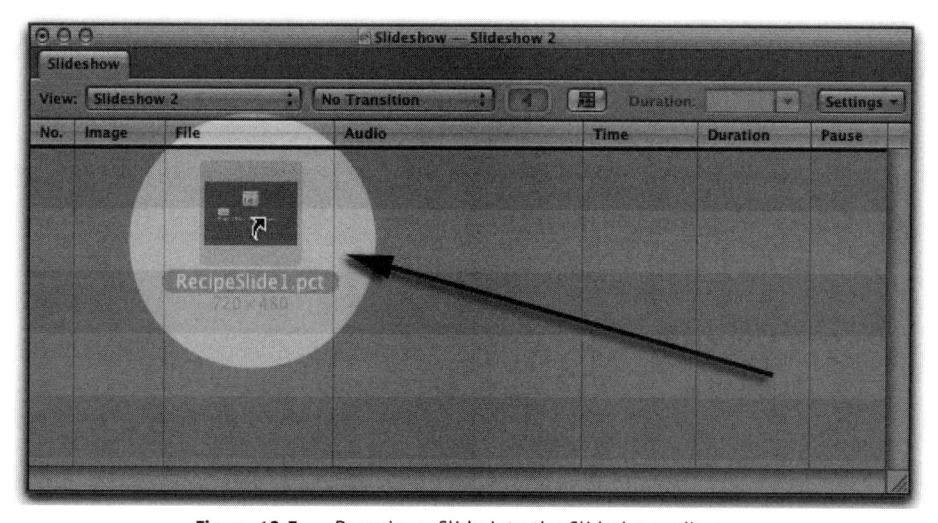

Figure 10-5 — Dragging a Slide into the Slideshow editor.

DVD Studio Pro 4

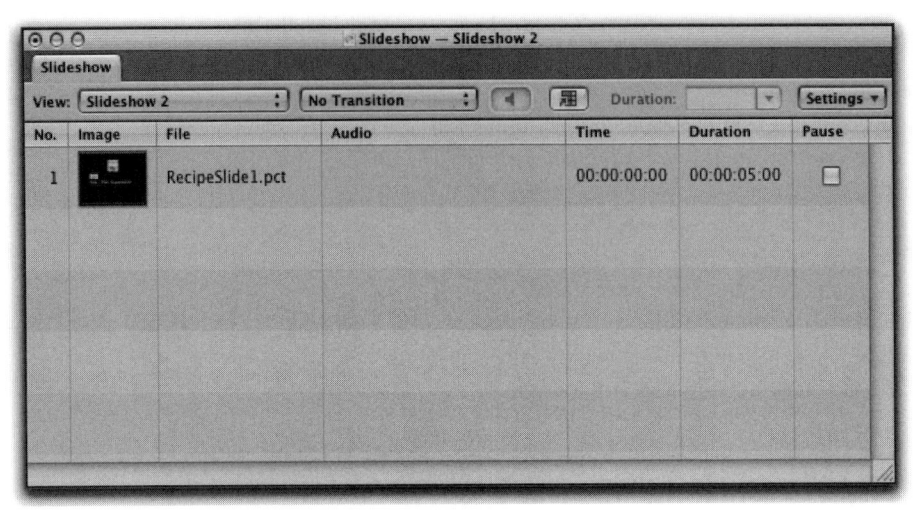

Figure 10-6 — RecipeSlide 1 event is now in the Slideshow list.

To add additional slides:

- Drag slides one at a time from the Palette, Finder, or Asset Tab, **or**...

- Drag multiple additional slides at one time, **or**...

- Drag a folder containing multiple slides...

Drag the desired slides onto the *Slideshow* entry in the *Outline Tab* or into the chosen Slideshow directly. Slides dragged into the slideshow will drop into the list at the location of the cursor when you let go. Existing slides will be pushed down into the list to compensate.

Continuing with our exercise:

- Drag in four more slides (RecipeSlide2 through 5) from the same source location (see Fig. 10-7). Note the selection line under *RecipeSlide1* (the arrow). This line indicates where the slide(s) being added will enter the list. If adding to an existing list of slides, slides will be added in at this location.

Once the cursor is released, the slides appear in the Slideshow event list in alphabetical order (with luck), at the drop point (see Fig. 10-8). Note that they don't always show up in the order you expect!

Also note that there are not yet any audio files attached to any slides. "Find Matching Files" only works for Tracks.

Adjusting durations of slides:

If you are not using audio with individual slides, you can adjust any slide event to an individual duration, and mix and match durations as you please. The "Slide Duration" pop-up (Fig. 10-9) controls the duration of any currently selected slide. It offers popular preset durations (1, 3, 5, 10 seconds) and the ability to enter a custom duration if you so choose.

To adjust the duration of a slide:

- Single slides can be set individually, as can groups of slides.

Authoring Slideshows

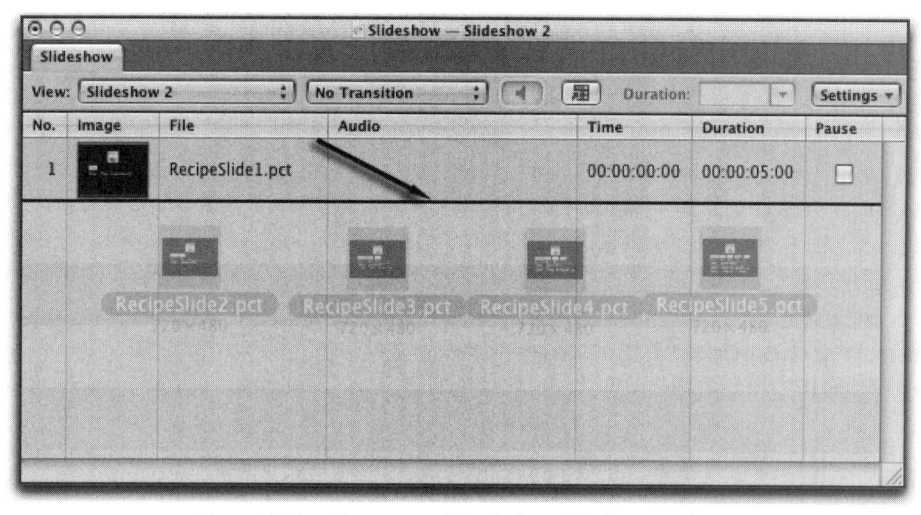

Figure 10-7 — Four more slides being added at one time.

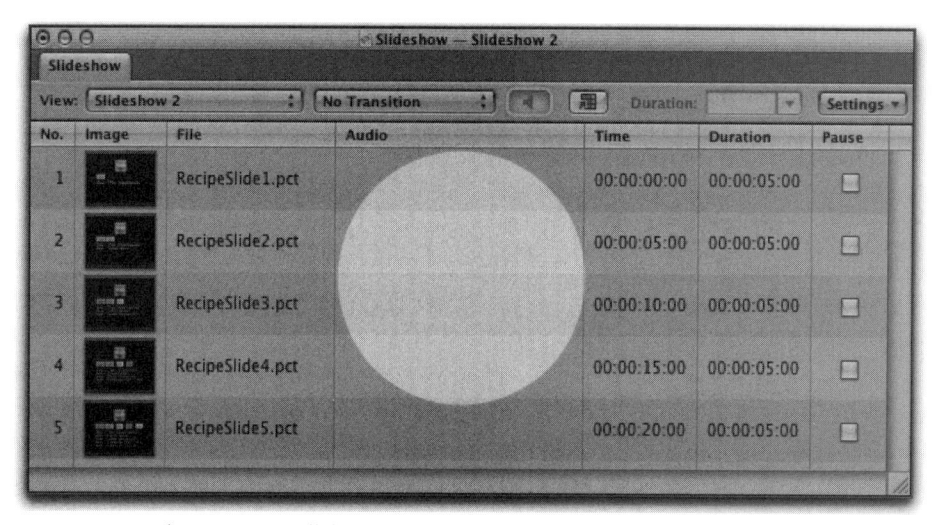

Figure 10-8 — All five slides added, but no audio yet—(note hilite).

- Click on one or more slides to select them (be sure to select the slide, not the audio field).
- Click on the Slide Duration pop-up, and select a value, **or...**

- Enter a custom value in the Slide Duration field, *then*
- Hit **ENTER** or **RETURN** to finish, **or** you can
- Add an individual audio file to a slide (that slide will take on the duration of the audio file).

303

DVD Studio Pro 4

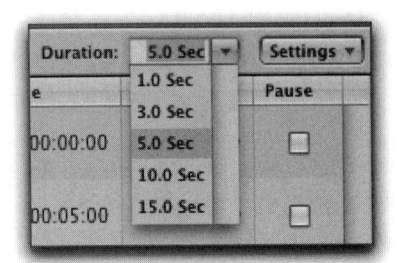

Figure 10-9 — Using the "Slide Duration" pop-up.

To adjust the duration of all slides or the slideshow:

- If you wish to adjust all slides, execute a Select All (⌘-**A**), then choose the duration in the pop-up, or enter a value.

- If you wish to adjust the run time of the slideshow with an overall sound file, see "Using Fit to Slides" or "Using Fit to Audio" just ahead in this chapter.

Setting Manual Advance

Setting the *Manual Advance* checkbox will set the Pause property for every slide in the Slideshow (see Fig. 10-10). This requires the user to click *Skip Next* or *Play* on the DVD Remote Control to advance to the next slide in the Slideshow. (Exactly which key can be pressed depends on the specific DVD player.) You can also select the Pause property for individual slides as well (e.g., to present an instructional slide that must be manually advanced) at any point in the Slideshow.

Adding Sound to Your Slideshow

If you are building a Slideshow with individual narrations or audio files for each slide, see "Adding an Individual audio File..."

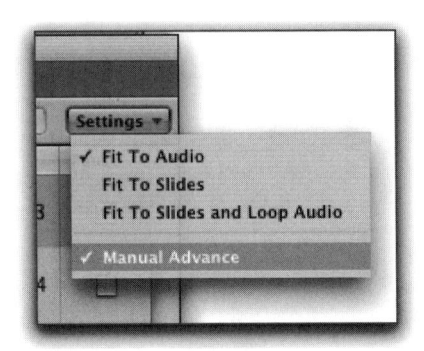

Figure 10-10 — Selecting "Manual advance" sets "Pause" on all slides.

If you want to know about using an Overall Sound file for the whole Slideshow, see "*Using an Overall Audio File...*" If you are going to use Audio files with your Slideshow, deselect *Manual Advance*!

> **NOTE!**
> You must be consistent in your use of audio file formats—a Slideshow can only use one format of audio throughout, either PCM (AIFF or WAV) or Dolby Digital (AC3).

Adding an Individual Audio File to a Slide

Adding an individual audio to an individual slide only requires dragging a suitable Audio file (PCM or Dolby Digital) to the audio column of that slide's entry in the Slideshow list. Figure 10-11 shows the audio file being added (note the highlight box around the selected slide). Figure 10-12 shows the Slideshow after the audio file has been added to the slide. Note that the Audio File column is selected, not the entire slide.

> **NOTE!**
> In Figure 10-12a that slide 1 has now taken on the duration of the audio file that was just added to it. Unlike in DVD Studio Pro 1, the slide's duration is not added to the audio file's duration.

Authoring Slideshows

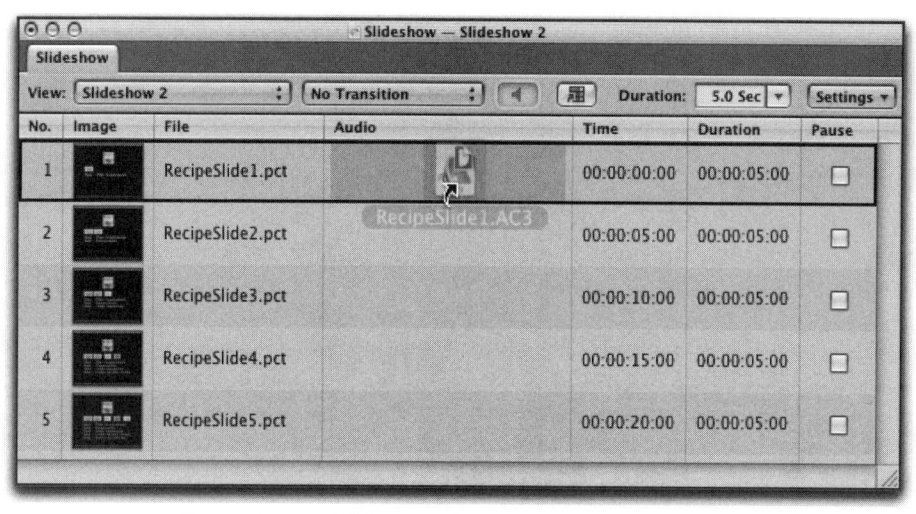

Figure 10-11 — Adding an individual audio file to one slide.

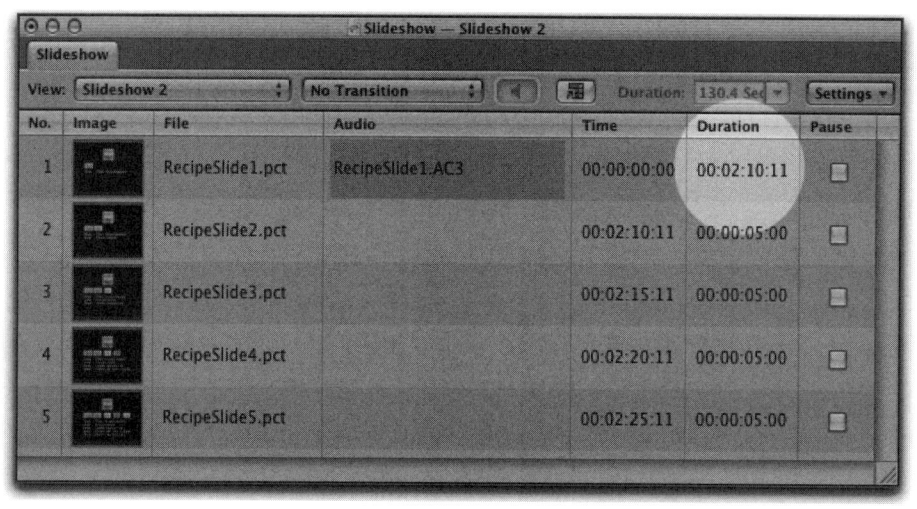

Figure 10-12 — Audio file added to Slide 1—note the new duration.

To delete an individual file, or multiple audio files:

- Click on one or more audio files with the cursor to select, *and*
- Press the **DELETE** key, **or...**
- Choose **Edit > Delete** in the menubar.

Using an Overall Audio File in the Slideshow

This is a feature long awaited in DVD Studio Pro! Adding an overall audio file means using one sound track under your entire slideshow, and having the slides change seamlessly and automatically above it.

305

DVD Studio Pro 4

You can even elect to use transitions between slides with this mode.

To Add an Overall Audio File to your Slideshow:

- Drag the chosen file to the Overall Audio File "drop bin" (see Fig. 10-13a and 10-13b). Also see the Slideshow Inspector General Tab.

Figure 10-13a — Adding an Overall Audio the file.

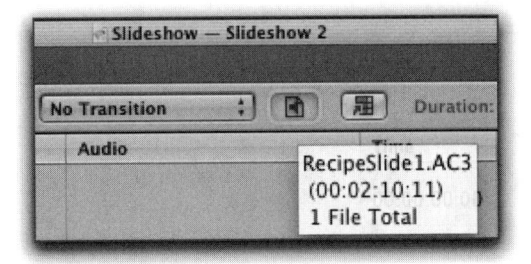

Figure 10-13b — Overall audio file in place

To Remove an Overall Audio File:

- Control-Click on the Drop Bin to remove the file (see Fig. 10-13c), **or**
- Drag the file out of the bin (see Fig. 10-13d and 10-13e), **or**
- Drop a new file in the bin. A new file will supersede the old file.

> **WARNING!**
> Adding an Overall Audio file to a Slideshow that uses Individual Audio files will **remove the individual files from the slideshow**. If you do this by accident, hit Cmd-Z (**Edit > Undo**) immediately!

Figure 10-13c — Control-click on the bin to remove.

Figure 10-13d — Drag the file out of the bin... and...

Figure 10-13e — POOF! It's gone.

That's all there is to it.

Authoring Slideshows

How does an Overall Audio File work?

When using an overall audio file, the Slideshow can be programmed to behave in two different ways:

Using Fit to Audio

Fit to Audio will space all slides evenly over the length of the Overall Audio File assigned to the Slideshow. If you want a nicely formatted Slideshow without chopping up the audio file, use this (Fig. 10-10).

Using Fit to Slides

Fit to Slides will add up the individual durations of all slides, and will trim the overall audio file to fit them (see Fig. 10-10). Note that the "trim" is not very graceful, as it does not fade out—the audio will just STOP when the total running time of the slides has been reached.

> **TIP!**
> For a more graceful presentation of slides, use Fit to Audio.

Transitions for Slideshows

The newest creative feature in DVD Studio Pro, Transitions, is very welcome in the Slideshow. The creative possibilities of DVD Studio Pro's Transitions are amazing and, best of all, they are created while you are authoring within the application—no need to go to an external Motion graphics application.

Before you attempt to add Transitions, take a moment to review the many possible transition types that exist in the preset list. You can find this list in the Slideshow Transition Inspector or the Slide Transition Inspector.

Adding Transitions to a Slideshow

To add a Global Slideshow Transition, one common to all slides, do this:

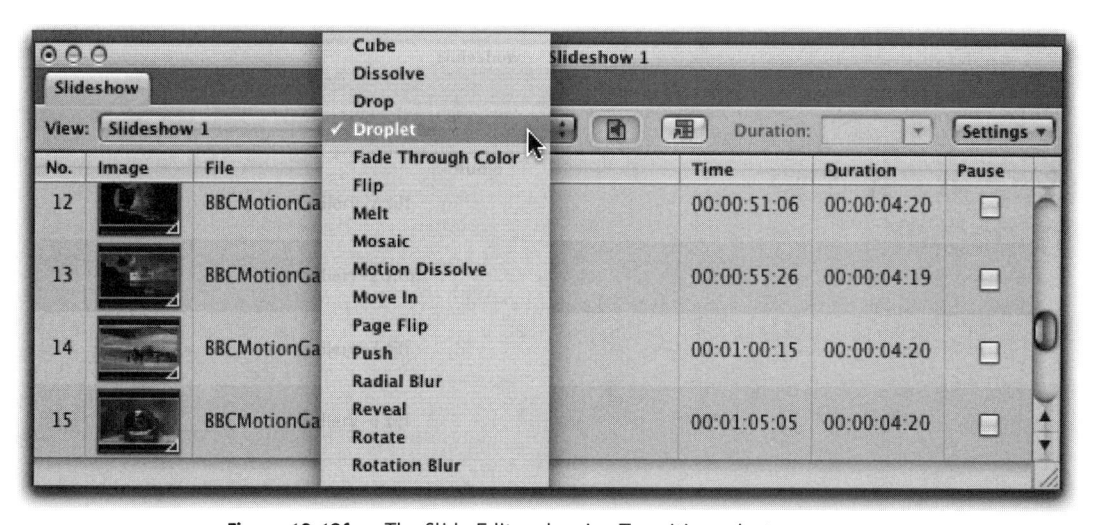

Figure 10-13f — The Slide Editor showing Transition selector pop-up.

307

DVD Studio Pro 4

- Select the desired Transition in the Transition pop-up of the Slideshow editor (see Fig. 10-13f), or the Slideshow Inspector's *Transition Tab* (see Fig. 10-14c).

You will note (if you look carefully in Fig. 10-13g) that the lower right corner of each slide thumbnail has a little triangle superimposed on it. This is the indication that this slide has a Transition—either an Individual one, or that it is subject to the slideshow's Global Transition.

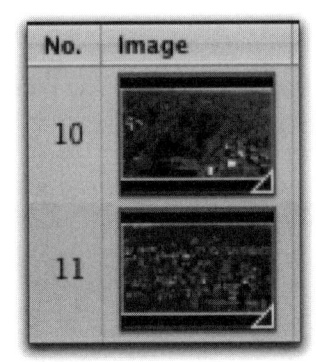

Figure 10-13g — The Slide thumbnail showing the Transition indicator.

We'll cover Transitions thoroughly in Chapter 17, because they are an Advanced Authoring feature applicable to more than just slideshows.

The Slideshow Inspector

Slideshows and Slides each have Inspectors with unique settings.

In DVD Studio Pro 4, the Slideshow Inspector now has three tabs, as follows:

Slideshow Inspector Top Area:

- *End Jump*—where the slideshow will jump to when finished.

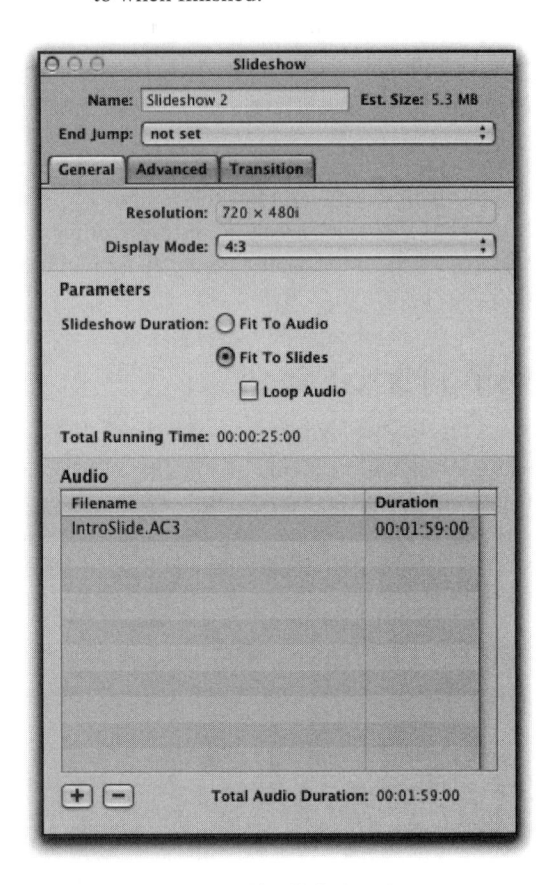

Figure 10-14a — The Slideshow Inspector General Tab.

Slideshow Inspector General Tab settings:

The Slideshow General Tab settings (see Fig. 10-14a) include:

- *Resolution*: displays format and size of slide
- *Display Mode*: selects between 4 × 3 and 16 × 9 modes

308

Authoring Slideshows

- *Parameters:*
    - *Slideshow Duration* controls (duplicates of those in Slideshow Editor Tab
    - *Fit to Audio*—adjusts Slides' durations to fit length of overall audio file. This is the preferred setting.
    - *Fit to Slides*—truncates audio to total of slide durations (please note this is not an elegant fade-out). If slides exceed length of sound,
        - *Loop Audio*—repeat overall audio file to fit slide durations.
- Audio
    - This field lists all audio elements used to make up an extended overall audio program. Add audio files here using the + and – icons. Non-PCM files can be selected and DVD Studio Pro will convert them to AIFF for the DVD.

Slideshow Inspector Advanced Tab:

- Playback Options
    - *Pre-Script*—a script to be executed prior to Slideshow play
    - *Remote Control*—overrides for the DVD Remote Functions (Fig. 10-14b):
        - Menu
        - Angle
        - Chapter
        - Audio
        - Subtitle

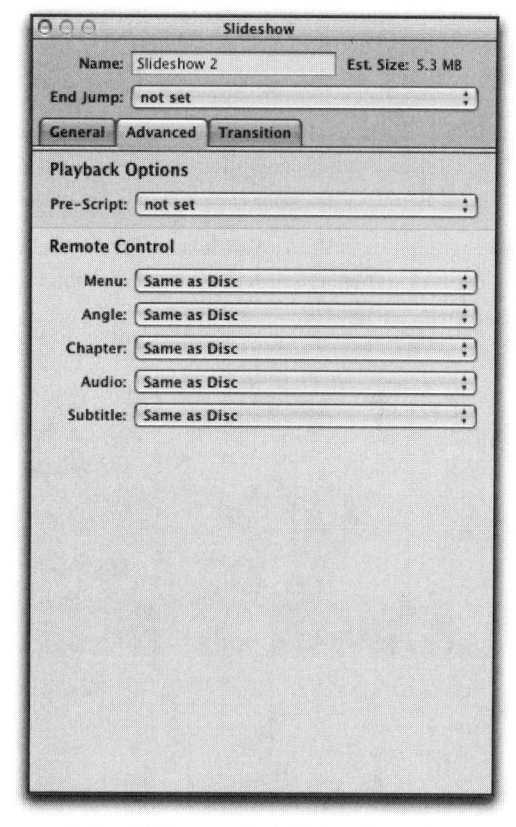

Figure 10-14b — The Slideshow Inspector Advanced Tab.

Slideshow Inspector Transition Tab:

The Transition Tab (see Fig. 10-14c) offers a *Transition*: pop-up to allow selection of the desired Overall Transition type, and a *Preview* button right next door, which will preview the chosen Transition in the center panel of the three-pane display at the top.

The three-pane image display shows the following:

- *Start:* The last frame of this Track, OR, a default seacoast view to hold the "A" effect position.

- *Transition:* When the Preview button is clicked, the current settings of the chosen Transition are previewed here, in real time.

- *End:* The frame that the Transition will end on, or a default nighttime shot to hold the "B" effect position.

The bottom pane will change with each Transition, as it displays the specific Options available for that Transition.

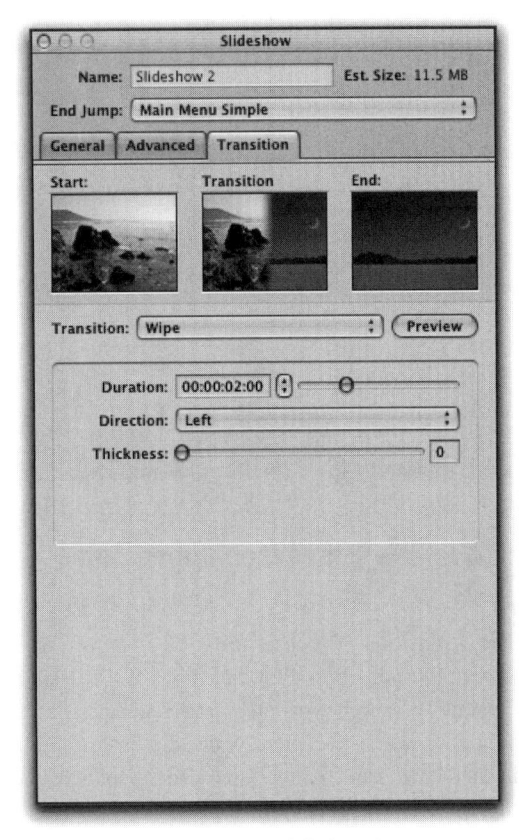

Figure 10-14c — The Slideshow Inspector Transition Tab.

The Slide Inspector

While the Slideshow Inspector is busy reporting on the Slideshow itself, in-depth information about the individual events is reported in the *Slide Inspector*, as outlined below:

Slide Inspector General Tab settings

The Slide Inspector General Tab settings include (see Fig. 10-15a):

- *Slide Duration*: As in the Slide Editor, allows selection of the selected Slide's duration, and Pause attribute ("Manual Advance").

- *DVD@ccess*: A URL, file or mail link that can be attached to this Slide. For information on how to use DVD@ccess, see "Chapter 17, Advanced Authoring."

  – *Name:* A reference name for this link.

  – *URL:* The actual DVD@ccess link used for this slide.

Slide Inspector Transition Tab

The *Slide Transition* Inspector (see Fig. 10-15b) offers a Transition control area exactly as the Slideshow Transition Tab, but this one specifically controls individual Slide Transitions, which may override the Slideshow Overall Transition.

Some Precautions about Audio for Slideshows

Using audio for slideshows is pretty straightforward except for one small thing: Once you have used an audio format in a slideshow (AIFF, for example), that

Authoring Slideshows

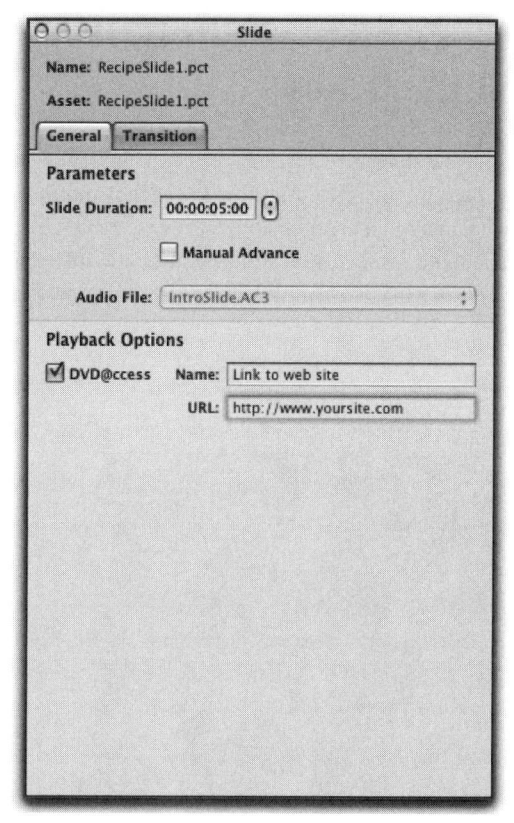

Figure 10-15a — The Slide Inspector General Tab (with DVD@ccess engaged).

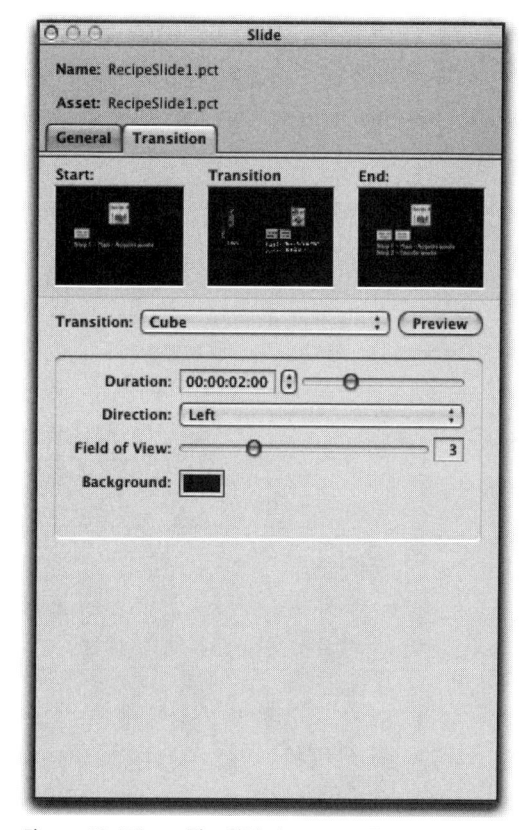

Figure 10-15b — The Slide Inspector Transition Tab.

Slideshow will not accept audio files of a different format. You cannot mix audio formats within a slideshow, so plan your asset preparation carefully, when creating materials for a Slideshow. If you violate this rule, you'll get a warning (see Fig. 10-16).

Convert to Track

DVD Studio Pro 4 includes a feature to convert your Slideshows into Tracks painlessly and easily (see Fig. 1-17a). On occasion, you may find that the material you have created in a Slideshow is better used from within a Track. Tracks have the advantage of allowing multiple audio streams, subtitle streams, angles, and even buttons over video.

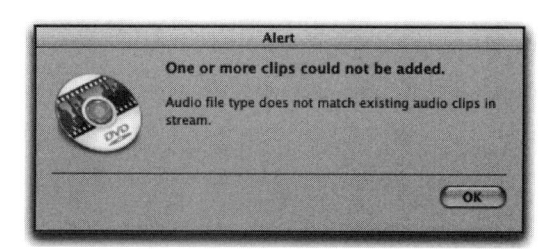

Figure 10-16 — Alert when mismatching audio in Slideshow.

311

DVD Studio Pro 4

To Convert a Slideshow into a Track:

- Bring the desired Slideshow into the Slideshow Editor.
- Click the "Convert to Track" button.
- The Track will immediately be created in the Track Edit Tab. It will inherit the name of the Slideshow that it was created from. (See Fig. 1-17b.)

Viewing and Simulating Slideshows

Once the slideshow has been created, the Viewer can be used to play the Slideshow to check its performance, in a rudimentary manner. However, the Viewer is lacking some advanced navigation controls. The Simulator offers a more accurate and functional preview.

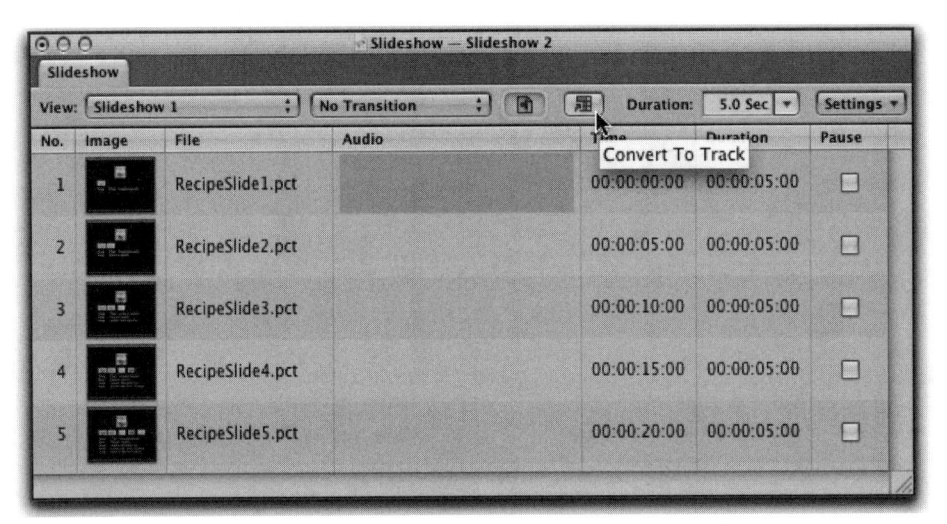

Figure 10-17a — Slideshow with five slides before conversion.

Figure 10-17b — Track made from Slideshow and Transitions.

Authoring Slideshows

To play the Slideshow in the Viewer:

- In the Outline Tab, Double-Click the Slideshow you wish to view.

- Click *Play* in the Viewer controls (Fig. 10-18a), **or**...

- In the Slideshow Editor, *Control-Click* on the slide you wish to begin playing at, and select "Play" from the pop-up.

Controls available while Viewing a Slideshow:

- Click the *Play* button to begin the show (see Fig. 10-18a).

- Click the *Pause* button to pause playback (see Fig. 10-18b).

- Click the *Play* button again to resume from a Pause.

- Click *Stop* (the square) when you wish to stop viewing;

- Click *Previous* (left triangle with line) for the previous slide.

- Click *Next* (right triangle with line) for the next slide.

Simulating a Slideshow

While you are Simulating, you can navigate through the slideshow using the skip NEXT and skip PREVIOUS buttons... If the slideshow has slides that are timed using individual audio files, it should automatically progress to the next slide once the sound file has completed playback. All other Simulator functions behave as you would expect.

To Simulate the Slideshow:

- *Control-Click* on a Slideshow in the Outline Tab, select "*Simulate*," **or**

- *Control-Click* on a slide, but select "*Simulate*," not "*Play*," **or**...

- Click the *Simulator* Tool in the toolbar.

If you begin Simulation from the toolbar, it will try to begin from the *Startup Action*, if it is set. If you haven't defined the Startup Action yet, try simulating from the Outline Tab instead.

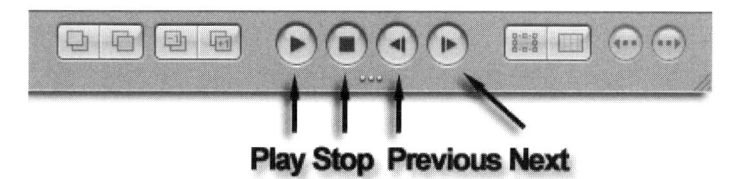

Figure 10-18a — Viewer Controls Stopped: Play | Stop | Previous | Next.

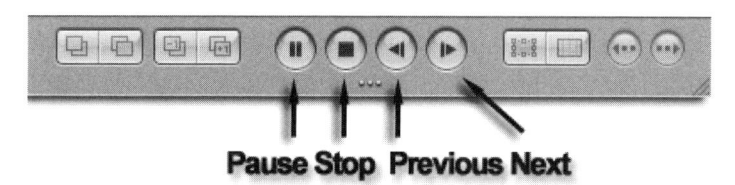

Figure 10-18b — Viewer Controls in Play: Pause | Stop | Previous | Next.

Keyboard Shortcuts (Slideshow Editor)

"Command" + "="	Shows large thumbnails
"Command" + "-"	Shows small thumbnails
"Command" + "A"	Selects All slides
"Spacebar"	Play/Stop Slideshow in Viewer Tab
"L"	Play/Pause Slideshow in Viewer Tab
"K"	Stop (and rewind) Slideshow
"–>" [Right Arrow]	Go to Next Slide
"<–" [Left Arrow]	Go to Previous Slide
"Command" + "–>"	Go to Next Slide
"Command" + "<–"	Go to Previous Slide
"Option" + "–>"	Go to Next Slide
"Option" + "<–"	Go to Previous Slide
"Control" + "–>"	Go to Next Slide
"Control" + "<–"	Go to Previous Slide

Authoring Slideshows

Summary

About Slideshows in General

Slideshows are sequences of up to 99 Still Images, each of which may be accompanied by a sound file. As an alternative, one sound file may be used as an Overall Sound File to run seamlessly underneath all slide images. In DVD SP2, this can be fit to the length of the image show, or the images can be "fit" to the length of the audio file. Slides may be timed to a certain length, or paused infinitely, which requires user interaction to move the Slideshow forward. Slides may not contain Buttons, but an interactive "pseudo-slideshow" may be created using a series of Menu containers to hold the image files (in the Menu backgrounds) with whatever Buttons may be desired.

What's Changed in DVD Studio Pro Slideshows?

A significant difference in DVD Studio Pro slideshows is that as of DVD Studio Pro 2, Slideshows can no longer use Video Clips, and it is no longer possible to use multiple audio programs underneath the slides.

DVD Studio Pro 3 added the killer *Transitions* that are available in not only Slideshows, but Tracks (see Chapter 7), and from Menu Buttons (see Chapter 9). See Chapter 17 for information on Transitions.

In DVD Studio Pro 3, the Slideshow Controls were consolidated into the top of the Slideshow editor tab, where they take up less room, and are easily accessed. They remain there in this version.

Chapter 11

Making Connections

DVD Studio Pro 4

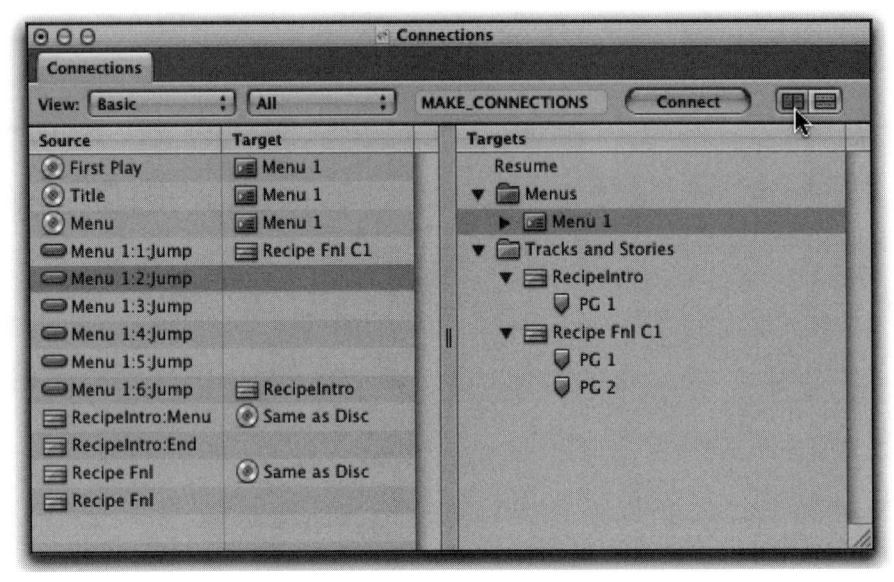

Figure 11-0 — The Connections Tab.

Goals

At the conclusion of this chapter, you should:

- Understand how the **Connections** window works
- Understand how to **connect project elements**
- Understand how to make **Control-Click connections**
- Understand how to **make connections in the Menu Editor**
- Understand how to **make connections from Buttons**
- Understand how to **make End Jump connections**
- Understand how **to define a Connection Source element**
- Understand how to **define a Connection Destination element**

Making Connections

Defining *Connections* is one of the most important aspects of authoring, because the connections you create are the only programmed pathways your DVD navigation can follow. Although you will see in Chapter 13, "Basic Scripting for DVD Studio Pro" that with Scripting we can often allow the DVD itself to make alternate or random choices as to navigation while it's playing, in actual fact, all of the choices the DVD is allowed to make are *predetermined* choices. So whether your Main Menu Button Number 1 goes to Track 1, or to Slideshow 2, a *Connection* has been made during authoring to allow that to happen.

Creating a Connection requires three things:

- There must be an element to make a Connection *from*. Each DVD project *Connection* must begin at some element and end at another element.

Making Connections

- There must be a *Source* to connect from within the element of origin; Sources include Buttons, Markers, and project actions like the End Jump Action.

- There must be a *Destination* element to make a Connection to (the Target).

Making Connections through Inspectors

One of the simplest methods of making Connections involves setting linkages using the Inspector. One of the easiest Connections to make is to set the "End Jump" property. Every Track and Slideshow has an *End Jump*, and a Target may be assigned to that action by simply clicking the pop-up to select the desired Target (see Fig. 11-1).

Figure 11-1 — Setting an End Jump action in the Track Inspector.

You will find the Target pop-up is available at any place where a connection may be made, with Targets listed that are appropriate to that source location.

In most cases, you should be able to Target every *Track, Menu, Button, Story, Slideshow, Slide,* or *Script* in your project from practically anywhere.

Making Connections from Menu Buttons

As you create buttons on your Menus, many of the connections will be automatically set by your choice of function in the Drop Palette that appears when you drag-and-drop an Asset or Folder onto the Menu Editor. Button Connections can also be made by Control-Clicking on a button and then setting the desired Target from the pop-up (see Fig. 11-2).

Figure 11-2 — Making a Connection from a Button by Control-Clicking.

Resume Function

Buttons also allow selection of the *Resume* Function, which returns navigation control to the last played location in a Track or Slideshow.

As you can see, keeping track of Connections could quickly become a nightmare. What we need is a central "clearinghouse" for displaying, and editing (or making) Connections. And we have that.

319

The Connections Tab

The Connections Tab is a central location where all of this "connecting" wizardry can take place, although as we've seen, it's not the only location. The advantage of the Connections Tab is that all of the linkages are available in one location to make, break, or modify.

The best way to use the Connections Tab is to have the Outline Tab and the Connections Tab both open to view at the same time. If by some chance both the Outline and Connections Tab are located in the same Quadrant, move the Connections Tab to a larger Quadrant so you can have more room to use for display—you will need it!

My personal preference is to have the Outline Tab in the upper left Quadrant, and the Connections Tab in the upper right one, since I think this gives the largest display area for the Connections Tab. Since DVD Studio Pro is SO customizable, you can put the Tabs wherever they feel comfortable for you, and save that layout as your default configuration.

Meet the Connections Tab

Once you Select a project Element (Track, Menu, Slideshow, Script, Story, or even the Disc itself), the Connections tab will display the Sources within that Element, and their current Connection state.

If the Target column is empty, that Source is not currently linked.

The Connections Tab shows three columns in its default view:

At left, the *Source* and (current) *Target* columns, and at right, the (possible) *Targets* Column (see Fig. 11-3).

> **NOTE!**
> Don't confuse *Target* (on the left) with *Targets* (at the right).

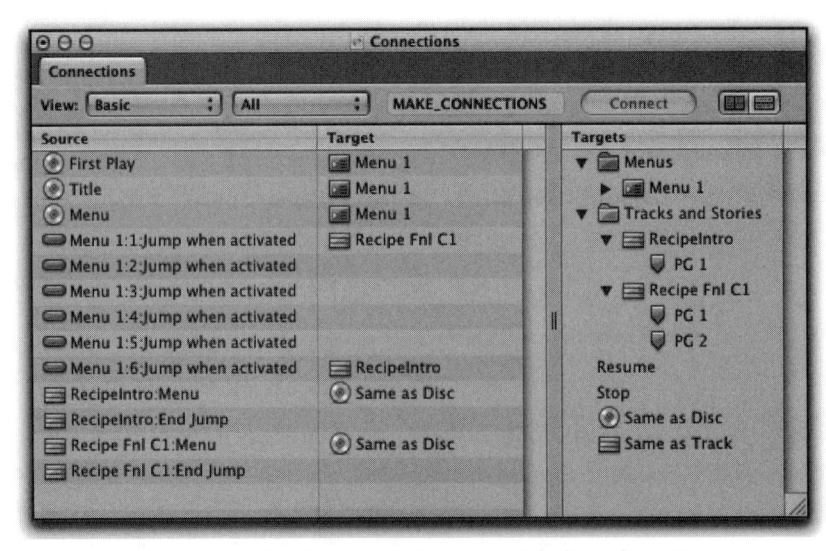

Figure 11-3 — The Connections Tab in its default configuration.

Making Connections

The Connections Tab Tools

Across the top of the Connections Tab are some of the now-familiar tools that you have seen in other Tabs. There is a *View*: pop-up, (as seen in Fig. 11-4) an,d next to it, a *view filter* pop-up that can be set to display only *Connected* sources, only *Unconnected* sources, or All (both)—see Figure 11-5.

Figure 11-4 — The View: pop-up tool.

Figure 11-5 — The Selection Filter pop-up (it's unnamed).

To the right of that is a field that displays the name of the currently selected Project element, so you can remember what you are connecting! Next is the *Connect* button shown in Figure 11-6, which becomes *Disconnect*, shown in Figure 11-7, when an already-connected Source is selected in the Source list.

Connections Tab Layout

Last, there are two buttons at extreme right, which change the Connections Tab configuration from divided Vertically to divided Horizontally. Vertical (Fig. 11-8) is the default for this Tab, and it puts the Sources on the Left and the destination Targets on the right side of the window. In Horizontal mode (Fig. 11-9), the destination Targets are displayed in the lower half of the window.

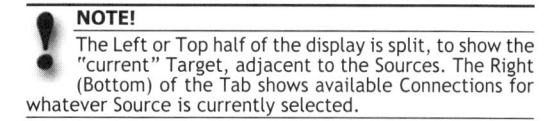

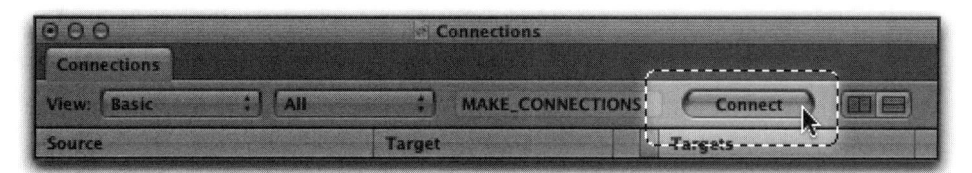

> **NOTE!**
> The Left or Top half of the display is split, to show the "current" Target, adjacent to the Sources. The Right (Bottom) of the Tab shows available Connections for whatever Source is currently selected.

Figure 11-6 — The Connect Button becomes...

Figure 11-7 — ...the Disconnect Button when a connected source is selected.

321

DVD Studio Pro 4

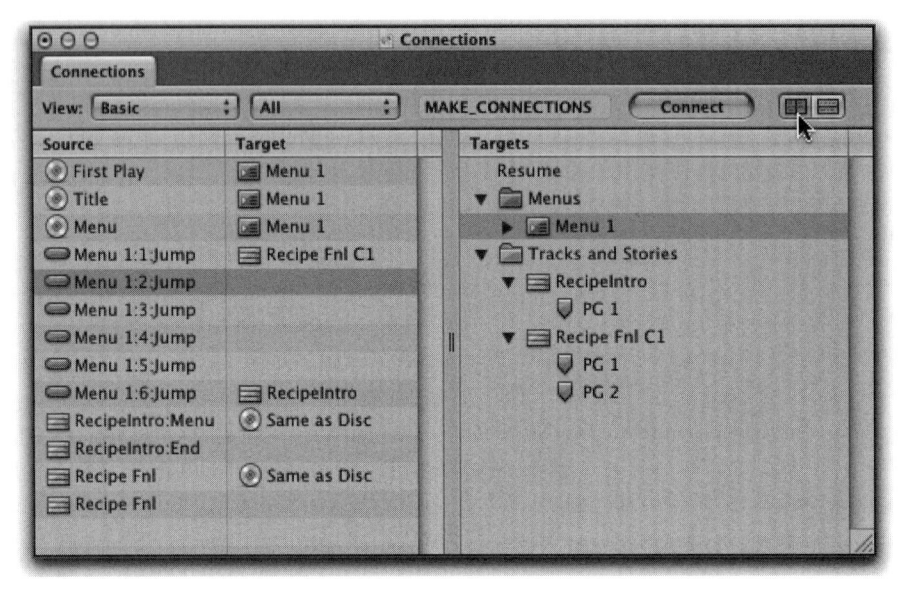

Figure 11-8 — Connections Tab Vertical Layout.

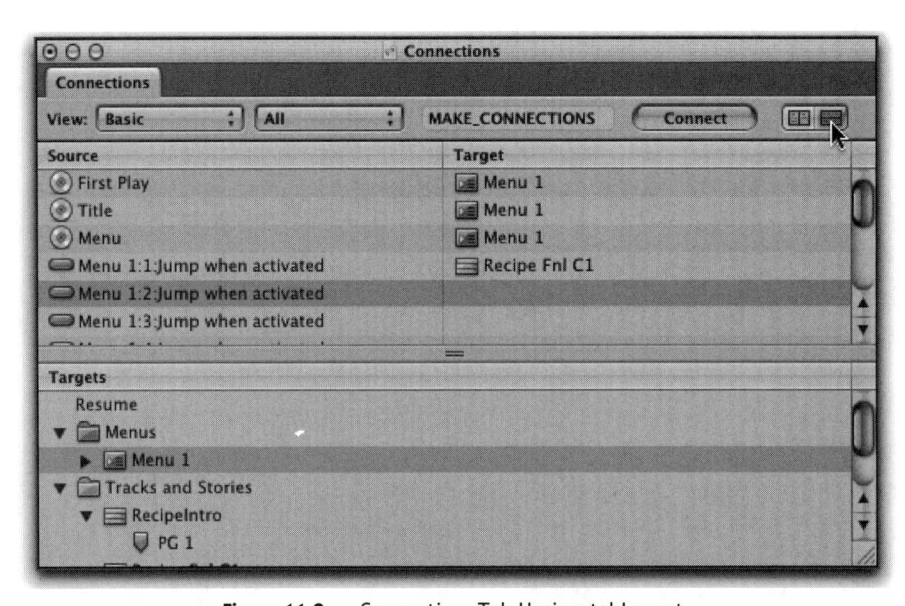

Figure 11-9 — Connections Tab Horizontal Layout.

Making Connections

Using the Connections Tab

To view and use the Connections Tab, use these steps:

- Select the Element you wish to connect *from*, in the *Outline* Tab;

- Select the desired Connections Tab *View*: Basic, Standard, or Advanced;

- Adjust the Filter pop-up to show *Connected*, *Unconnected*, or *All*.

The contents of the Connections Tab Source column will change depending on the currently selected Element. Menus have different Connections possibilities from Tracks, and the same goes for Slideshows, Stories, and Scripts as well. Every selected element displays only the connections appropriate to its current configuration; i.e., if you have a Track with Markers, the display will reflect the number of markers defined.

Once a Source is selected, and the depth of view adjusted with the View and Filter pop-ups, you will see any existing Connections in the (current) Target column immediately adjacent to the Sources column. The other half of the window (Right or Bottom) lists the available Target possibilities (destinations) for that Element's sources.

> **NOTE!**
> Select to view the Disc Connections with *View: Advanced*, *All* to see everything in the Project; select DISC with *View: Advanced, Unconnected* to see what's in the project that has not yet been connected.

Basic, Standard, Advanced— the View: Pop-up

The *View*: pop-up allows you to control the number of Sources displayed for the currently selected Element.

- *Basic*—is the simplest view, and shows the fewest connections. For a Track, Story, or Slideshow, this means the Menu Button action, and the End Jump action. For a Menu, this means Buttons. For the Disc, this means the Title and Menu Buttons, and the First Play.

- *Standard*—The Standard view adds the most common possibilities: For Tracks, Stories, and Slideshows, Pre-scripts are listed, as well as Menu and End Jump function for each Marker, Slide, or Story Entry Point. For Menus, Pre-scripts are listed along with a Menu Timeout Function. This Jump is processed after the Menu has been on screen for the amount of time specified in the Menu Inspector General Tab.

- *Advanced*—the Advanced setting adds the remaining functions not yet shown in Basic or Standard. This includes the Additional Remote Control functions for the Disc, and the Return function for Menus. *Advanced* also adds the Additional Menu Controls found on a few DVD player remotes: buttons for Angle, Audio, Subtitle, and Chapter Menus; DO NOT confuse these functions with standard DVD remote Control function for Angle, Audio, and Subtitle STREAM selection.

323

DVD Studio Pro 4

> **DVD Reality Check!**
> FEW, if any, DVD players have these buttons on their remote. You may find that the better method of accessing these kinds of Menus is the traditional method—use a Button from the Main Menu to take you to a Setup Menu for Angles, Audio, and Subtitles, or a Chapter Menu to select Chapter points to begin playing from.

Next, Previous Jumps—Special Cases

The Next and Previous Jump settings are only visible in the *Advanced* view, for Tracks or Slideshows. Here's why it's important to remain aware of these settings:

While your Track or Slideshow is playing, the Next and Previous buttons on the DVD remote allow you to skip to the next Marker, or Slide. This works fine up until you get to the very last Marker or Slide before the end. Due to the quirks of the DVD Spec, your Next button will stop working at the last Marker in the Track before the end marker. The same phenomenon occurs in slideshows, at the last Slide.

One way to work around this is to use the Next Jump command, which will allow a linkage to be made that can activate a Jump when there are no more Markers or Slides.

There is a catch, though, if you do use these connections:

1. The Menu command can only be set for the first Marker, and
2. The DVD player will no longer display a play time for this Track or Slideshow.

There are a few interesting workarounds, to solve this annoying issue:

1. In a Track, drop a black still image in at the end of the track, and set a Chapter Mark at the beginning of that slide—make it short (15 frames, if you can); the Next Jump will skip to that slide, and then exit immediately through the End Jump of the Track.

2. In a Slideshow, duplicate the last slide, without a Pause. That will force the Next button to work, leading you to the End Jump of the Slideshow. Since the image doesn't change, it will appear as if it has just "skipped" through the last slide, which is exactly what you want!

All, Connected, Unconnected—the Filter Pop-up

The unlabeled Filter pop-up allows you to further restrict the Source display to only those Sources that are Connected, Unconnected, or All (the default).

> **TIP!**
> Choosing *Disc, View: Basic, Unconnected* will quickly point out any missing links.

The Recipe4DVD Sample Projects

Perhaps the best way to see how to use the Connections Tab is to explore a project.

In the DVD that comes with this book is a folder of sample projects that help you explore DVD Studio Pro 4. Look for a folder called "** Projects." In that folder, you will find a project called "Recipe Web Project." Feel free to open that project into DVD Studio Pro 4 if you'd like to explore along with us, or you can just read about the steps. We'll explain everything as we go along.

Making Connections

Connections are extremely simple to make; you can:

- Use the Connect Button to link a Source and a Target, or

Making Connections

- Click on a Source Element, then Control-Click in the Target Column, and select the desired Target from the pop-up menu, or

- Drag a Target element to connect it with a Source element.

To Make a Connection Using the Connect Button:

- Select the desired Element in the Outline Tab, then

- Select the desired Source in the "Source" Pane, then

- Select the desired Target in the "Targets" Pane, and

- Click the "Connect" button (which will now be active) (Fig. 11-10) or

- Select the desired Element in the Outline Tab, then

- Select the desired Source in the "Source" Pane, then

- Double-Click the desired Target in the "Targets" pane.

To Make a Connection using Drag-and-Drop:

- Select the desired Element in the Outline Tab, then

- Select the desired Source in the "Source" Pane, then

- Click on the desired Target in the "Targets" Pane, and

- Drag it to the "Target" field next to the chosen Source (Fig. 11-11).

> **NOTE!**
> If you do NOT see the black box selection rectangle, your chosen Target is not valid for that Source!

To Make a Connection using Control-Click:

- Select the desired Source Element in the Outline Tab, then

- Control-Click on the empty Target field adjacent to it, and

- Select the desired Target from the pop-up menu (Fig. 11-12).

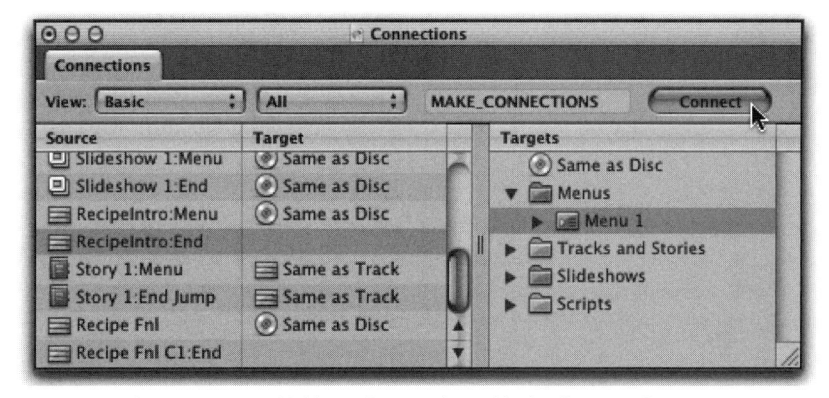

Figure 11-10 — Making a Connection with the Connect Button.

325

DVD Studio Pro 4

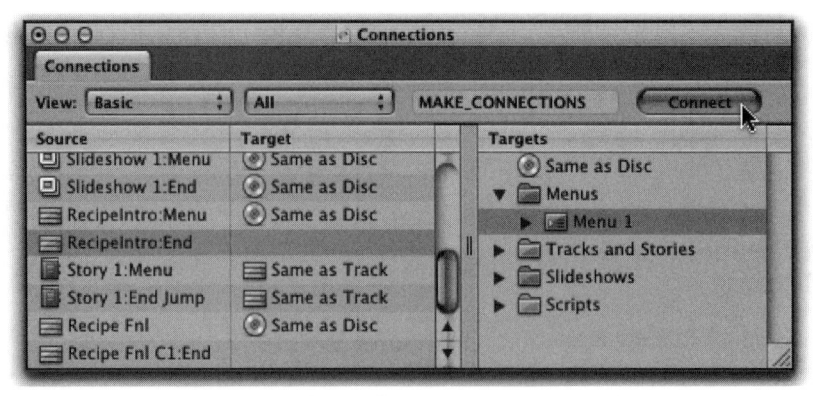

Figure 11-11 — Making a Connection using Drag-and-Drop.

Figure 11-12 — Making a Connection using the Control-Click menu.

To Make a Connection using the Keyboard:

> **POWER USER TIP!**
> I can use the keyboard to connect? Yes, you can... (and this is a *real* Power User Tip!)

- Select any Source within the Connections Tab Source column; (you can use the Arrow Up and Down keys to scroll through the list).

- Holding the Control Key down and using the Arrow Keys will scroll you through the *opposite* side of the Tab (in this case, Targets) to select a Target. This side will show a light gray highlight.

- Once you have highlighted the desired Target, hit Return to Connect; If already connected, Return will Disconnect them.

It works the other way too (those clever Apple engineers!):

- Select any Target within the Connections Tab Targets column; (you can use the Arrow Up and Down keys to scroll through the list).

- Holding the Control Key down while using the Arrow Keys will scroll you through the

Making Connections

opposite side of the Tab (in this case, Sources) to select a Source. This side will show a light gray highlight.

- Once you have highlighted the desired Source, hit Return to Connect; If already connected, Return will Disconnect them.

> **NOTE!**
> The Highlight Colors between the two sides are different! Check out the highlights colors very carefully while doing this.

I call the side you click first the *Primary* highlight zone. The Purple highlight occurs in the Primary highlight zone (and that zone scrolls with Arrow Keys).

I call the opposite side of the Tab the *Secondary* Highlight zone. The light gray highlight occurs in the Secondary Zone (that zone scrolls with Control-Arrow keys).

To reverse the primary and secondary selection, just Click in the other side of the Tab. The highlight colors will reverse.

> **POWER USER TIP!**
> If you need to scroll down a list of Targets or Sources while connecting, use the *Enter* key to make the Connection, instead of the Return key. Using *Enter* will make the connection, and *scroll down automatically to the next Target or Source in line*, depending on which side your Primary Highlight is (the purple one). You can then scroll down your list of connections quick as a rabbit, using the keyboard and not the mouse.

Breaking or Modifying Connections

To Break a Connection using the Disconnect Button:

- Select the desired Element in the Outline Tab, then

- Select a *connected* Source in the *Source* Pane, then

- Click the *Disconnect* button (which will now be active—see Fig. 11-13).

To Break a Connection using the Delete Key:

- Select the desired Source Element in the Outline Tab, then

- Select *a connected* Source in the "Source" Pane, then

- Use the Delete Key.

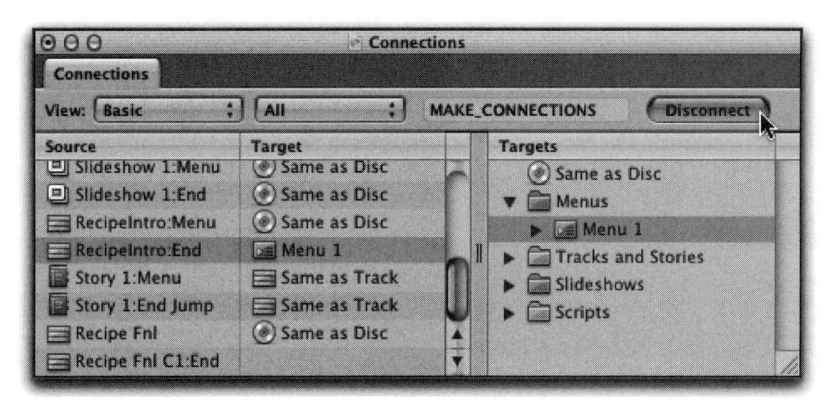

Figure 11-13 — Breaking a Connection with the Disconnect Button.

327

DVD Studio Pro 4

To Modify a Connection:

- Select the desired Element in the Outline Tab, then
- Select *a connected* Source in the "Source" Pane, then
- Select and drag a new Target into the "Target" column adjacent to the Source column, **or**
- Select the desired Source Element in the Outline Tab, then
- Select *a connected* Source in the "Source" Pane, then
- Control-Click and select a new Target from the pop-up menu, **or**
- Use the Keyboard scrolling technique above to make selections;
- Selecting a different Target for an already connected source will MAKE that connection if you hit Return or Enter.

Setting the First Play (IMPORTANT!)

Setting the First Play is a vital task—If you don't set the First Play, the finished disc will not automatically begin to play, which could cause some concern among DVD enthusiasts who are already used to DVD's automatic start behavior.

What Is the First Play?

Contrary to some reports, the First Play isn't really an entity in itself—it is a POINTER to a project Element, which becomes the First Play(ed) Element—see how it works? (First Play is shown in Fig. 11-14.)

As with everything else in DVD Studio Pro, setting the First Play is easy—just Display the Disc Inspector and click on the First Play pop-up to set a destination. All legal destinations will be displayed for you.

Make your selection from the pop-up target list.

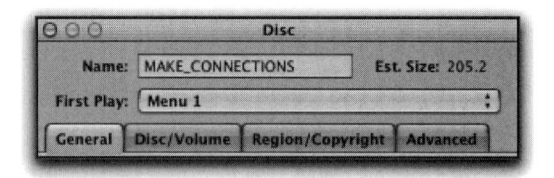

Figure 11-14 — The First Play Connection.

> **TIP!**
> To set the **First Play** quickly, Control-Click in the Outline Tab and set the First Play from the pop-up menu. You may now also Control-Click on an Element Tile in the Graphical View and set the First Play attribute from there (see Fig. 11-15).

Figure 11-15 — The First Play Connection.

Checking the Connections

The Simulator is the method you should use to check the accuracy of your Connections. The Viewer does

Making Connections

not process links, so Simulating is the best way to verify your DVD is on the road to proper operation.

See Chapter 14, "Building and Formatting" for information on using the Simulator. There are some very nice improvements and additions to the Simulation capabilities of DVD Studio Pro 4.

Avoiding Remote Control Key Problems

The DVD Remote Control is part of every DVD player, and manufacturers are required to have certain keys on their remote controls.

The required controls include: Transport Controls (Play, Pause, Stop, Rewind, FFwd, Next, Previous) Menu Navigation Arrows and Enter key, Title (or Top Menu), Menu, Angle, Audio, Subtitle, and Return.

Some (but not the majority of) DVD Players also include additional buttons for: Angle MENU, Audio MENU, Chapter MENU, and Subtitle MENU.

While DVD Studio Pro gives you complete flexibility to determine what Remote Control keys do within your project, do NOT confuse the Angle/Audio/Subtitle/Chapter *MENU Select keys* (which are optional) with the Angle/Audio/Subtitle *Stream Select keys* (which are required on all DVD remote controls).

An alternate method of making these Connections would be to include buttons on your Main Menu to connect to a Setup Menu for Audio, Angle, and Subtitle selections, and a Chapter Menu for Chapter Selection. In DVD Studio Pro, you can make a Chapter Menu painlessly by linking a Movie with embedded Markers in the Menu Editor (see Chapter 9, "Authoring Menus").

Likewise, it is a simple matter to create a Setup Menu where the Buttons themselves can make an Audio, Angle, or Subtitle selection at the same time as initiating playback of the Track. Alternatively, you can always use Scripting (See Chapter 13) to set a GPRM to keep track of your Angle, Audio, or Subtitle Stream selections. There is ALWAYS more than one way to do practically EVERYthing in DVD Studio Pro.

Understanding Sources and Their Connections

The following list organizes the displayable Sources into their display Groups: Basic, Standard, and Advanced, explains the action of each Connection, and shows alternate places where they may be connected.

Basic Sources in Detail

The following connections must be made prior to building, or your DVD will likely fail to navigate properly. *First Play* is VERY important!

Disc (Basic)	Function	Also can be set in:
First Play	Sets the disc's first played Element	Disc Inspector (Default: Menu 1)
Title Call	Sets the destination when TITLE is pressed while a Track or Slideshow is playing	Disc Inspector General Tab
Menu Call*	Sets the destination when MENU is pressed while a Track or Slideshow is playing	Disc Inspector General Tab

*This setting may be overridden by similar settings on the Elements themselves.

Menu (Basic)	Function	Also can be set in:
Button: Jump	Sets the destination of a Button activation	Control-Click on Button, or use Button Inspector, or Drag Element to Button

Track (Basic)	Function	Also can be set in:
End Jump	Sets the destination when end marker is reached	Track Inspector Top Area
Menu Call	Sets the destination when MENU is pressed	Track Inspector General Tab
Marker: Button: Jump	Sets the destination when a Marker button is activated	Subtitle Inspector Button Tab

Story (Basic)	Function	Also can be set in:
End Jump	Sets the destination when end marker is reached	Story Inspector End Jump (default: Same as Track)
Menu Call	Sets the destination when MENU is pressed	Story Inspector General Tab (default: Same as Track)

Slideshow (Basic)	Function	Also can be set in:
End Jump	Sets the destination when end marker is reached	Slideshow Inspector End Jump (sometimes set by default)
Menu Call	Sets the destination when MENU is pressed	Slideshow Inspector General Tab

Making Connections

Standard Sources in Detail

The following Sources and Connections will be listed in the Standard Source group, which will also include the Basic Sources listed above.

Menu (Standard)	Function	Also can be set in:
Pre-Script	Specifies a script to run before displaying this Menu	Menu Inspector Menu Tab
Menu Timeout	Sets Target when Menu Timeout function expires	Menu Inspector General Tab

Track (Standard)	Function	Also can be set in:
Pre-Script	Specifies a script to run before displaying this Track	Track Inspector General Tab
Marker X: Menu Call	Sets the destination when MENU is pressed	Marker Inspector General Tab

*This setting overrides the Menu setting in Disc and Track.

Marker X: End Jump	Sets the destination to go to when this Marker's section has completed playback	Marker Inspector: End Jump

*The Marker End Jump will be executed at the frame prior to the next Marker.
NOTE—This setting will cause a non-seamless playback WITHIN the Track.

Story (Standard)	Function	Also can be set in:
Pre-Script	Specifies a script to run before displaying this Story	Story Inspector General Tab
Marker X Entry: Menu Call	Sets the destination when MENU is pressed	Story Marker Inspector: Remote

*This setting overrides the Menu setting of Disc and Track.

Marker X Entry: End Jump	Sets the destination to go to when this Marker's section has completed playback	Story Marker Inspector: End Jump

*The Story Marker End Jump will be executed at the frame prior to the next Marker.
NOTE—This setting will cause a non-seamless playback WITHIN the Story.

Slideshow (Standard)	Function	Also can be set in:
Pre-Script	Specifies a script to run before displaying this Slideshow	Slideshow Inspector General Tab
Slide X: Menu Call	Sets the destination when MENU is pressed during this slide	None
Slide X: End Jump*	Sets the destination to jump when next Slide is reached	None

*Using this function defeats a seamless continuous slideshow.
(These Slideshow functions are not listed in the Manual for Version 1,0, but DO exist.)

331

Advanced Sources in Detail

The following Sources and Connections will be listed in the advanced Source group, which will also include the Basic and Standard Sources listed above.

Disc (Advanced)	Function	Also can be set in:
Angle Menu Call*	Sets the destination when ANGLE MENU is pressed on DVD Remote Control	Disc Inspector Advanced Tab
Audio Menu Call*	Sets the destination when AUDIO MENU is pressed on DVD Remote Control	Disc Inspector Advanced Tab
Chapter Menu Call*	Sets the destination when CHAPTER MENU is pressed on DVD Remote Control	Disc Inspector Advanced Tab
Subtitle Menu Call*	Sets the destination when SUBTITLE MENU is pressed on DVD Remote Control	Disc Inspector Advanced Tab

*This setting may be overridden by similar settings on Tracks and Slideshows.
NOTE—Many DVD players DO NOT HAVE THESE MENUS BUTTONS—use carefully.

	Function	Also can be set in:
Return**	Sets the destination when RETURN is pressed on DVD Remote Control	Disc Inspector General Tab

**This setting may be overridden by similar settings on Menus.

Menu (Advanced)	Function	Also can be set in:
Return*	Sets the destination when RETURN is pressed on DVD Remote Control	Menu Inspector Menu Tab

*This setting overrides the Disc setting.

Track (Advanced)	Function	Also can be set in:
Angle Menu Call*	Sets the destination when ANGLE MENU is pressed on DVD Remote Control	Track Inspector General Tab
Audio Menu Call*	Sets the destination when AUDIO MENU is pressed on DVD Remote Control	Track Inspector General Tab
Chapter Menu Call*	Sets the destination when CHAPTER MENU is pressed on DVD Remote Control	Track Inspector General Tab
Subtitle Menu Call*	Sets the destination when SUBTITLE MENU is pressed on DVD Remote Control	Track Inspector General Tab

*This setting overrides the Disc setting. DEFAULT: "Same as Disc."
NOTE—Many DVD players DO NOT HAVE THESE MENUS BUTTONS— use carefully.

Making Connections

Track (Advanced)	Function	Also can be set in:
Next Jump**	Sets the destination to go to when NEXT is pressed after the last Marker is passed	None
Prev Jump**	Sets the destination to go to when PREV is pressed before the first Marker	None

**Read "Next and Prev Jumps" earlier in this chapter—Important information.

Story (Advanced)

There are no Story-related Sources added in the Advanced Group— See Basic and Standard.

Slideshow (Advanced)	Function	Also can be set in:
Angle Menu Call*	Sets the destination when ANGLE MENU is pressed on DVD Remote Control	Track Inspector General Tab
Audio Menu Call*	Sets the destination when AUDIO MENU is pressed on DVD Remote Control	Track Inspector General Tab
Chapter Menu Call*	Sets the destination when CHAPTER MENU is pressed on DVD Remote Control	Track Inspector General Tab
Subtitle Menu Call*	Sets the destination when SUBTITLE MENU is pressed on DVD Remote Control	Track Inspector General Tab

*This setting overrides the Disc setting. DEFAULT: "Same as Disc."
NOTE—Many DVD players DO NOT HAVE THESE MENUS BUTTONS—use carefully.

Next Jump**	Sets the destination to go to when NEXT is pressed during the last Slide	None
Prev Jump**	Sets the destination to go to when PREV is pressed during the first Slide	None

**Read "Next and Prev Jumps" earlier in this chapter—Important information.

Understanding Targets and How They Work

The list of Targets displayed in the Connections Tab is the same list for all Sources; however, not all of them are usable for all Sources.

Here is what happens when you select a specific Target:

- Menu—Selecting a Menu as Target will cause that Menu to open, with the default Button Selected. If there is a Pre-script assigned to this Menu, it will execute when the Menu is navigated to. Selecting a Button within that Menu will cause that Button to be highlighted when the Menu opens, overriding the default Button.

- Track—Selecting a Track will cause the specific Track to begin playing from Chapter 1. If there is a Pre-Script attached, it will execute when the Track is selected. Selecting a Marker within the Track will cause the Track to begin playing from that location, *and it will continue to play to the Track End*, unless a remote control function (like Menu or Title) causes playback to end prematurely. At the End of the Track, the Track's End Jump will be processed.

> **NOTE!**
> Jumping to a Marker in a Track gets you into the Track, but not out before the end marker.

- Story—Selecting a Story will cause the specific Story to begin playing from Story Entry Marker 1. If there is a Pre-Script attached to the Story (NOT the Track), it will execute when the Story is selected. Selecting a Story Entry Marker within the Story will cause the Story to begin playing from that location, *and it will continue to play to the End of the Story*, unless a remote control function (like Menu or Title) causes playback to end prematurely. At the end of the Story, the Story's End Jump will be processed, which quite probably is different from the End Jump of the Track that contains the Story.

Jumping to a Story defines an In point, possibly a series of additional segments to play, and generally an out point other than the end of the Track. Stories are very powerful, as they can re-use content without requiring the assets to take up additional space in the build.

- Script—Selecting a Script as Target will immediately execute that Script and transfer control to the instructions within that Script.

The following Targets only appear for certain Sources:

- Resume—This command will execute a "resume play" function if a resume pointer has been set by the DVD (meaning you have previously played some content in a Track or Slideshow). It may have unpredictable results if used from a Menu before any Tracks or Slideshows have been player. Test this before you deliver a project!

- Stop—Does exactly what it says. This command will make the DVD layer stop playing when it is executed. This is not to be used lightly. This command may only be assigned to the End Jump settings of Markers and Slides. I cannot think of a single good reason to use it there, unless it is a specific Presentation DVD where you have no need for navigation to continue on past that slide or chapter.

Making Connections

- Same as Disc—Allows the specific Source to perform the behavior programmed into the same connection in the Disc. This can only be assigned to sources that specify Remote Control actions, like Menu, Audio Menu, Angle Menu, etc.

- Same as Track—Allows Story Sources to perform the same as their counterpart Sources in the Track. Can only be assigned to Story Sources like Menu, End Jump, etc.

Summary

Each DVD project you create is a fluid collection of Elements: Tracks, Menus, Slideshows, and Scripts. You envisioned that they would perform in a certain manner when you designed this DVD project. However, your DVD will only perform in the manner you have planned for it if you make the proper *connections* between the project elements.

Chapter 12

Creating Subtitles

DVD Studio Pro 4

Industry, which saw DVD as the next generation in Home Video distribution (and they were right!).

To facilitate the worldwide enjoyment of films in territories other than those that could speak the film's original language, Hollywood has long used the translating and *Subtitling* of the spoken dialog from the original language of the film. Subtitling means adding (usually by superimposing) viewable graphics that contain the written equivalent of the spoken dialog. This is accomplished in film releases by printing the Subtitle graphics onto a copy of the finished print, where they become a permanent part of the film.

You may have already seen subtitles in action if you are a fan of Asian cinema, especially martial arts movies, a great number of which have been subtitled into English or other non-Asian languages on top of their original Cantonese or Mandarin spoken soundtracks. Home video releases of these movies, even on VCD or DVD, may in many cases still carry these original subtitles, permanently burned into the video encodes. (Good thing, too—my Mandarin is pretty rusty!)

Goals

The goal of this chapter, is for you to:

- Understand the role and usage of Subtitles
- Understand how to create them using the Track Editor
- Understand how to properly Author Subtitles
- Understand the difference between Subtitles and Closed Captions
- Understand how to Import Subtitle Files
- Understand how to Create individual subtitles
- Understand the limitations of using subtitles

While this works fine for film releases, it is expensive to do, and requires keeping track of the correct print to be sent to the correct territory. Oh—one more thing: Subtitling only works for one language at a time, making subtitled prints difficult to distribute in territories other than the ones they are subtitled for!

Encoding an MPEG file from a Subtitled movie would not work—you would be stuck with the language that was on the print when it was encoded. What Hollywood needed was an easy and cost-efficient method to create multiple Subtitles in DVDs that could accommodate the need for distribution in multiple territories. The DVD Forum created just such a tool when they codified the graphical Subtitle function for DVD video discs.

Some Subtitle Basics

While the DVD Specification was being formulated, much input was provided from the Hollywood Film

Creating Subtitles

Uses for Subtitles

Of course, the most common use is to translate and display the dialog in a different language. But Subtitles can also be used to display production notes (as in the *Ghostbusters Collector's Edition*), commentary tracks (as in *The Abyss Collector's Edition*), or even graphical watermarks or logos. Because these graphics may be changed every few frames (within limitations), it is even possible to create rudimentary animation, or add graphical diagrams to commentary tracks, like the Telestrator™ tool used on television to diagram football plays or military strategies.

Video Commentaries

In fact, probably one of the most unique uses of Subtitle streams to date has got to be the video commentary track created for *Ghostbusters Collector's Edition*, and continued in *Men in Black I and II* (see any of those DVDs, for a display of this feature). The video commentary uses the Subtitle Stream to display a shadow image of the individuals who you hear in the audio commentary track (like *Mystery Science Theater 3000*, if you've seen that show), and in *Men in Black*, you can watch the commentators pointing out things in the film, and see the Telestrator™ effect drawing arrows and circles in the Subtitle track to identify where they are on the screen! It was very innovative...for its time.

Buttons over Video

Other interesting uses of Subtitles includes *Buttons over Video*, where an interactive button is created over the encoded video in a track, as opposed to in a Menu. An excellent example of this technique can be seen in New Line Cinema's InfiniFilm DVD releases, where interactive buttons are displayed briefly at the beginning of each chapter, allowing selection from among some "behind-the-scenes" video clips. You could almost think of this as a Movie with *hyperlinks*, reminiscent of a web page. When the behind-the-scenes video clip is finished, it navigates back to the Chapter where it was selected, allowing the movie viewing to continue. This is a very cool feature, although your DVD projects may not yet need this kind of feature.

Some Demonstrations

If you have a copy of Jim Taylor's *DVD Demystified 2nd Edition* around, now might be a good time to break out the DVD and check out the Subtitle demonstration section, to see what can be done with Subtitle streams. The *3rd Edition* should be out about the time you are reading this book.

Naming Conventions

Throughout this chapter, Subtitles will be referred to as "Subtitle events." Apple refers to them as "Subtitle Clips" in the DVD Studio Pro manual. Both words mean the same thing—a yellow subtitle area, which exists on one of the 32 Subtitle streams in a Track. I think of them as *events*, to distinguish them from video *clips*. That's the real difference—solely one of semantics.

What Are DVD Subtitles?

Subtitles are graphical bitmaps that can be superimposed on top of the MPEG video content playing in a Track. These bitmaps can be used in a number of different ways within DVD Studio Pro, usually as dialog translations (Subtitles). They can be turned on when desired, and turned off when not needed (see Figs. 12-1 and 12-2a).

DVD Studio Pro 4

DVD by providing dialog translations in many languages. This is one method of overlaying text on top of video in a DVD, but there is another method: Closed Captioning.

Closed Captions versus Subtitles

Subtitles should not be confused with *Closed Captions*, a method of embedding text into the scan lines of the video picture, and decoding them from and displaying them on top of the video signal in real time (see Fig. 12-2b). Closed Captions can be authored using DVD Studio Pro, but must be created in another application (see our DVD for a demo version of *Mac Caption*). Viewing the Closed Captions requires a decoder of some kind, usually built into the Television receiver or Video Monitor. While DVD settop players can play back Closed Captioned DVDs, many computers that can easily play DVDs

Figure 12-1 — A snippet of MPEG without a Subtitle.

Figure 12-2a The same snippet with a DVD Text Subtitle.

The DVD Spec allows for inclusion of up to 32 low-bandwidth Subtitle streams in each presentation (Track) of your DVD. These can be used to enhance and extend the multi-language capabilities of the

Figure 12-2b — The same snippet with a Closed Caption.

Creating Subtitles

with Subtitles cannot decode and display DVD Closed Captions—it's a function of the software DVD player. You may wish to bear this in mind when deciding on whether to use Closed Captioning or Subtitling on your DVD titles. Subtitles are very easy to play back, while Closed Captions require a CC decoder. (See Chapter 17, "Advanced Authoring" for information on using Closed Captioning in DVD Studio Pro.)

How DVD Subtitles Work

Subtitles in DVDs are displayed using a DVD playback capability called the *Subpicture*. The Subpicture is actually a graphical bitmap layer that is superimposed on top of the MPEG Video playing in the background layer of the picture. Using a switchable subpicture stream superimposed on top of the actual video means the Subtitle may be either on or off, and different subtitles can be used to provide language translations to serve many different territories of the world.

In actual fact, each DVD Track can contain up to 32 Subtitle Streams and 8 Audio Streams. The Subtitle Stream is limited in the number of colors it can display, specifically 4. The smaller bandwidth required for Subtitles means a large number (32) of subtitle streams can be stored and displayed efficiently in a single Track.

Subtitling Options in DVD Studio Pro

Depending on the specific method used to create each Subtitle Event, you can control the start and stop time of the Subtitle, a fade-on and fade-off time, the text, the font color, and the font style.

There are a few different ways to create Subtitles:

- **Individual Subtitle entry**—DVD Studio Pro can create Subtitles one at a time, by creating an event in a Subtitle Stream and entering Text or a graphic directly into that event. Each Subtitle event can control the color and font attributes of the text of that subtitle, as well as its position on the screen.

- **Import a Text File**—Multiple Subtitles in a stream can be created by importing a text file that contains the timecode start and end times for each Subtitle, as well as the specific text for that Subtitle. The format for this text file is described later in great detail in the DVD Studio Pro User Guide.

- **Import a Subtitle File (script with matching graphics)**—Subtitles may also be created by a third-party Subtitling company, and those files used to create a series of DVD Subtitle events. The difference in this technique is that third-party Subtitle files use a Script file in a specific format, and a folder of graphics files, which actually contain the rendered subtitle bitmaps. The Script file tells DVD Studio Pro which graphic file appears at which timecode location, and the color and position.

Subtitle files can be imported in the following formats:

- STL—the Spruce Technologies format
- SON—the Sonic Solutions format
- SCR—the Daiken Labs subtitle format

The Apple DVD Studio Pro manual contains extensive information on creating STL format scripts.

341

Importing SPUs from DVD Studio Pro 1 Projects

SPU files created in DVD Studio Pro 1 may be used with DVD Studio Pro 2 and above, but not directly. Although an SPU file cannot be imported, a project that *contains* an SPU can be imported into DVD Studio Pro 2, 3, or 4. The imported SPU will be converted into individual Subtitle stream events, each with a TIFF file providing the subtitle information for that event in graphical form.

- Graphic Subtitles can be created using Subtitle Events to display a specific bitmap graphic (see Fig. 12-3). To create this graphic, you can use the rules for creating a Subpicture Overlay for a menu. Up to four colors or grayscale levels may be used to define the graphics. This is the same technique discussed in Chapter 9, "Authoring Menus."

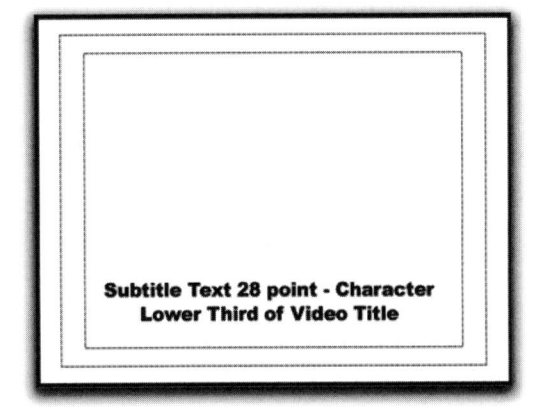

Figure 12-3 — A sample of a Graphic Subtitle.

- *Watermarks*—Instead of using text as the Subpicture Overlay with an Interactive Marker, a graphical file or logo can be used to superimpose a watermark or "bug" on top of the MPEG video contained in the Track (see Fig. 12-4). While technically not a Subtitle per se, this form of display may be helpful to some DVD Producers as a content-branding tool.

Figure 12-4 — A sample of a Watermark.

> **IMPORTANT!**
> Subtitles cannot run continuously underneath Chapter Markers; it's a violation of the DVD spec. While previous versions of DVD Studio Pro were able to run a continuous Subtitle underneath a Chapter Marker, causing problems, in Version 2.04, DVD Studio Pro changed the way it treats Subtitle events, making this less of an issue. The approach since DVD SP 2.04 automatically divides the Subtitle at each Chapter point. This provides the effect of an unbroken Subtitle while complying with the rules of the DVD spec.

Subtitle Tools in DVD Studio Pro 4

Creating Subtitles in DVD Studio Pro 4 involves three tools:

- *The Track Editor*—All Subtitles are entered in the Track Editor, on one of the 32 available Subtitle Streams. The Track Editor

Creating Subtitles

allows the creation of individual or multiple Subtitle events and the importation of any Subtitle Files or Text files. All Subtitles in DVD Studio Pro are visible as events in one of the 32 Subtitle streams. Regardless of the manner in which they are created, attributes of Subtitles may be modified using the Subtitle Inspector:

- *The Subtitle Inspector*—The Subtitle Inspector allows customization of the subtitle attributes for the currently selected Subtitle (you will see it in the figures that follow).

- *The Viewer*—The Viewer displays the Subtitles in the currently selected Subtitle stream over the Video and Audio streams currently selected with the Track Visibility buttons (review Fig. 12-2).

There are also some important characteristics that are set in various Panes in the Preferences. We'll cover them a bit later.

Checking Subtitles

Subtitles can be checked/verified in two different ways:

- *Preview*—The Viewer Tab can be used to play a Track and display the Video, Audio, and Subtitle stream currently selected. To switch a Subtitle stream during playback, read on:

- *Simulator*—The Simulator can be invoked to view a Track with Subtitles. The currently selected Streams can be changed while the Track is Simulating by using the Stream Selection tools in the Simulator display (see Chapter 14, "Building and Formatting" for more on using the Simulator display).

Switching Subtitle Streams in Preview

In a finished DVD, switching Subtitle streams can be accomplished with the Subtitle key on the Remote Control. This will cause the currently selected Subtitle stream to switch to the chosen new stream at the earliest opportunity. The behavior of the *Preview Tab* is a little different from a finished DVD disc, though.

According to Apple's DVD Studio Pro manual, switching Subtitle streams in Preview can only be accomplished at the following times:

- *The end of the currently playing subtitle event*. Streams cannot be switched, and Subtitles cannot be turned on or off until a subtitle event has concluded.

- *The commencement of a new subtitle event*. Streams can be switched mid-stream, but an event on a new stream will not be displayed until the start frame of that event passes the playhead after that stream has been selected.

In testing these rules, it seems that they are not followed rigidly, and you might expect subtitles to turn off when commanded, even if that command is issued in the middle of a subtitle event. The speed of the Macintosh you are using will directly affect how quickly these commands may be processed during preview. Expect slower machines (733, 867, 1 GHz G4s, etc.) to be somewhat sluggish in preview displays, compared to faster G4s and G5s.

DVD Studio Pro 4

Setting Subtitle Preferences

Certain preferences can be set that control the font, color, duration, and fade-in and fade-out times of Subtitles created either individually or by importing a file. In some cases, these settings may be overridden (most commonly by using an STL format Subtitle file that contains commands that directly specify the font and color setting, among other things).

To preset Subtitle preferences:

- *Text:* The DVD Studio Pro Font Preferences pane has a pop-up selector for *Subtitle Text Settings.* These settings may be used to preset the Font, Size, Justification and Vertical Placement of Subtitles created by hand or by importing a file (see Fig. 12-5).

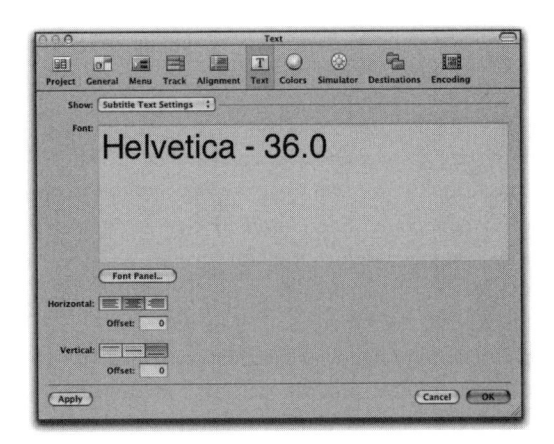

Figure 12-5 — Text Preferences Pane—Subtitles.

- *Color:* The Color Preferences pane has a pop-up selector for *Subtitle* that allows setting the type of color mapping used by the subtitles, as well as the color choice. (See Fig. 12-6.)

NOTE!
Color mappings for Subtitles use a slightly different scheme than menus. The primary color is the Text color (BLACK), which defaults to displaying White on-screen, but can be changed easily enough.

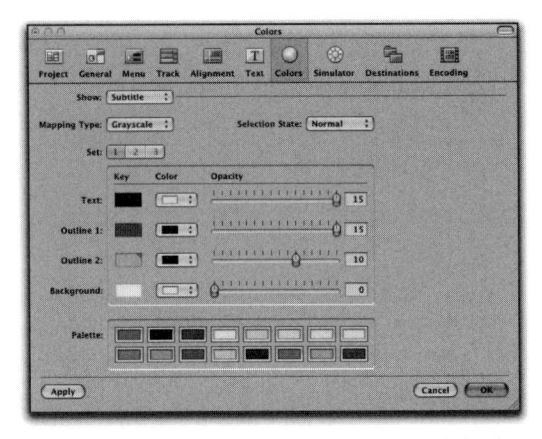

Figure 12-6 — The Colors Preferences Pane—Subtitles.

- *Timing attributes*: The Duration of a Subtitle as well as the fade-in and fade-out length (if any) can be set in the Subtitle section of the *General Preferences Pane.* (See Fig. 12-7.) Fade-in and -out lengths do not apply to subtitles created for Buttons over Video.

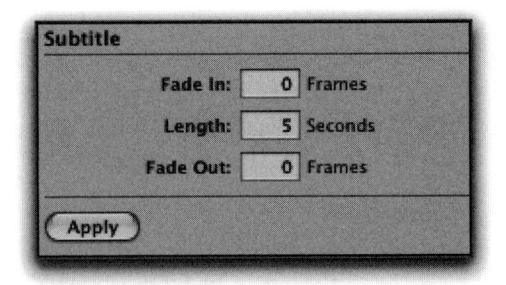

Figure 12-7 — The General Preferences Pane—Subtitle settings.

344

Creating Subtitles

Applying Settings Globally to an Entire Stream

Certain settings may be applied to all Subtitle events in that stream by using the *Apply to Stream* button for that particular feature.

In the Subtitle Inspector General Tab, the following settings may be applied:

- *Fade In and Fade Out timings* may be applied to the entire stream (see Fig. 12-8).

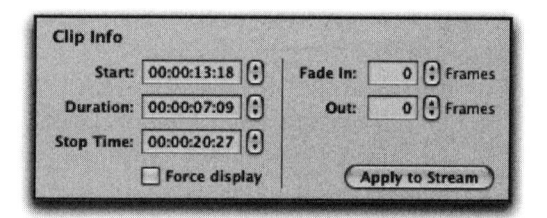

Figure 12-8 — Applying fade-in and -out settings. Make the desired selection adjustments and then click "Apply to Stream."

- *Justification and Vertical placement and offsets* may be applied to the entire stream (see Fig. 12-9):

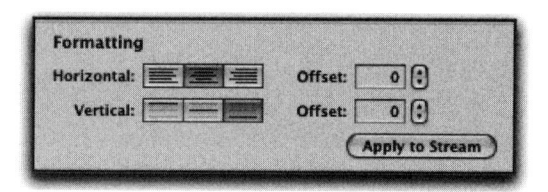

Figure 12-9 — Applying formatting settings. Make the desired selection adjustments and then click "Apply to Stream."

- *Color Settings* may also be applied to the entire stream (see Fig. 12-10):

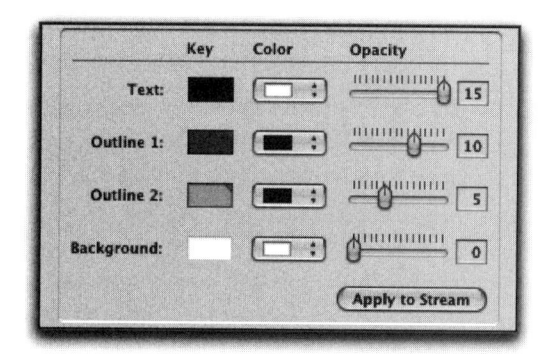

Figure 12-10 — Applying Color Settings. Make the desired selection adjustments and then click "Apply to Stream."

About the Subtitle Inspector

As with the other Inspectors, the various areas of the Inspector separate its functions.

Top Area

The Top Area only contains the field for Text entry. Once text is entered in this field, by hand or by import, it may be edited as needed (see Fig. 12-11).

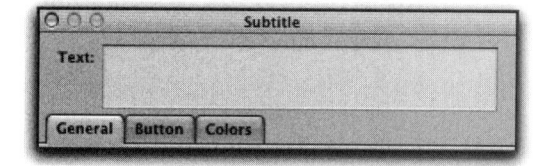

Figure 12-11 — Subtitle Inspector Top Area.

General Tab

- *Start:* The time where this Subtitle event begins. Enter a new value in this field, or drag the head of the Subtitle in the Track Editor. Changing the start time affects the

345

Subtitle's Stop Time, but not its duration. It just moves the Subtitle in the timeline.

- *Duration:* The length of the Subtitle event. If you change the duration, the Subtitle's stop time also changes to reflect the difference. You can also drag the head or tail of a Subtitle event in the Track Editor to change its duration.

- *Stop Time:* The time where the Subtitle event disappears. Enter a new value here or drag the Subtitle event's tail edge in the Track Editor. Changing the stop time also affects the Subtitle duration.

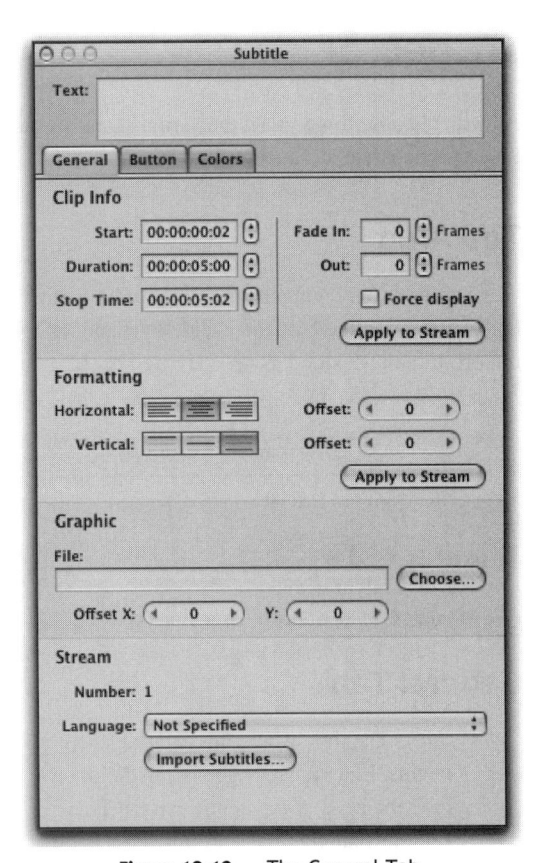

Figure 12-12 — The General Tab.

- *Fade In:* The length of time (in frames) that the Subtitle takes to fade in. The fade in will begin at the Subtitle's Start time. A value of zero (0) will cause the Subtitle to "pop on." (See Fig. 12-12.)

> **NOTE!**
> A Subtitle using Buttons over Video cannot use the Fade setting.

- *Out:* The length of time (in frames) that the Subtitle takes to fade out. The fade out finishes at the Subtitle's stop time.

> **NOTE!**
> Fades longer than 15 frames may not produce smooth fades, due to the limited number of opacity levels available.

- *Force display:* Selecting this checkbox forces Subtitles to appear, whether or not Subtitles have been disabled. Subtitles so set will only appear if their subtitle stream is currently selected.

- *Apply to Stream:* Clicking this button will apply the fade-in and fade-out settings to all Subtitle events in this stream.

Formatting

- *Horizontal:* Click the left, center, or right Icon to set the horizontal justification of the text.

- *Horizontal Offset:* Fine-tunes the horizontal position of the text, by pixels, from the selected horizontal setting. A value of 0 positions the text at its justification setting. Positive values move the text to the right and negative values move it to the left.

- *Vertical:* Click the top, center, or bottom Icon to set the vertical placement of the text.

Creating Subtitles

- *Vertical Offset:* Fine-tunes the vertical position of the text, by pixels, from the selected horizontal setting. A value of 0 positions the text at its justification setting. Positive values push the text down, and negative values pull it up from the default position.

- *Apply to Stream:* Clicking this button will apply the formatting settings to all subtitle events in this stream.

Graphic

- *File:* Displays the name of the graphics file used by this Subtitle. Click the *Choose*: button to select a file, or enter the path and name for a graphics file.

- *Offset X and Y:* Use to fine-tune the graphic's position.

- *Offset X* fine-tunes the horizontal placement. A value of 0 positions the graphic at its default position. Positive values push the graphic to the right and negative values pull it to the left.

- *Offset Y* fine-tunes the vertical placement. A value of 0 positions the graphic at its default position, while positive values push the graphic down and negative values pull it up.

Stream

- *Language:* Selects the stream's language tag, which applies to all subtitle events in the stream. You can also set this language tag from the stream language pop-up in the Track Editor.

- *Import Subtitles:* Click this button to import a subtitle file. This file may be a single file or graphics file in STL format.

> **NOTE!**
> Importing a subtitle file will delete any existing subtitle clips from the stream.

Button Tab

The Button tab in the Subtitle Inspector (shown in Fig. 12-13) is used to configure *Buttons Over Video*. This Tab is inactive unless you have created one or more buttons within a Subtitle event.

- *Name:* Buttons are named in the ascending numerical order of their creation, starting with "Button 1." You should consider renaming buttons to more accurately reflect their function.

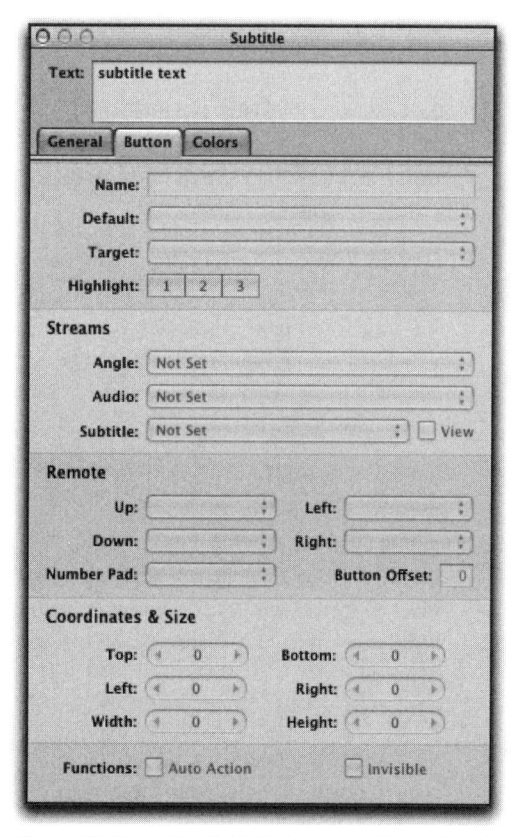

Figure 12-13 — The Subtitle Inspector Button Panel.

347

- *Default:* The Default button will be the one highlighted when the subtitle appears during playback of the title. As in earlier versions, there still does not seem to be a method to jump to a specific targeted button within this subtitle marker.

- *Target:* This pop-up menu is used to define a Button's target when activated. This may also be set using the Connections tab or by Control-clicking the button.

- *Highlight:* This selects the highlight set (1, 2, or 3) to use for this button.

- *Angle/Audio/Subtitle Streams:* Allows a button to select specific angle, audio, and subtitle streams to play back in a Track target, just as in Menus. You must select the "view" checkbox for a subtitle stream to be enabled as visible. However, selecting a Subtitle stream without the View checkbox would allow "forced" Subtitles to be made visible by the target Track.

- *Remote:* These controls define the button navigation relationship for this Subtitle. This relationship, as in a Menu, controls which button the Remote Control arrow keys will navigate to when pressed.

- *Number Pad:* Defines which buttons are directly accessible by a DVD player's numeric keypad. Choose All, None, or a button number from the pop-up menu. Buttons greater than the selected value are blocked from direct access, normally left at "All" unless you have a specific need to the contrary.

- *Button Offset:* The same as in a Menu, this allows the User to enter a number on the remote that is different from the actual button number. For example, in a series of 6-button chapter menus, a Button offset of 6 might be added to the second Chapter Menu, allowing numerical selection of buttons numbered 1 through 6 by entering 7 through 12 on the Remote. Whatever offset value is added to the Subtitle will affect all buttons, just as in a Menu.

- *Coordinates & Size:* These values set the location of each side of the button area, plus the button's height and width. This will precisely position the button's active area and size. All values are referenced from the upper-left corner of the menu, which is at pixel 0, line 0. You can also drag the button and each of its edges with the cursor.

- *Auto Action:* This attribute will cause the button command to execute immediately when the button is navigated to. Note that this is NOT supposed to mean if the button is selected as the default button, but rather if the button has been navigated to from another button.

> **NOTE!**
> Only the "Activated" color will appear when the button is navigated to. The "Selected" color will be ignored.

- *Invisible:* Applies to overlay-based buttons. Disables the normal, selected, or activated state highlights. This is useful when you want to have a subtitle with Easter Egg type invisible buttons, or buttons with text and no highlights.

Colors Tab

The Colors tab in the Subtitle Inspector (see Fig. 12-14) provides controls to set the display colors in a manner very similar to the Colors tab in the *Menu Inspector*. Both use the DVD Subpicture to display

Creating Subtitles

the pixels described in the subpicture overlay file, or subtitle text.

- *Mapping Type:* Select the color scheme used to create the overlay graphic. Chroma uses black, 255 red, 255 blue, and white. Grayscale uses black, 66% dark gray, 33% light gray, and white. This setting only applies to graphics files imported into the Subtitle.

- *Selection State:* Only the Normal setting is available unless you have created "Buttons over Video" in this Subtitle. If so, you can select the button state you want to configure, and that state's colors and transparencies are displayed.

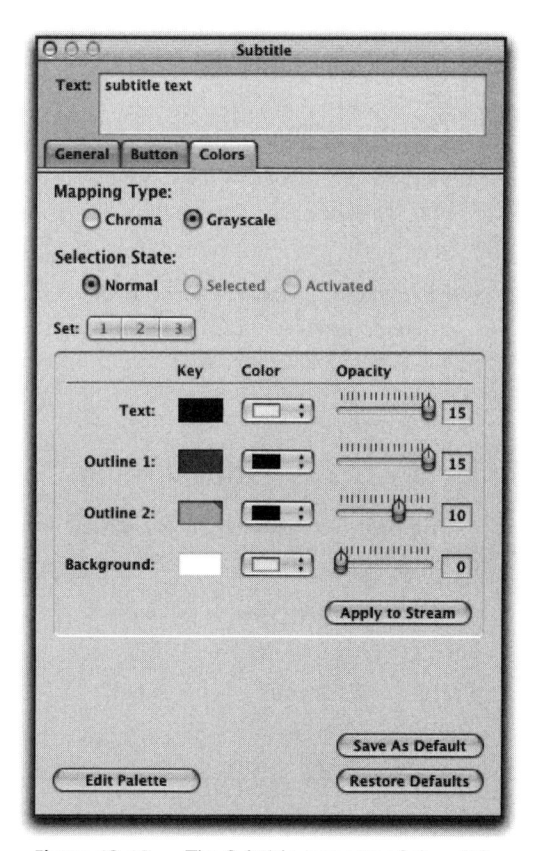

Figure 12-14 — The Subtitle Inspector Colors Tab.

- *Set:* Select the highlight set (1, 2, or 3) to configure. All new buttons use Set 1. This setting is accessible only if this Subtitle has buttons over video.

- *Key:* This area shows the key color for each of the four color controllers. Which set is displayed is determined by the Mapping Type setting.

- *Color:* Selects the color for each of the four color controls (Text, Outline 1 and 2, and Background) from the current 16-Color Palette. If this subtitle has Buttons over Video, you need to set all three Button states.

- *Opacity:* This function sets the opacity for each of the four color controls. You can drag the slider or enter a numerical value. 15 is completely opaque, 0 is completely transparent.

- *Apply to Stream:* This button applies the color and opacity settings to all subtitle clips in this stream.

- *Save As Default:* Click to save this color configuration as the default to be used on all new Subtitles you create.

> **NOTE!**
> These color settings are separate from those for the Menu Editor.

- *Restore Default:* Use this button to restore the default Subtitle color configuration.

- *Edit Palette:* Click to display the *Color Palette* dialog, which can be used to edit colors in the current 16-color palette.

349

DVD Studio Pro 4

Creating a Subtitle

To create a Subtitle, you must first create a Track element. Once the Track has been created, you have the framework required with which to create the subtitle events.

> **NOTE!**
> You will always create a Subtitle event in a Subtitle stream area, not in a Video or Audio stream area.

Creating the Subtitle Event

Creating a Subtitle begins by determining which of the Subtitle streams of the desired Track will be used for the subtitles you will create. You will also need to determine where in the Track's timeline the Subtitle event will be placed. Once you have those two items determined, you can create the Subtitle event:

To Create a Subtitle Event, use one of these methods:

- Double-Click in the selected Subtitle stream, at the appropriate location, **or**

- Locate the Playhead to the point where you wish the Subtitle to appear, and choose **Project > Timeline > Add Subtitle** at Playhead from the menubar, **or**

- Type Command-Shift-Tilde (⌘-**Shift-~**) (or **Shift-~**), **or**

- Locate the Playhead to the point where you wish the Subtitle to appear, then *Control-Click* and select *Add Subtitle at Playhead* from the pop-up menu, *or*

- *Control-Click* in the desired Subtitle stream, and select *Add Subtitle* from the pop-up menu.

Each of the above methods will create an empty (textless) Subtitle event with the default duration and fade times as set in the General preferences Pane, and with the Colors set in the Color Pane.

> **TIP!**
> If the Viewer Tab has not automatically opened to reveal the first frame of the Subtitle event, click the Viewer Tab to reveal it.

To modify the Subtitle position:

If the subtitle is not located at the correct frame of the video, you can move it using a few different methods:

- Click on the Subtitle event, and drag it forward or backward in the timeline to a new position. The new start timecode will not be displayed until you release the cursor, **or**...

- In the Inspector General Tab, modify the *Start*: field timecode by directly entering a new value, or by clicking on the up or down arrows, **or**

- Position the Playhead to the desired location, then Control-Click on the Subtitle and select *Move Subtitle Clip to Playhead in the pop-up*.

To Modify the Subtitle duration:

- Trim the end of the Subtitle event itself, using the trim cursor.

- Enter a new *Stop Time* in the Subtitle Inspector.

- Enter a new *Duration* in the Subtitle Inspector.

> **NOTE!**
> Trimming the beginning of the event, changes its Start time!

350

Creating Subtitles

Selecting Subtitles

In order to enter text, or modify Subtitle attributes, you will need to be able to select a specific Subtitle from a stream.

To Select a Specific Subtitle:

- Click on the desired Subtitle in the Subtitle stream area;
- In the Viewer, click on the right or left arrow tools;
- In the Timeline, use the Up and Down arrow keys to step through each subtitle event;
- Select **Project > Subtitle > Next** or **Subtitle > Previous** in the menubar.

Once a Subtitle is selected, its information is available in the Subtitle Inspector. The currently selected Subtitle is the bright yellow one; unselected Subtitles are a more pale yellow, and transparent.

If a Subtitle event contains text, that text is visible in the event in the timeline. Depending on the current zoom setting of the timeline, you may or may not be able to read the entire Subtitle text.

Working with Text in Subtitles

For Subtitles created individually, text can be entered two ways:

- Enter the desired Subtitle text in the Subtitle Inspector Top Area text field, **or**
- Enter text directly into the Subtitle field of the Viewer, if it is visible. If not visible, double-click the desired Subtitle event to force the Viewer display, and open the Subtitle field for entry.

Subtitle text can be edited as you would expect within any Mac text processor. You can cut, copy, paste, delete, and so on.

There are some limitations, however, to Subtitle creation:

Each Subtitle is one text area. You can position this text area anywhere on top of the video background, but you cannot place it in two different areas, nor can you create two Subtitle events on top of each other in one stream. In other words, the Subtitle event is an invisible field that will contain and display your text. If you need to interpose text with graphics, or place text in two or more locations on the screen, you might wish to use a graphic file containing the text. We'll cover that in just a bit.

To reposition the Subtitle text field:

When the field is not being edited, it can be moved around the screen to customize its position:

- Click on top of the text, and drag it to a new location—notice the *Offset* values in the Inspector will change when you do this, **or**
- Use the *Formatting* controls to change the Horizontal or Vertical positioning or text justification, and then fine-tune it with the offset controls as needed.

To change the Font attributes:

To change Font attributes (see Fig. 12-15), open the Fonts Window using one of these methods:

- Select **View > Show Fonts** in the menubar,
- Select **Format > Font > Show Fonts** in the menubar,
- Type *Command-T* (⌘-**T**), **or**
- Click *Show Fonts* in the Toolbar, if it is visible.

351

DVD Studio Pro 4

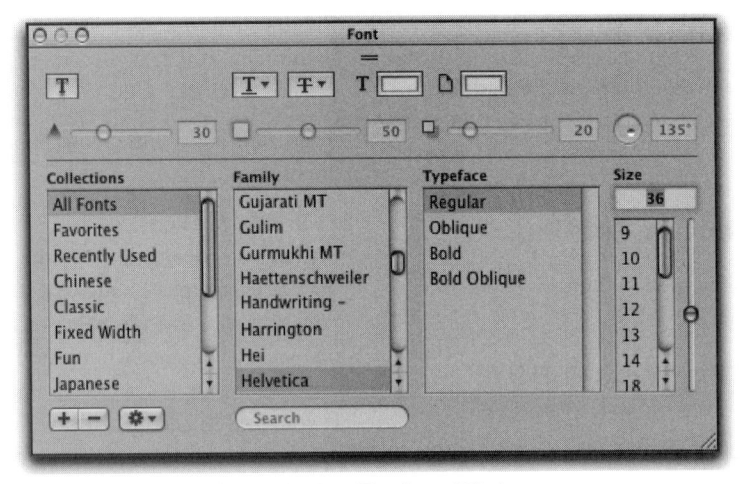

Figure 12-15 — The Fonts Window.

Once the *Fonts Window* is displayed, you may change the current font selection, which will affect the next text entered in the Subtitle editor. If you wish to modify text already written, select the text first, then modify the text attributes in the Font Window.

- You can change the Font Family and the Typeface by clicking on a selection in the Fonts Window.

- You can change the Font size by entering a new numerical value, selecting a preset size in the list, or scrolling the scrollbar on the right edge.

Cautions about using Fonts

You can use any font installed on your system to create subtitle events, but be aware that if you should move the project to a different system that does not contain a font used in a Subtitle, a substitution must occur to a font that *is* resident on that system.

Likewise, you may use two-byte fonts for foreign languages to create Subtitles (assuming you can write and read that language), but you may also encounter difficulties if you move the DVD project to a system that does not contain those fonts.

To avoid these problems, you will want to be sure to *Build* your project while you have access to those fonts. Once the Subtitles are built, they are converted into graphic bitmaps, and the original fonts are no longer needed. The Subtitles will appear correctly on any DVD player once they are built.

To change the Font Color:

- In the Subtitle Inspector Colors Tab, set the *Text*: color pop-up to the color and opacity you choose.

You will note there are four controllers there, and four color pop-ups, as well as four opacity settings.

If you look carefully at the text in the subtitle, you will see that there are gradations around the edges, where the *Outline 1* and *Outline 2* colors have control. This allows you to create a contrast border around the Text color, to improve readability of the Subtitles on top of difficult backgrounds. You can experiment with the settings of Outline 1 and 2 to see which contrast settings work well for you.

Creating Subtitles

The *Background* setting is almost always set to zero (0) because the *Background* color (usually white) will map to every pixel that isn't in the text, which means it will cover the entirety of the image in the MPEG video—experiment with it—you won't want to use it much!

Remember that choice of font, size, and typeface can make your Subtitles readable, or unreadable, depending on how you set them. At a minimum, 24-point type size is readable, but you might consider a larger size if 24 doesn't work well.

Positioning of the Subtitle Text

In the General Tab of the Subtitle Inspector, you have formatting control of the Horizontal justification and vertical placement. You will be pleased to note that these controls are aware of, and respect, the Title Safe zone limitations. So even if you choose Left Justification, your text will be justified *inside* of the Title Safe zone—a really nice feature!

Using a Graphic File for a Subtitle

For complex images, mixtures of text and graphics, or even simple watermarks, a graphic image may sometimes be the best choice for a Subtitle.

Subtitle graphics files are created in the same method as menu Subpicture overlays files. Grayscale values of 100% (Black) 66%, 33%, and 0% (white) are used to allow mapping of the display colors.

Once you have created a suitable graphic file, you can add it to a subtitle using one of these methods:

- Select an existing Subtitle or create a new one and select it.

- In the General Tab of the Subtitle Inspector, click the *Choose:* button in the *Graphic* area.

- Select the graphic file you desire. Once selected, the filename and path will be visible in the *File:* field, or

- Locate a suitable graphic in the Finder, Palette or Asset Tab.

- Drag the file directly onto a subtitle event, **or**

- Drag the file into the desired Subtitle stream, at the place where it should begin.

- Adjust the Start time to fine-tune the placement.

To remove the graphic file:

- Highlight the filename in the File field; hit the *Delete* Key.

> **NOTE!**
> A graphic file may be added to a Subtitle that already has been created using text. In some cases, this may create overlapping text, so be certain to preview this Subtitle event carefully.

Advanced Subtitling

Chapter 13 of the DVD Studio Pro User Manual outlines methods to be used to import single files that contain text entries to create Subtitles, or Subtitle files that can access prebuilt graphics files.

It also contains a very complete description of the syntax used in creating an STL format Subtitle file.

Please review Chapter 13 of the DVD Studio Pro User Manual for more information on these subjects and instructions for creating Buttons over Video.

353

Summary

Subtitles are graphic elements that are superimposed over the MPEG video in the DVD. Although they mostly display translated text, they may also be used to display commentary information, director's notes, production comments, and so on.

Subtitles may be created one at a time or may be imported from text files that contain the Subtitle text and timecode, or timecode and a reference to a graphics file that was created to display that Subtitle or graphic.

The Spruce STL format is highly flexible and very usable. It is explained in some detail in Chapter 13 of the DVD Studio Pro 4 User Manual, beginning on page 456.

Chapter 13

Basic Scripting for DVD Studio Pro 4

DVD Studio Pro 4

Goals

By the end of this chapter, you should know:

- What is Scripting?
- What is a Script?
- What kinds of Scripts are there?
- How do scripts work?
- What is a GPRM?
- What is a SPRM?
- How do the GPRMs work?
- How do the SPRMs work?
- How to create a Script
- How to edit a Script
- How to link Scripts to Elements
- How to link Elements to Scripts
- About GPRM Partitions (new in DVD Studio Pro 4)
- How to create and use GPRM Partitions

Scripting = Enhanced Interactivity

When the DVD format was created, a powerful control language was included in the format, allowing the enhanced nonlinear interactivity that has become the hallmark of DVD Video projects. While basic DVD interactivity can be created by linking project elements together while authoring—jumping from Menu Buttons to Tracks and Markers, for example— enhanced interactivity in DVDs can be created by Scripting the various DVD project elements to exert control over the DVD player.

Scripting makes use of the DVD Spec's navigation and presentation *Commands, System Parameters*, and General *Parameters*, which are built into the DVD Specification and every DVD player.

Just as making links between Tracks, Menus, and Slideshows defines code to create those links, Script commands you create define micro-code instructions that will be created when your project is built. When this microcode is executed by the DVD player, it can perform amazing functions, from remembering and interpreting user responses to generating random values, and even automatically navigating through pre-programmed sequences of content. DVD Players are actually amazingly smart devices, despite today's modest cost.

What Is Scripting, Exactly?

Scripting is DVD Studio Pro's method to create command elements that can control DVD player navigation and presentation. Scripting Commands can access 8 of the defined General Parameters and all of the 24 defined System Parameters in the DVD player. You'll find there are only 21 System Parameters currently in use, and you may actually use only a few of them regularly. (More on Parameters in just a bit.)

Scripting is *not* Java, Jscript, C++, Perl, HTML, XML, VBscript, or anything you may already be familiar with, but it is somewhat reminiscent of some other languages you may have seen. In some ways, it resembles high school algebra ("add 6 to GPRM0," "multiply GPRM1 by 1024"), and in other ways, simple logical expressions ("if GPRM0=1, *then* to Jump Slideshow 1"). The important thing is: DVD Studio Pro Scripting is designed to make it EASY for you to create sophisticated DVD behavior, without having to learn the DVD Command language!

Basic Scripting for DVD Studio Pro 4

The Script Editor included in DVD Studio Pro will actually do the hard work for you. All you have to do is point, click, and drag to create a Script! Occasionally you may need to type something, perhaps to add a comment, but not very often.

General and System Parameters—Scripting Tools

To understand Scripting, we need to understand the DVD player's Register Memories—let's begin with GPRMs.

Understanding General Parameters—GPRMs

Each DVD player contains 16 General Parameter Register Memories (GPRMs, or "germs"). Similar to RAM in your computer, these are generic memory locations the player can use like a scratchpad to read and write information while it is operating. DVD Studio Pro scripting allows you to access 8 of these GPRMs, GPRM 0–GPRM 7.

The remaining GPRMs are unavailable for scripting purposes. All GPRMs default to a value of 0 initially. GPRMs have no specific preset function—any GPRM can be used for a particular function in one disc, and then may be used for a completely different function in another disc. You will find that you will use them a lot in Scripting. They are extremely powerful, and very versatile.

GPRM Issues to Be Aware Of

GPRMs can store numerical values between 0 and 65535 (the largest value possible in 16 bits), but you will want to be careful if you are planning on performing math operations on GPRM values.

Overflow and Underflow

The DVD command set has no provision for overflow or underflow conditions caused by math operations on GPRMs, so you will find that operations that result in values less than zero (0) or greater than 65535 will "wrap around" that end of the numerical scale, and your resulting operations may not be what you expect!

As an example, subtracting 4 from 3 would normally result in –1, but in the GPRM, without underflow protection, the –1 will instead become 65535! To prevent this behavior from being a problem, you can program your Script to check for this condition before it happens.

GPRM Volatility

Like the RAM in your Computer, GPRM contents are "volatile." If the player power goes off, or the disc is stopped, the contents of the GPRMs will generally revert to zero. This is disconcerting if the user stops your disc in the middle of a highly interactive section, utilizing lots of GPRMs to track what they have seen. Those GPRM values will disappear, making continued operation of the disc unpredictable, or impossible. This should be kept in mind when you are creating a DVD that relies heavily on scripting to keep track of GPRM values. Locking the DVD remote to prevent selecting *Stop* isn't the way. But perhaps an advisory screen letting the user know that stopping the DVD will cause problems continuing the program might be the way.

Understanding System Parameters—SPRMs

A DVD player continually tracks its ongoing operation in a group of memory registers called System Parameter Register Memories (SPRMs or "spirms"). The status of the player's operation (i.e., which track is playing, what audio stream is currently selected, the last menu button selected, last chapter marker passed, and so on) can be read by script commands that access the SPRMs. Unlike the generic GPRMs, SPRMs have very specific predefined functions.

Here is a list of the System Parameters and their uses:

- *SPRM 0:* Menu language description code; DVD player setup by user
- *SPRM 1:* Current audio stream number; set by user or program
- *SPRM 2:* Current subtitle stream number; set by user or program
- *SPRM 3:* Current angle number; set by user or program
- *SPRM 4:* Current playing title number (Title); set by DVD player
- *SPRM 5:* Current Video Title Set (VTS) number; set by DVD player
- *SPRM 6:* Current title PGC number; set by DVD player
- *SPRM 7:* Current part of title number (chapter); set by DVD player
- *SPRM 8:* Current highlighted button; set by viewer in DVD player
- *SPRM 9:* Navigation timer; set by program, then times out
- *SPRM 10:* Timer target; DVD track to play when SPRM 9 times out
- *SPRM 11:* Audio mix mode for Karaoke; set by program or DVD player
- SPRM 12: Country code for parental management
- *SPRM 13:* Parental management level currently in effect
- *SPRM 14:* Player video configuration (Aspect Ratio, Letterbox Mode)
- *SPRM 15:* Player audio configuration (DTS, AC-3, MPEG, PCM, SDDS)
- *SPRM 16:* Initial language code for audio, player setup by user
- *SPRM 17:* Initial language code extension for audio
- *SPRM 18:* Initial language code for subtitle, player setup by user
- *SPRM 19:* Initial language code extension for subtitle
- *SPRM 20:* Player region code
- *SPRM 21:* Reserved – not currently defined or in use
- *SPRM 22:* Reserved – not currently defined or in use
- *SPRM 23:* Reserved – not currently defined or in use

The DVD Studio Pro Scripting language allows you to write values to three of these parameters (SPRM 1, 2, and 3) by using the *Set System Stream* command. The usefulness of a fourth, SPRM 8, can be exploited using the *Jump Menu* command with a specific Button as Target. SPRM8 always contains the value of the last Menu button selected by the user multiplied × 1024 (because it's a binary bit value). Divide the value of SPRM 8 × 1024 to determine the number of that button. You can use that knowledge to target a jump to the appropriate button on that menu or any other one.

Other SPRMs

The remaining SPRMs cannot be written to by the scripting language. Many of these other SPRMs con-

Basic Scripting for DVD Studio Pro 4

tain values set by the player owner during setup, or by the manufacturer. The player's Region Code (SPRM 20) cannot be changed by the owner, or by scripting, but it can be read by Script commands, so you can determine for what region the player was originally manufactured. Check the DVD Studio Pro Manual for a complete listing of SPRMs and their values, some of which may look peculiar because of the binary bitmapping used.

How Do I Use a Script?

One of the first things to learn about scripting is what you can DO with the script.

Scripts can move you around the disc:

- Navigate to a MENU (Jump To "MenuName").
- Navigate to a BUTTON (Jump To "MenuName::ButtonName").

Navigate to a TRACK (Jump To "TrackName").

- Navigate to a MARKER (Jump To "TrackName::MarkerName".)
- Navigate to a STORY (Jump To "TrackName::Story::EntryPoint").
- Navigate to a SLIDESHOW (Jump To "SlideshowName").
- Navigate to a SLIDE (Jump To "SlideshowName::SlideName").
- Navigate to a SCRIPT (Jump To "ScriptName").

Scripts can change what's being seen or heard, and can control GPRMs:

- Set an Audio Stream (1-8; Immediate, or GPRM-based).

- Set a Subtitle Stream (1-32; Immediate, or GPRM-based) (On/Off).
- Set a Video Angle (1-9; Immediate, or GPRM-based).
- Set a GPRM (Variable) (mov, add, etc., Value to "TargetGPRM").
- Perform Conditional Branching ("Jump If…" commands).
- Generate Random numbers to create random play!
- Perform mathematical operations (+, −, *, ÷, divide modulo).
- Perform Binary operations on bits (AND, OR, XOR).
- Determine DVD operation/status (read SPRM commands).

…and SO much more…

Once you have decided WHAT you can use a script for, it's time to decide how you are going to GET to and from your script; you can:

- Navigate to a Script from a Button (JUMP WHEN ACTIVATED).
- Navigate to a Script from a Track (JUMP WHEN FINISHED).
- Navigate to a Script from a Menu (TIMEOUT ACTION).
- Navigate to a Script from a Marker (Chapter End Jump).
- Call a Script from another Script ("Jump to Script").
- Call a Script before playing a Track, Story, Slideshow, or Menu.
- Call a Script in response to the Title, Menu, or Return Controls.

359

…and of course, even more options exist.

If this is your first time attempting to script in DVD Studio Pro, please do yourself a favor and read Chapter 14 of the DVD Studio Pro 4 Manual. It contains four nice tutorials that will help you get started with Scripting, and some reference material that we will not duplicate here. You will find that we've built the first one for you, the ever-popular "Play All Script," and included it on the DVD accompanying this book.

When to Use Scripting

One of the first things to do with your project is determine if you need to use a script, or if you can instead utilize the built-in navigation of DVD Studio Pro. Many functions that appear to require scripting may be easily programmed using the built-in navigation instead! You may find, for example, that you can easily use an End Jump Action pointing to a Button to target a specific Button for highlighting, without using a Script. This was not possible in DVD Studio Pro 1 without using Scripting!

It should be obvious to you by now that Scripting is the means to enhance or upgrade the interactivity of your DVD and really only needs to be used when the normal interactivity doesn't suffice.

Here are a few typical times when you might use DVDSP Scripting:

- When specifying a specific button to return to on a menu (instead of letting the player's SPRM 8 decide)
- To perform Conditional branching in conjunction with a test (i.e., "Play SlideShow 'Catalog' if GPRM0=1")
- To generate or modify a random number "ran GPRM0, 255" (usually used with conditional branching—"Jump if")
- To initialize (set) a GPRM value, or perform a function on a GPRM (mov, swap, add, subtract, multiply, divide, divide modulo)
- To perform Binary logic operations on a GPRM value (AND, OR, XOR)
- To create and use a "Play All" type of project. This uses a GPRM (variable) and conditional branching…
- To "password protect" some content. This requires setting and testing GPRMs (variables).

Of course, there are many more times when you might wish to use scripting—these are just a few of the most-used basics.

One of the best ways to really learn how Scripting works is to Simulate your DVD project with the *Info Pane* revealed. When this pane is open, you will be able to watch the real-time interpretation of the script commands, including seeing the evaluation of formulas, the changing of the GPRM (variable), and SPRM values, etc. Watching the Simulation display is an invaluable lesson in how Scripts control the innermost functions of the DVD, and I urge you to use it a lot while you are learning about Scripting.

What Kinds of Scripts Are There?

There are two types of scripts: **Pre-Scripts**, and the other kind, which I like to call **Inline Scripts**.

Inline Scripts

An Inline Script is a script that exists by itself as an individual Element, which may be accessed "**In the line** of normal navigation" by a Menu, Button, Track, Story, Slideshow, or even another Script. The action

Basic Scripting for DVD Studio Pro 4

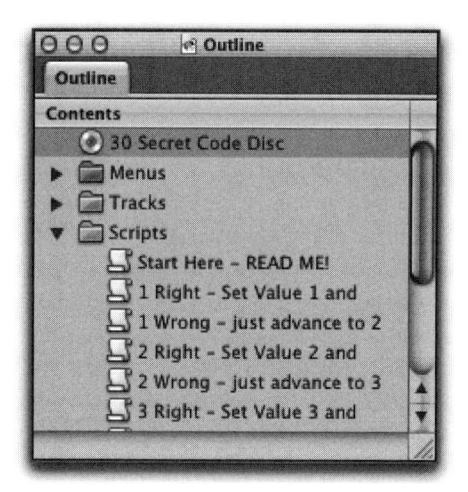

Figure 13-1 — All Scripts are accessible in the Outline Tab.

or timeout properties of any other Element can be used to navigate to a Script. (See Fig. 13-1.)

Figure 13-2 shows the Connections Tab for a typical DVD Studio Pro project using a lot of Scripting. All of the scripts shown here are being used as Inline Scripts, either connected to directly *from* another element, or connecting directly to another project Element. Notice that the *Enter Code #1* Menu does have a place where a Pre-Script could be attached, but it is currently not used. Notice also that the *First Play* and the remote control functions for the *Title* and *Menu* buttons are routed to the Script named "Start Here -READ ME!"

Pre-Scripts

Pre-Scripts are Scripts, just the same as inline Scripts, but they are attached to a project element and may execute prior to the playback of that element (under some circumstances). The most noticeable difference between the two is that a Pre-Script can only be accessed by navigating to the root of the Element to which it is attached. Pre-Scripts can be attached to a *Track, Story, Menu*, or *Slideshow* by selecting the desired Script in the *Pre-Script* property of that Element's Inspector (Fig. 13-3), or by using the *Pre-Script* link in the Connections Tab (see Fig. 13-4). Examine the Connections Tab or the Inspectors of various elements to see where Pre-Script links are. Be sure to have the Connections Tab's Standard or Advanced view open, or you won't see the Pre-Script link.

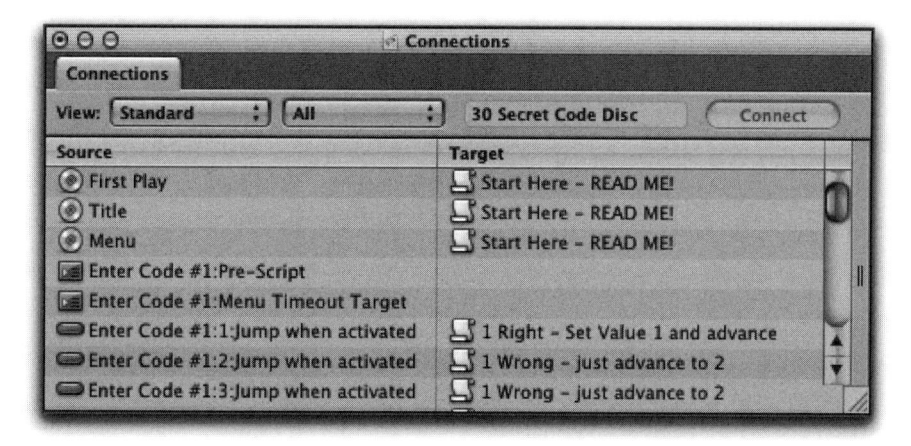

Figure 13-2 — Inline Scripts in the Connections Tab.

361

DVD Studio Pro 4

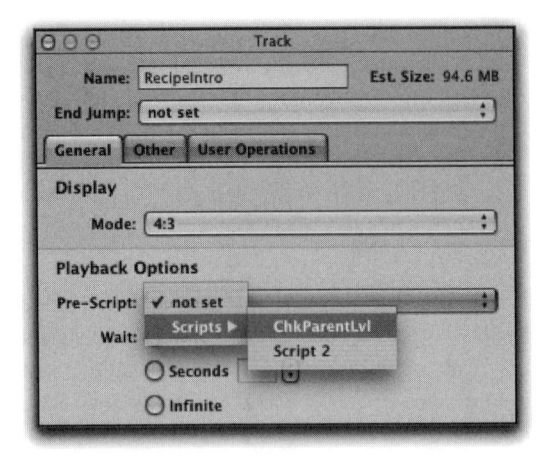

Figure 13-3 — Setting a Pre-Script in a Track Inspector.

Figure 13-4 — Linking a prescript to a Track using Connections Tab.

> **REMEMBER!**
> The Pre-Script Property must be attached to, and will *only* execute when you navigate to the *root* of Element where you want the Script to execute. This means a Pre-Script can be activated when navigating to a Menu, but *not* a Button; by a Track or a Story, but *not* a Marker; by a Slideshow, but *not* a Slide.

Why Use a Pre-Script?

You might use a Pre-Script to check for a setting every time an element is accessed, or to set a specific stream setting before playing a Track. For example, you might Script the "Set System Streams" command to set the Audio Stream setting to select a Surround Sound stream if the user has selected 5.1 Surround from an Interactive Menu, or Stereo if they selected Stereo. The uses for Scripts and Pre-Scripts are endless.

General Notes on Using Scripts

When a Script is executed, it can do any number of things, but the last thing it must always do is *exit* to another project Element—a Script needs to have an "escape clause"—a command to jump to some other Element where navigation may continue. If not, the script (and your DVD!) will *stop dead*. Oops! (Not a good thing.)

> **REMINDER!**
> Always check to be sure every Script you create has a way to exit, to allow playback to continue.

Making a New Script

To Make a Script (in general):

- Create a Script Element; rename it if you wish.
- Select it and add more Command Lines if needed.
- Use the Script Command Inspector to modify command line;
- Repeat as needed, to create the number of commands needed.

To Create a New Script Element

In order to create a new Script element, you may use any of the following methods:

- Choose **Project > Add to Project > Script** from the Menu, (see Fig. 13-5) **or**
- Click the *Add Script* button in the Toolbar (if visible) (see Fig. 13-6) **or**
- Type *Command-Apostrophe* (⌘ - ') on the keyboard.

Basic Scripting for DVD Studio Pro 4

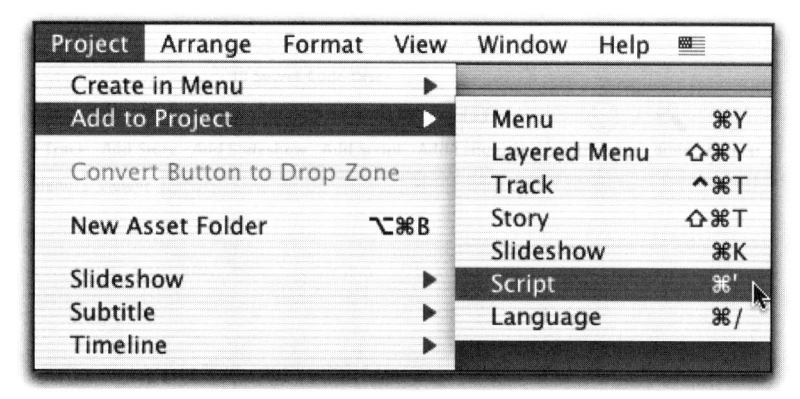

Figure 13-5 — The "Add Script" Menu Command in action.

Figure 13-6 — The "Add Script" Tool in the Toolbar, with Tool tip shown.

Scripts that you create will be numbered sequentially in each project, beginning with Script 1. All Scripts you create will be organized in the *Outline Tab*, in the Scripts Folder, regardless of whether they will be used as Inline or Pre-Scripts.

Renaming a Script

- In the *Outline Tab*, click on the Script to select it, click again on the name to open the name field, then type a new name to replace it.

- In the *Inspector*, click in the Name field and rename as needed.

Duplicating a Script

It's often faster and easier to edit a script that's already written.

To duplicate a script:

- Select the desired Script in the Outline Tab.

- Select **Edit > Duplicate**, or type *Command–D*, ⌘**-D**, or

- *Control-Click* on the desired Script, select *Duplicate* in the pop-up Menu.

363

Saving a Script

Scripts may be saved in two different ways, and can be imported into other DVD Studio Pro projects. Scripts can be saved as:

- Scripts files (with a .DSPScript extension), or
- Description files (with an .IDesc extension).

To save a Script:

- *Control-Click* on the Script in the Outline Tab, and select *Save Script* in the pop-up Menu.
- Select **File > Export > Item Description** in the Menubar to save a Description file.

Loading a Script or Script Description

To reload a Script file (.DSPScript) into a project:

- *Control-Click* in the Outline Tab, select *Load Script* in the pop-up Menu.

To reload a Script Description file (.IDesc) into a project:

- Select **File > Import > Item Description**; select the Description file in the Dialog box.

Files loaded using either technique will appear in the Outline Tab and can then be edited as needed.

Using the Script Editor

Once you have a new Script Element created, it's time to access the Command Lines with the *Script Editor*, as shown in Figure 13-7. Using the *Script Editor* is simple. In each new Script, the first Command Line is created by default, containing a NOP (No Op) Command. More Command Lines may be added, up to the limit of approximately 128 command lines. (Complicated commands may use more than one command step, so you may have less than 128 available.) The *Script Inspector* will display the number of command steps remaining.

Selecting a line in the Script Editor Tab allows you to use the *Script Command Inspector* to modify the Script command in that line. Each Command Line can contain only one Command.

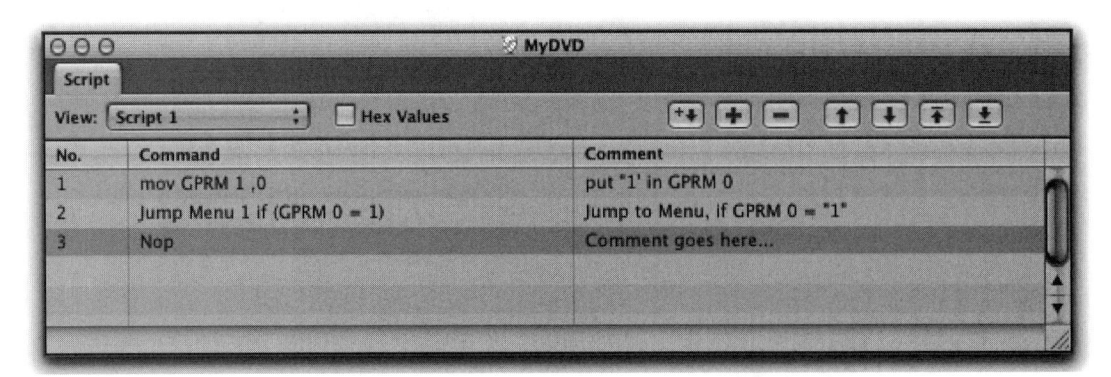

Figure 13-7 — Command Line Tools in the Script Editor.

Reordering Command Lines

The order of the Command Lines (as well as the commands you create in them with the Script Command Inspector) will determine how (and if!) the script functions in the finished DVD.

The order in which the Command Lines appear within the Script is *critical*, because Scripts execute in "top down" fashion—that is, the Script Editor evaluates commands in sequential order beginning at Line 1, then working down through the list.

Figure 13-8 — Command line reordering tools.

The sequencing of the Command lines may be modified using the re-ordering tools at the top of the Script editor (shown in Fig. 13-8).

The three tools on the left side will (from the left):

- Insert a line below the currently selected one (or type *Shift-Command-=* (**Shift ⌘ =**).
- Add a new line at the bottom of the list—*Command-=* (⌘=).
- Delete the currently selected line (*Delete Key*).

The four tools on the right side will (from the left):

- Move the currently selected line up one position (the line above will slip down to take its place).
- Move the currently selected line down one position (the line below will slip up to take its place).
- Move the currently selected line to the top of the list (all command lines above it will ripple down).
- Move the currently selected line to the bottom of the list (all command lines below it will ripple up).

> **NOTE!**
> Every Command Line must be processed and evaluated, but not every Command in a line may actually execute. *Conditional* Commands exist, which will only execute if a specific Condition being evaluated is "true." These commands allow you to create "If . . .Then" command constructions in your DVD, and are quite popular.

Script Command Syntax

Building scripting commands consists of creating a properly formed *syntax*, just like when creating a computer program. Preparing a proper Script Command is easy in DVD Studio Pro. By using the pop-ups available in the Script Command Inspector to generate the commands automatically, you create perfect Scripting Commands by avoiding typing errors. In fact, you will find it's almost impossible to type anything into a command, except perhaps an immediate value for a variable. Even though DVD Studio Pro makes is virtually impossible to make a bad command, you must be sure you are using the correct commands in the correct order to accomplish the desired behavior in the finished DVD.

Using the Script Inspector

There are two Inspector displays possible when dealing with Scripts: the *Script Inspector* and the *Script Command Inspector*.

- Selecting the Script Element itself in the Outline Tab will display the Script Inspector (Fig. 13-9), which only gives the name of the Script, and a set of fields in which to rename the GPRM variables from their default names. (This same set of fields is available in the *Disc Inspector, Advanced Tab*.) Only ONE set of names is permissible

365

DVD Studio Pro 4

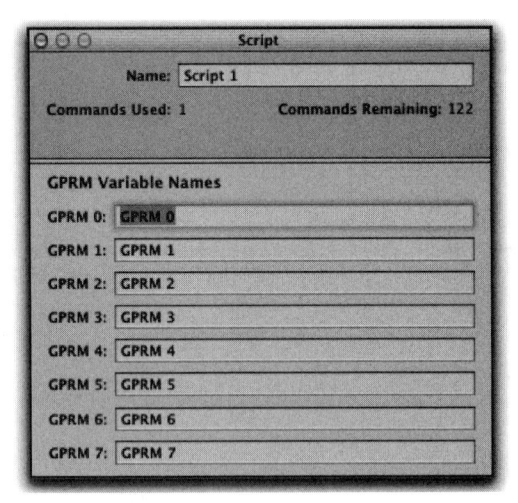

Figure 13-9 — The Script Inspector.

per project—the GPRM names are *global* definitions that apply to every script within that project.

- Selecting a Command Line in the Script Editor Tab will display the *Script Command Inspector* (Fig. 13-10). Use the Script Command Inspector to view and edit the specific content of each Command Line.

Be sure to notice the "Start at Loop Point" feature in the Jump section, allowing you to bypass a Menu opening and land directly on the Loop Point—this is new in DVD Studio Pro 4.

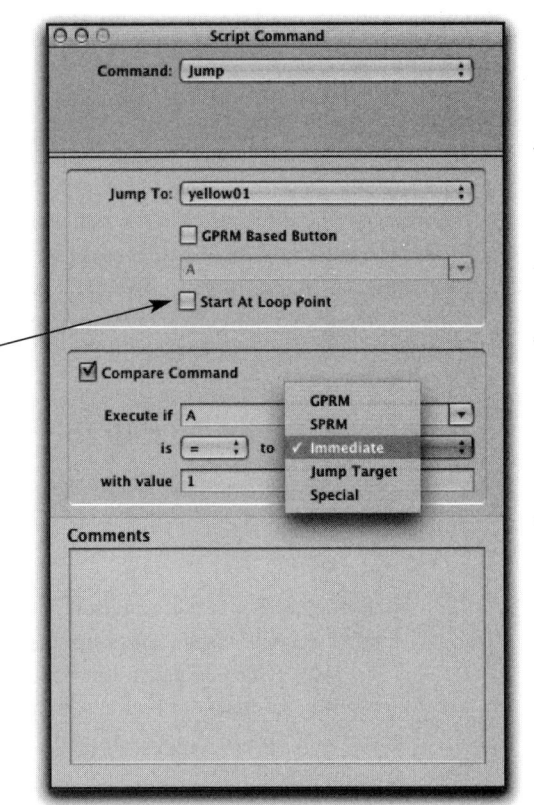

New!

First Pane Command Selector

Select from the 10 Command Types

Second Pane Command Modifier

Fine-tune the command here. Set options, or activate

A GPRM-based operation.

Third Pane Compare Command

Sets up the criteria for evaluation of the Comparison test. If true, the command executes.

Comments Pane

Add comments in this pane to document the script's logic: "Why is this command here?" "What does this command do?"

Figure 13-10 — A Script Command Inspector.

366

Basic Scripting for DVD Studio Pro 4

Editing in the Script Command Inspector

Once a Command Line is selected, the *Script Command Inspector* is used to create the proper Command Syntax. The Script Command Inspector contains three command panes, and a comments pane.

The First Pane—Command Selection

The top pane of the Script Command Inspector (Fig. 13-10) contains the *Command:* Pop-up, listing the 10 Command types available. Here you can select the basic command type you need. Seven of the ten Commands in the *Command:* pop-up will display their options in the second pane of the Script Command Inspector. *NOP, Exit, and Exit Pre-Script* commands do not have options. Both *Exit* commands may be made Conditional, if desired.

The Second Pane—Command Options

The second pane of the Script Command Inspector will display options appropriate for the command selected (except NOP). Adjusting those options in the Inspector creates the precise Command needed.

The Third Pane—Compare Command

The third pane of the Script Command Inspector will always contain the Compare Command, which is available for every command except NOP. The Compare Command can be activated (using the checkbox) and programmed to make that command line a Conditional Command. See the section on Conditional Commands, further on.

The Fourth Pane—the Comments Field

Instead of marking empty Script lines with a special Comment character (as the "#" in DSP1 scripts), *Comments* may now be embedded into each Command Line, in the Comments pane of the Inspector. Be sure to comment your scripting code as you create it, or you may find yourself forgetting just why you needed to make that "Jump Track 2" command Conditional. You'll be glad you did it!

Conditional versus Unconditional Commands

DVD Commands created in Scripting may be Unconditional or Conditional. Unconditional Commands execute immediately when they are encountered during DVD play. Conditional Commands are one of the most powerful features of the DVD command language—its ability to make decisions based on the status of SPRMs or GPRMs, or the user's behavior. This kind of command is pretty simple to program, but very powerful in use.

To make a Command Conditional (in general)

- To activate the Compare Command, check the Compare Command checkbox in the third Pane, then specify the test to use.

- If the Compare Command test equates to *True*, then the Command in that line will be executed.

- If the Compare is *not true*, that command is ignored. The Script will continue down the command list, and attempt to execute the next Command Line in sequence.

DVD Studio Pro 4

- The Script will continue to evaluate and/or execute commands until it exits or runs out of command lines prematurely before an exit (whoops—watch out for that!)

> **NOTE!**
> If a Script runs out of valid command lines and does not somehow return back to another Element in the DVD, it is a flawed script, and your DVD is broken! Be sure each Script has a valid "escape path" and can return control to another Element.

To make a Command Conditional (detailed overview):

- Activate the Compare Command checkbox in the Inspector.
- Select an "*Execute If*" source—(which GPRM will you test?).
- Select an "*is*" option—(determines the kind of test to perform).
- Select a "*to*" target—(which Element type will you test against?)
- Select a "*with value*" option—(which Specific Element to test?)

Setting up a Conditional Command:

- Activate the Compare Command checkbox (see Fig. 13-11).

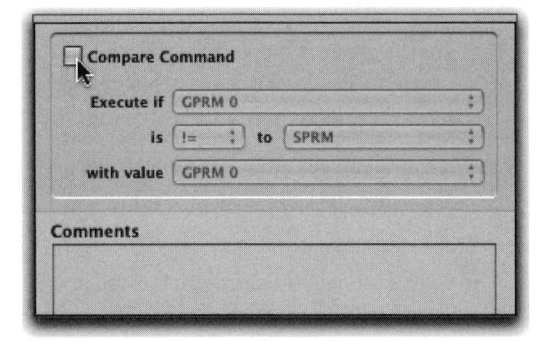

Figure 13-11 — First, activate the Compare Command.

Figure 13-12 — Select a source GPRM holding the test value.

- Select an "Execute If" source—choose the GPRM you will test.

Select one of the 8 GPRMs as the target to test. You will want to have set a value into this GPRM somehow, before you test it. This value can be a simple numerical value or could even be a *Jump Target*. (See Fig. 13-12.)

- Determine the kind of compare test to perform (see Fig. 13-13):

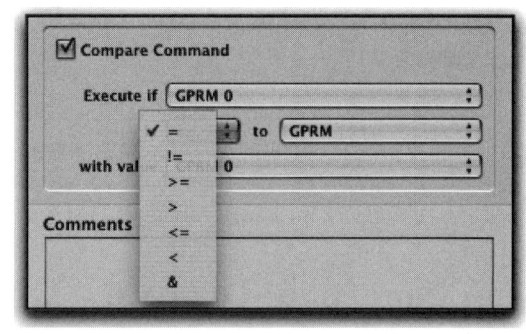

Figure 13-13 — Select a Comparison type with the "is" option.

Comparison tests may be made with these operators:

- = *(equal)*: The command executes if the two values are the same.
- != *(not equal):* The command executes if the two values are not equal.

368

- **>=** *(greater than or equal to):* The command executes if the selected GPRM is greater than or equal to the selected target.

- **>** *(greater than):* The command executes if the selected GPRM is greater in value than the selected target.

- **<=** *(less than or equal):* The command executes if the selected GPRM is less than or equal to the selected target.

- **<** *(less than):* The command executes if the selected GPRM is less than the value of the selected target.

- **&** *(Binary 'and'):* Performs a bit-wise "and" operation between the two values and executes the command if the bits set to 1 in the selected GPRM are also set to 1 in the selected target.

> **NOTE!**
> It helps to have at least a basic understanding of working with binary numbers to be able to use these bit-related operations.

- Select an Element type to test against (see Fig. 13-14).

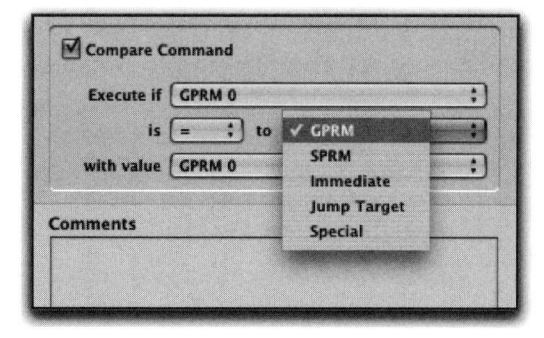

Figure 13-14 — Select an Element type to compare against.

You can select which type of element you wish to compare against your selected GPRM source. Select a type in this pop-up first, and then select a specific element of that type in the pop-up below, "with value."

- *GPRM:* Allows you to compare to any of the eight available GPRMs.

- *SPRM:* Allows you to compare to any of the SPRMs. (See "Understanding SPRMs" earlier for a list of SPRMs and their functions.)

- *Immediate:* Allows you to enter a decimal value, from 0 to 65535, to compare.

- *Jump Target:* Allows you to select from all available project elements (Menus, Tracks, Stories, Slideshows, and Scripts) to compare.

- *Special:* Choose Current Item, Last Item, or Last Track to compare.

    - *Current Item* is this script, unless this is a pre-script. If this is a pre-script, the Current Item is the project element (menu, track, slideshow, or story) the script is assigned to.

    - *Last Item* is the project element that started this script running.

    - *Last Track* is the last track that was played, even if this script was started by a button on a menu.

- Selecting "with value" option chooses a Specific Element to test (see Fig. 13-15).

This pop-up is context sensitive. The available options change based on the Element type selection ("to") made in the previous step. GPRMs are shown in Figure 13-14, but each Element type has its own specific list of possible selections.

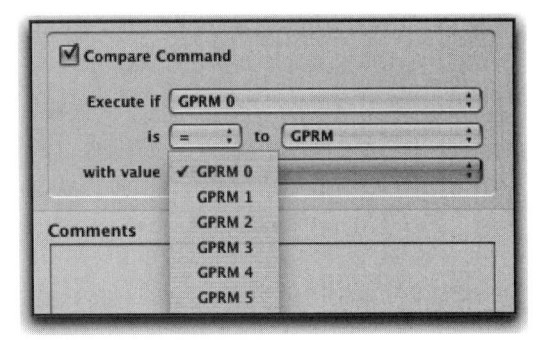

Figure 13-15 — Select a specific target for the comparison—"with value."

Thinking Logically about Comparisons

Newcomers to scripting are often challenged by the Conditional Command their first few times. The key concept is this:

The Compare test you define will be evaluated to see if it is true. There are MANY ways to establish the test, more than just checking to see if something is equal to 1 or zero. For example:

- You could use a Compare to check for a GPRM value which signifies where in the Menu structure you are—1 means Menu 1, 2 means Menu 2, etc. This could then be used to direct navigation precisely back to a specific Menu Element; or

- You could check for the value of SPRM 7, to see which was the last chapter marker passed—that value could be used to JUMP back into the Track to that chapter point, if you didn't want to use "resume"; or

- You could check for the value of SPRM 8 to see what was the last button pressed on a menu and set it to the next button value (these are in steps of 1024, by the way), thereby advancing the menu button highlight to the NEXT button in series;

… the list goes on—there are thousands of ways to use Scripting, GPRMs, and SPRMs to do exciting things in your DVD.

Scripting Commands in Detail

DVD Commands generally correspond to two types: *navigation* commands and *presentation* commands. Navigation Commands make the DVD change <u>where</u> it is playing, and Presentation Commands make it change <u>how or what</u> it is playing, by adjusting audio, subtitles, angles, etc. The few commands that don't fit this description perform general functions, like changing the mode of a GPRM, or exiting.

The following Script Commands are available (see Fig. 13-16).

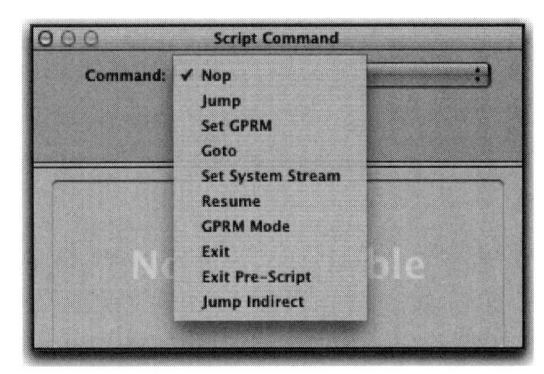

Figure 13-16 — Command select pop-up (top of Command Line Inspector).

- NOP
- JUMP
- Set GPRM
- Goto [another script line]
- Set System Stream
- Resume

Basic Scripting for DVD Studio Pro 4

- GPRM Mode
- Exit
- Exit Pre-Script
- Jump Indirect

The NOP Command

Every new command line contains a NOP command by default. The NOP Command ("No Operation") is nothing more than a placeholder—it keeps the command slot occupied until it can be replaced by a real command. Leaving a NOP command in your DVD project is a sloppy technique, as many DVD players react badly when they encounter this command, sometimes stopping immediately during playback!

The JUMP Command

JUMP commands are *navigation* commands, pure and simple. They cause the DVD to navigate immediately from the current location to the new location specified in the "Jump To" command target in Pane 2. This new location can be a Menu or Button, a Track, Story or Marker, a Slideshow or Slide, or even another Script. The rare exception seems to be targeting a Button over Video within a Track element.

Jump Commands may be made *Conditional* if the Compare Command is activated, and a comparison programmed (see Fig. 13-17). If the Compare Command equates to "True," then the Jump Command will be executed. If <u>not</u> true, the command is skipped, and the Script continues to the next Command Line, looking for another valid command.

Jump Command options

JUMP commands (*not* "JUMP Indirect") must directly specify the destination they are jumping to, unless the "GPRM based" option is activated. In this

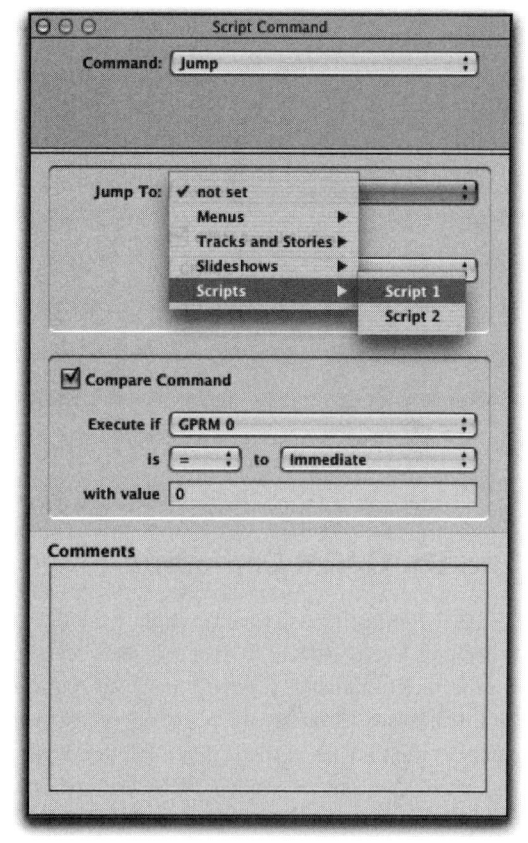

Figure 13-17 – A Jump Command with Target pop-up revealed.

case, the value in a GPRM may be used to point to a Button within a Menu or a Marker within a Track or Story. The value specified should respect the limit allowed for Buttons (1–36 in 4 × 3 menus, 1–18 in dual aspect ratio menus) or Markers (1–99). Marker numbers in excess of 99 will be ignored, and navigation will land on Marker 1. Setting "Jump To" to a Button or Marker within an Element will be ignored if the "GPRM Based…" selection is invoked. Only the root element will be targeted, and the GPRM will determine which Button (on a Menu) or Marker (in a Track, Story) will be the target (see Fig. 13-18).

371

DVD Studio Pro 4

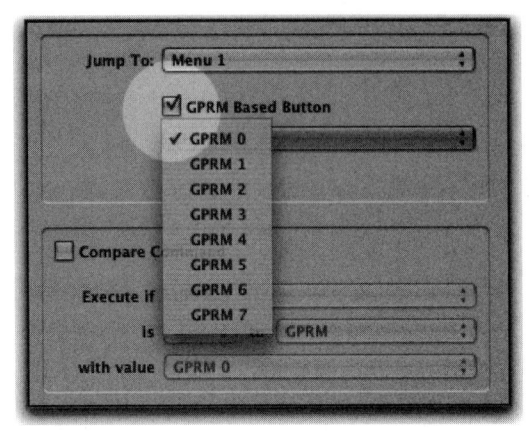

Figure 13-18 — Setting up a GPRM-based Menu Button Jump.

The Set GPRM Command

The *Set GPRM* command sounds simple, but it belies a complex set of functions that can be accessed within the Command (Fig. 13-19). *Set GPRM* Commands have 11 different types of operations that can be performed, using five different Source Types, and writing to one of eight GPRM destinations, called "Targets."

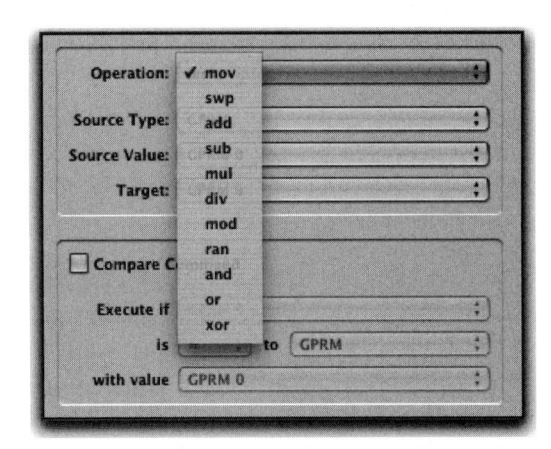

Figure 13-19 — Set GPRM Options.

The 11 types of Set GPRM *Operations* function as follows:

- *mov:* Moves the specified source value (S) to the target (T), replacing any current value in the target. **T = S**.

- *swp:* Swaps the source value and the target value. The current source value is written to the target GPRM, and the target's current value is written to the source location. The SP2 manual says this is the only operation that writes to the source location. **S -> T, T -> S**

- *add:* Adds the source and target values together and writes the result in the target location. **(T = S+T)**

- *sub:* Subtracts the source value from the target value and writes the result in the target location. **(T = T-S)**

- *mul:* Multiplies the source value by the target value and writes the result in the target location. **(T = T*S)**

- *div:* Divides the source value by the target value and writes the result in the target location. **(T = S/T)**

- *mod:* Divides the source value by the target value and writes the remainder (modulo) in the target location. **(T = S/T modulo)**

- *ran:* Generates a random value between 1 and the source value and writes the result in the target location. **(T=Rnd [1-S])**

- *and:* Performs a binary bit-by-bit "and" operation between the 16 individual bits of the source and the target values, and writes the result in the target location. An AND means that the bits that are set to 1 in both source AND target are left at 1. Bits that are not set to 1 in both values are set to the value of 0.

Basic Scripting for DVD Studio Pro 4

```
Example:   S=1111000001010101
           T=1010111111111111
T [and]    S=1010000001010101
```

- *or:* Performs a binary bit-by-bit "or" operation between the 16 individual bits of the source and the target values and writes the result in the target location. An OR means that the bits that are set to 1 in either the source OR target are left at 1. Bits that are set in both source <u>and</u> target are set to the value of 1.

```
Example:   S=1111000001010101
           T=1010111111111111
T [or]     S=1111111111111111
```

- *xor:* Performs a binary bit-by-bit "exclusive or" operation between the 16 individual bits of the source and the destination values and writes the result in the destination location. EXCLUSIVE OR means bits that are set to 1 exclusively in source <u>or</u> target (but <u>not</u> <u>both</u>) are left set to 1. Bits that are set to 0 in both, or are set to 1 in both of the values, are set to 0.

```
Example:   S=1111000001010101
           T=1010111111111111
T [xor]    S=0101111110101010
```

The five types of *Sources* are as follows:

- *GPRM:* Allows you to select any of the 8 GPRMs as the source.
- *SPRM:* Allows you to select any of the 21 usable SPRMs as the source. See "Understanding SPRMs " for more on SPRMs.
- *Immediate:* Allows you to enter a decimal value, from 0 to 65535, as the source.

- *Jump Target:* Allows you to select from all available project elements (Menus, Tracks, Stories, Slideshows, and Scripts) as a source.
- *Special:* Allows you to choose from Current Item, Last Item, and Last Track as the source.
- *Current Item* is this script, unless this is a pre-script. If this is a pre-script, the Current Item is the project element (menu, track, slideshow, or story) the script is assigned to.
- *Last Item* is the project element that started this script running.
- *Last Track* is the last track that was played, even if this script was started by a button on a menu.

The Goto Command

The Goto command allows the Script to navigate to another line number within the same script (see Fig. 13-20). Used with Conditional Compares, this can be a powerful method of making one script do multiple things, depending on a GPRM value, for example, or the state of a certain SPRM. A Script can be programmed with several "modules" of code, the Compare can determine an external condition, and the Goto will specify which module to execute, based on that condition.

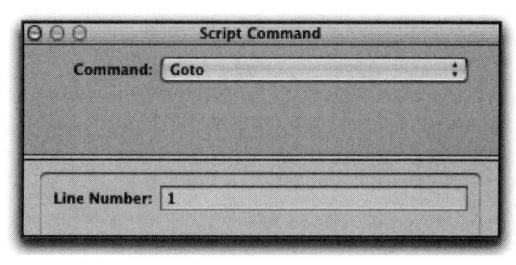

Figure 13-20 — Goto Command.

The Set System Stream Command

The Set Stream Command can be used to change the currently playing Video Angle, Audio Stream, or Subtitle Stream number, and to turn Subtitles on or off (see Fig. 13-21). This command contains a checkbox for each of the streams, and a pop-up selector for each. It can be used as an Immediate Command, where the specified stream is selected directly in the pop-up, or as a GPRM-based command, where the pop-up specifies a GPRM that contains the desired value for each stream that is to be modified (see Fig. 13-22). Of special note is Subtitle stream, where the View checkbox is used to turn the selected subtitle stream on or off.

Resume Command

The RESUME command allows return to the last Element that was being played prior to selecting the MENU button. It directs the DVD to the "resume" pointer location, stored at the time the MENU button was pressed. Typically, this is used from a Menu.

> **NOTE!**
> Be careful in using the Resume command from within a Pre-Script. *Resume* returns navigation to the last Element played. In the case of a Pre-Script, navigating to an Element that has a Pre-Script attached will redefine the "last Element Played" to be the Element attached to the Pre-Script. *You may not go where you expect to go!*

GPRM Mode

GPRMs have two different modes: Register and Counter. This command allows selection of the desired mode for a specific GPRM. Select the desired mode and the Target GPRM (see Fig. 13-23).

- *Register Mode*—A value stored in the Target GPRM remains there until it is replaced by another operation.

Figure 13-21 — The Set System Command, showing Immediate Mode.

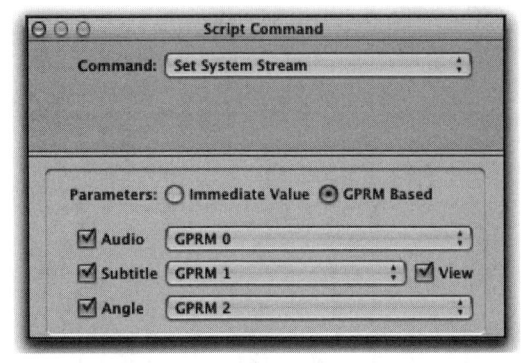

Figure 13-22 — The Set System Command, showing GPRM Mode.

Figure 13-23 — Set GPRM Mode command, showing selector pop-up.

Basic Scripting for DVD Studio Pro 4

- *Counter Mode*—A value stored in the Target GPRM will decrement by one each second, until the value reaches 0. This can be used for more sophisticated time-out routines.

Exit Command

The Exit command stops a script from executing, and if encountered during DVD playback, will stop the DVD Title from playing back. This command could be used to prevent DVD playback if certain conditions are not met, like failure to correctly enter a security code or a mismatch in region coding, but remember: An *Exit* is not a graceful ending.

Exit Pre-Script

This command allows a Pre-script to exit its execution, and pass control back to the root Element it is attached to. That Element will then play normally.

> **IMPORTANT!**
> Unlike Inline scripts, a NOP command encountered as the last command of a Pre-Script will be treated like an *Exit Pre-Script* command, and control will be passed to the attached root Element.

Jump Indirect

Jump Indirect executes a *Jump* navigation command, but instead of directly specifying the destination, it specifies a GPRM where the destination location has previously been stored (see Fig. 13-24a). You will need to store a *Jump Target* into a GPRM before this command can be used.

New Scripting Tool in DVD Studio Pro 4

New! Scripting is much the same as in DVD Studio Pro 2 and 3, but one new feature has been added: *GPRM Partitioning*.

Figure 13-24a — Jump Indirect Command showing GPRM selector.

What's a GPRM Partition?

In DVD Studio Pro 4, GPRMs can be divided into various subgroups in the Advanced Tab of the *Disc Inspector*. The options available allow for each of the 8 available GRPMs to be split into 1, 2, 4, 8, or 16 partitions (see Fig. 13-24b).

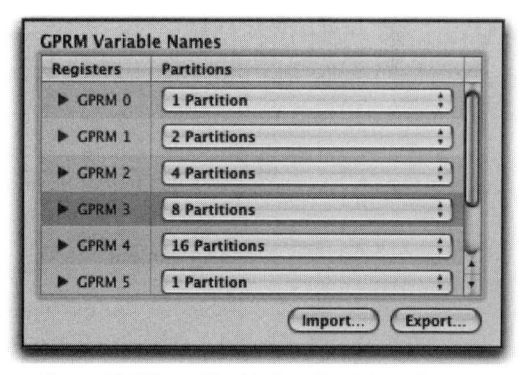

Figure 13-24b — The Various Partition schemes.

Each partition scheme equally divides the available 16 bits among themselves equally. You can create:

375

DVD Studio Pro 4

```
 1 16-bit partition  (max. decimal value for 16 bits = 65,535)
 2  8-bit partitions (max. decimal value for  8 bits = 255)
 4  4-bit partitions (max. decimal value for  4 bits = 16)
 8  2-bit partitions (max. decimal value 2 bits = 4)
16  1-bit partitions (max. decimal value 1 bit = 1)
```

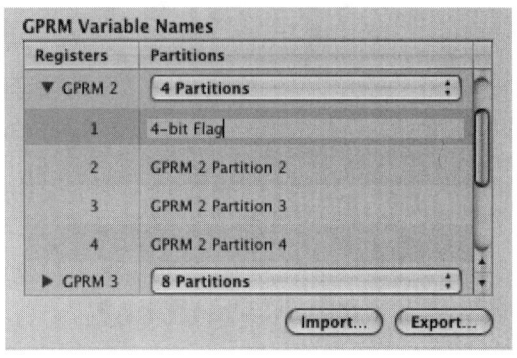

Figure 13-24c — Naming a Partition.

Each Partition can be named individually, as shown in Figure 13-24c.

Why Use a Partition?

It makes binary operations much easier. One of the most common uses of a GPRM in DVDSP is to set a "flag" to denote a state; usually "on" or "off"—the common Play All script uses this technique. While this can be done in any number of ways, the simplest method is to set a GRPM value to something other than zero, usually "one" (1).

Each GPRM can hold a 16-bit binary value, and while that means that each GPRM can be used to hold a decimal value of from 0 to 65,535, these registers are actually comprised of 16 binary digits (bits), so in a GPRM, the decimal value of 65,535 would look like this in Binary:

1 1 1 1 1 1 1 1 1 1 1 1 1 1 1 1

while the value of the number 1 would look like this in Binary:

0 0 0 0 0 0 0 0 0 0 0 0 0 0 0 1.

Experienced and advanced DVD authors know they can use each of these 16 bits to represent an on or off state, which means that using a GPRM to represent only zero or one is a waste of 15 perfectly good bits that could be used for something else (like 15 other "flags").

In previous versions of DVD Studio Pro, the programming required to separate the GPRMs into multiple values was complex and time-consuming, involving the use of Binary AND functions, and lots of comparison commands. DVD Studio Pro 4 has made this SO much easier to do, since each GRPM may now be split into various combinations of binary digits (*bits*) from one 16-bit register all the way down to 16 1-bit registers.

Practical examples of partition use

In the traditional *Play All* script, one GRPM is used simply to hold a one-bit value to indicate that the Play All function has been activated. It's a waste of a perfectly good GPRM to only use it to hold one bit, but up until DVD Studio Pro 4, that's what we had to do, if we wanted to keep authoring simple.

Now, a GPRM can be partitioned various ways, and one of these can be used as the 1-bit "*Play All* flag," while the remainder of the bits are available to serve other purposes.

While the Play All is admittedly a very simple use of the partitioning function, a more exotic use of this might be a 16-element "*Play List*" script, where one GPRM can be partitioned into 16 1-bit flags to keep track of whether or not each of up to 16 tracks have

376

Basic Scripting for DVD Studio Pro 4

been played. This can be of great use in a "random playlist" function, as it can be used to prevent a track from repeating before all of the tracks have been played.

This can be accomplished by Scripting the DVD, so that at the completion of the playback of a given track (1-16), the GPRM Partition being used as the "I've Been Played" flag for that track would be set to 1. To make it simple, each 1-bit Partition would be named to reflect the name or number of the Track for which it is the "Been Played" flag. Once all tracks have been played, the flags can all be reset to zero.

Practical Uses of Scripting

The above detailed explanations tell you what the commands CAN do, but I am sure you want to know HOW to use them.

The next section will dissect a working script, so you can see the logic that goes into that particular script.

To help you learn more about scripting, there are some additional resources for you on the DVD accompanying this book. We've included some scripting projects taken from Gnome Digital's very successful *Pro-Pack 2—Scripting for DVD Studio Pro* training CD, just revised to include new things from DVD Studio Pro 4.

In the Pro-Pack 2 Folder, you will find assets and projects, as well as descriptions of the logic behind the scripting. Each time you build a scripting project, you can learn a new scripting technique!

Anatomy of a "Play All" Script

If there's one script I have been constantly asked about since the dawn of Scripting, it's the Play All script. This Script carries a certain fascination for DVD authors everywhere. It always seems to be the first one they want to write (or need)—so let's create that function, but let's dissect it, so you know what it's doing.

What Does a "Play All" Script Do?

The name says it all. Typically, a DVD project will have a certain number of Tracks, each individually accessible from a Menu Button. Normally, each Track plays, and then returns back to the Menu, to await another Button selection. The mission of the Play All script is to play *all* of the Tracks, in a continuous, predetermined nonstop sequence.

If you take a look at the *PlayAllTutorialSP4* sample project (it's in the Pro-Pack 2 folder in the book's DVD—see Fig. 13-25a), you will see that the Menu has four numbered buttons, and there are four tracks. This is a perfect candidate for a Play All—all we need is some place to create a button for the Play All. Of course, in DVD Studio Pro, we can just add a cool Shape with Text to a new menu button, but in this Pro-Pack project, we already built the Play All button for you.

The Logic of It

The Play All is a script that seems to contradict the rules of DVD—on the one hand, we set up a nice menu that links each button to the corresponding Track, and then program the End Jump of each track to return to the same button on the Menu. Well, here's the puzzle: How the heck do we get the tracks to *ignore* the End Jump instruction, and play the next track in sequence instead? The answer is simple—we use Scripting to create a Conditional Branching situation. In other words, instead of jumping back to the menu, we'll jump first to a *Script*, to check for some condition to see if we're really supposed to go back

377

DVD Studio Pro 4

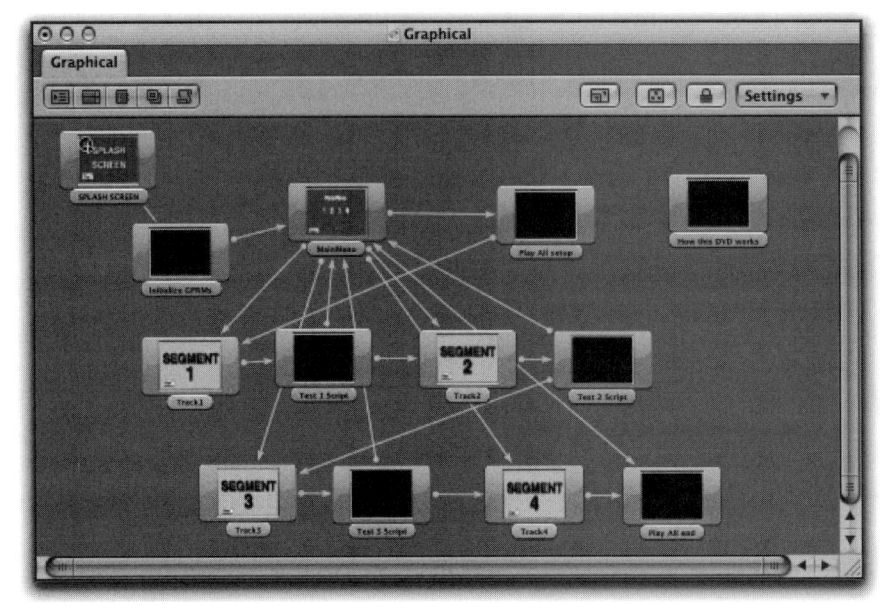

Figure 13-25a — The Play All Tutorial project—already built for you.

to the menu, or if we are supposed to continue to the next Track, as part of the Play All. Piece of cake!

Setting up the Script Conditions

To make this script work properly, we'll have to assume a few things:

- We'll use GPRM 1 to keep track of our Play All status;
- If GPRM 1 = 1, we're doing a *Play All*.
- If GPRM 1 = zero, the project is playing Tracks normally.
- We'll link a *Play All button* to a script to set GPRM1 to 1.
- After that, the Script will jump to Track 1 and start playback.

These conditions allow us to test after each Track to determine:

- If GPRM 1 = 0, the Track's own button was pressed,
- If GPRM 1 = 1, the Play All button started us off.

So, if GPRM 1 = 1, we can be sure we're in *Play All* mode.

- In Play All mode, we continue to play Tracks and test at the end.
- When we have played the last track, or in the case of an unusual occurrence (like the Menu or Title button being pressed), we have to restore the normal mode.

The "Play All" Script Logic Diagram

It's usually easier if you can see it (see Fig. 13-25b).

In the logic diagram (Fig. 13-25b), all action starts at the Menu.

378

Basic Scripting for DVD Studio Pro 4

The Play All Script - Logic Diagram

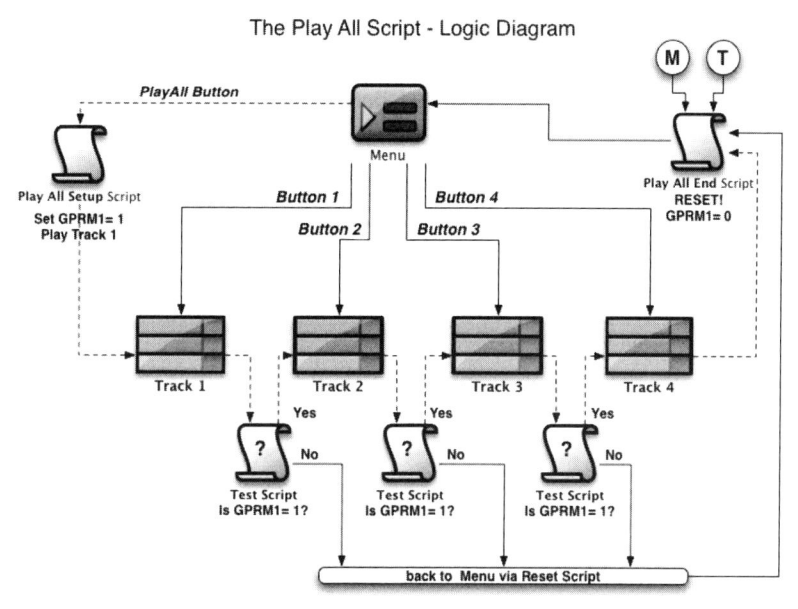

Figure 13-25b — The "Play All" script logic diagram.

- When a Track Button is selected, we play that Track.
- At the end of Track playback, we navigate to the test Script.
- The Test Script checks—is GPRM 1=1?
- If no, we came from a Track button—go back to Menu.

You can see this Script in Figure 13-26. Notice that the first command is Conditional, and only executes if GPRM1 = 1. If that command fails, control passes to the next command, which is Unconditional, and *Jumps* back to Menu 1.

However, if we click the *Play All* button instead of a Track button, here's what happens differently:

- Playback begins from the Setup script, which sets GPRM 1=1.
- Navigation passes to Track 1.

- At the end, again we check: is GPRM1=1?
- This time, GPRM 1 *does* = 1, and so the Conditional command will be TRUE, and we Jump to the next track, repeating this behavior until we get to Track 4.
- After Track 4, there's no need to test—just go to the Menu.

In Figure 13-25b, dashed lines indicate the path of the "Play All" navigation, while solid lines indicate normal navigation. (I've left off the normal Track returns to the menu for clarity.) This script may be extended to any number of Tracks, but each Track should test, and nav should always return to reset the GPRM 1 value when all tracks have played. If either the MENU or TITLE button are engaged, nav jumps to "Play All End" Script, cancels Play All, and goes to the Menu.

The Script Inspector for the test command sequence is shown in Figure 13-26.

379

DVD Studio Pro 4

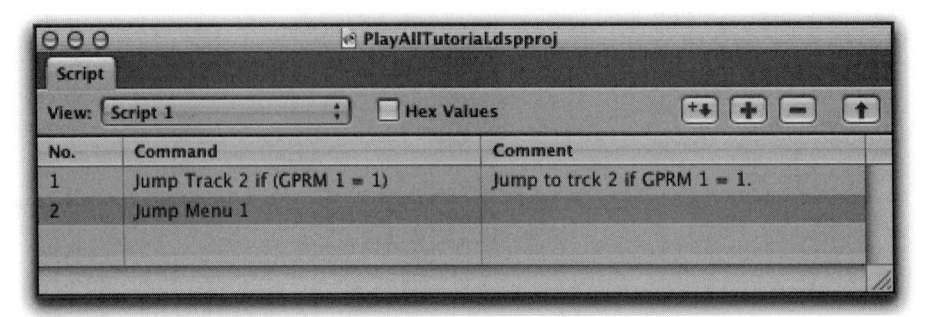

Figure 13-26 — The Conditional Jump Script for Track 1.

So, for a Play All, you need:

1. A GPRM to use, and a Setup Script to initialize it to start Play All.
2. A test Script for the end of each Track.
3. An End script, to turn the Play All Flag (GPRM0) off (make it 0).

Figure 13-27 shows the Script Command Inspector for the Conditional Command from Figure 13-25a.

By the way, we've included this diagram in the DVD so you can see it in larger size. It's called PlayAllLogicSP4.tiff. This diagram was created using OmniGraffle Pro, an exceptionally cool layout program, and using a killer set of Stencils for DVD Studio Pro that you can use to flowchart your DVD projects while you are planning. The stencils were the brainchild of Andrew Porter—way to go, Andrew!

You can download this and other OmniGraffle stencils here: **http://www.omnigroup.com/applications/omnigraffle/extras/content.html**.

Knowing that's a lot to type, try this URL, courtesy of our friends at TinyURL.com: **http://tinyurl.com/dweyz**. Isn't that better?

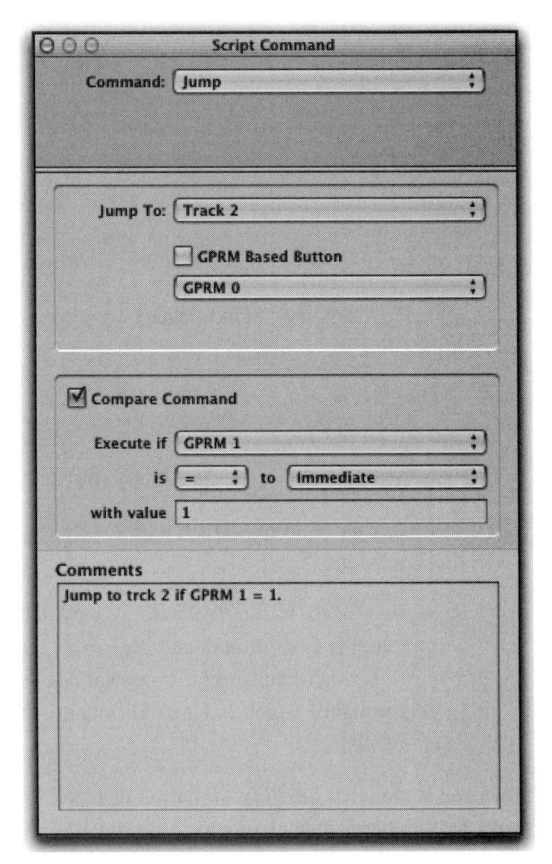

Figure 13-27 — The Conditional Jump Command, completed.

OK—it's also in the book DVD as well—even easier for you! Where? Here: ****Goodies > for OmniGraffle**

380

Basic Scripting for DVD Studio Pro 4

Summary

Scripting is DVD Studio Pro's method to create command Elements that can control DVD player navigation and presentation.

Scripting makes use of the DVD Spec's navigation and presentation commands, System Parameters (SPRMs), and General Parameters (GPRMs), which are built into every DVD player. Scripting Commands in DVD Studio Pro can access 8 of the 16 defined General Parameters and can read the 21 of the 24 defined System Parameters currently in use. Scripting can write new values to SPRMs 1, 2, and 3, allowing for control of which video, audio, and subtitle streams are presented in the DVD.

Enhanced interactivity in DVDs can be created by Scripting the various DVD project elements to control DVD player navigation in response to memorized user selections or conditions stored in GPRMs. GPRMs can be used to track the operation of the DVD, by storing SPRM values at certain times, and Scripts can make decisions on what content to present, based on this stored data.

Scripts can be created and used as stand-alone Elements, called Inline scripts, or they can be attached to certain "root" Elements within the DVD as Pre-Scripts, which will execute when the root element is navigated to.

Scripting allows a great deal of control over the DVD's operation, and can be used to add random playback behavior, prebuilt playlists, and a greater level of user interactivity, including the ability to remember users' menu entries and act upon them at a later time.

The DVD Studio Pro 4 Manual contains a few sample script projects you can create: A Random Play Script, a Play All Tracks script, and a script to determine the Parental Control level of the DVD player. Building these projects will help you learn scripting.

In addition, included in the DVD of this book are several additional Scripting Projects, excerpted from Pro-Pack 2, *Scripting for DVD Studio Pro 1–4*. These projects will give you additional practice in scripting, and will expose you to some new scripting ideas as well.

Chapter 14

Building and Formatting

DVD Studio Pro 4

Figure 14-0 —Quantum DLT-4000.

Goals

The goal of this chapter is to understand the purpose and function of Building your DVD, and how to output (Format) the finished DVD, whether burning it to DVD-R (see Fig. 14-1), or writing it to DLT (Digital Linear Tape—see Fig. 14-2).

Check Your Project Thoroughly before Output

Once your DVD project has been completely Authored, you will undoubtedly want to output the DVD to some tangible form for delivery.

Prior to outputting, you will want to be sure that all of the navigation connections have been checked by you during Authoring, and that project element playback is proper and in sync. After all your hard work making the DVD, it doesn't hurt to give it another once-over to be sure you haven't missed anything. Long before you actually deliver a burned disc, or a DLT master, DVD Studio Pro 4 offers a number of methods to fine-check your project to verify its proper operation.

- **Preview**—Checking the playback behavior of the DVD tracks and slideshows using the Viewer can provide verification of video and audio sync, proper timing of Slideshows, and proper playback of Slideshow audio files.

- **Simulation**—Checking the playback behavior of the DVD using the project assets, prior to building (multiplexing). Once you are confident the project is functioning as you intended, you can Emulate.

- **Emulation**—Once it has been built, the VIDEO_TS folder can be checked in Apple DVD Player to verify its performance. This is the closest you will get to the finished DVD without burning a test disc.

Once you are confident the project is functioning as you intended, the final steps to output include:

- **Burn**—The quickest way to make a DVD-R disc.

- **Build**—Creates a *DVD Volume* folder, which contains the VIDEO_TS folder, with DVD

Figure 14-1 — Some of the DVD Burners available for DVD Studio Pro. L: Pioneer DVR-S201 Authoring; R: Pioneer A03/04/05/06 General Media.

Building and Formatting

Figure 14-2 — Some of the DLTs available for DVD Studio Pro. L: DLT-4000 20/40 GB; R: DLT-8000 40/80 GB.

files organized into the proper DVD file structure. Before building, be sure to set project properties that affect the Build, i.e., First Play action, menu languages, and so on.

- **Format**—Writes a built DVD Volume to DVD-R, or DLT. Prior to Formatting, be sure to set any remaining properties, like DVD-ROM folder inclusion, Copy Protection, Region Coding, and any other properties that affect the final Format.
- **Build and Format**—Does both of the above in one step.

Testing the DVD before Output

Using the Viewer

While authoring, double-clicking on a Track or Slideshow element will bring that Element into the Viewer for casual review. I say casual, because the Viewer isn't equipped for more than a cursory playback of the raw media elements and lacks the bells and whistles of the Simulator. But as a handy first check, it's easy to use the Viewer to determine audio-video sync in Tracks and Slide timings in slideshows.

The Transport controls at the bottom of the Viewer Window provide the bulk of the motion control available, although certain Keyboard shortcuts are available:

Spacebar or L—Play/Pause

K = Stop

Arrow Keys—Step Forward (right) or Step Backward (left)

What the viewer cannot do is prove connections between the Track or Slideshow and other elements within the DVD; for that, you need the assistance of the Simulator.

Using the Simulator

The Simulator is a beefed up Preview window, with lots of capabilities. It acts like a DVD player and can test not only how your media plays, but also how well the connections in your DVD function.

To Simulate the entire project beginning at the First Play:

- Select `File > Simulate` from the menubar.

385

DVD Studio Pro 4

- Click the *Simulator* Icon in the Toolbar (if visible).

- Type *Command-Option-0* (zero) (**⌘-Option-0**).

To Simulate the project beginning at a specific item:

- *Control-Click* on the desired item in the Outline Tab; select Simulate in the pop-up.

- *Control-Click* within a Track or Slideshow; select Simulate.

- *Control-Click* within a Menu; select Simulate.

To Stop Simulating:

- Click the Close button in the Simulator's upper left corner to quit.

- Type Command-W (the Mac Close Window command) (**⌘-W**).

Setting the Simulator Preferences

The Simulator Preferences Pane allows the Simulator to be preset to mimic the characteristics of a DVD player in any region, and with any combination of Audio, Subtitle, and Menu languages (see Fig. 14-3).

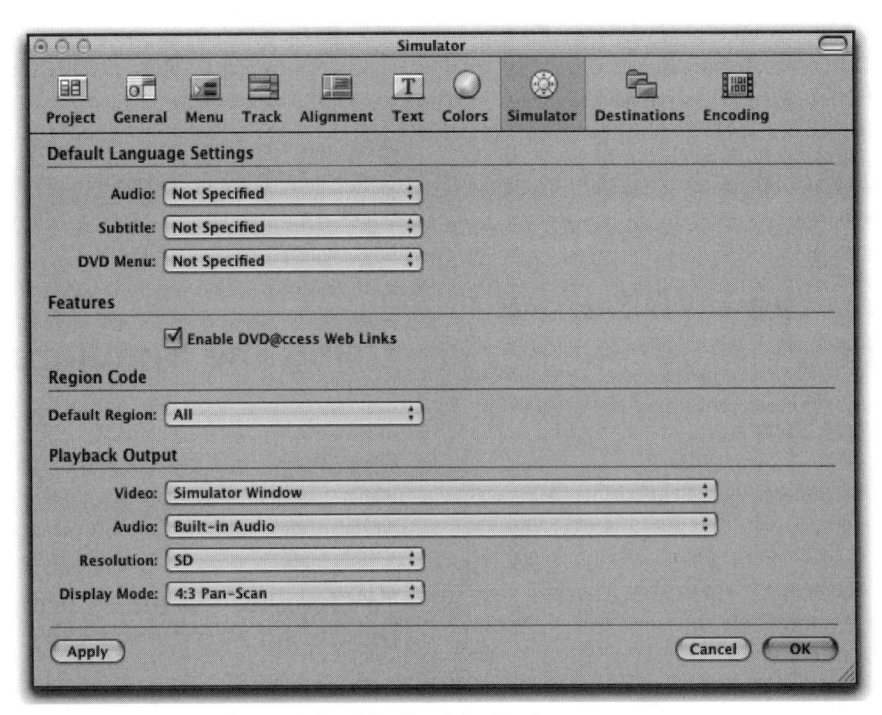

Figure 14-3 — The Simulator Preferences pane.

Building and Formatting

> **NOTE!**
> If you wish to debug DVD@ccess functions, be sure to turn on the "Enable DVD@ccess web links" preference before simulation.

Default Language Settings:

- *Audio*—Selects a preferred Audio Language Stream, when available

- *Subtitle*—Selects a preferred Subtitle Stream, when available.

- *DVD Menu*—Sets a preferred Menu Language, when more than English Menus exist. English is the default when not specified.

Features:

- *Enable DVD@ccess Web links*—When enabled (checked) allows the Simulator to process DVD@ccess web links authored into the current project. Can test web URLs, local file access, email links.

Region Code:

- *Default Region*—Allows selection of one specific region, or "All" (regions) to emulate default behavior. This can be used to test region-specific behavior such as alternate element playback, or conditional branching scripts that depend on a specific region code.

Playback Output:

- *Video*—Select between the traditional *Simulator Window* and *Digital Cinema Desktop Display* that can use a large-format monitor to display image resolutions up to 1920 × 1080.

- *Audio*—Select between Built-In Audio and an external Audio decoder (Dolby Digital or DTS) connected via S/PDIF using Optical or Digital cabling.

- *Resolution*—Select from SD, HD 720, or HD 1080, as appropriate (uses the pop-up Menu at lower left).

  - *4 × 3 Letterbox*—Simulates a 4 × 3 display, letterboxes 16 × 9 content

  - *4 × 3 Pan & Scan*—Simulates a 4 × 3 display, Pans & Scans 16 × 9 content

  - *16 × 9*—Simulates a 16 × 9 display

The Simulator Window

The Simulator Window consists of the following:

- *The Display Window* (see Fig. 14-4a and 14-4b)—Where the DVD video will be displayed

- *The Remote Control Panel*—Emulates a DVD player remote control

- *The Information Drawer*—Shows value of GPRMs (General Parameters), SPRMs (System Parameters), and more.

The Display Window

The right side of the Simulator is the Display window, where the DVD will be previewed under control of the Simulator preferences. This window is "live" and menu buttons can be clicked on to activate them.

The Remote Control Panel

On the left side of the Simulator, the Remote Control Panel operates like the buttons on the DVD remote control of your set-top DVD player, allowing you to

DVD Studio Pro 4

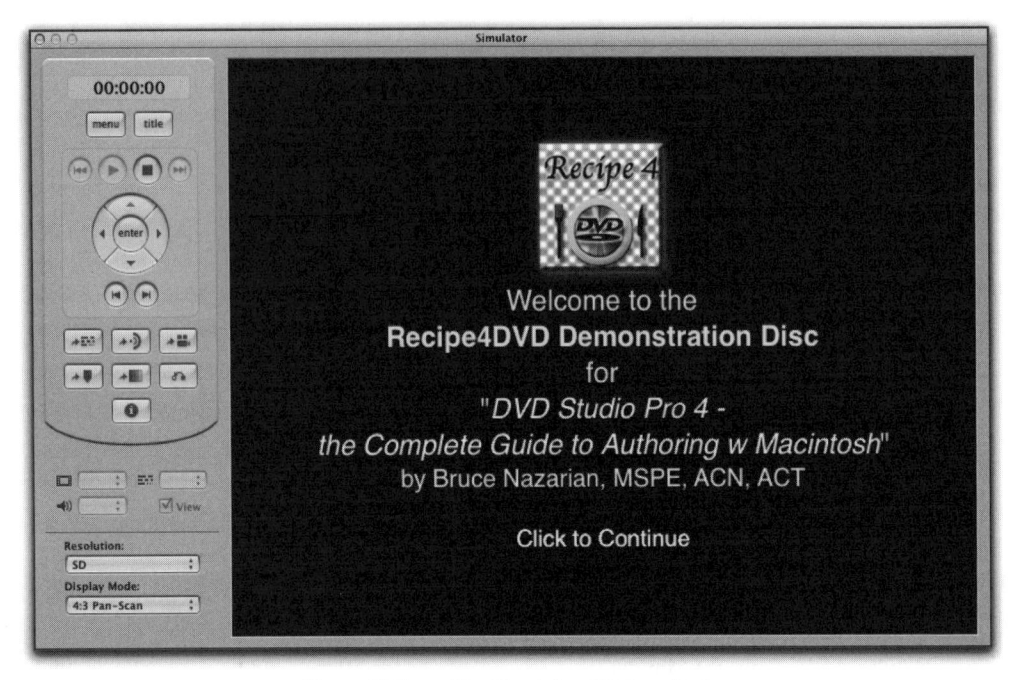

Figure 14-4a — The Simulator Window 4 x 3.

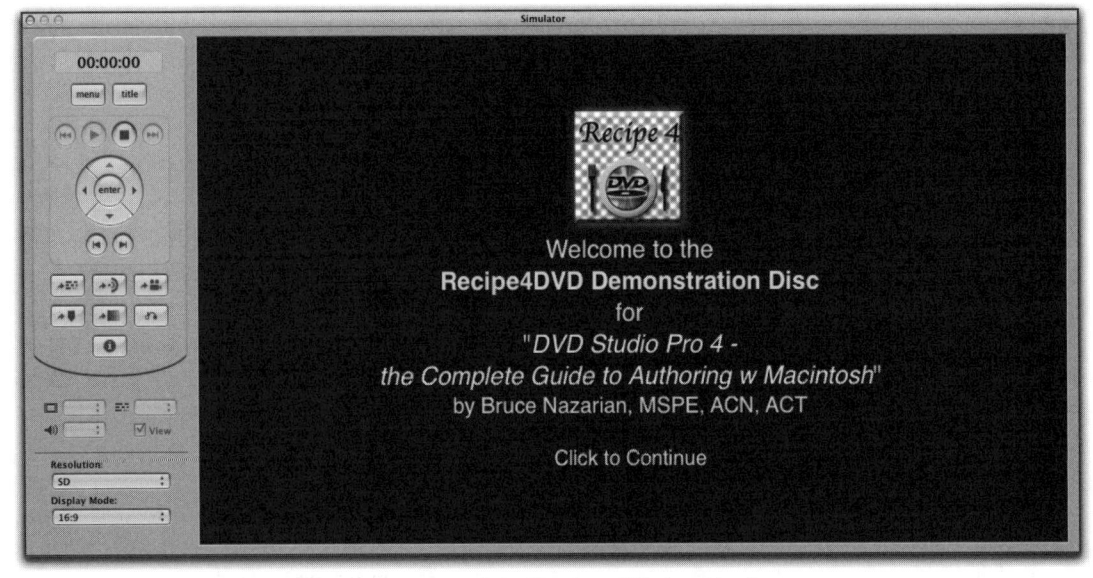

Figure 14-4b — The Simulator Window 16 x 9.

Building and Formatting

navigate through your DVD-in-progress as if it was playing from a burned disc. (See Fig. 14-5.)

The Remote Control Panel is divided into four areas of interest (from the top down): Transport Control, Menu Control, Stream Control, and Resolution Control (this last one is new in DVD Studio Pro 4).

Transport Control—Top Area

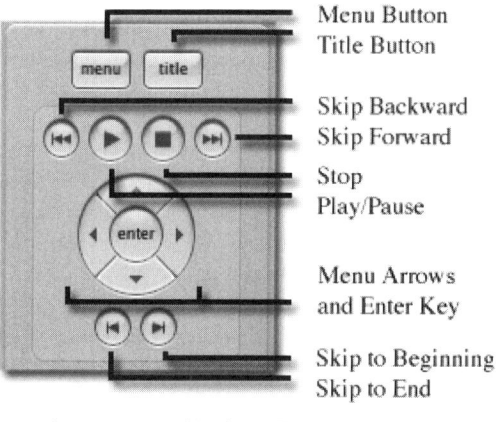

Figure 14-5 — Simulator Transport Control.

- *TC Display*—Shows current play time location within the current Element (not shown—at top above the Menu and Title buttons).

- *Menu Button*—The Menu button executes any Menu call connection attached to this Track, or executes the Resume-to-play function when sitting on a menu (once a Track has been played)

- *Title Button*—The Title button executes any Title connection programmed into the Disc Connections. In the Simulator, it does NOT execute any return-to-play function, as many DVD players WILL do.

- *Chapter Skip Reverse**—Skips to the previous chapter.

- *Play/Pause*—Play (when Stopped), Pause (when playing).

- *Stop*—Stops Simulation playback—doesn't close Simulator.

- *Chapter Skip Forward**—Skip to Next Chapter (not the End Mark).

- *Menu Navigation Controls*—Arrows and Enter Key for Menu navigation. Keyboard arrows and Enter Key may also be used.

- *Track Skip to Top**—Returns to beginning of current track.

- *Track Skip to End**—Skips to end of current Track.

* These keys were mislabeled in the DVD Studio Pro 2 Manual.

Menu Control Area

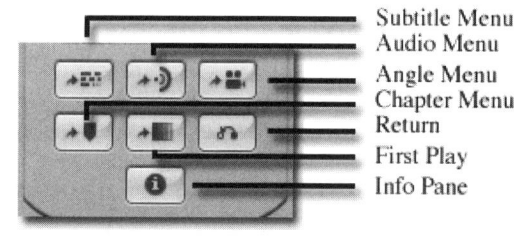

Figure 14-6 — Remote Control Panel Menu Control area.

> **NOTE!**
> Do not confuse these with the Stream Selection controls.

- *Subtitle Menu Select*—Jump to the Element connected to this Element's "Subtitle" link in the Connections Tab.

- *Audio Menu Select*—Jump to the Element connected to this Element's "Audio" link in the Connections Tab.

389

DVD Studio Pro 4

- *Angle Menu Select*—Jump to the Element connected to this Element's "Angle" link in the Connections Tab.

- *Chapter Menu Select*—Jump to the Element connected to this Element's "Chapter" link in the Connections Tab.

- *First Play Select*—Return to the First Play action.

- *Return Button*—Jump to the Element connected to this Element's "Return" link in the Connections Tab.

Connections made to Subtitle Menu, Audio Menu, Angle Menu, Chapter Menu, and Return can be different for each element.

- *Info Button*—Opens the Information Drawer, detailed below (see Fig. 14-8), to display information about the DVD in real time while Simulating.

Stream Selection and Display

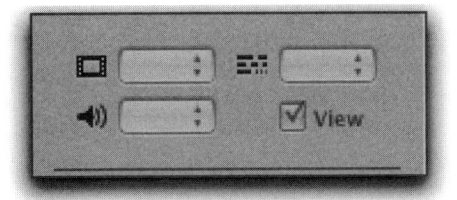

Figure 14-7a — Stream Selectors area.

Left side (top to bottom):

- *Video Stream*—Select an alternate Video Angle, if any exist;

- *Audio Stream*—Select an alternate Audio Stream, if any exist;

Left side (top to bottom):

- *Subtitle Stream*—Select a Subtitle Stream, if any exist;

- *View*—Check "√" to make selected Subtitle visible in Simulator.

Resolution and Display Mode Selectors:

DVD Studio Pro 4 adds new HD DVD features, and the ability to preview and simulate in widescreen and 4 × 3 HD resolutions. The Simulator window now has tools (at the bottom) to allow selection of these new resolutions.

Figure 14-7b — Resolution pop-up.

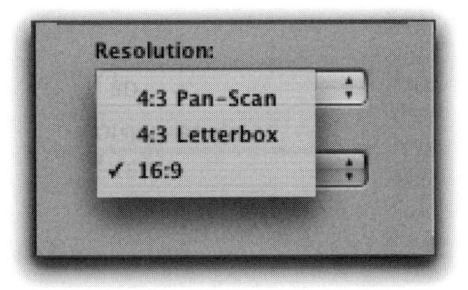

Figure 14-7c — Display Mode.

Resolution (upper pop-up):

- *SD*—Select Standard Definition (4 × 3) resolutions

- *HD 720*—Select HD 1280 × 720 (16 × 9)

- *HD 1080*—Select HD 1440 × 1080 (4 × 3) or 1920 × 1080 (16 × 9).

Building and Formatting

Display Mode (lower pop-up):

- *4:3 Pan-Scan*—Simulates a 4 × 3 display, Pans & Scans 16 × 9 content within the frame.

- *4:3 Letterbox*—Simulates a 4 × 3 display, letterboxing 16 × 9 content within the frame.

- *16:9*—Simulates as a 16 × 9 display.

Information Drawer

The Simulator's Information Drawer displays DVD registers and Item property values in real time during Simulation (see Fig. 14-8).

- *Item Properties*—Displays current values of Items

- *Registers*—Displays current values of SPRM and GPRM registers

- *SPRM GPRM checkboxes*—Check to turn on display of selected registers (In DVDSP 2.0, works upside down)

- *Hex*—Check to display numerical values in Hexadecimal (base 16) 0–9, A (10), B (11), C (12), D (13), E (14), F (15).

The Simulator Display Is "Live" and Active

When the Simulator is displaying a menu, or live clickable button over Video, the buttons visible in the display may be directly clicked upon, as if the Simulation was really being played on a Computer (see Fig. 14-9). This is in addition to the Menu navigation arrows and Enter key in the Transport Control Panel.

You can verify this function by using your mouse to pass over Menu buttons displayed during simulation—your mouse cursor will turn into the Pointing Hand (see Fig. 14-9b), indicating the presence of a button *hot zone*, whenever your cursor is within the active area defined by a button.

> **NOTE!**
> Even if a button is defined as invisible, or lacks a highlight (see Chapter 9), you should be able to sense the invisible button area with your mouse—keep this in mind when designing Easter Eggs (invisible buttons with content links).

Emulating a Project with Apple DVD Player

If your project passes Simulation with flying colors, one additional means exists for advanced testing—*Emulation*. Playing your built DVD volume in the Apple DVD player can verify that it will actually perform properly *after* it has been built, but without having to burn a disc.

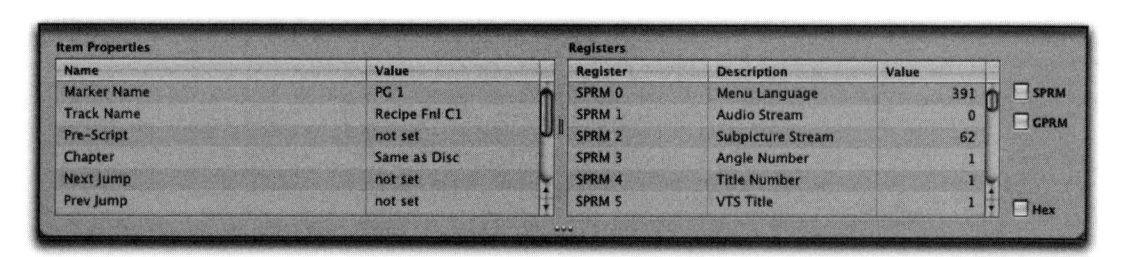

Item Properties		Registers		
Name	Value	Register	Description	Value
Marker Name	PG 1	SPRM 0	Menu Language	391
Track Name	Recipe Fnl C1	SPRM 1	Audio Stream	0
Pre-Script	not set	SPRM 2	Subpicture Stream	62
Chapter	Same as Disc	SPRM 3	Angle Number	1
Next Jump	not set	SPRM 4	Title Number	1
Prev Jump	not set	SPRM 5	VTS Title	1

☐ SPRM ☐ GPRM ☐ Hex

Figure 14-8 — The Information Drawer.

DVD Studio Pro 4

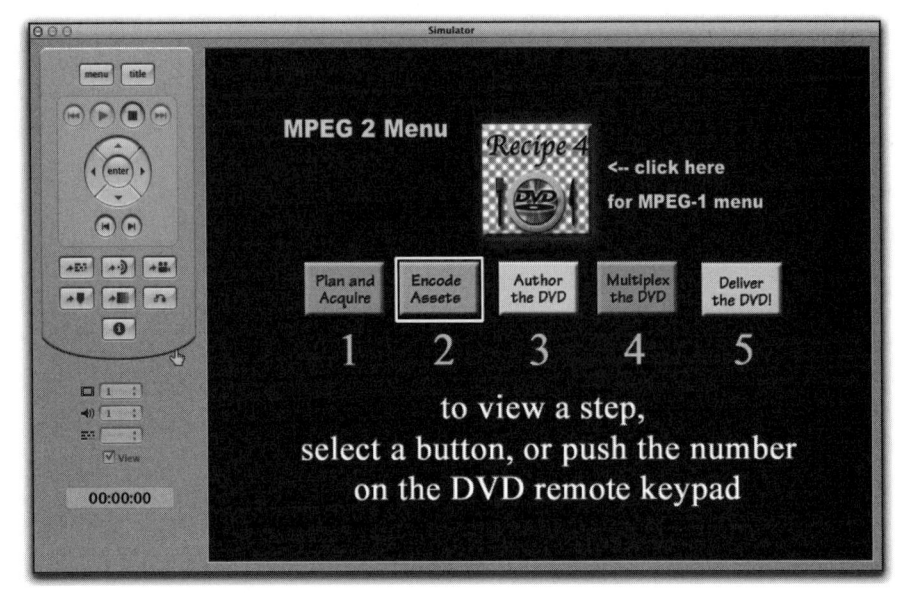

Figure 14-9a — A live menu highlight in the Simulator Window.

Figure 14-9b — The Button Cursor you see on a Computer.

Build before Emulation

To view your project in Emulation, you must build it to your hard drive, and then load it into Apple DVD player (see *Building...* later on).

To open your built DVD volume in Apple DVD Player:

- Select `File > Open DVD Media...` from the menubar (see Fig. 14-10a), **or**

Figure 14-10a — Opening a VIDEO_TS folder.

- Type *Command-O* (⌘-**O**).

In the dialog box that appears, navigate into the DVD Volume folder you built to find the VIDEO_TS folder. You *must* select the VIDEO_TS folder, and no other folder, or DVD playback will not occur.

> **NOTE!**
> You may need to click the DVD Remote Control's Play triangle to commence playback once the DVD Volume has loaded into Apple DVD player.

Emulating with DVD@ccess Links

If your DVD contains DVD@ccess links (see Chapter 17), you will want to enable those links during emulation.

To enable DVD@ccess links in Apple DVD Player:

- Select **DVD Player > Preferences...** in the menubar.

- In *Disc Setup*, check *Enable DVD@ccess web links*.

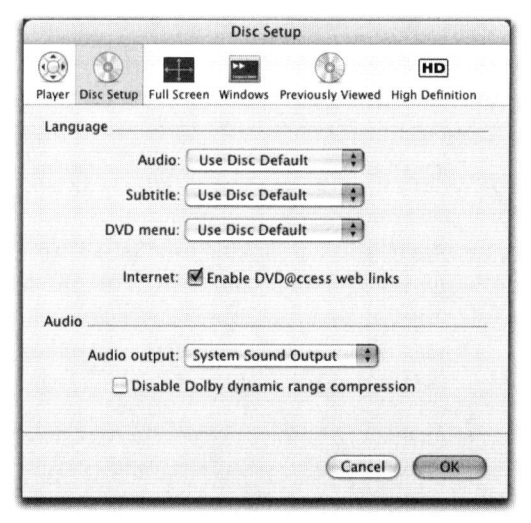

Figure 14-10b — "Enable DVD@ccess Web Links" in Apple DVD Player.

> **TIP!**
> If you are planning on emulating a number of times while you are making revisions, consider adding the Build Folder to the Favorites display. It will be much easier to find it each time.

Setting Disc Properties before Building or Formatting

Disc properties that affect output options need to be properly set prior to attempting to build, format, or both. Certain very important properties, like the First Play action, can break an otherwise perfectly functioning disc.

Start by verifying that the Disc properties have been set correctly and completely. If the Disc Inspector isn't visible:

- Click on the *Disc* Icon in the Outline Tab,

- Select **View > Show Inspector** in the menu bar, or

- Type *Command-Option I* (⌘-**Option-I**), or

- Click on the Inspector tool in the Toolbar (if visible).

Top Area Settings

- *Name*—The default name UNTITLED_DISC is easily replaced with a name of your own choosing. Only alphanumeric characters, 0–9, A–Z, and the underscore character "_" are valid in a disc name. Other characters are ignored; lowercase characters are converted to UPPERCASE automatically when entered.

- *First Play*—This is the first project element that will be played automatically when the disc is inserted into any DVD player. Without this being set, the DVD will not begin playing properly. This may be set to any Track, Slideshow, Menu, or Script.

393

DVD Studio Pro 4

> **TIP!**
> Many DVD players do not like to encounter a script or menu as the first play item, so consider creating a short (1 second) Track of Black video, and using that as the First Play target.

> **VERY IMPORTANT!**
> Always be sure to set the First Play action!

The Disc Inspector General Tab

All of the settings in the General tab except the DVD-ROM settings must be set before building a project (see Fig. 14-11).

- *DVD Standard*: Selects the DVD resolution to be used in this project, SD or HD.

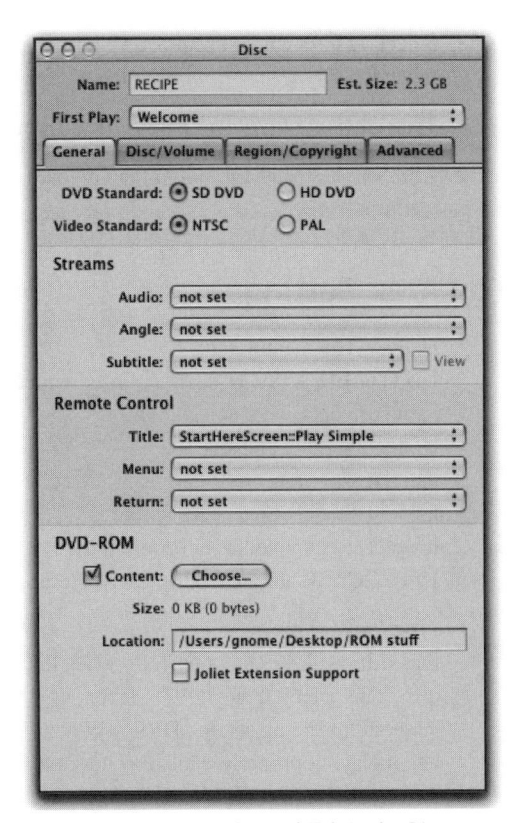

Figure 14-11 — General Tab in the Disc Properties Inspector.

> **NOTE!**
> Once HD is selected for a project, it cannot be downgraded to SD.

- *Video Standard*: Selects the video format used in this project, NTSC or PAL. This must be set before you assign any assets to the project.

Streams

These settings control any Stream selection overrides you wish this DVD to activate.

- *Audio*: The default setting, *not set*, allows the DVD player to control the audio stream setting. You can choose an Audio Stream that will serve as the initial active stream, overriding the DVD player's setting.

> **NOTE!**
> The word *Empty* in parentheses after a stream number indicates that that stream is currently not used.

- *Angle*: The default setting, *not set*, allows the DVD player to control the video stream. You can choose a Video Stream that will serve as the initial active stream, overriding the DVD player's setting.

- *Subtitle*: The default setting, *not set*, allows the DVD player control the subtitle stream setting. You can choose a Subtitle Stream that will serve as the initial active stream, overriding the DVD player's setting. You can also force the Subtitle Stream to appear by selecting its *View* checkbox.

Remote Control

- You can set the *Title, Menu*, and *Return* remote control settings, which correspond to a DVD player's remote control buttons. See Chapter 17 for details on these controls.

394

Building and Formatting

DVD-ROM

See "Creating a Hybrid DVD Disc" later in this chapter for more information about using these settings and creating DVD-Video+DVD-ROM discs:

- *Content*: Selecting the *Content* checkbox enables the DVD-ROM content settings (Location and Joliet Extension Support). These settings convert this disc into a DVD Hybrid. Click *Choose:* to select the DVD-ROM folder location.

- *Location*: This setting shows the path to the folder containing the DVD-ROM contents for the disc. Clicking *Choose*: opens a dialog box to use to select this path. You may also type it in manually.

- *Joliet Extension Support*: Checking Joliet Extension Support allows long filenames to exist in the DVD-ROM content. For greatest compatibility with replicators, you might consider limiting filenames used in your DVD-ROM content to conform to 8.3 (DOS) standards.

> **NOTE!**
> Some replicators' software verification may report filename errors when DVD-ROM files exceed the 8.3 filename spec permissible in DOS. This should not be a problem, unless their software *truncates* file names to force compliance. Ask your replicator!

The Disc Inspector Disc/Volume Tab

These settings must be verified before you attempt to format your project (see Fig. 14-12).

Disc Information

- *Layer Options*: Choose either single- or dual-layer for the project. Note that in newer

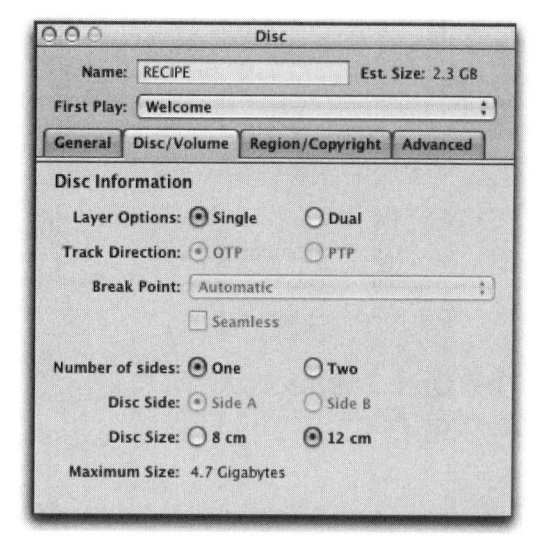

Figure 14-12 — Disc size settings in the Disc Inspector.

G5 systems, you will likely have a Double-Layer-capable DVD burner.

- *Track Direction*: Use this control to select either OTP (Opposite Track Path) or PTP (Parallel Track Path). In OTP, the first layer plays from the center to the edge, then the laser refocuses and plays back from outer edge to the center. In PTP, both layers play in the same direction The advantage of the OTP method is that it minimizes the pause that occurs during the layer break.

- *Break Point*: Use this pop-up menu to choose the dual-layer break point, the point of division between the first and second layer of the disc. You can select any marker on any track, but ideally the layers should be balanced as to size. When left as Auto, DVD Studio Pro automatically chooses the break point. (see Chapter 8 for more information on setting Layer Break Markers).

- *Number of sides*: Select if this is a one- or two-sided disc.

395

DVD Studio Pro 4

> **NOTE!**
> A two-sided disc requires two projects—one for each side.

- *Disc Side*: Which side of the double-sided DVD is this project? Select Side A or Side B.
- *Disc Size*: Choose either the standard 12 cm or the smaller 8 cm disc size. This setting affects the choices you have available in the Disc Media setting.

The Disc Inspector Region/Copyright Tab

The only setting in the Region/Copyright tab that you must set before building a project is the *Macrovision* setting (see Fig. 14-13).

Setting Playable regions

Playable Region Codes: Select the checkbox next to the regions in which you want to be able to play the disc (all are enabled by default). All DVD players (stand-alone and computer) must be assigned a region code. This code divides the world into six regions, with an additional region for discs played on airplanes. These region codes allow you to have some control over where a disc can be played.

Copy Protection settings (CSS, Macrovision)

This setting in the Disc Properties is the global Macrovison switch for all Tracks. Macrovison may be modified on a Track by Track basis once this switch is enabled, but unless this switch is enabled, Macrovision cannot be turned on at the Track level.

- *Copyright Management*: Select the checkbox to enable either digital-based (CSS) or analog-based (Macrovision) copyright protection of the disc. Selecting this checkbox enables the Copy Generation setting.

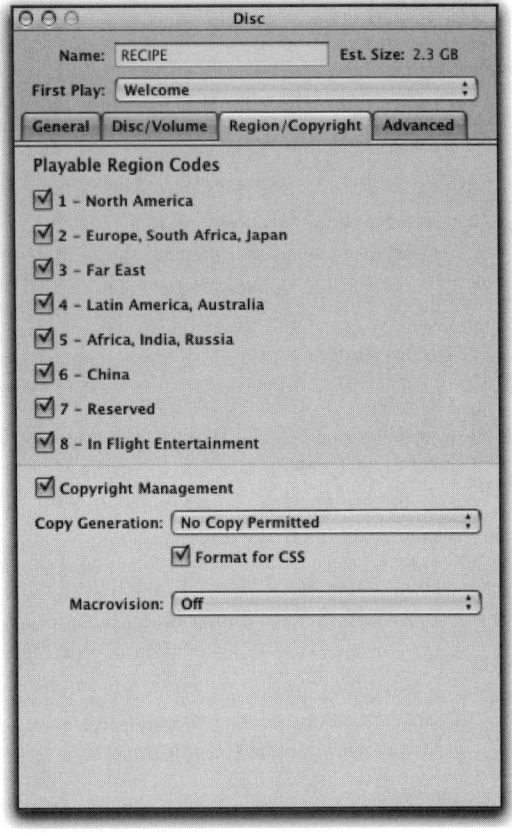

Figure 14-13 — Region Codes and Copy Protection settings.

> **NOTE!**
> This setting affects the number of bytes in each sector of the disc. With copyright management disabled, each sector has 2048 bytes. When enabled, each sector has 2054 bytes.

> **WARNING!**
> In most cases, it is NOT POSSIBLE to write to DVD-R (General) with a 2054-byte sector size selected! *Be warned!*

- *Copy Generation*: This pop-up menu specifies whether copies are permissible, or copying is denied.
  - *Copying Permitted*: The disc is not protected against copying. The Format

Building and Formatting

for CSS checkbox and Macrovision pop-up menu are disabled. The only difference between this setting and leaving the Copyright Management checkbox unselected is the number of bytes in the sectors. *THIS IS AN UNUSUAL SETTING!*

- *One Copy Permitted*: Allows users to make a copy of the disc, but no additional copies from that copied disc (the Copy Generation status of the copied disc switches to No Copy Permitted). You can make as many copies as you want from the original disc; you just can't make more copies from the copies of the original. The Format for CSS checkbox and Macrovision pop-up menu are disabled.

- *No Copy Permitted*: Enables the Format for CSS checkbox and Macrovision pop-up menu so that you can choose the type of copy protection to use.

- *Format for CSS*: Only available when Copy Generation is set to *No Copy Permitted*. The Content Scrambling System (CSS) provides digital-based copyright protection. Selecting Format for CSS alerts the replication facility to apply CSS encryption to the disc (if it is licensed and authorized to do so). *CSS REQUIRES DLT or DVD-R (A) output!*

- *Macrovision:* This must be set before you create the VIDEO_TS folder in the build process. Macrovision provides analog-based copyright protection. See Chapter 17 for more information on Macrovision.

The Disc Inspector Advanced Tab

These settings (see Fig. 14-14) apply to specialized features used on advanced projects. You may leave

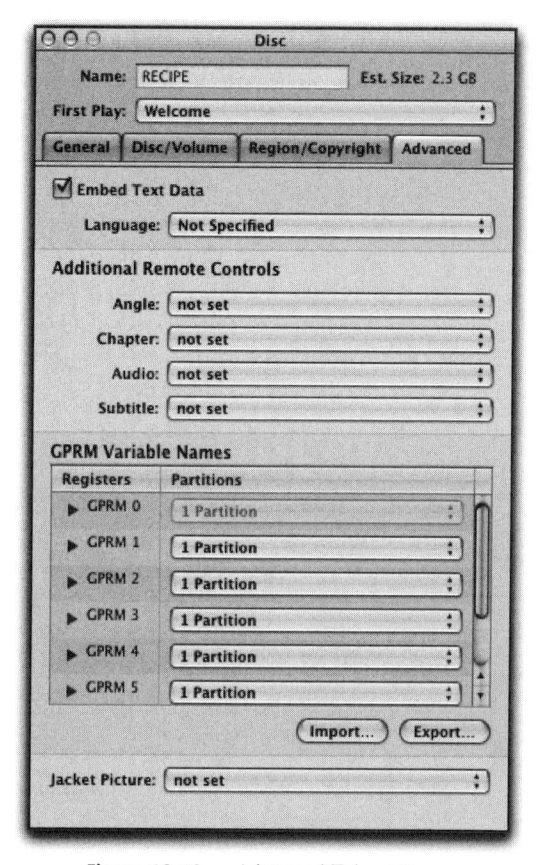

Figure 14-14 — Advanced Tab settings.

the default GPRM names in place, if you choose to do so, and have used those names in your scripts.

- *Embed Text Data*: This checkbox is automatically selected if you use the DVD@CCESS feature within your project. You can also select it if you want to add the names you assigned to the tracks, slideshows, and menus to the DVD disc. This text can be displayed on DVD players designed to support the Text Display extensions of the DVD-Video spec (version 1.1). (There are some limitations on the length of names that can be used—see Chapter 17).

397

> **REALITY CHECK!**
> Today, very few DVD players (mostly SONY ones) actually display this information. It is also used by PowerDVD on PCs.

- *Language*: If the Embed Text Data checkbox is selected, you can choose which one DVD player language setting in which the text names will be visible. Choose *Not Specified* to have the text appear regardless of the DVD player's language setting. (This does NOT translate the Characters you author into the names into a different language!)

- *GPRM Variable Names*—For use with Scripting, these fields allow you to set meaningful names for the 8 General Parameters (variables) used to hold values during script execution. These names are Global throughout the entire DVD project. Changing a name in this field will change it throughout all of the existing scripts.

- *GPRM Partitions*—GPRMs may now be partitioned into subunits, to allow for easier Scripting when using bit values to track player functions. See Chapter 13 for more information on using GPRM partitions.

Outputting a Finished DVD Project

Two distinctly different operations are required to output the finished DVD project—*Building and Formatting*:

Building

Building the DVD changes the populated skeleton of your DVD project into a finished *DVD Volume*, with the files arranged properly according to the rules of the DVD Specification. The elementary files you began with have now been converted into IFO, BUP, and VOB files, and arranged into the familiar VIDEO_TS folder that contains all of the *Video Title Sets* that comprise this particular DVD.

A Video-only DVD volume

Figure 14-15 shows what a built DVD Volume looks like for a project that only contains DVD-video content (no ROM):

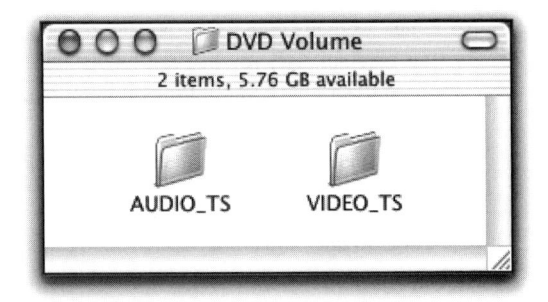

Figure 14-15 — A Built DVD-Video Volume.

The AUDIO_TS folder is always empty in a DVD-Video project; that folder is for DVD-Audio data. DVD Studio Pro only creates DVD-Video data. The AUDIO_TS folder is created to provide the most compatible DVD possible. Many players do not react well if the AUDIO_TS folder is missing, so it is always built, and you should always leave it in the DVD Volume folder, and include it in any copies you make.

Inside the VIDEO_TS folder lies the heart of the DVD (see Fig. 14-16).

The finished DVD files are organized into VTS_XX sets starting with the VIDEO_TS set. Each *Title Set* corresponds to a Track, Story, or Slideshow in DVD Studio Pro. Note that each set contains an IFO, a BUP, and one VOB file per GB of DVD media in that Track, Story, or Slideshow. The IFO is the navigation infor-

Building and Formatting

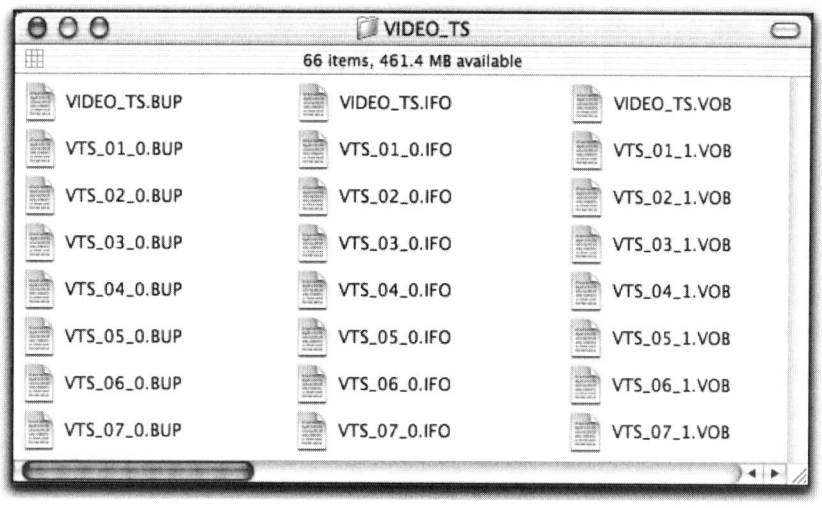

Figure 14-16 — The VIDEO_TS folder revealed.

mation and Menu data for that Title Set, while the VOB (Video Object) contains the built DVD data for that Element. The BUP file is a Backup for the IFO, used in case the IFO is not readable for any reason.

A Hybrid DVD—(Video + ROM Data)

In a DVD Hybrid disc, the DVD-Video data are still contained in the VIDEO_TS folder—that's standard. The Disc still contains an AUDIO_TS folder for compatibility. The additional data are included in an area of the disc called the ROM ZONE (or "Others Zone"). These data are readable by a computer, but not usable in a DVD set-top player connected to a TV Set. As of this writing, no DVD settop players can process DVD-ROM readable material, but it's not for a lack of trying. Perhaps in a few years we will see set-top DVD players that can surf the web, or read DVD-ROM files, but not yet.

The specific figures above show a DVD-Video-only project, which is quite a common configuration. If there were any DVD-ROM-based files included in the finished DVD, it would look like that shown in Figure 14-17 (as an example).

Formatting

The DVD project may be formatted to DVD-R, or DVD+R using the Apple Superdrive, or an external DVD-R drive (using inexpensive DVD-R General Media)—or you may choose to use an external Pioneer DVR-S201 (a SCSI drive, which uses Authoring disc media instead). (You should be aware that DVD-R and +R discs may not be completely compatible with all DVD players.) For a list of known compatible DVD players, you can browse the exhaustive and authoritative study of DVD-R compatibility at **http://www.dvdtoday.com**.

Another output method is DLT (Digital Linear Tape), which has traditionally been the delivery method of choice for DVDs that will be replicated (see Chapter 16, "Duplication and Replication").

Lastly, you may output your DVD project to a DVD-RAM disc if your Mac is equipped with a DVD-RAM Drive (very few are these days). While

399

DVD Studio Pro 4

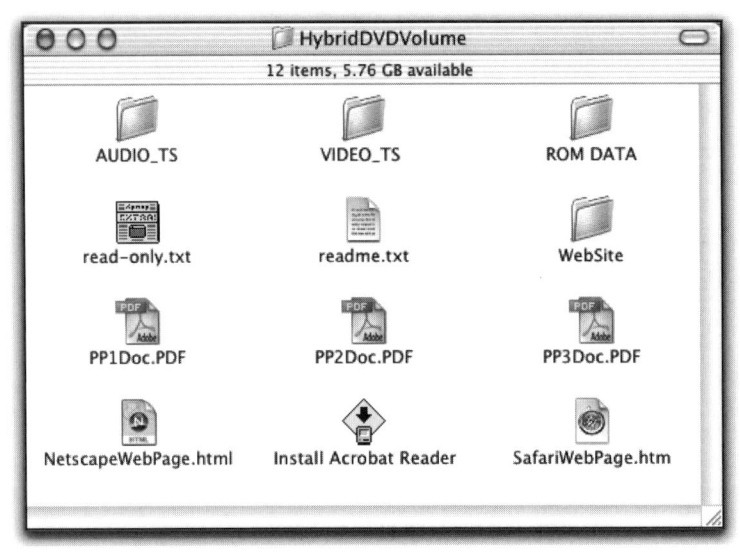

Figure 14-17 — The Hybrid DVD Volume exposed!

DVD-Video projects *may* be written to DVD-RAM media, that media is not as readily compatible with DVD settop boxes as DVD±R or DVD±RW media. In point of fact, it may well be that you can only play a DVD-RAM disc on a computer suitably equipped with a DVD-RAM drive, or perhaps a DVD-RAM-based settop DVD recorder.

A Disc Image file may also be written to Hard Drive.

Burn, Build, Format, or Build and Format?

There are four different commands available in the File Menu that allow you to output the DVD project: *Burn*, *Build*, *Format*, and *Build and Format* (see Fig. 14-18).

Burn

Burn is a quick convenient command that does one thing—it Burns your current project (in whatever state it is) to a DVD±R/RW disc. This process is really streamlined, as it offers no options for DLT or writing a disc image.

To Initiate a Burn:

- Select **File > Burn...** from the menubar
- Click on the *Burn* tool in the toolbar (if visible)
- Type *Command-Shift-F* (⌘-**shift-F**).

A few assumptions are made, to expedite this command's operation—it will build the DVD volume into the folder already specified in the *Destinations Preference* pane. It will also take the current DVD name from the *Disc Name* property. Once it's ready to begin burning, *Burn* will look for the first General Media compatible drive it can find on your system and expect to find a blank DVD recordable disc in it. If not, it will open the drive drawer and prompt for you to enter a blank disc of any acceptable format (see Fig. 14-19).

Building and Formatting

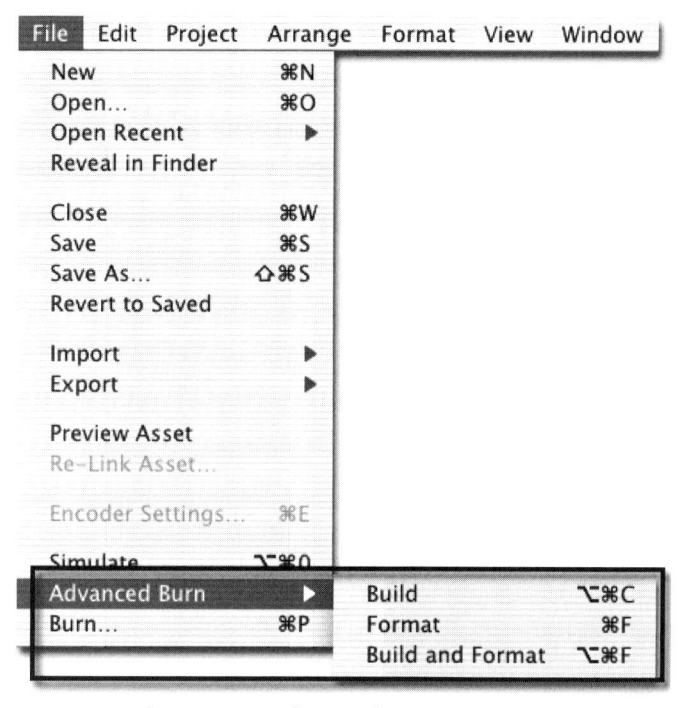

Figure 14-18 — Commands to output a DVD.

Figure 14-19 — Dialog asking for a bank DVD-R disc for burning.

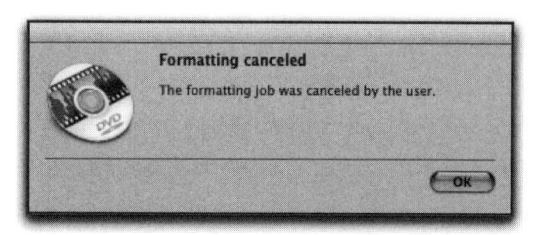

Figure 14-20 — Dialog Cancelling Formatting.

This operation now works with DVD-R, +R, –Rw, +Rw, and on some systems, +DL media. If you do not wish to burn at that time, hit *Cancel*, and the Burn operation will terminate, and you can return to authoring (see Fig. 14-20).

When finished, the system will report formatting success and return control of the application to you. If you wish, you may burn another disc at that time.

Build

Building the DVD will only create the DVD Volume folder on a local hard drive. This folder may then be viewed using the Apple DVD player, for further testing of the DVD project before committing it to an expensive piece of burnable media. Before you build, exhaustively test your DVD using Simulation in DVD Studio Pro to minimize wasted effort and use-

401

less builds. One mistake in authoring is enough to send you back to square one, and force a rebuild—a DVD can't be *almost* perfect, it has to be *completely perfect* before you can release it from your control!

To Build the DVD volume, do one of the following:

- Select `File > Advanced Burn > Build` from the menubar; **or**

- Click on the *Build* tool in the toolbar (if visible), **or**

- Type *Command-Option-C* on the keyboard (⌘**-Option-C**).

Once you select the *Build* command, you can select an existing folder to use, or create a new folder to build your DVD Volume into. This is not a good time to find out you don't have enough hard disk space.

How Much free Hard Disk space is needed?

The *Choose Build Folder* dialog box will be asking for a location where it can *build* the DVD Volume—the Folder where the multiplexed DVD data will be written. This will require free hard disk space equal in size to the finished DVD project. If you're building a DVD-5, you'll need around 5 Gigabytes of free space—around 9 Gigabytes for a DVD-9. If you are making a smaller disc, obviously you only need a smaller amount of space. Practical experience says that "more is better." A few hundred gigs is not an unreasonable drive size for DVD authoring.

Maximizing the Build speed

Be aware that during the Build, a lot of digital data are going to be transferred from one location to another on your hard drive. One of the effective ways to speed this process up (besides having a faster Macintosh CPU) is to configure the hard drive storage of your project in an efficient manner.

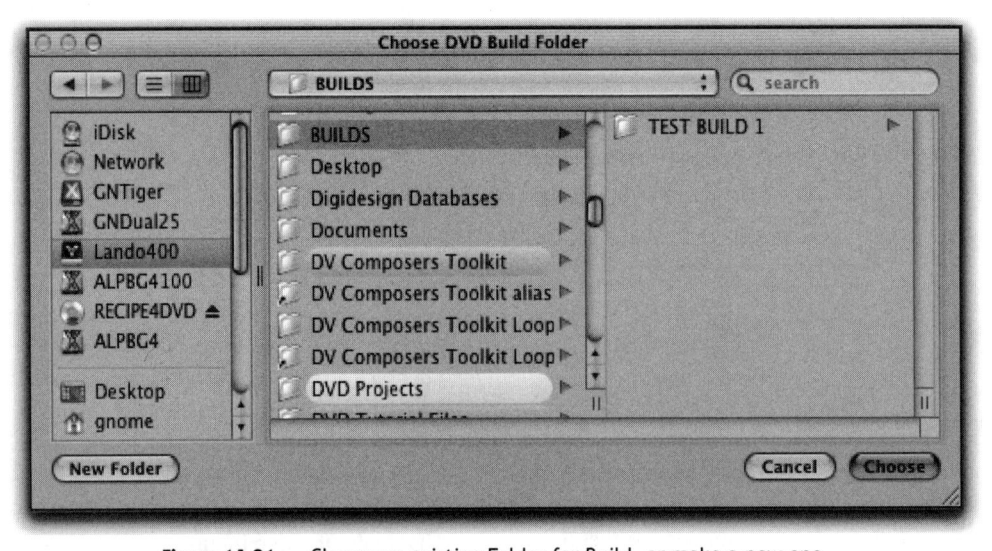

Figure 14-21 — Choose an existing Folder for Build, or make a new one.

Building and Formatting

Configure the right drive setup

Store all of your project assets on one drive, while Building to a different drive. By configuring your authoring system in this way, the flow of data is always one-way; from the source drive to the build drive. This will speed up the flow of data, and your build will be completed faster.

Use fast disk drives

We've talked a lot about FireWire, but don't forget that internal ATA or SATA and external SCSI drives can frequently be accessed faster than FireWire drives. On processes like *Building* that are very disk I/O intensive, this speed advantage will pay off for you.

Shut down unneeded processes

Quit programs you don't need running while building—they will just steal CPU time away from the build.

Keep your hard disk space optimized

This will also speed up reading and writing of your Build. Alsoft's *Disk Warrior* is your friend!

Build Messages

Once the Build Folder has been selected, the process of Compiling will commence. Watch the Log Tab for details on the Build in progress, including the success message. When a Compile (the Build) has completed successfully, you will see this dialog (see Fig. 14-22).

Once your Build is successful, you may now Emulate that DVD Build using Apple DVD Player before going on to burning a DVD.

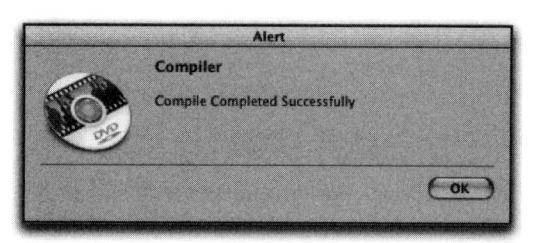

Figure 14-22 — A Successful Build completed!

The Apple DVD player in OS X has an "Open DVD Media" function (⌘-**O**) to allow you to play the Build folder as if it was a DVD disc.

Format

Formatting means writing the built *DVD Volume* (which we assume you have tested and passed as OK by this time) to one or another of the available devices on the authoring system. These devices may include a DVD±R/RW +DL Superdrive (General Media) or a DLT (Digital Linear Tape) drive. Both of these are typical choices, but DVD-RAM is another possibility, as is the venerable Pioneer DVR-S201 DVD-R (Authoring Media) drive. The DVR-S201 requires a SCSI interface in your Mac, as does a DLT drive.

DVD Studio Pro 4 also allows you to Format *any* finished DVD volume on your hard dive, eliminating the need to rebuild, even partially. Once your DVD project works correctly, you can Format as many times as you like, in as many different formats as you like, either burning to DVD±R/RW, or writing to DLT.

It is also possible to Format a *Disc Image* file to your hard drive, which may be mounted and played in the Apple DVD player, or loaded into Toast Titanium and burned to any device usable with Toast, including (but not limited to) DVD-R/RW, DVD+R/RW, DVD+R DL, DLT, and DVD-RAM.

DVD Studio Pro 4

If you wish, you may choose to include DVD-ROM data during the formatting stage, and the information within the selected DVD-ROM folder will appear on the root level of the finished DVD disc (root level means at the same level as the VIDEO_TS folder). This applies to a Disc Image, a DVD-R, or a DLT.

> **REMEMBER!**
> DVD-ROM data cannot be processed by a set-top DVD player!

To Format an already built DVD volume, do one of the following:

- Select **File > Advanced Burn > Format** in the menubar; **or**
- Click on the *Format* tool in the toolbar (if visible), **or**
- Type *Command-F* on the keyboard (⌘-**F**).

When the Formatting dialog appears, sections of the General Tab allow selection of specific functions, as outlined in Figure 14-23:

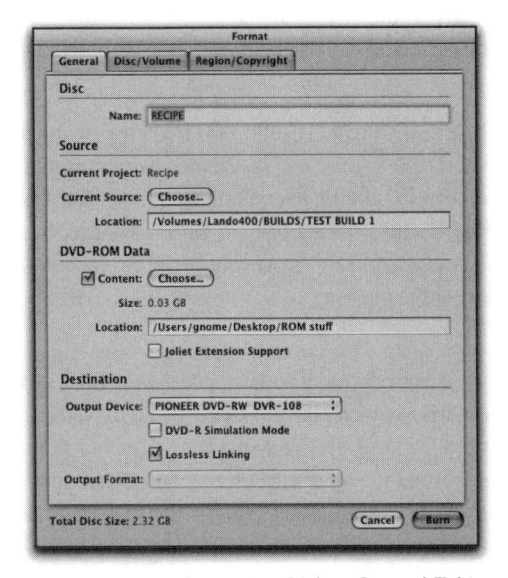

Figure 14-23 — Formatting Dialog (General Tab).

- Disc
  - *Name*: Defaults to the current name of the Disc, but you may also select a new name. Enter a new name in the *Name*: field by replacing the old one.

- *Source*
  - *Current Project*—Indicates the name of the current project.
  - *Current Source*—Click the Choose button to select which DVD volume you wish to Format.
  - *Location*—Will display the filepath to the Build folder chosen to be formatted.

- *DVD-ROM Data*
  - Check the DVD-ROM *Content*: checkbox to enable the DVD-ROM function (if it has already been enabled in the Disc Properties, it will already be checked).
  - Click the *Choose...* button to locate the DVD-ROM folder you wish to add to your disc (see Making a DVD Hybrid Disc, later in this chaptr).
  - *Location*—Will display the filepath to the Build folder chosen to be formatted.

- *Destination*
  - *Output Device*: This pop-up will list all devices found on your system and currently available to write to, both DVD burners, and DLT drive(s), also *Hard Drive*.
  - *DVD-R Simulation* mode will turn OFF the burning laser in a DVD-R burner, so don't select this unless you are just checking your system.

Building and Formatting

- *Lossless Linking* is a way to prevent a buffer under-run (write error) in DVD burners, but it's been said that this feature can affect readability of DVD-R discs, so you may want to avoid using this. It defaults to on. Consider turning it off if you have problems playing DVDs written from DVD Studio Pro.

- *Output Format* will change depending on which device you select to format to:

  - If Output Device = a recordable DVD drive, nothing will appear;

  - If Output Device = DLT, you can choose between DDP 2.0, 2.1, and *CMF*; DDP 2.0 is the accepted standard for DLT masters. Check with your replicator before using DDP 2.1.

  - If Output Device = Hard Drive, only *.img* will be available. Formatting to Hard Drive can only create a *DVD Disc Image*. If you wish to create a DVD Volume folder, use *Build* instead, not *Format*.

Format Options in the Disc/Volume Tab

Disc Information

The Disc information presented in the Disc/Volume Tab of the Format Dialog should reflect settings you have already set in the Disc Properties (see Fig. 14-24). NOTE that new to DVD Studio Pro 4 are the settings for red Laser or Blue Laser devices and media. As of this writing, there are not any significant Blue laser devices available for the Macintosh.

Volume Information

Volume Information automatically Defaults to a date/time stamp set at the moment the Format Command was selected. This is taken from the current system time and date.

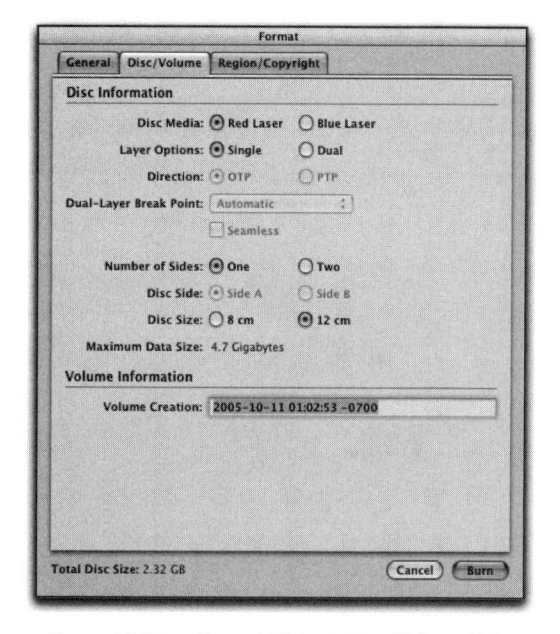

Figure 14-24 — Format Dialog's Disc/Volume Tab.

Format Options in the Region/Copyright Tab

Playable Region Codes

The Disc information presented in the *Playable Region Codes* area of the Format Dialog should reflect settings you have already set in the Disc Properties. They are presented in this dialog for your convenience in overriding the settings for this particular Format operation.

405

DVD Studio Pro 4

Copyright Management

The Disc information presented in the *Copyright Management* section of the Format Dialog should reflect settings you have already set in the Disc Properties. They are presented in this dialog for your convenience in overriding the settings for this particular Format operation (see Fig. 14-25).

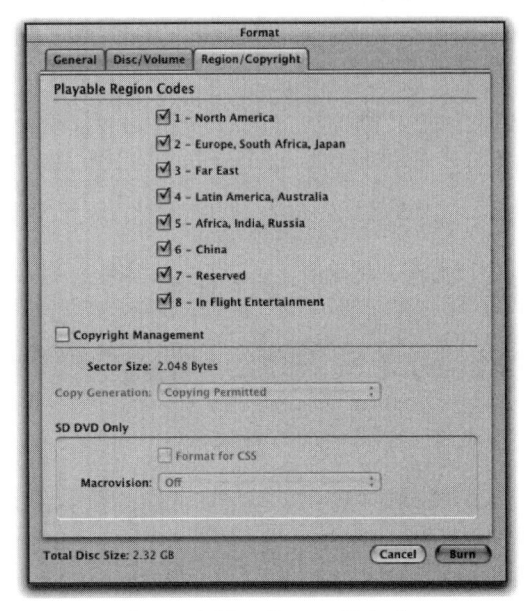

Figure 14-25 — Format Dialog's Region/Copyright Tab.

Build and Format

To Build and Format the DVD in one integrated operation:

- Select **File > Advanced Burn > Build and Format** from the menu, **or**
- Click on the *Build and Format* tool in the toolbar (if visible), **or**
- Type Command-Option-F (⌘-**Option-F**) on the keyboard.

Selecting this command will essentially perform both the *Build* and *Format* operations listed above, but in an uninterrupted sequence, accomplishing both operations. A *DVD Volume* will be built to the current source location, and the built DVD Volume folder will be *Formatted* to the device selected in the Destination section of this dialog. If DVD-ROM data is enabled, it will be added in the usual manner.

Once again, you may go through the Format dialog Tabs and set the precise setting you need for this particular output (Fig. 14-26).

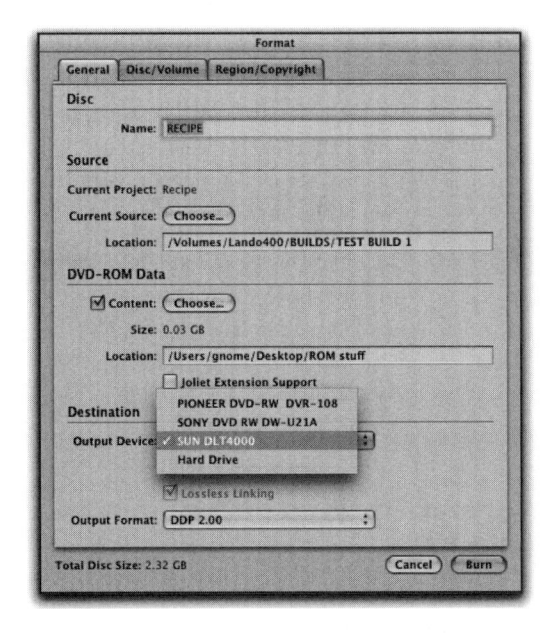

Figure 14-26 — The Formatting Dialog presented for "Build and Format."

Re-Using a Build Folder

If you attempt to build into the same folder more than once, you may get the following dialog warning (see Fig. 14-27).

Choosing *Reuse* will allow DVD Studio Pro to only rebuild those portions of the VIDEO_TS it needs to fix—usually the IFO files, and occasionally a Menu

406

Building and Formatting

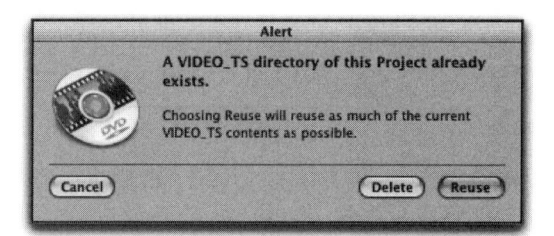

Figure 14-27 — If a VIDEO_TS already exists, you'll see this.

VOB file. This can save a large amount of time when fine-tuning the same build again and again.

Choosing *Delete* will immediately proceed to delete the contents of this build, and will force the recreation of all of the VOB files previously build—this can add a significant amount of time to your build, especially if not much has changed in your project except a few navigation links.

Keep an eye on the behavior of the finished DVD project, though, because in past versions of this process, the Reuse function has been known to cause some strange performance issues. Since that was reported in DVD Studio Pro 1, however, we should assume it's fixed in DVD Studio Pro 4 and give it the benefit of the doubt.

In either case, once you select a choice, DVD Studio Pro will begin Building the data, creating the DVD files and folders that make up the final DVD.

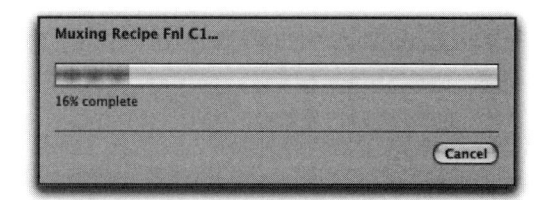

Figure 14-28 — The Build progress display.

Once the Build portion is complete, the dialog will switch to showing a Recording progress, while it occupied with the Format operation, whichever version it is doing.

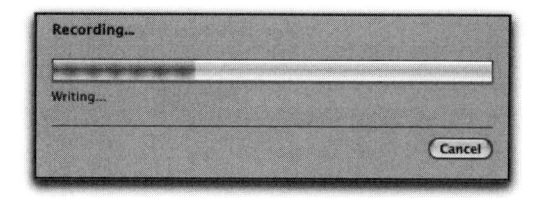

Figure 14-29 — The Format progress display.

While the DVD is being built, the Log Tab will continually display messages which track the progress of the Build and Format. When the DVD is completely built, the LOG window will display messages confirming the DVD's compliance with the DVD Video spec.

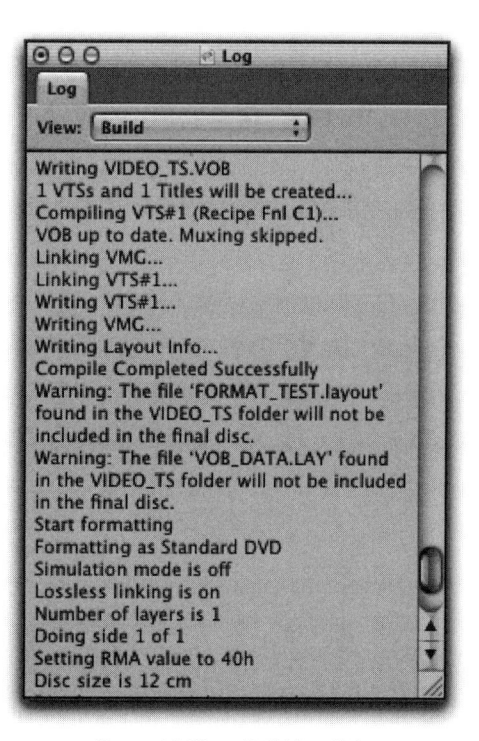

Figure 14-30 — Build Log Tab.

Creating a DVD Hybrid Disc

A DVD Hybrid disc is one that contains both DVD-Video information, and computer-readable information in the ROM zone of the DVD. (The Apple manual calls this a DVD-ROM disc.) These discs are playable in both set-top DVD players and computers with DVD players, but the ROM zone data is only accessible to the computer. A set-top DVD player will be able to process all of the DVD-Video content, but will ignore the ROM data. The method of creating a DVD-Hybrid disc varies, depending on which version of DVD Studio Pro you are using. When preparing your DVD-ROM files, pay attention to the naming convention (see *Joliet Files*, later in this chapter).

Making a DVD Hybrid Disc in DVD Studio Pro 4

We have already seen that both the Disc Properties Window and the Format Dialog display contain settings where the DVD-ROM function can be enabled.

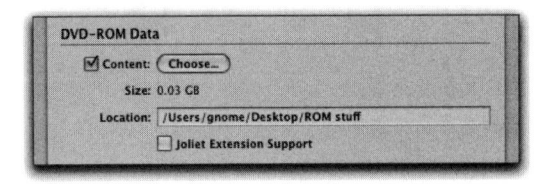

Figure 14-31 — The Format Dialog showing DVD-ROM Data setting.

Adding DVD-ROM data to any DVD-Video project is very easy in DVD Studio Pro; just activate the DVD-ROM feature in the Disc Properties, or the Format dialog, and *Choose...* (select) the folder containing the DVD-ROM files for that project. They will be added to the root level of the finished DVD.

NOTE!
Some DVD players have problems accepting a large number of files in the root directory—be frugal with your ROM content.

Making a Hybrid Disc Using DVDSP V1.5

- **CREATE** your DVD-Video project as usual. If you are using the **DVD@ccess** function, be sure to make your links while you are Authoring.

- **Before you BUILD** the DVD-Video volume onto your Hard Drive, be sure to set the DVD-ROM folder in the DISC Property Inspector.

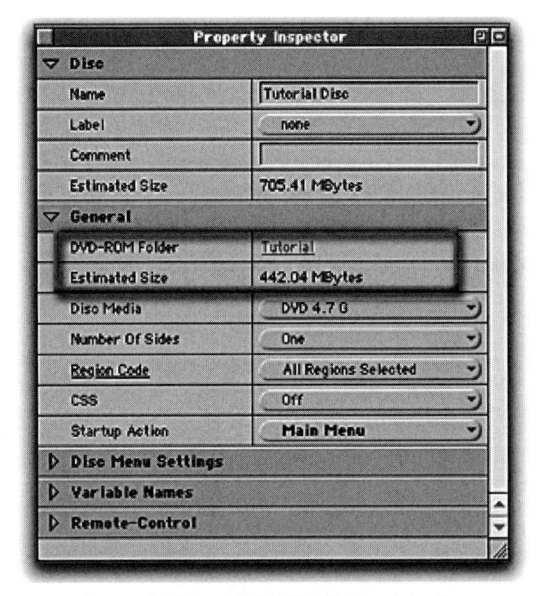

Figure 14-32 — DVD-ROM folder Selector in DVDSP 1.5.

- **LOAD** the ROM material for your finished disc into the "DVD-ROM Folder" BEFORE you build. This folder will not appear in the build, but its CONTENTS will have been

Building and Formatting

moved to the root level directory of your finished DVD.

- **BUILD & FORMAT** a DVD-R or a DLT Master tape using DVD Studio Pro's own output capabilities. The finished disc will include all contents that were in the "DVD-ROM Folder" specified in the Disc Properties.

Creating a DVD-Hybrid in DVD Studio Pro 1.2

To create a DVD disc, which can also deliver computer-readable material in the DVD-ROM zone of the disc (especially useful for creating File-enabled discs using DVD@ccess), follow these steps:

Making a Hybrid Disc Using DVDSP V1.0-1.2 and Toast Titanium

- **CREATE** your DVD-Video project as usual. If you are using the **DVD@ccess** function, be sure to make your links while you are Authoring.
- **BUILD** the DVD-Video volume onto your Hard Drive.

Figure 14-33 — Basic DVD Volume Folder.

- Bring the ROM material into the root level of your DVD folder.

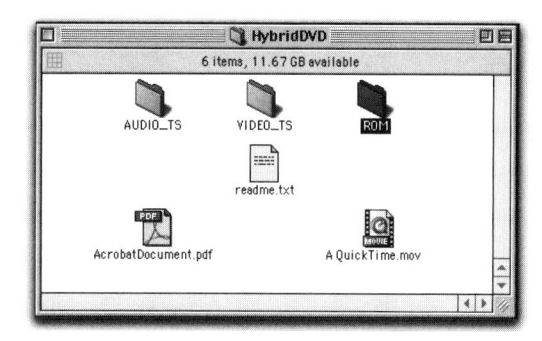

Figure 14-34 — A HybridDVD Folder.

- **BURN** a DVD-R with Roxio TOAST 5.0.1 or better…or
- **WRITE** a DLT Master tape with TOAST 5.0.1 or better.

Making a Hybrid Disc Using DVDSP V1.0-1.2 without Toast

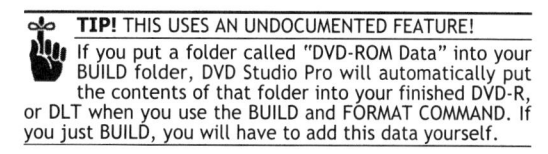

> **TIP!** THIS USES AN UNDOCUMENTED FEATURE!
> If you put a folder called "DVD-ROM Data" into your BUILD folder, DVD Studio Pro will automatically put the contents of that folder into your finished DVD-R, or DLT when you use the BUILD and FORMAT COMMAND. If you just BUILD, you will have to add this data yourself.

- **CREATE** the DVD-Video project as usual. If you are using the **DVD@ccess** function, be sure to make your links while you are Authoring.
- **CREATE an EMPTY** Build folder on your Hard Drive (be sure you have enough room). Inside of it, create a folder named "DVD-ROM Data," exactly. DO NOT create any other folders.

409

DVD Studio Pro 4

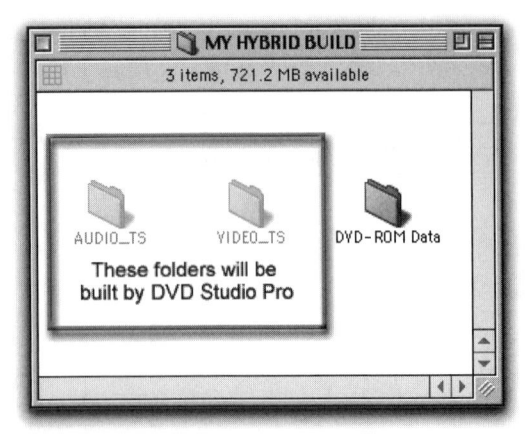

Figure 14-35 — Using the prebuilt BUILD folder in DVD Studio Pro 1.2.

- **LOAD** the ROM material into the "DVD-ROM Data" folder. This folder will disappear during the build, and its CONTENTS will be moved to the root level directory of your finished DVD.

- **BUILD** the DVD volume onto your Hard Drive as usual, being sure to specify the Hybrid Build folder you created in step 2 as the "target," or

- **BUILD & FORMAT** a DVD-R or a DLT Master tape using DVD Studio Pro's own output capabilities.

DVD Hybrid Pitfalls!

IMPORTANT!
Be careful of the following problem areas:

Joliet Files

- For absolute compliance with some replication verification software, be sure your filenames comply with ISO9660 standards (a.k.a. 8.3 file names—MYTEXT.DOC, for example). These short file names will pass through DVD Mastering controls without flagging errors upon analysis. Longer names may be flagged as errors, even if they are not!

DVD@ccess Installers

- If you DON'T use the BUILD & FORMAT command, be sure to include the DVD@ccess INSTALLERS onto your DVD disc, or the recipient will not be able to utilize the DVD@ccess function on their machines. The Installers are located in the RESOURCES folder of the DVD Studio Pro Folder in OS 9. They are located inside the DVD Studio Pro application bundle in OS X. Command-click the application Icon to open it up and locate the resources folder inside.

TIP!
DVD Studio Pro will AUTOMATICALLY put these into a finished disc if you use the BUILD and FORMAT command. If you just BUILD, it won't.

DVD@ccess on Windows

- Be aware that as of the writing of this documentation (October 2005), it has been reported that there are difficulties in using DVD@ccess on Windows-based machines running Win 95, 98, ME, and Win 2000. This is being looked into for a fix. Reports indicate that DVD@ccess links do work in Windows XP. Check the DVD Studio Pro support forums or *late breaking news* for further information on using DVD@ccess.

Building and Formatting

Hybrid Discs Viewed on Settop Boxes

- Be aware that when you create a DVD-Hybrid, you will need to remind your DVD viewers that the ROM material is NOT viewable from on a set-top DVD player—they cannot access the DVD-ROM material without a computer. It's always a good idea to plan your navigation with this in mind as well—always create clear, understandable instruction screens that explain what's supposed to happen, or how to return to the DVD-Video portion of the program. You would never want to leave the DVD viewer in the dark about the operation of the disc.

Important Limit on DVD-R Disc Size!

(Thanks to replication consultant Jon Wenger for this important info!)

> "There is a limit of slightly under 4.7 Billion Bytes of data that can be contained in a General Media DVD-R disc (DVD-5). That's about 4.37 Gigabytes of data, measured the same way as your hard drive."
>
> If you are sending out your DVD-R (G) for replication remember this hard limit: check for a volume size of no greater than 4,699,979,776 bytes. If you are under that you are OK."

It is unlikely that this is still an issue in DVD Studio Pro 4, but it's good to have this definitive size information available.

Roxio Toast Titanium— the Swiss Army Knife of Optical Media

Figure 14-36 — Toast Icon.

You will benefit from having Roxio TOAST TITANIUM software around, especially if you are going to create a DVD-Video or DVD Hybrid disc on DVD+R/RW, DVD-RAM, or DVD-RW instead of DVD-R. It can do many things to assist in keeping a productive DVD workflow.

IMPORTANT SUPERDRIVE WARNING!!

If you have a system with an original Apple DVR-103 or DVR-104 Superdrive, or have installed an aftermarket Pioneer A03 or A04, **BE SURE YOU UPGRADE THE DRIVE FIRMWARE BEFORE YOU EVER TRY TO USE 4X MEDIA!!**

Inserting 4X DVD-R General media in an unmodified DVR–103, A03, -104 or A04 MAY BURN OUT THE LASER MECHANISM!

If you are NOT sure which drive you have, or which firmware is proper, CONSULT the APPLE.COM SUPPORT AREA for information on this VITAL UPGRADE! Search on "Superdrive" to find the appropriate Document for your system, PowerMac or iMac, and your Operating System OS X or OS 9. (You're not *still* running OS 9 are you? Say it isn't so!)

411

Summary

Although you may preview and Simulate your DVD endless times, to make it a functional DVD, you will need to Build the data structure, and Format it to some device—usually a DVD-recordable disc, but also possibly a DLT. Burn, Build, Format, and Build, and Format are simple commands that can output your finished DVD to your hard disk drive or to an internal or external DVD burner or DLT drive.

Burn is the simple path to a DVD-R disc—it Builds and Burns in one integrated operation, but with few options.

Build allows you to specify where you will write the DVD Volume folder.

Format allows you to write a built DVD Volume to any device usable on your system.

Build and Format combine these two operations into one selection. Building may also be used to include DVD-ROM data in the finished DVD.

Using the DVD-ROM folder parameter, you can designate the folder that contains data to be copied to the DVD-ROM zone in the finished DVD.

Formatting can also introduce copy protection into the finished DVD. CSS, the digital copy protection form, can be output only if you are using DLT or DVD-R Authoring Media.

Macrovision adds an analog copy protection to the video content, but again requires DLT or DVD-R Authoring Media.

Region coding can be used to restrict the world regions in which the DVD may be playable, but again, really requires DLT or DVD-R Authoring Media output.

Chapter 15

Graphics Issues for DVD Images

DVD Studio Pro 4

Goals

By the end of this chapter, you should be able to show how to create proper DVD Graphics, including stillframe DVD Menus using Adobe Photoshop, how to create proper PICT files, and how to use basic Photoshop Techniques for DVD Studio Pro.

This document will reveal the pitfalls that await you in creating graphical menus, as well as some of the standard techniques used for creating motion menus.

The Menu Editor in DVD Studio Pro has radically changed the way in which beautiful DVD menus can be created, minimizing the need for external applications, but professional DVD authors will benefit from a clear understanding of the underlying principles of graphics for Video usage. (See Fig. 15-1.)

Basic DVD Image Concepts (Important Stuff!)

DVD stillframe menu images, and image files for slideshows can be very easily and efficiently created in Adobe Photoshop, or other pixel-based graphic applications, like Corel Paintshop Pro. All you need to know are a few very basic survival techniques. Here they are!

Image Size

DVD Images require that the dimensions of the finished image be correct for the specific Video Format used in the DVD.

$$\text{NTSC} = 720 \times 480 \text{ pixels}$$
$$\text{PAL} = 720 \times 576 \text{ pixels}$$

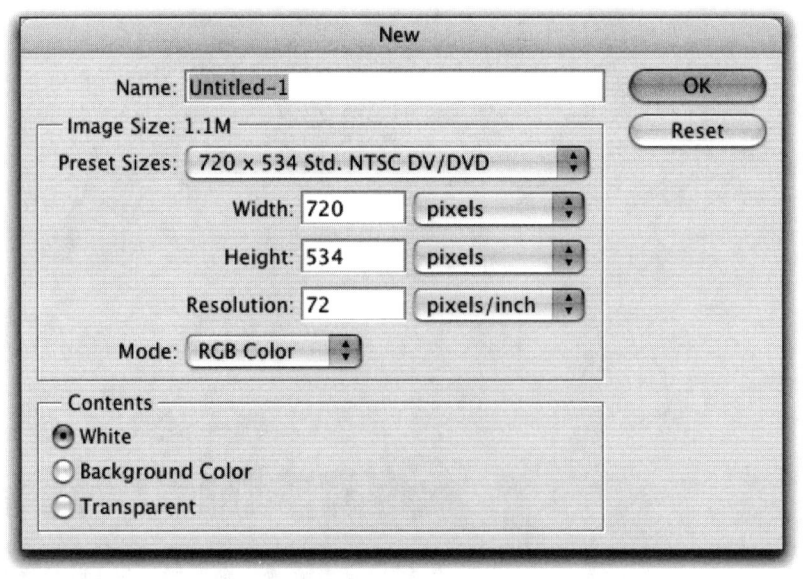

Figure 15-1 — A new DVD-sized document in Photoshop.

Graphics Issues for DVD Images

But more important than the final dimension is the method of menu assembly, i.e., HOW you get to those dimensions.

Creating a DVD image directly at 720 × 480 is WRONG! This will not take into consideration the difference between the pixel shape on a computer screen (they are square) and on a video monitor (those pixels are taller than they are wide). This means that we must ADUST the size of the finished artwork, while preserving the proper Aspect Ratio for display on the Video Monitor. You can do this by starting the composition at 720 × 540, and converting to 720 × 480 using the "Image Size" command in Photoshop (see Figs. 15-2 and 15-3).

In addition to the issue of pixel size, DVD menus must be created with a healthy understanding that the color you see on the computer screen is NOT the same color you will see in the finished DVD Video screen. Serious differences exist between the Computer screen and the Video screen, and we must account for them in the image design.

DVD Image Size Basics

Before starting your Image, be aware of the Video Format in which you are working—the dimensions for PAL images are significantly different than those for NTSC (see Figs. 15-4 and 15-5). Create your graphics with this in mind—we'll use NTSC standard in this document for simplicity.

DVD Image Issues

To make a DVD graphic of the proper size (regardless of whether it is a PICT, TIFF, BMP, or TARGA file or a layered PSD file), there are several VERY important issues that need to be kept in mind, all of which relate to the difference between VGA displays and Video monitors:

- **Pixel shape is wrong**
    - Pixels on computers are square
    - Pixels on Video Monitors are taller than square

540 Square Pixels = 480 Tall Pixels

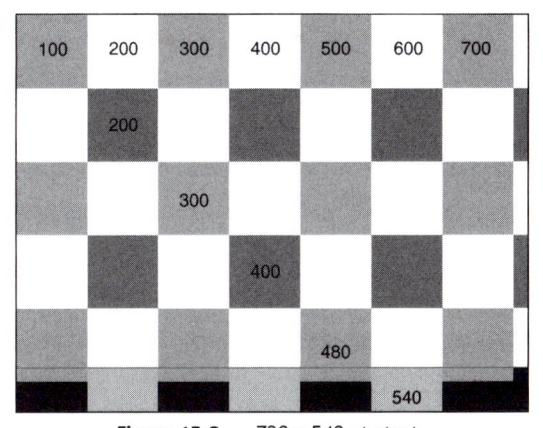

Figure 15-2 — 720 x 540 at start.

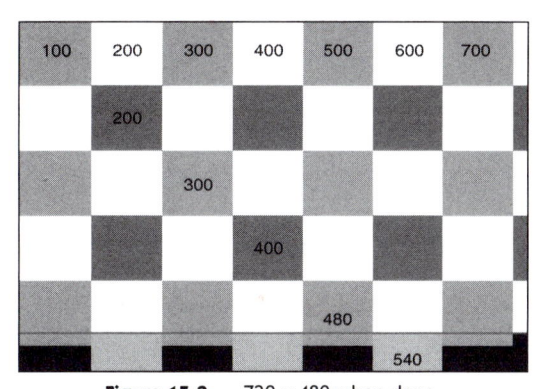

Figure 15-3 — 720 x 480 when done.

415

DVD Studio Pro 4

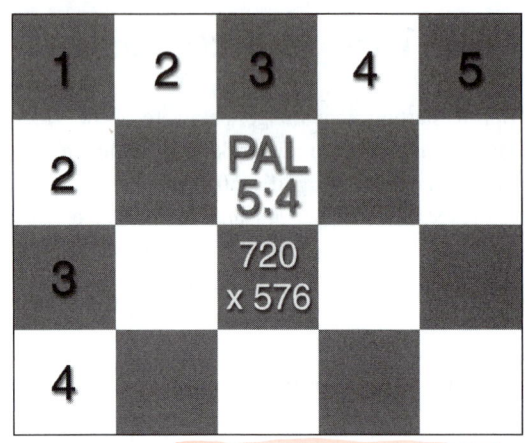

Figure 15-4 — 720 x 576 PAL finished Image size begins at 768 x 576.

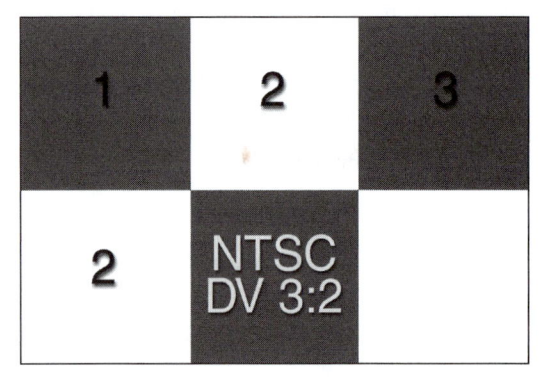

Figure 15-5 — 720 x 480 NTSC finished Image size begins at 720 x 540.

- **Color space is different**
  - The colors you see on your computer monitor are not accurate representations of the colors you will see on a Video Monitor.
  - RGB can create colors than cannot exist in NTSC or PAL.
- **Screen real estate is not all usable**
  - While it may be possible for you to view the entire Image screen on a Computer

DVD playback, that is not always the case on a Video Monitor, due to "Overscanning"—a function on televisions that renders a small portion of the sides of the Video image unviewable.

- **Font choice needs careful consideration**
  - Due to interlaced imaging on Video Monitors, Serif fonts (Times, New York) do not image as cleanly as Sans Serif fonts (those without serifs) like Helvetica and Arial.
- **Graphic elements need consideration**
  - Elements with single-pixel lines or borders do not image well, and will "jitter" due to video field interlacing.

In other words, understand that you're making *video* screens, and not computer display screens. As a result, the rules of broadcast design apply to DVD image design just as if you were creating television broadcast graphics.

Easy DVD Image Rules

There are some easy rules to follow to make proper DVD Images:

- Begin the Photoshop document at 720 × 540 pixels (see Fig. 15-6).
  - You will see this has some alternatives—read on.
- Use 72 DPI resolution.
- Use the RGB Color space.

The **720 × 540 image size** will provide a proper graphical 4 × 3 Aspect Ratio on the square pixel screen of a computer, allowing you to properly visualize the geometry of your DVD Image. (There's more on this in a minute… read further!)

416

Graphics Issues for DVD Images

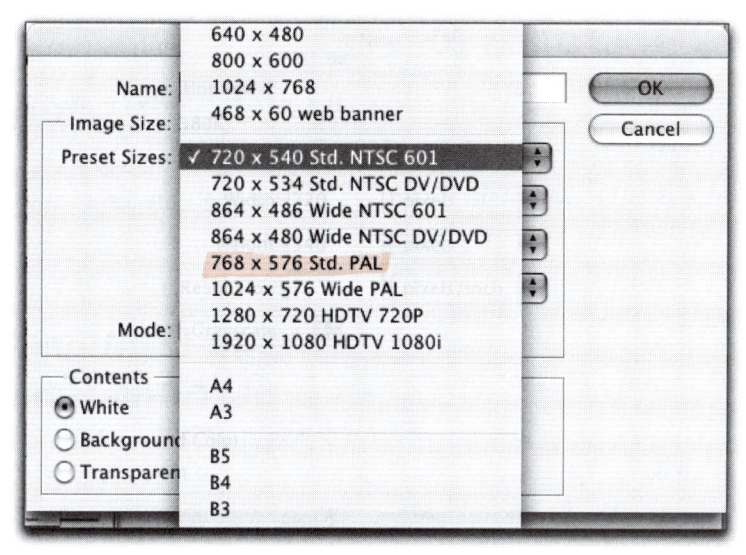

Figure 15-6 — Beginning your DVD graphic in Photoshop.

72 DPI is the resolution of a standard computer screen, and a TV set as well. There is nothing to prevent you from creating your Image in a higher DPI resolution (150 or 300 DPI), but you MUST reduce the DPI resolution to 72 DPI if you expect the finished files to operate properly within DVD Studio Pro. This is an ABSOLUTE requirement.

RGB is the proper Photoshop Color Space in which to create. We will STILL have to be conscious of our image, while composing as the RGB color space allows millions more colors than can be accurately displayed in the NTSC Video world! Yow!

Alternate Image Sizes

There exists a controversy over exactly WHICH size is the proper size to begin a DVD Image document. Adding fuel to this fire is Adobe's recent inclusion of DOCUMENT PRESETS in the "New…" dialog that provide not one but TWO different NTSC screen sizes (ignoring, for a moment, widescreen Images).

It is argued that the 720 × 534 dimension is MORE correct for a DVD Image than the 720 × 540 NTSC 601 image. Interestingly, note that there IS NO such controversy for PAL, which is properly fixed at the opening image size of 768 × 576 (which will later be reduced to 720 × 576 for a DVD Image). In many ways, PAL authors have an easier time than NTSC authors do—there's only one PAL setting. (See Fig. 15-7.)

Figure 15-7 — Which preset to choose 720 x 540, 720 x 534, or…???

417

DVD Studio Pro 4

Yet a Third Approach to Image Dimensions!

As if two different approaches to Image dimensions weren't enough, there is a THIRD size approach you should be aware of, and it has some pretty solid logic behind it.

Instead of starting at 720 × 540 and crushing the picture to 720 × 480 when finished, some Image specialists suggest starting at 640 × 480 and expanding the horizontal dimension to 720! (See Fig. 15-8.) There is a preset for this dimension in Photoshop 7, and you can very easily create this dimension in Photoshop 5 or 6. (Just remember all of these dimensions MUST be 72 DPI in the RGB Color Space.)

In theory, the logic behind 640 × 480 is pretty hard to fault—in creating Image images that are 540 tall, and "crushing" the vertical dimension down to 480 in order to achieve the proper height, we run the risk of accidentally squeezing horizontal line elements into odd pixel heights, or smaller pixel heights than originally created.

But in fact, it's better to start at true 4 × 3, 720 × 540, and scale down to 720 × 480. Scaling up from 640 × 480 to 720 × 480 may very well cause more problems than scaling down from 720 × 540 to 720 × 480.

Image Composition Limitations ("Safe Zones")

While we have figured out how to properly display the Aspect Ratio while creating the image, not all of the available final image space can or should be used. Due to some inherent limitations of Video playback, the outermost edges of the Graphic image should be avoided when designing.

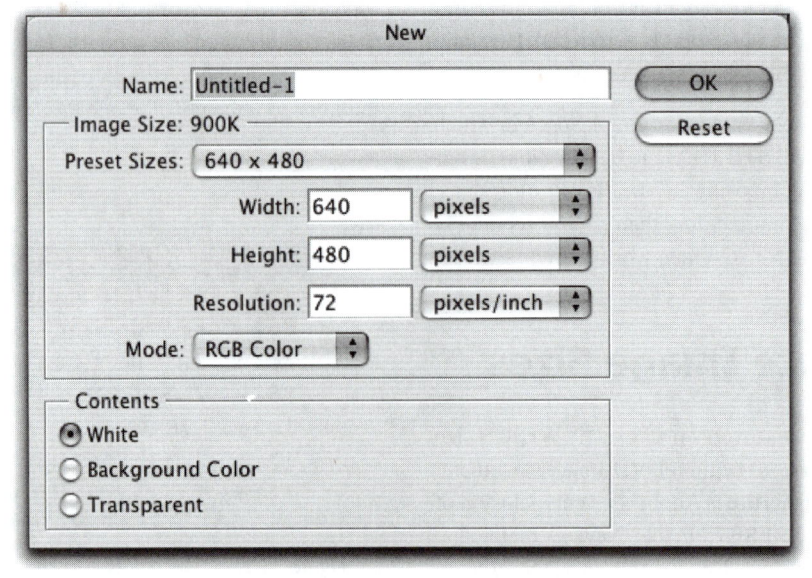

Figure 15-8 — Why not 640 x 480? Read on....

Graphics Issues for DVD Images

The traditional method of broadcast image composition establishes two "safe zones" within the full image size. The "**Action Safe Zone**" is the maximum area usable for "live matter" as our printer friends call it—this means anything you want to see in your Graphic should be composed within this reduced image space, typically 5% in from each edge. A further reduced space, called the "**Title Safe Zone**," is typically 10% in from each edge. The traditional rule is that all titles and text should exist within this smaller zone.

While this may not be of concern for DVDs that play back on computers (which are capable of imaging the entire Graphic screen), we must compose Graphics for the worst-case playback—namely, video monitors or TV sets that "overscan," pushing the outer edges of the Graphic image outside of the viewable area.

Figure 15-9 — Safe Zones defined.

In the image sample (Fig. 15-9), the red zone is the region to be avoided while composing the image. The black area describes the outer limit of the "Action Safe" zone, and the gray area ("Title Safe") is completely legal for both image and text. This file is included in the Goodies folder of the Book's DVD, and can be used as a guide to Menu image composition.

DVD Image Color Rules—RGB versus Video

In DVD image design, certain video color issues need to be addressed. You cannot arbitrarily use just any color in your Graphic design, as the NTSC color space has far fewer legal colors than your computer.

One such limitation concerns proper values for black and white. In video, black saturation is limited (compared to RGB values), and white saturation is limited to the value, which generates "100 IRE."

Creating proper, video-legal video blacks and whites involves limiting the Photoshop Values of black and white used in your DVD Graphic. You can do this in the Photoshop Color Picker (see Figs. 15-10 and 15-11).

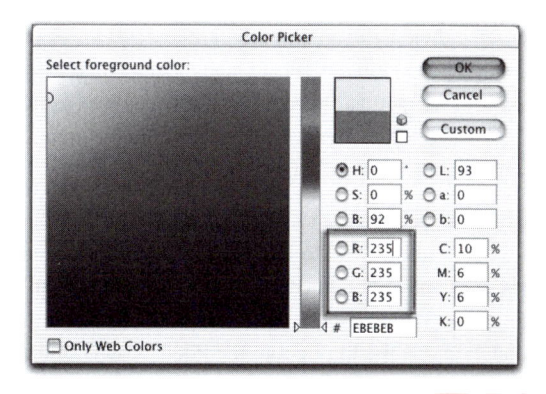

Figure 15-10 — White is limited to RGB values of 235, 235, 235.

419

DVD Studio Pro 4

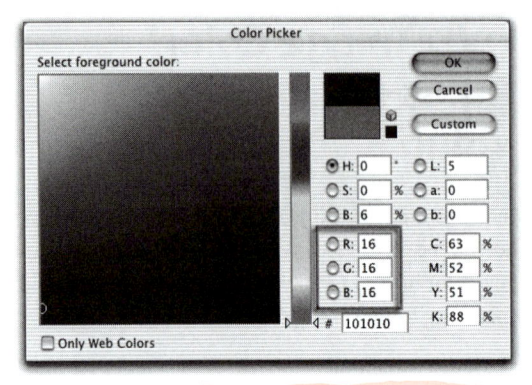

Figure 15-11 — Black is limited to RGB values of 16, 16, 16.

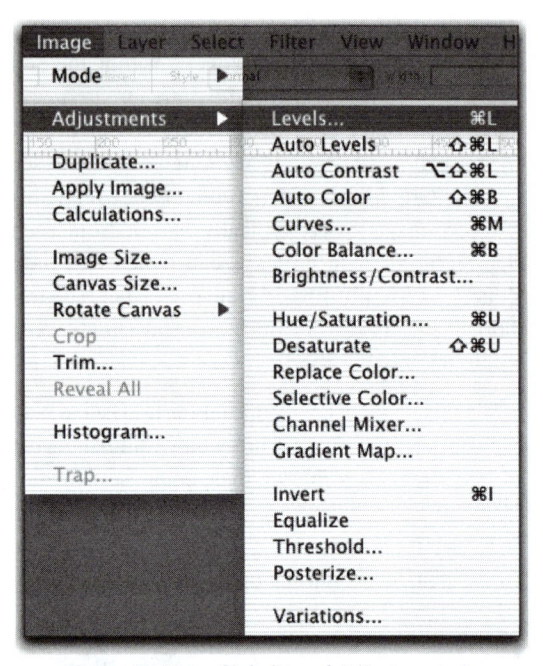

Figure 15-12 — Global Level Adjustments in Photoshop.

A Global Method to Control Black/White Levels

While it is possible to adjust the levels of black and white individually, a more efficient method involves changing the global values of black and white for the entire document at one time. This can be done using the **Image > Adjustments > Levels...** command in Adobe Photoshop (see Fig. 15-12).

The **Image > Adjustments > Levels** command sequence allows you to globally limit the lower and upper boundaries at one time—in a way, it's like running the same color selection operation on ALL blacks and white throughout the Photoshop Graphic document at the same time.

Once you select the command (don't select **Auto Levels** by mistake!), you can set the lower boundary to 16 and the upper boundary to 235, and the entire document will be affected (see Figs. 15-13 and 15-14).

In an unmodified Photoshop document, the maximum black and white levels appear unlimited, as in Figure 15-13.

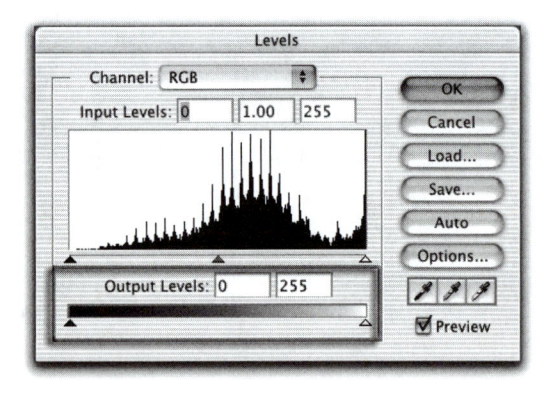

Figure 15-13 — Levels as they normally appear (0-255).

Adjust the Black Triangle to set the Black point to 16 and adjust the White triangle to set the White point to 235. This will limit the black and white levels to those usable in video and DVD.

Graphics Issues for DVD Images

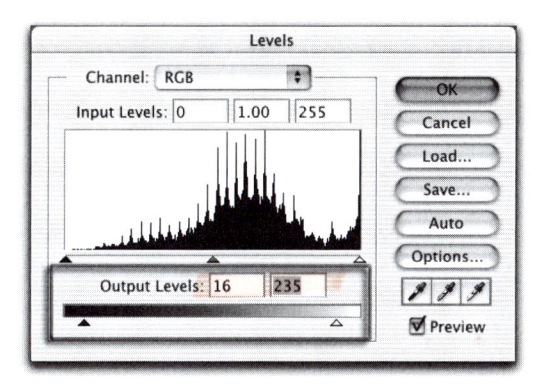

Figure 15-14 — Levels after being limited for Video Compliance (16-235).

This is a much easier method to use than laboriously setting the black and white levels for every layer you create in Photoshop.

How to Reveal Graphic Image Flaws and Limitations

Graphical elements in DVD Images are subject to interlacing issues when displayed on video monitors. For this reason, the best approach in creating your DVD Image graphics is to view them on a video monitor while you are creating them. When possible, mirror your desktop to NTSC video to preview while composing!

Several options exist for this:

Software display using FireWire to output to video:

Echo Fire from Synthetic-Aperture, Inc.

http://www.synthetic-ap.com/

Hardware display using a video capture card:

Aurora Igniter

http://www.auroravideosys.com/

Digital Voodoo D1 Desktop

http://www.blackmagic-design.com

A secondary Video card with an NTSC or PAL output.

ATI Rage has a dual head card, capable of S-Video.

A second video card can be added to most computers.

TIP!
How about Video out from your Mac G5 to NTSC for $20 (see Fig. 15-15)?

I'm not high, it's true—you can use a simple $19 DVI-to-Video adapter from Apple, part number M9267G/A to convert the second monitor output from a G5 to an NTSC signal—the adapter does both Composite and s-Video. If you remember to set the resolution properly, you will get a pretty decent output to video, good enough to check your colors and composition!

Hardware display using Mac Video card and Apple adapter: http://www.apple.com/store/ search on M9267G

Figure 15-15 — The $20 video output solution for G5s.

The advantage of displaying your Image in real-time is that you can immediately see when graphical elements are not working properly. Graphical elements

421

like fonts, lines, and shapes are all affected by the action of interlaced video scanning, and will not display the same on a video monitor as they will on a computer's VGA display. The colors are also quite different, as we discussed above, and by mirroring to an external monitor, you can immediately see the final composition of your DVD Image.

Image Size And Shape Limitations

To avoid interlacing jitter, limit line width to 3 or more pixels.

This also means when creating shapes, limit the width of border elements (box outlines, circles, ovals, etc.) to 3 or more pixels. This will prevent jittering of graphic elements once converted to MPEG, and displayed on a video monitor.

When creating Subpicture Overlays for Menu highlights, pay special attention to the shapes you choose. Unlike the background pictures, Menu Highlights are not converted to MPEG, and are not displayed in an anti-aliased manner. Try to avoid using circles, ovals, and other rounded shapes, as the sharp edges of the non-anti-aliased pixels will look bad in the finished DVD menu (see Figs. 15-16a and 15-16b).

Anti-Aliased

Figure 15-16a — Anti-Aliased images have smooth sides.

not Anti-Aliased

Figure 15-16b — Non-Anti-Aliased images have jagged sides.

Font Size and Style Issues

When designing DVD Images, it's good to remember that DVDs are many times watched from across the room on a television set or video monitor, rather than being watched from up close on a Computer screen. While it may be easy to see a small font up close, it is difficult to see small font information when the DVD is being viewed from a distance.

For this reason, choice of font *size* is important, and choice of font *face* and *style* will contribute to good readability, and again, lack of jitter.

As a general rule, use Sans Serif Fonts (like Arial, Helvetica, and similar) instead of complex, fancy or Serif fonts (like Times, New York, or any script font). (See Figs. 15-17 and 15-18.).

Sans-Serif (Arial - Helvetica)

Figure 15-17 — Sans Serif font faces don't have tiny points.

Serif (Times - New York)

Figure 15-18 — Serif Font faces are not the best choice for DVD.

Using Layer Styles (Layer Effects) with DVD Studio Pro

One popular method of adding production value to DVD Studio Pro menus is to use Photoshop's Layer

Graphics Issues for DVD Images

Styles (called "layer effects" in Photoshop 5 and under).

The Layer effects/styles add drop shadowing, texturing, beveling, embossing, and a wide variety of other graphics "assists" to ordinary Photoshop layers, which can take simple text and move it from humdrum to quite striking, as shown in Figure 15-19.

Figure 15-19 — Plain Text—flat and lifeless.

Adding Drop Shadowing to the text will make it stand out from the flat background layers (see Fig. 15-20).

Figure 15-20 — Text with Drop Shadow—adds some depth and class.

Adding Beveling adds even more depth to the text, and some 3-dimensionality as well (see Fig. 15-21).

Figure 15-21 — Shadowed Text Beveled—adds 3-D texture to the 2D text.

While DVD Studio Pro 2 has some nifty new text capabilities in the Menu Editor, it does not offer the depth of expression possible with the Type handling capabilities of Photoshop. It's great for a lot of things, but sometimes you need, well, something *more*!

Rendering Layer Styles for Layered DVD Menus

While the extra image impact provided by layer styles is striking, a small problem exists when using these effects with a menu destined for DVD Studio Pro. While Photoshop can display these layer effects in their natural state, DVD Studio Pro cannot. A step called **rendering** must first be executed, as shown in the steps below. Rendering will consolidate the Pixels of the Type or Art and any Layer Styles on to that layer into a single whole image, that can be visible in DVD Studio Pro.

To Render Layer styles for DVD Studio Pro:

Start with a layer with live Layer Styles and/or live Type (see Fig. 15-22):

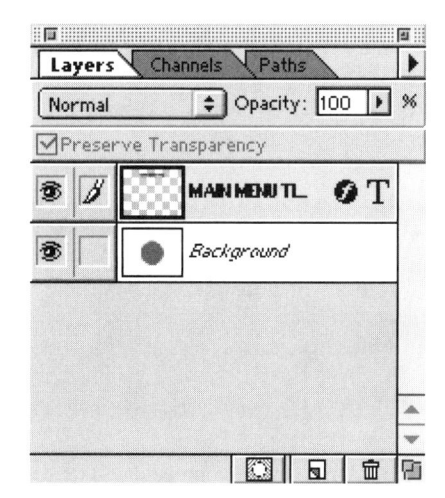

Figure 15-22 — The original Photoshop with live type and layer effects.

Add a new empty layer underneath the layer you wish to render down, using the **Layer > New** command, or the new layer tool (see Fig. 15-23):

DVD Studio Pro 4

Figure 15-23 — Adding a layer below the one you wish to render.

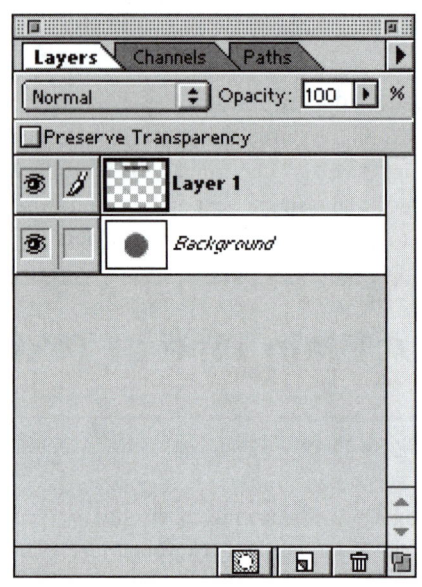

Figure 15-25 — The Styled Type Layer after being "merged down."

Click on the upper layer (the one you want to Render) to select it, and use the **Layer > Merge Down** command (shortcut: Cmd-E) shown in Figure 15-24:

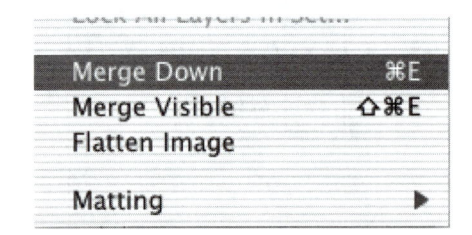

Figure 15-24 — Photoshop Merge Down command.

Using the "Merge Down" command will render all of the type and the Layer Styles down into a single layer containing all of the text and the layer effects as a single unified layer (see Fig. 15-25).

Keeping the Colors within Video Standards

Rendering the Layer Styles solves the problem of seeing the effects in DVD Studio Pro's Menu Editor, but there's a big gulf between what the menu looks like in Photoshop and how it will appear on a Video monitor. We've already handled the Black and White levels—now it's time to work on the *colors*.

Using the NTSC Color Filter

The NTSC Color filter (**Filter > Video > NTSC Color**) can be used to reduce oversaturated colors in the Photoshop document for proper display in the Video colorspace. This filter will, for example, reduce the saturation of RED in your menu to an

Graphics Issues for DVD Images

acceptable level for video playback. Try an experiment with the MyMenu.psd file—there is a layer in there called Red 255 Dot. If you run the MTSC color filter on this layer, it will reduce the saturation of the red color from 255 to 181, making it less "strident" in the final Video Playback.

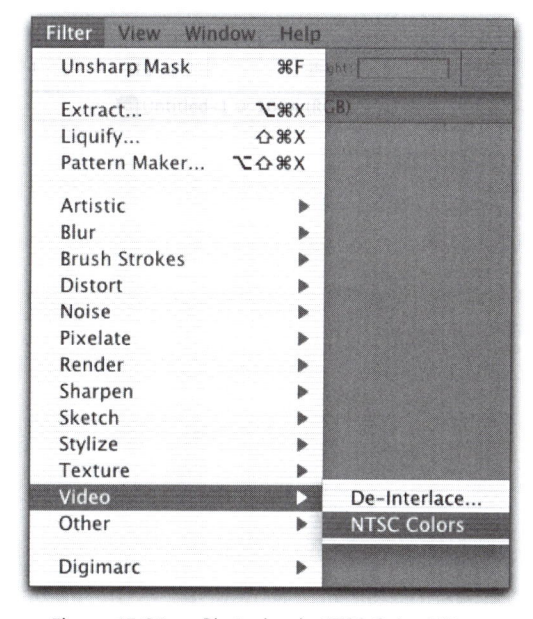

Figure 15-26 — Photoshop's NTSC Color Filter.

Notice there is no ellipses (the three periods) after the NTSC Colors command in the Menu shown in Figure 15-26, because selecting this command causes it to activate immediately on the currently selected layer.

> **TIP!**
> If you're making a TIFF or PICT, consider using this filter on the flattened image, instead of every layer of your .PSD file.

Completing the Photoshop DVD Graphic Image

Regardless of whether you are creating a layered PSD file, or flattening it into a TIFF or PICT file for easier use (and possibly to add audio to), the image size must be properly adjusted for DVD playback. This is accomplished using the **Image > Image Size** command in Photoshop.

Although the new DVD Studio Pro Menu Editor can scale image sizes, setting the proper size is easy, and important. Since you are already this far in Photoshop, might as well go all the way. To resize, be sure to unlock the "Constrain Proportions" checkbox, or the Width will be changed at the same time as the Height, and this is NOT what we want.

To resize the finished image in Photoshop (see Fig. 15-27):

- Use "Image Size" command.
- Adjust image to 720 W × 480 H.
- Don't constrain the proportions (just adjust the *height*).

Save as a TIFF or PICT for easiest use as a still menu or slide, or layered PSD for Layered Menus. (See Chapter 9, "Menus and Buttons.")

Pay attention to these parameters while resizing.

Motion Menu Basics

Motion Menus are those that contain a moving image in the background picture. Any image application that can perform video compositing is capable of creating a Motion Menu. In current practice, this menu may be created by a number of different applications:

DVD Studio Pro 4

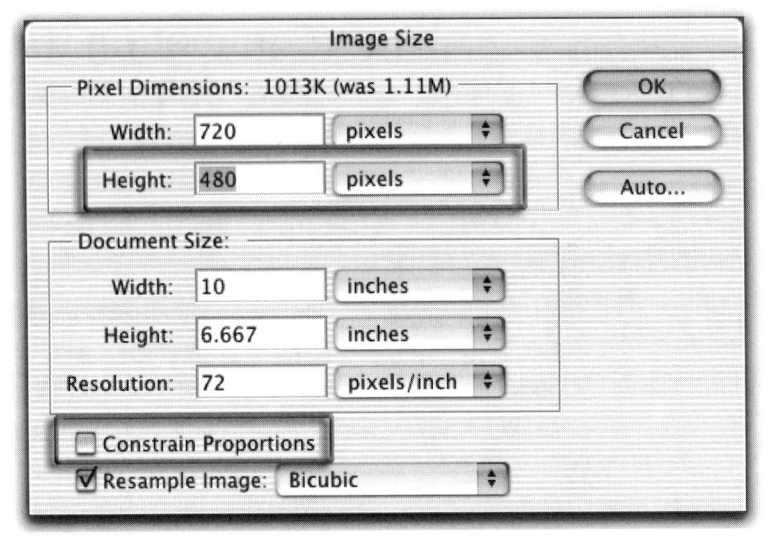

Figure 15-27 — Uncheck "Constrain" then modify Height.

- **Final Cut Pro**
- **Motion**
- **Adobe After Effects**
- **Avid**
- **Media 100**
- **Combustion**

Believe it or not, even the QuickTime Pro Player is capable of creating a simple but usable Motion Menu—you can do this by importing (adding) several motion clips into the QT Player, and resizing them and moving them around. Once they have been set on the background in the proper positions, you can export the entire composition as its own self-contained (or reference) QuickTime movie (see Fig. 15-28).

Most of the serious Motion Menu work is done in After Effects, or other similar Composition tools. The key is to get the sizes right, just as it is in Photoshop.

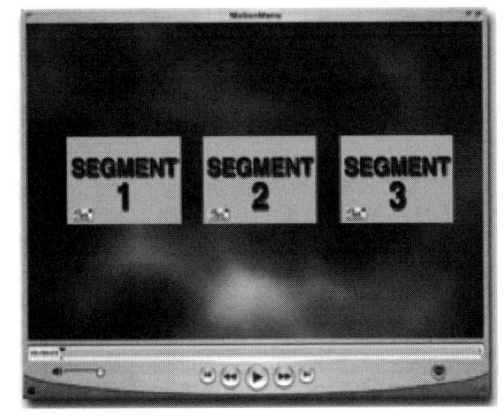

Figure 15-28 — QT Pro Player assembling a quickie Motion Menu.

Motion Image Sizes— 720 x 486, 720 x 480

Motion Menus are subject to image size and resolution issues in the same manner that still menu images are, but because Motion Menu assets are encoded

Graphics Issues for DVD Images

into MPEG from their original Video files, there may be a few different dimensions than their still image counterparts. For example, you may be creating a Motion Menu from digitized footage captured in Motion JPEG at 720 × 486. This image size will be converted into a 720 × 480 MPEG asset when it is encoded by cropping six lines from the original × 486 image size. Each encoder crops with its own method.

Creating a simple Motion Menu in Final Cut Pro can be done easily in the DV codec, from DV source footage. One advantage of this is that the image size is already 720 × 480, which makes the conversion to 720 × 480 MPEG that much easier. The DV codec is also partially compressed to begin with (around 5:1) so it makes the final MPEG encode even simpler.

> **NOTE!**
> The DV codec is spectacularly BAD for doing text work, as it chews up the text edges and makes them very raggedy. Consider using Animation codec at 100%, or uncompressed.

Creating Motion Menus in Adobe After Effects

DV is not the only codec that can be used to create a Motion menu, however. Many of the nicest Motion Menus around begin life as 2D motion compositions in Adobe After Effects. These motion sequences are more typically output using the Animation codec, usually at 100% quality. This particular codec outputs a very high-quality source movie, which can be encoded into a very high quality MPEG image. This codec also handles text elements much better than the DV codec.

In most cases, an After Effects motion composition will begin in the same aspect ratio as a Photoshop file—namely, 720 × 540, because of the square pixels on the computer video monitor.

As you might expect, somewhere along the line, the aspect ratio must be corrected to the proper one for encoding into MPEG-2 for DVD.

There are several ways to do this, but one particularly effective workflow has been documented by noted DVD Author and Producer Ralph LaBarge of Alpha DVD, and is reproduced here by courtesy:

To Create and Export a DVD Motion Composition with After Effects

1. For a 4:3 motion menu, set the composition up as a 720 × 540 resolution, 30 fps, in the Animation codec.

2. Make sure you have After Effects set to Best Quality for each layer.

3. Run Effects-Video-Reduce Interlace Flicker on each layer in After Effects.

4. Run Effects-Video-Broadcast Colors on each layer in After Effects.

5. When you output from After Effects, make sure you are saving in an uncompressed format.

6. Resize the animation to 720 × 480 as you are saving it in After Effects.

7. Use a software MPEG-2 encoder (e.g., Bitvice, Digigami, Main Concept, Compression Master) to convert the uncompressed file into an MPEG-2 video stream.

> **NOTE!**
> Don't forget to create a subpicture overlay file while you are assembling this motion composition, if you plan on using it as a menu.

DVD Studio Pro 4

About Transitions

Some of the coolest DVDs use video elements to carry you elegantly from a Menu button to the selected video, replacing the rather stark "freeze and cut" that happens when you click a typical menu button (see Fig. 15-29).

Transitions, as these elements are called, are easy to create in After Effects, Final Cut Pro, other NLEs, and Apple Motion. The most effective transitions match the outgoing frame of the Menu to the beginning of the Transition element, creating the illusion of a seamless transition. In this case, illusion is everything. The Transition itself may be as simple as a fade to black, or white, or a complex effect element. Taste and creativity pay off handsome dividends here.

About Transitions in DVD Studio Pro

To make your creative efforts even more rewarding, DVD Studio Pro has included built-in Transitions since Version 3. You can use these Transitions on Menu Buttons, in Slideshows, and on Tracks. There are lots of these, and they are as simple to use as just selecting one. You will find them in the Transition Inspector for Tracks, Slideshows, and Menus/Buttons.

Creating Motion Menus in Nonlinear Edit Tools

A motion menu can be easily created in Final Cut Pro or another Non-Linear Edit (NLE) system by using the Motion Effect tools available (see Fig. 15-30). For example, in Final Cut Pro, you can create a mul-

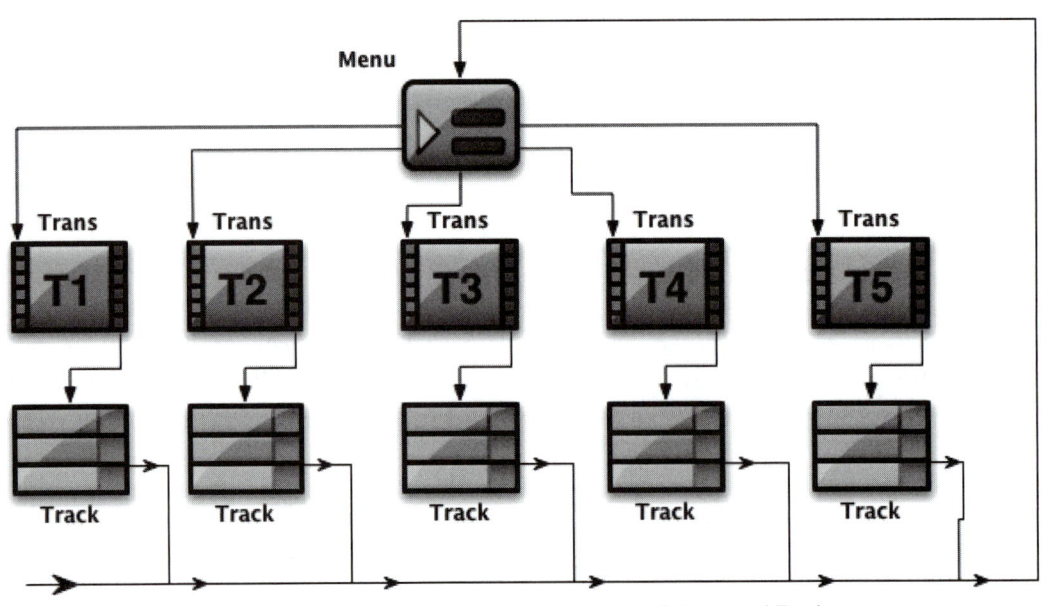

Figure 15-29 — Menu with Transitions in between Buttons and Tracks.

428

Graphics Issues for DVD Images

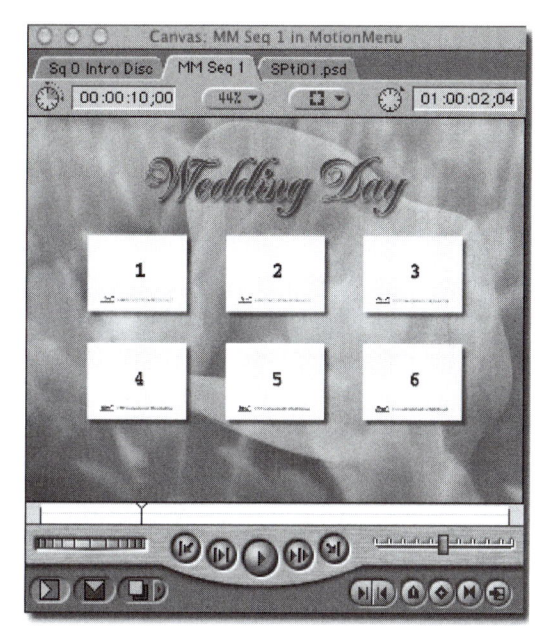

Figure 15-30 — A Final Cut Pro motion menu sequence.

Assembling a Motion Menu Sequence in Final Cut Pro

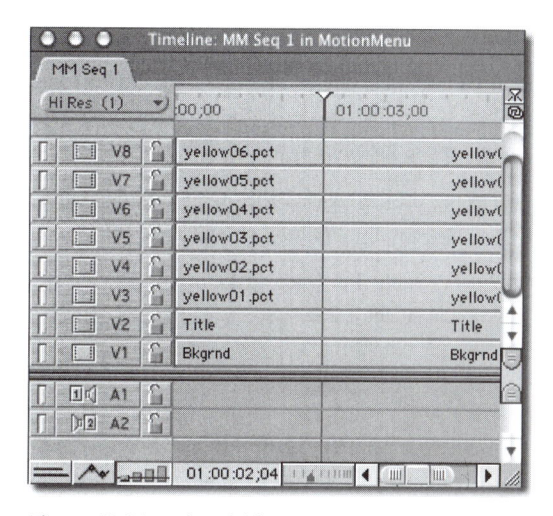

Figure 15-31 — A multi-layer Motion Menu sequence.

tiple layer composition with 4, 6, or 8 Motion Buttons created and overlaid onto a still-frame graphic backplate or a motion background file.

There are hundreds of collections of both graphics and motion clips available for your perusal and purchase.

A very complete catalog of motion clips is at Central Stock (**http://www.centralstock.com/**). This is an extensive online catalog of motion graphics, and stock footage—many of the elements here are great for creating Motion Menu Backgrounds. Another good source of unique images is the relatively new FPS collection (**http://www.fps.tv**).

Assembling a Motion Menu sequence in Final Cut pro is straightforward—the key is assembling the layers of the composite in the proper order and sizing the video clips using motion effects.

In the sample shown in Figure 15-31, there are a total of eight video layers, arranged in specific order from the bottom up.

The bottom layer is a graphic "backplate"—a foundation on which we can composite the additional video clips (or still-frames) desired to create the finished menu design. In this case, I borrowed one of the nice Photoshop 720 × 480 graphics from our own Wedding-Pack One, which contains two layers—the "Bkgrnd" and the "Title." Final Cut Pro intelligently splits the file into video layers for us automatically when it is imported and placed into the FCP timeline. What a great time-saver!

429

On top of the graphics backplate, and the title, we've added six placeholder graphics, used to shape the sequence template.

Customizing This Simple Menu Sequence

You can use the power of Final Cut Pro to easily replace these placeholder graphics by using the following procedure (see Fig. 15-32):

1. Import the desired clips for your new menu (be sure they're long enough).

2. For each clip, Open that Clip in the Viewer, and select it all.

3. Click on the desired layer (placeholder) to replace.

4. Select "**Edit > Paste Attributes**" and replace ONLY the "Content."

5. Be careful about scaling the time of the source content.

6. Click "OK" to perform the Paste (replace the content).

7. Repeat as many times as needed to replace the desired clips.

8. If you make a mistake, either "Undo" or do the Paste again.

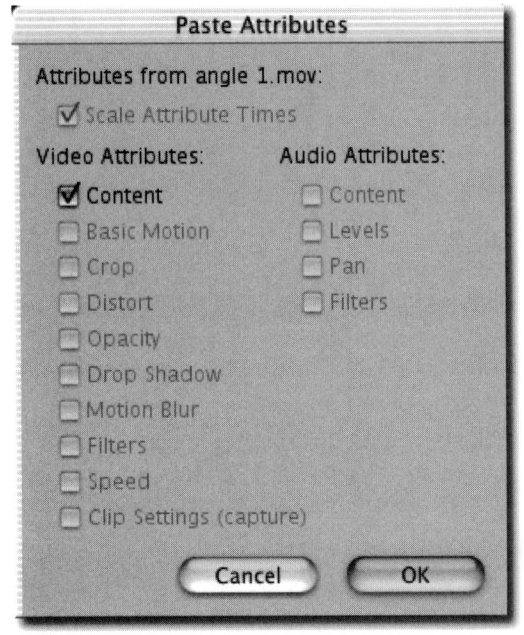

Figure 15-32 — The Final Cut Pro "Paste Attributes" Command.

Encoding Motion Menus into MPEG Once Completed

On the Macintosh, you can use *Compressor*, or the *QuickTime MPEG-2 encoder* (once you have installed DVD Studio Pro), Heuris MPEG Power Professional (www.heuris.com), or the new Bitvice Encoder (www.innobits.com) to software encode these motion menus into their finished MPEG form. All of these applications share one common trait: They will accept a QuickTime movie as the source file for your motion menu encode.

Graphics Issues for DVD Images

Motion Menu Overlay Pictures

Motion Menus cannot use the "switched Photoshop layer" highlight scheme using a layered Photoshop PSD file that you saw in Chapter 9 earlier, but they CAN look into a Photoshop document and use one layer as the Subpicture Overlay.

If you're creating a Motion Menu from scratch in Adobe After Effects, Final Cut Pro, AVID, Media100, etc., making the Overlay File is a piece of cake—just be sure you have a graphic that is the proper dimension for the video format you are working in—NTSC (720 × 486 or 720 × 480) or PAL (720 × 576).

If you are working in an intermediate dimension (720 × 486 or 720 × 540 for example), be sure to adjust the image size properly before separating out the overlay file.

> **NOTE!**
> If you are working in D1 resolution (720 x 486), you may find it easier to create the Subpicture Overlay file LATER, after the MPEG encode has taken place.

How? It's actually quite easy. Given the sophisticated graphics capture routines built into Mac OS X, you can import the asset in question and display it in the Menu Editor Screen, where a graphics capture can be made.

- Shift-Command-4 makes a PDF of the selected screen image, or
- Use SNAPZ PRO 2 to grab a screenshot with great control.

Remember, you're only looking to grab an accurately scaled GUIDE image with this technique, not a high-quality graphic still for the menu. (You *can* export a 72 DPI TIFF from the Marker Inspector!)

Capturing a Menu Image from a Screen Image

If your Motion Menu is already encoded into MPEG-2 form, and imported it into DVD Studio Pro, you can use this tip: Once imported, create a Menu Item, using any of the methods you already know.

Now, open the Menu Editor window up all the way, and resize it as large as possible. It may be easier to do this if you tear-off the Menu Tab.

Figure 15-33 — The open DVD Studio Pro Menu Editor.

You can tell when you have the Menu open far enough—you will see gray around the edges of the menu image, as shown in Figure 15-33.

You can select the right aspect ratio by using the Settings pop-up, upper right corner. Set it to "square pixels" to display a true 720 × 480 image, or "rectangular pixels" to display 720 × 540.

Remember, your overlay file must be 720 × 480, 72 DPI.

Once you've grabbed a screen shot, you'll need to clear away the extra pixels. We can use a feature of Photoshop to help eliminate these extra pixels simply.

Trimming the Screen Capture in Photoshop

We need to extract the Live Menu picture from your captured image—we'll be copying out a 720 × 480 selection (or 720 × 576 for you PAL users out there). It's very easy to do:

Load the Screen Capture into Photoshop (see Fig. 15-34).

Then select the Rectangular Marquee tool (see Fig. 15-35)…

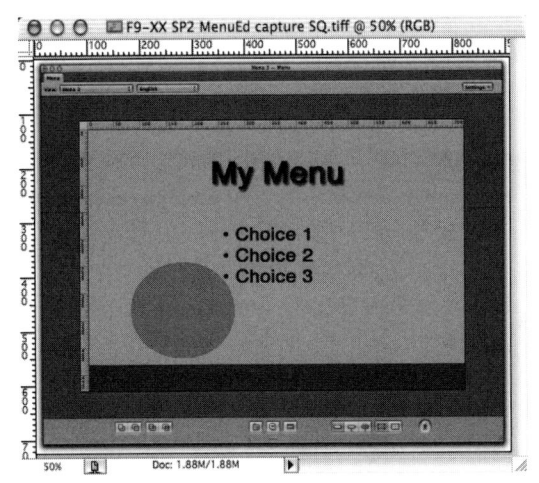

Figure 15-34 — The Captured Menu image, now in Photoshop.

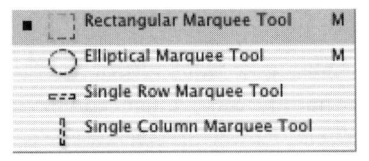

Figure 15-35 — Rectangular Marquee Tool in Photoshop.

…and set the Marquee Tool Style to *Fixed Size* (see Fig. 15-36).

Adjust the Fixed Size to 720 px × 480 px (720 × 576 for PAL); (px means "Pixels"—be sure Photoshop doesn't set it to Inches; see Fig. 15.37)!

Clicking will now create a Fixed Size Selection of 720 × 480 pixels; move this around until it is exactly aligned with the image of the menu (see Fig. 15-38).

Now COPY this selection, capturing the precise 720 × 480 (or 720 × 576) image to the Mac clipboard; use Cmd-C (⌘-**C**) (see Fig. 15-39).

Figure 15-36 — Select Fixed Size form the Marquee tool options.

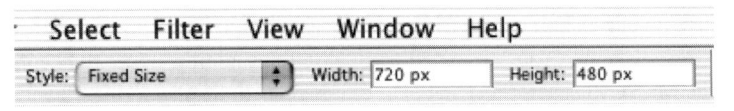

Figure 15-37 — Marquee size set to 720 x 480.

Graphics Issues for DVD Images

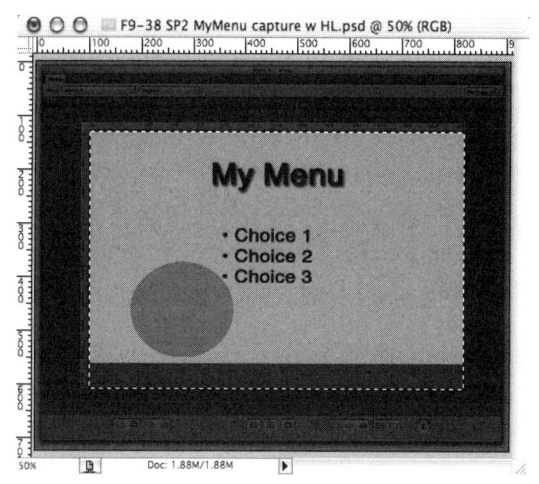

Figure 15-38 — The Menu Capture, with a Fixed Size selection.

Figure 15-39 — Copy the selected region with the "`Edit > Copy`" Command.

Immediately initiate a new Photoshop document (Cmd-N) (⌘-**N**) (see Fig. 15-40).

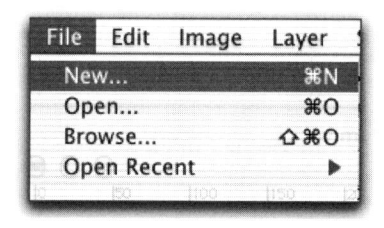

Figure 15-40 — Make a new PS file with the "`File > New`" command.

This new Photoshop document will be exactly the size as the last COPY you performed (a neat trick) (see Fig. 15-41).

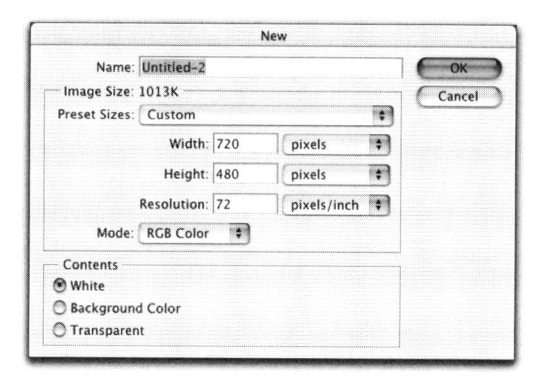

Figure 15-41 — The new Photoshop doc will be the size of the Copy.

Paste the copied image into the new Photoshop document.

De-Interlacing

If you have captured an image that suffers from Interlacing artifacts (jagged lines resembling comb teeth), you can eliminate them with the De-Interlace filter (`Filter > Video > De-Interlace`) (see Figs. 15-42a and 15-42b).

Once De-Interlaced, this new Photoshop document will be a good guide for creating your Subpicture Overlay file. If you need to make overlay elements that are NOT rectilinear, then you may need to adjust the image size. To create Highlight elements that are round or ovoid, expand the image size back out to 720 × 540 (768 × 576 PAL) and create the image highlights (using the resized captured image as your guide). Once you have completed the creation of the Subpicture Overlay hilites, convert the image size back down to the proper dimension and save the file under a unique but appropriate name.

433

DVD Studio Pro 4

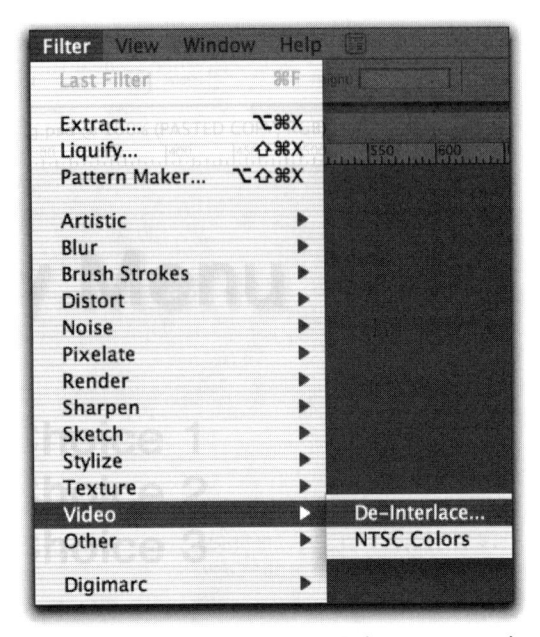

Figure 15-42a — Photoshop De-Interlace... command.

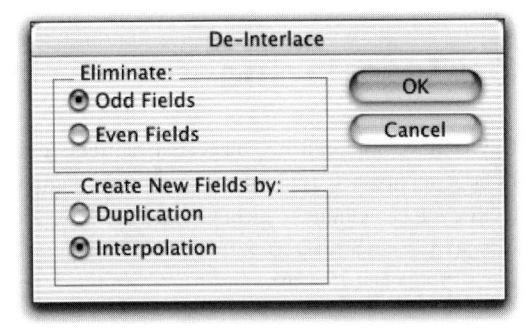

Figure 15-42b — Default Photoshop De-Interlace Filter settings.

Remember, you can save this Subpicture Overlay file as a TIFF or PICT file, or as a .PSD (with all Highlight elements contained on ONE layer).

Creating the Highlights in the Overlay File

The Subpicture Overlay file works by providing a bitmap of pixels for the Menu Editor to use as a reference in determining which pixels to light up in response to Button Selection or Activation modes.

You create a Highlight by designing pixels of a particular shape and color on a white background. In DVD Studio Pro, these pixels may either be Grayscale (i.e., 100% Black, 66% Gray, 33% Gray), or you can use Red 255, Blue 255, and Black.

See Figure 15-43 as an example:

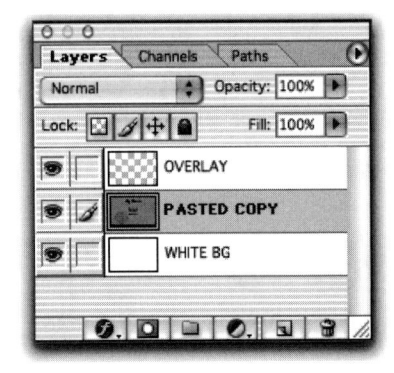

Figure 15-43 — Trimmed Menu as a guide for a new Subpicture Overlay.

Use the captured menu image with reduced opacity as a guide. Put the guide image on top of a white background, and then reduce the Layer Opacity of the guide until it "ghosts."

Create the overlay elements in a new layer on top of the reference image. This is similar to the "onion-skinning" technique used by artists and cartoonists for years.

When the Overlay image is completed, you can save the file as a layered PSD and use only the overlay layer, or turn off all layers except the finished overlay layer, and save the file as a PICT file. Remember to adjust the Image Size to 720 × 480 or 720 × 576.

434

Graphics Issues for DVD Images

Summary

Graphics images play an import role in DVD menus and slideshows, and even occasionally as the source for a Track.

Graphic images are subject to problems because of the differences in aspect ratio between the square-pixel screens on our computers and the nonsquare pixels on both NTSC and PAL video monitors, where we watch the majority of our DVDs.

While DVD Studio Pro 4 includes a dandy Menu editor, advanced authors may wish to take advantage of external graphics applications like Adobe Photoshop with which to properly prepare complex graphics.

To effectively create graphics for DVD in Photoshop, one needs to pay attention to the pixel differences, white and black levels, and color saturation issues. In addition, for layered documents being created to take advantage of DVD Studio Pro's Layered Menus, Photoshop *Layer Styles* (effects like Drop Shadow and Enbossing) will need to be rendered in order for them to appear within DVD Studio Pro.

Chapter 16

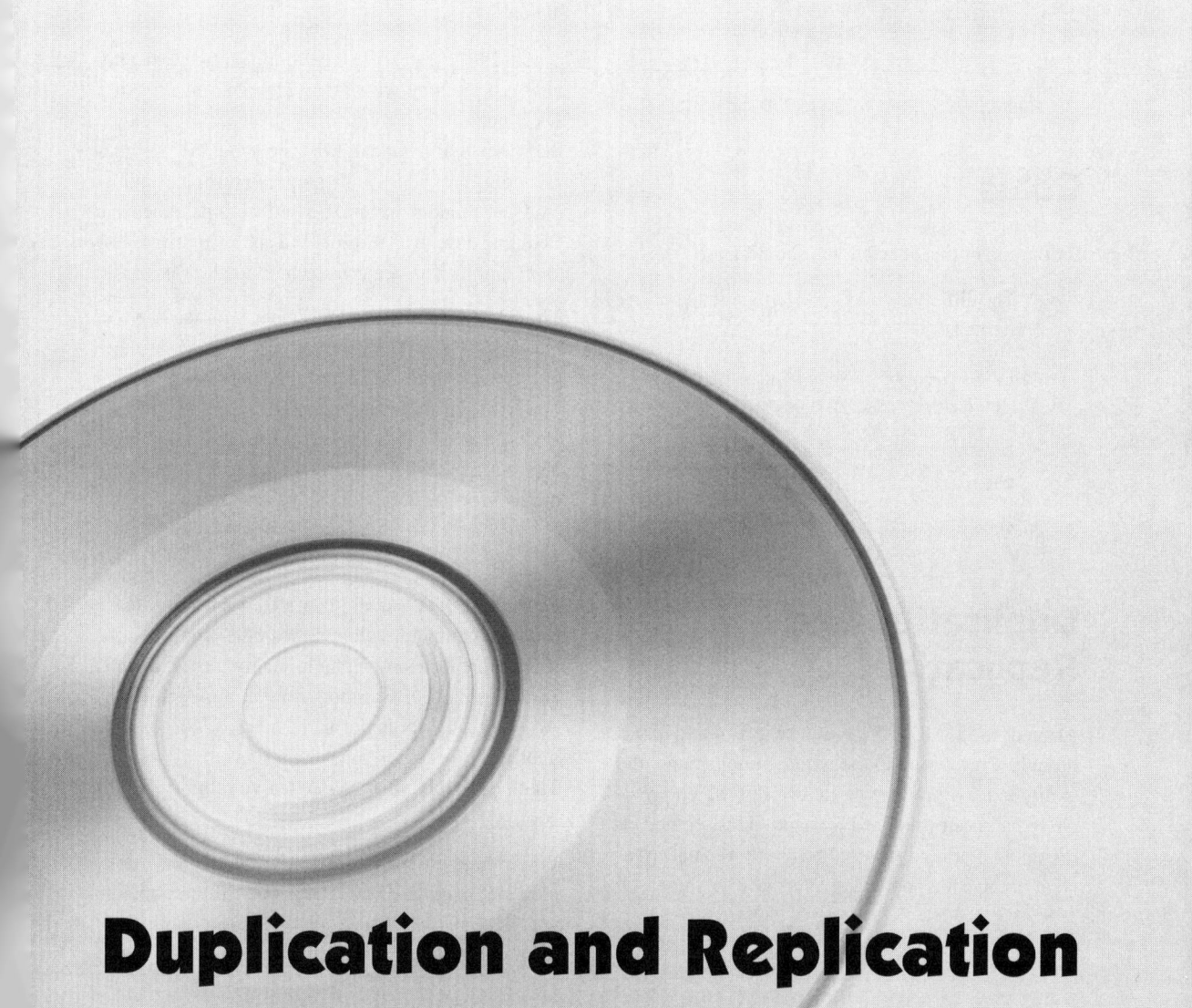

Duplication and Replication

DVD Studio Pro 4

Figure 16-0 — Primera Bravo DVD Duplicator.

Goals

By the end of this chapter, you should know:

- The difference between Duplication and Replication
- How to converse with a Duplicator or Replicator
- How to intelligently place an order for either one
- Which method to choose to fill your needs

Duplication and Replication

Around 1455 A.D., the Gutenberg printing press brought a revolution to the world—publishing—the ability to mass produce written works, which previously had been produced by hand. This method of distributing printed material has been a mainstay of what became the publishing industry ever since.

Fast-forward more than 500 years, to around 1985, when the desktop publishing revolution begins. Harnessing computer word processing programs and (relatively) inexpensive laser printers, publishing had finally come to the masses, indeed (and it only took a little more than five centuries)!

Just <u>10 short years later</u>, DVD surfaced on the radar screen of history, a true testament to how blazingly fast innovations in technology are being created these days.

Since 1997, the DVD revolution has irrevocably changed the way we watch video, and the desktop DVD authoring revolution has irrevocably changed the way we distribute it, bringing about the "democratization" of video publishing. No technical industry I can think of has progressed from its expensive fledgling days to an affordable desktop industry in so short a time. That's a great benefit to all of us.

The DVD revolution follows closely on the heels of the DV video revolution, which brought cost-effective video editing programs to the desktop of today's powerful personal computers. The content being created on the tens of thousands of previously unaffordable editing systems is, in no small part, providing the fuel for the DVD publishing revolution.

Today, the costs of desktop DVD publishing have dropped so dramatically that practically any household of median income can afford an inexpensive personal computer equipped with the requisite software to create a simple, high-quality movie, and the hardware and software to burn that movie to DVD. Add to that affordable blank media, and personal publishing becomes fun, inexpensive, and popular. That's probably what has gotten you this far in this book.

While there are many different ways to create more than one original DVD disc, you will undoubtedly use ONE of these methods to create quantities of identical DVDs, if that is your goal. (It usually is.)

438

Duplication and Replication

Here are the options for copying and distributing DVD discs you create:

"One-offs"—Burning One Disc, or Perhaps a Dozen or Two

Burning a single DVD is what I like to call "personal publishing." Today, almost nothing stands in the way of documenting personal milestones and achievements with startling digital clarity, and creating a "personal DVD" experience from that content. It's so amazingly affordable, it's practically unbelievable. Yet, it's here, and Mac programs like iMovie and iDVD are the proof of that pudding.

DVD Studio Pro 2 takes this personal publishing one giant step further, bringing professional DVD authoring features to thousands of small businesses and advanced hobbyists. Creating one DVD today is a piece of cake—creating 10 of them just takes a little bit longer.

Advances in high-speed (4x) DVD writers and media have reduced the time required to burn a full DVD-5 from one hour to a mere 15 minutes. The next revolution in write speed (8x) will cut that in half again. Imagine—a complete DVD burned in less than 8 minutes!

The only real difficulty with one-offs is that they are very time-intensive (even at 15 minutes per disc); they tie up the host system completely while burning (or you run the risk of buffer under-run, turning out a very nice drink coaster or Christmas tree ornament), and they usually demand personal attention—*yours*. Someone has to remove the burned disc and feed a new blank into the burner; guess who? For more than a few discs, *duplication* is a better solution!

Duplication—Making More than 10, Less than 100

Duplication is simple—take your original DVD, burn another identical copy. Presto! *Duplication!* This technique works fine for many DVD Producers, especially if your target goal is less than 100 discs.

While the disadvantages are slow production time (even the newest 4x drives takes 7–8 minutes to burn a complete DVD), the advantages are a minimal investment in Mastering costs (your original DVD-R), and the ability to have your finished product immediately.

When you need to duplicate dozens of discs with professional quality, there are even automated desktop DVD duplication systems available. If you need Professional graphics, some can even produce full-color *printed* DVDs by the dozens or hundreds, with little or no human intervention. For a few thousand dollars, you can be your own DVD Publisher.

Replication—When You Need More than a Few Hundred

Replication is a bit more complex—in order to create hundreds or thousands of identical manufactured DVDs, a certain procedure must be followed precisely. This procedure consists of preparing a finished DVD Pre-Master (usually on DLT, but today it can be delivered on DVD-R), converting that data to a "Glass Master" through precision laser burning, and then manufacturing the production materials (stampers) that will actually be used to turn simple polycarbonate beads into thousands of identical finished DVDs through injection molding. The equipment required to build a single replication line starts at around USD $1 million, and that doesn't take into account the thousands of dollars of additional equipment required for Glass Mastering, plating, QC, and other operations, which are integral to good replication.

Duplication versus Replication—How Do I Decide?

An excellent question! The factors that will make you choose between Duplication and Replication are actually pretty simple: Cost, Complexity, and Compatibility. With DVD Studio Pro 2, you will find there's only a little extra work involved in making a Replication Master.

Cost Issues for Replication

Replicated DVDs require a fixed expense for "Glass Mastering" (usually a few hundred dollars) but enjoy a relatively low per-unit cost, even when you factor in the costs of designing and printing the Disc Face Art, inserting the jacket wrappers, and loading any printed material that goes inside.

As a general guideline, it's not unusual to find DVD-5 discs being offered at less than USD $1 per disc (and this figure may even include Glass Mastering costs, if the replicated quantity is sufficient).

These costs generally are quoted to include:

- Manufacture of the finished discs,
- Insertion of the DVD disc into the selected packaging,
- Insertion of any printed jacket wrappers,
- Insertion of any printed insert material inside the package,
- Screen or offset printing of the disc face,
- Shrink-wrapping of the assembled, packaged DVD, and
- Boxing of finished product.

Ideally, you should receive boxes of finished goods ready to ship to your retail clients. You can select a different delivery configuration; just speak with your replicator. You don't have to have them shrink-wrapped and boxed if you are using them for promo handouts, for example. However, in most cases, this is the standard method.

Replicators usually like to produce at least 1,000 copies of your DVD, and you may find that the prices they quote you for quantities below that level are prohibitive, to say the least. At first glance, this might seem to be another deterrent to choosing Replication, but when you look around at duplication costs, you may not think so. Don't forget that when opting for Replication, you will be liable for the entire manufacturing cost upon completion of manufacture, and usually prior to delivery!

When evaluating the *total* cost of *replication*, be sure you factor in any external hidden costs, like graphics design fees and printing costs for the packaging covers.

Cost Issues for Duplication

On the Duplication side, costs begin with the cost of the duplication service, which includes the blank DVD disc, and sometimes a Jewel Box package. Opting to have the disc face printed may add to the cost of the service, as well as the raw media cost. Surface-printable recordable discs are more expensive than their generic branded, nonprintable counterparts. Likewise, you can have an insert created for your duplicated DVD, although it may usually be a short run of booklets or insert cards created and executed using desktop publishing. Interestingly, an entire cottage industry of suppliers who offer prepared Jewel Box insert cards and ready-to-print sticky disc labels has arisen.

Duplication and Replication

Compatibility

Playback compatibility is the hot issue. There's little point in making 100 copies of your disc if those copies are flawed and unreadable. Compatibility involves achieving a proper combination of disc, media, and player, and sometimes this magic combination seems as unattainable as making gold from base metal.

Replicated discs are the ideal format to distribute—the entire DVD business was built on the standards promulgated for replicated discs, and every DVD player should be able to read a replicated disc *without fail*, or it shouldn't have a DVD logo! Likewise, you should find that most if not all replicators should provide quality services all through the DVD production chain—mastering, metallizing, and molding.

Duplicated discs are reliant on one form of DVD-recordable or another; either DVD-R (DVD "dash" R) or the more recent DVD+R (DVD "plus" R). Either way, the compatibility of a duplicated disc is very dependent on which DVD media it is recorded on, and which player that disc is then played on. There are many workable combinations of this equation, but there are also some nonworkable combinations as well. The goal is to obviously come out on the good side of this equation, but it's easy to wind up with unplayable coasters, even after the most sincere efforts.

Complexity

Complexity refers to some of the advanced features of DVD you may choose to use—specifically, CSS (Content Scramble System) encryption, Macrovision copy protection, and Region Coding. In almost all cases, CSS cannot be added to disc masters delivered on DVD-R (G), but it is possible to deliver a DVD-R (A) Authoring disc that includes CSS, Macrovision bits, and Region Coding bits set. This is a format called CMF (Cutting Master Format) and may not be accepted by all replicators, so check first.

Rather than expend time, energy, cost, and stress trying to work around these issues and delivering on DVD-R, do yourself a favor and acquire a DLT drive, as we discussed earlier. DLT delivery solves all of these problems and delivers a more secure replication master.

Duplicators Galore!

One-to-One Burner—One Drive

With only one drive, this little duplicato (Fig. 16-1) first reads your original DVD master into its internal hard drive, then ejects it, then burns one DVD clone after another, until it senses a new master DVD has been loaded. It will then load that disc image, and commence cloning that disc. Advantage? Smart! A number of jobs can be mixed in the same load, and can run unattended—yes, both DVD and CDs can be copied.

Figure 16-1 — Micro Orbit—self loading one-drive auto-duplicator.

441

DVD Studio Pro 4

One-to-One Burners—Two Drives

A lot more sophisticated than sitting in front of your computer, these little duplicators will faithfully clone your original DVD, one at a time (see Fig. 16-2).

One dedicated read drive, one dedicated write drive. Advantage? Cost is low, and it's a little faster than the one-drive burner.

Figure 16-3 — Orbit II from Microboards.

example of this kind of Duplicator is the cute-but-powerful Orbit II from Microboards (see Fig. 16-3). Advantage? Large feed stack (50 discs), inexpensive price, does DVD and CD, and unattended operation. Orbit ProModel exists, which has 150-disc feed stack.

Multiple-Burner Straight Duplicators

Figure 16-2 — Microboards QD-DVD one-to-one duplicator.

Single-Burner Automated Duplicators

These are duplicators that burn one-disc-at-a-time, but automate the feed and removal steps. A good

These are duplication systems that can burn many discs at one time, writing the same information to multiple DVD burners—2, 3, 4, 8, 10—even as many as 16 simultaneous burners (see Fig. 16-4). Advantage? Throughput! In one period of burn time, multiple discs can be created. These systems currently use 4x DVD burners, for about an 8-minute

Figure 16-4 — Small (L) and large (R) tower duplicators.

Duplication and Replication

burn cycle. Wow! Sixteen finished DVDs per 8-minute cycle!

Single-Burner Robotic System with Printers

An interesting choice for a beginning DVD production system has got to be the Primera Bravo—a slick combination of robotics for automatic loading and unloading, full color 2400 DPI disc printing, and high-speed (4x DVD) disc burning (see Fig. 16-5). This system is available for both Mac and PC, and has quite a few proponents. What's the advantage? There are several—low cost, high performance, easy setup, and easy to use.

Figure 16-5 — Primera Bravo DVD Production System.

Multi-Burner Robotic Systems with Printers

As the prices go up, duplicators become duplication systems, complete with robotics for feed and removal, printers, even remote host job control from a PC or Macintosh system. Some systems can even automate the printing of the disc face. Advantage—they do it all! Disadvantage? These systems can get pricey, fast!

Figure 16-6 — Rimage Protégé II DVD Production System.

The systems shown in Figures 16-5 and 6-16 are the step below Replication—if you need more discs, let's talk about…

Replication by the Numbers

As we said earlier, Replication is a series of steps that must be followed faithfully and accurately if the digital data from your DVD Master is going to be transferred error-free to thousands of clones.

Many folks who are new to DVD have never had the chance to tour a DVD replication facility, and may not know the steps that go on behind the scenes. Replication is very interesting, quite and complex.

Nine steps to producing a DVD:

1. Physical Formatting
2. Glass Mastering
3. Metallization
4. Electroplating
5. Molding
6. Sputtering

443

7. Bonding

8. Labeling

9. Packaging

1—Delivering the DVD Master—Physical Formatting

DVD-5s can be delivered on DLT (with Disc Descriptor Protocol—DDP), and some replicators even allow DVD-R delivery. Each Layer of the finished DVD requires its own master element, hence the need for two DLT master tapes for a DVD-9. This is also another reason why DVD-9 masters are almost always delivered on DLT, rarely on DVD-R. There are few, if any, reliable methods of splitting the DVD layers into deliverable halves on DVD-recordable or DVD-rewritable discs. If desired, CSS encryption can be added in Replication, but only if the master was created with CSS sectoring (2054 bytes instead of 2048). Because of this nonstandard sectoring, CSS encrypted masters are generally delivered on DLT, but rarely they may be written to DVD-R (A) Authoring media, in a format called CMF.

Preparing a proper DLT Master is essential. The data from your finished DVD is written to DLT using Disc Descriptor Protocol (DDP) to "wrap" the data with information that informs the replicator's computers about the data size and organization on the tape. DDP also provides a method of *error-checking* to be sure the data have been transferred properly from your computer to theirs. This is essential!

Delivering your DVD master on DVD-R disc eliminated the DDP protection from your delivery chain (unless you are using Authoring Media, and a Pioneer DVR-S201 drive and writing a CMF master). Only this drive and DVD-R (A) Authoring media can provide an error-protected delivery chain to your replicator.

The latest twist on delivery is via secure high-speed FTP connection to the replicator. WAM! Net can provide this service when time is of the essence. It's pricey, but very fast, and secure.

2—Preparing the "Glass Master"

It's not really made of glass, but it's built on a pristine optical glass platter to facilitate movement through the creation process. The actual Glass Master is a faithful bit-for-bit recording of your DVD master onto a photoresist substance mounted on the glass platter (see Fig. 16-7). The Master image data are prepared by loading it into the Laser Beam recorder's playback system (Hard Disk), and a Glass Platter coated with the Mastering Material is prepared in the Laser Beam recorder. The DVD image data are then written in a very precise manner onto the photo-resist material covering the Glass Platter. Each Layer of the finished DVD requires its own Glass Master, which is why each layer is delivered on a separate DLT tape. This is a critical step.

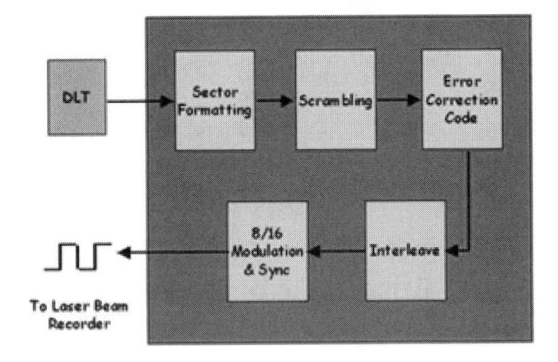

Figure 16-7 — Preparing the Master DVD image for burning Glass Master.

Duplication and Replication

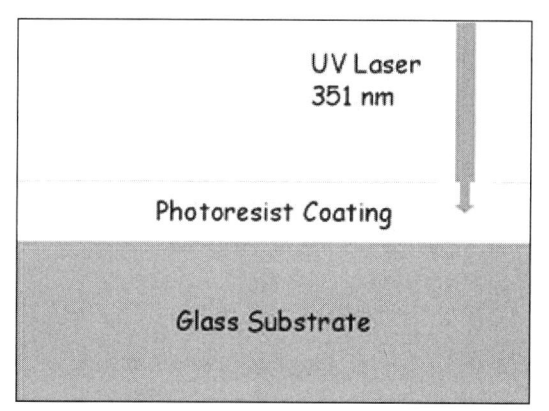

Figure 16-8 — Recording the Glass Master from your DVD data.

3—Metallizing the "Glass Master"

Once the Glass Master has been recorded (see Fig. 16-8), the next step is to create a more physically robust image of these data, to allow for the creation of the *stampers*. If you look at Figure 16-9, you will see the metallization layer fills in the pits burned by the laser, and creates a "negative image" of the data on the Glass Master. It also provides a surface that is far more durable than the developed photoresist material, allowing for a further layer of plating to occur in the next step.

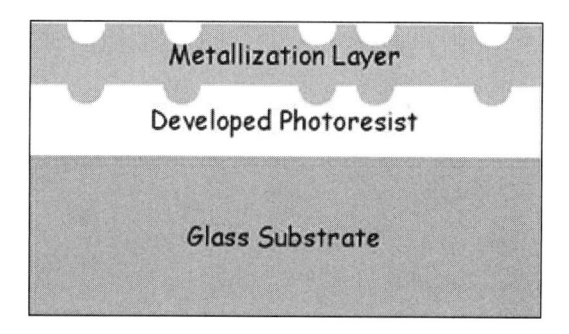

Figure 16-9 — Metallizing the recorded Glass Master.

4—Plating—Growing the "Stamper"

By repeating the process of adding metal to the existing DVD disc image, the electroplated Stamper is born (see Fig. 16-10). These Stampers can be mounted into the Injection Molding machines that actually press the polycarbonate beads into DVDs. Because each DVD is actually two layers of polycarbonate bonded together, each "layer" of data must become its own stamper, and the individual layers are created separately, then assembled according to the needs of the DVD.

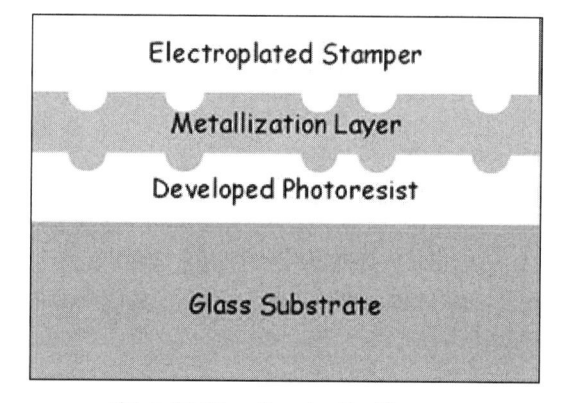

Figure 16-10 — Forming the Stamper.

5—Molding—The "Stamper" Goes to Work

On the replication "line," polycarbonate beads are injected into the molding press, where the Stampers are used to impress the final DVD data image on the polycarbonate (now liquefied) under extremely high pressure (~30 tons!). One finished DVD layer emerges from the process every 5–10 seconds, depending on the specific press used. (See Fig. 16-11.)

DVD Studio Pro 4

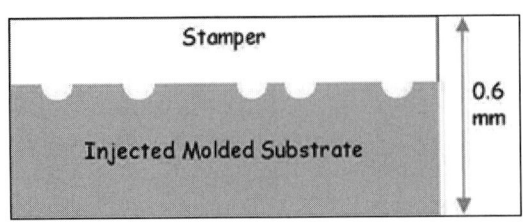

Figure 16-11 — Injection Molding the DVD from fluid polycarbonate.

6—Sputtering—The DVD Gets a Reflective Layer

Once the injection molding process has created a substrate with DVD information impressed into it, a very fine layer of metal is added to it to enhance the reflectivity of the information (see Fig. 16-12). Sputtering is a process where a metallic source is vaporized, and the metallic vapor deposits onto the polycarbonate substrate.

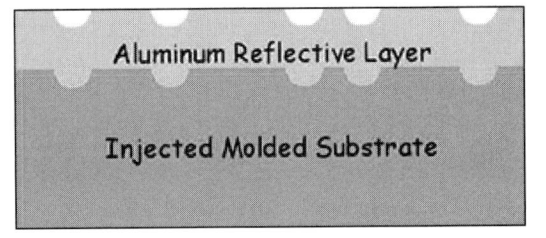

Figure 16-12 — Plating the DVD impression.

7—Bonding

Each DVD substrate produced by the injection molding system is actually only one half of a finished DVD, .6 mm in height. To produce a proper finished DVD, two halves must be bonder together to produce the full 1.2 mm height disc (see Fig. 16-13). This is how we get dual-layered DVDs.

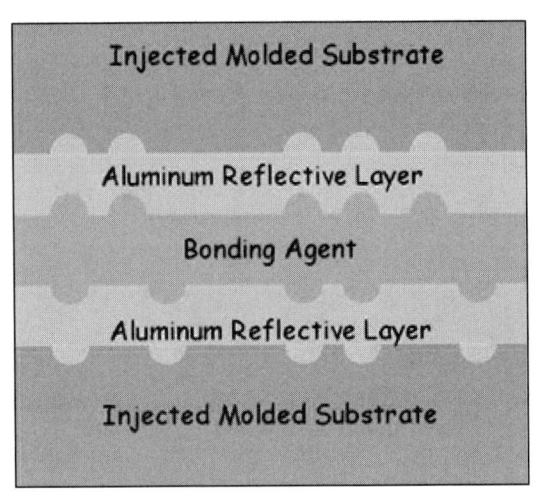

Figure 16-13 — Bonding the DVD layers together.

8—Labeling

For DVD-5s and DVD-9s, virtually the entire disc face surface is available for labeling, by a variety of means:

- Screen Print: Spot or process print as used by the CD industry. One to five colors can be applied with caution.

- Reverse Screen Print : Printing on internal surface of unused side for "Wet Look" appearance.

- Offset Printing: Full color picture printing similar to magazine print. Low ink mass and shrinkage is beneficial for DVD balance and tilt.

- Flexographic Printing: New disc printing process similar to offset printing but with spot print capability.

- Pit Art: Stamper produced for molded artwork/printing.

You will notice the deliberate omission of "sticky labels" in the above information. Sticky labels are the

Duplication and Replication

bane of DVD authors. They work fine on CDs, but are to be avoided at all costs on DVDs!

Single-layer DVS (DVD-5, each side of a DVD-10)

Single-layer DVDs include one substrate with data, bonded to one blank substrate (see Fig. 16-14).

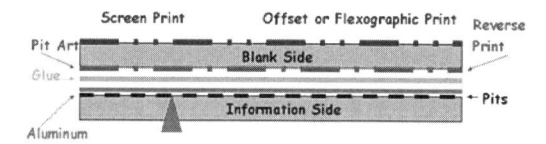

Figure 16-14 — Single-layer DVD—readable information on one layer only.

Dual-layer DVD (DVD-9, each side of a DVD-18)

Dual-layer DVDs include two substrates with data, bonded together in such a way as to allow the laser to focus on either layer 0 or layer 1, as required. The laser does not flip-flop between the two layers, but rather makes one strategic jump at the layer break—the end point of layer 0, to begin playing layer 1. (See Fig. 16-15.)

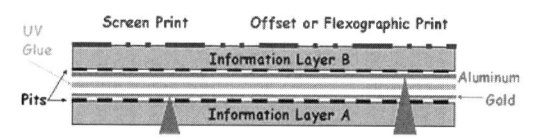

Figure 16-15 — Dual layer DVD—readable information on both layers.

9—Packaging

Finished DVDs can be packaged for distribution in a variety of different forms. The traditional AMARAY-style case is used a lot, as it is the officially chosen DVD distribution package style. DVDs can also be packaged in CD-style Jewel Boxes, paper or cardboard packages (as inexpensive CD shipping sleeves), and even plastic and cardboard combinations (like the Snapper package used on Warner Brothers DVDs). There are even more extravagant packaging styles available, like the crystal clear Super Jewel Box. Expensive, but elegant.

Before you ask for a quote from the replicator, you should have a good idea of which package style you wish to use, and how you are going to have the graphics designed for it.

In each case, a conversation with the replicator will determine which packaging options are available through that replicator, and more importantly, what the physical printing and design specs are for each type of package.

Distribution

Traditional Distribution

How you deliver the finished DVDs to your customers will vary as your customer type varies. If you have a distribution contract with a video distributor, the discs may be warehoused at the replicator and drop-shipped to retail as needed to replenish store stock. Or, discs may be consigned to retail with a return policy if they remain unsold.

Direct-to-Customer

Today, it is very simple to create an Internet store where you can sell directly to your end customer. If you choose to do this, remember the cost of manufacture and shipping will be borne by you in its entirety, whereas some traditional distributors may take on this expense as part of your distribution deal.

DVD Studio Pro 4

Distribution on Demand

In addition to the tried-and-true methods listed above, there are even new methods of distribution being created. One of these is Customflix, where your DVD master is prepared and stored, and purchases are fulfilled on a one-by-one basis, with each disc being manufactured as it is sold! This minimizes your investment in inventory, freeing up your capital for other things. The Customflix service includes online cataloging, as well as complete manufacture, online e-commerce sales and billing, fulfillment, and full reporting to you.

We've included some information on this new form of DVD publishing-on-demand in the DVD with this book.

Duplication and Replication

Summary

Whether burning a one-off DVD-R, duplicating a few dozen discs, or replicating thousands, one of these solutions will provide the proper method for distributing your DVD to its intended market.

Remember that each method has its advantages and its disadvantages, which should be weighed carefully before determining the appropriate choice.

Perhaps the biggest quandary in DVD production these days is the question of playback compatibility of DVD-recordable discs (both DVD-R and DVD+R). While DVD-R General Media discs are the normal format used in DVD Studio Pro authoring, there are many formulas of DVD-R (G) available from many manufacturers, and some of them work better than others.

Duplication can provide a more time-effective method for making dozens of discs than burning them one at a time, but you are still working with DVD-Recordable in some form, and therein lies the compatibility issue: Not all DVD-R (or +R) discs will play on every DVD player.

Replication provides the most cost-effective method of large-scale distribution, but requires a significant upfront cost to cover the glass mastering and a commitment to a minimum number of replicated DVDs—usually 1,000. Below quantities of 1,000, the per-disc cost may mushroom, as the cost of glass mastering is amortized over fewer and fewer discs, making each cost more. Be sure to check with your chosen replicator to see if there is a charge for glass mastering, or a minimum replication run of DVDs required to eliminate that cost. Some replicators throw in mastering if you replicate as few as 1,000 units, but others may require replication runs of 2,000 to 2,500. Avoid misunderstandings. Ask!

Chapter 17

Advanced Authoring

DVD Studio Pro 4

Goals

By the end of this chapter, you should understand:

- *DVD@ccess*
- User Operations
- Display Condition
- Mixed Angles
- Hybrid Discs
- Closed Captioning vs Subtitling
- Previous and Next Buttons

DVD@ccess— Interactivity Beyond DVD-Video

DVD Studio Pro contains a unique tool called *DVD@ccess* that allows a computer playing your DVD-Video disc to interact with web pages and files, given the proper environment on the computer.

What Does DVD@ccess Do?

DVD@ccess can be used to link a URL action to specific Elements of your DVD project—a Menu, a Track Marker, or a Slide.

You can use these URL actions to:

- Open Internet Web pages via a Web Browser,
- Connect to web pages or compatible files on the DVD disc itself, or
- Initiate an email!

What Do I Need to Use DVD@ccess?

A computer that has DVD-Video playback capability is a requirement. *DVD@ccess* works on Macs and (many) PCs. To connect to Web pages on the Internet, a live Internet connection is required. Web Browser software is required by *DVD@ccess* to make a connection. If you create a *mailto:* link, email software will be required to process this link.

To Configure DVD@ccess on a Macintosh:

End users running a Macintosh can use the Apple DVD Player to process the *DVD@ccess* links. Macintosh computers can use Apple DVD Player V2.4 or above in OS 9, or any version of Apple DVD Player in OS X. To use *DVD@ccess* links, you need to activate the *DVD@ccess* Preference in Apple DVD Player. In OS X, you can find it in **DVD Player > Preferences > Disc Tab**.

To Configure DVD@ccess on a PC:

PC users must install a small piece of software called *DVD@ccess.exe* to enable the communications between the DVD player software and the PC's Web Browser. This Installer is automatically added to the DVD-ROM zone of any *DVD@ccess*-enabled disc you build and burn with DVD Studio Pro 2, but is not added if you just format a disc.

You can add the *DVD@ccess* installer for PC if you copy it out of the DVD Studio Pro application bundle, but **DO THIS CAREFULLY!** The DVD Studio Pro application can be damaged, or the installation corrupted if you add or remove important files.

Advanced Authoring

To Locate and Copy the DVD@ccess installer:

- Locate the DVD Studio Pro 2 application bundle.
- Control-click on the DVD Studio Pro 2 application Icon;
- Select "*Show Package Contents.*"
- Open the *Contents* and *Resources* folders.
- Select the *DVD@ccess* folder;
- Option-Drag it to a location outside of the DVD Studio Pro 2 package. Option-Drag makes a copy, leaving the original in place.
- Add that Folder to the DVD-ROM folder you prepared for this DVD.
- Be sure to activate the DVD-ROM function and point to the folder that includes the *DVD@ccess* installer and the ROM files (see Fig. 17-1).

Setting the DVD@ccess Property

You may only set *DVD@ccess* links in these Elements:

Track Markers

Any (or all) of the up to 255 Marker locations may be assigned a *DVD@ccess* link. The *DVD@ccess* link is assigned in the *DVD@ccess* property of the Marker Inspector General Tab. During playback of that Track, the *DVD@ccess* link will be executed when the Marker that it is assigned to is played. If you wish to make it appear as if the track has a *DVD@ccess* link, assign the URL action to the first Marker in the Track. (Stories and Story Markers cannot have *DVD@ccess* links.)

To assign a DVD@ccess link to a Marker:

- Select the desired Marker (*DVD@ccess* in available on all Markers).
- Activate the *Marker Inspector, General Tab*;

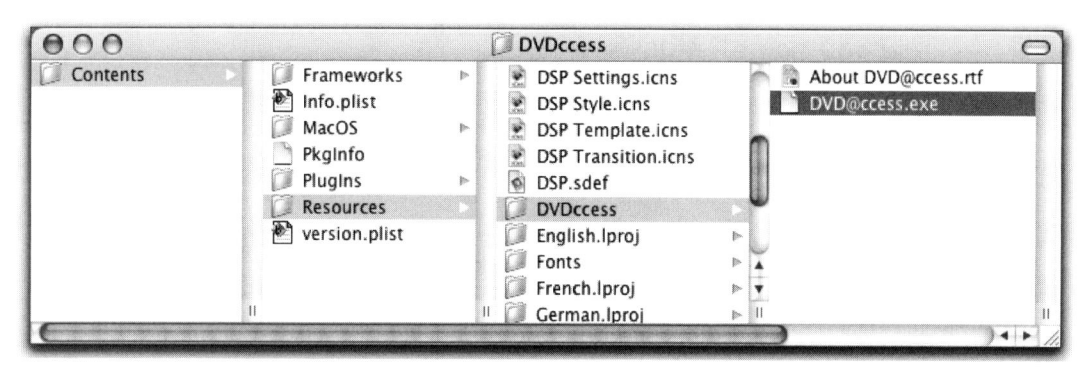

Figure 17-1 — OS X Contents folder showing DVD @cess installers.

453

- Activate the *DVD@ccess* checkbox.
- Enter the *DVD@ccess* URL according to the method outlined below.

Menus

Menus can have a *DVD@ccess* link attached, which will execute when the Menu appears on screen. While buttons may not have *DVD@ccess* links attached, you can emulate that behavior by linking a button to another menu with an information screen, and attach the *DVD@ccess* link to the Menu that is connected to the button. The effect will be identical.

To assign a DVD@ccess link to a Menu:

- Select the desired Menu (*DVD@ccess* in available on all Menus).
- Activate the *Menu Inspector, Menu Tab*;
- Activate the *DVD@ccess* checkbox.
- Enter the *DVD@ccess* URL according to the method outlined below.

Slides

Each slide of a slideshow may be assigned a *DVD@ccess* link. During playback of that Slideshow, the *DVD@ccess* link will be executed when the Slide that it is assigned to is displayed.

To assign a DVD@ccess link to a Slide:

- Select the desired Slide (*DVD@ccess* is available on all Slides).
- Activate the Slide Inspector;
- Activate the *DVD@ccess* checkbox.
- Enter the *DVD@ccess* URL according to the method outlined below.

Slideshows cannot be assigned *DVD@ccess* links, only Slides. If you wish to make it appear as if a Slideshow has a *DVD@ccess* link, assign the URL action to the first slide.

Setting the DVD@ccess Property

In each element, the *DVD@ccess* property is set using this dialog, which can be found in the Playback Options section of the relevant Tab (see Figs. 17-2a, 17-2b, and 17-2c).

To activate a DVD@ccess link:

- Check the *DVD@ccess* checkbox to activate the function.
- You may enter a *Name:* to help you remember what this link does;
- Enter the appropriate URL text in the *URL:* field.

Proper DVD@ccess Syntax Is Important

Please note carefully that the URL syntax is VERY particular for each different function, and you MUST enter the URL accurately in order for the function to work properly. URLs in many cases are CaSe SeNsItIvE.

To Link to Web pages live on the Internet:

- Enter a fully resolved URL name in the proper format, as shown:
  http://www.recipe4dvd.com/[index.html]

454

Advanced Authoring

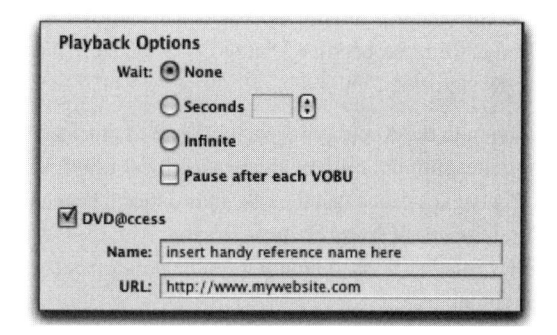

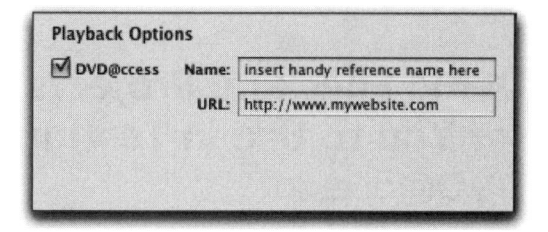

Figure 17-2a, b, and c — The DVD@ccess fields: Marker, Menu, slide.

To Link to local HTML pages or files on the DVD Disc:

To properly set the *DVD@ccess* link for local files, you will need to decide on a name for your DVD before you set the *DVD@ccess* links. *Please note that to properly utilize this feature, the URL or filename cannot contain a space character " "*.

Links to local pages should specify the filename in a complete path: **file:///Volumes/DISC/FOLDER/FILENAME.ext** (for OS X, OS 9, Windows).

DISC is the name of the DVD itself, FOLDER is the name of the first folder containing the file, and FILENAME.EXT is the file itself.

If your file is contained inside of one or more subfolders, your filepath should name those folder(s): **file:///Volumes/DISC/FOLDER1/[F2]/[F3]/.../FILENAME.ext**

F2, F3 represent subfolders, and /.../ indicates where to add any additional subfolders into the path.

The Apple manual no longer shows the following syntax, but it does still seem to work (in DVD Studio Pro 2.0): **file://localhost/Volumes/DISC/FOLDER/FILENAME.ext (OS X)**

The other syntax (file:///Volumes/) is much easier to write.

To create an email:

Enter the following text syntax in the URL field: **mailto:username@domain.com (or .net, .org, .edu, etc.)**

Limitations on local file access:

To open a "local file" from the DVD disc, the user's computer must have an application that can process the particular type of file requested in the URL. In addition, since the Browser software is used to launch the helper application in many cases, the Browser should be properly configured to process the expected file types.

For example:

- HTML pages can be read using only the installed web browser;
- PDF documents can be read by the Browser using a Plug-In, or may be passed along to the Acrobat Reader. In both cases, Acrobat Reader must be installed in order for PDF documents to be properly read.

455

- Text files, Word Docs, Excel Spreadsheet, each have their own particular helper application setting that should be checked.

These local page links can be very effective in creating Hybrid DVD discs for industrial use, training, or distance learning.

Testing Your DVD@ccess Links

Before deploying a DVD project with DVD@ccess links, it's important to test the links to be sure they are functioning correctly.

Testing Using the Apple DVD Player

Before attempting to test a DVD@ccess-enabled project using the Apple DVD player, enable the *DVD@ccess preference* in the Disc Tab.

To test the project using Apple DVD Player:

- Build the project and format it to your hard disk as a *disc image*.
- Mount the Disc Image file (.img) by double-clicking it, then
- Select the "`File > Open VIDEO_TS Folder...`" command, and
- Open the VIDEO_TS folder in the mounted DVD volume.

You can then play the disc as if it were burned to DVD-R.

Testing Using Simulation

Before testing DVD@ccess links in Simulator, be certain to check "Enable DVD@ccess Web Links" in the Simulator Preferences Pane. This allows the Simulator to process the DVD@ccess links for Web pages and email links.

Although the Apple DVD Studio Pro 2 Manual says that the simulator is not able to test local file links, it's my experience that the files may, indeed, be read. Build the project and Format it to your hard disk as a Disc Image file, then mount it on the desktop before Simulating.

Mounting a .img File on the Mac Desktop

To mount a disc image, just double-click on the .img file and the DVD Volume will mount on the Macintosh desktop. You may then Simulate, or open the VIDEO_TS using Apple DVD Player.

Demo Files and Projects for You to Use in Testing DVD@ccess

One of the extra features included in this book's DVD is a special folder full of Custom Web pages (HTML) including their support graphics.

The folder "Web" includes pages which the *DVD@ccess* enabled projects will use to demonstrate the DVD-to-web enablement that is possible using the *DVD@ccess* feature of DVD Studio Pro.

We've also included several projects (called "File-enabled) that show you how to use the *DVD@ccess* function in DVD Studio Pro to create a self-contained hybrid DVD that plays web pages from the ROM portion of the same DVD disc on which the video is contained. Other projects ("web-enabled") show you how use *DVD@ccess* to link your DVD Video to practically ANY page available on the

Advanced Authoring

World Wide Web, or available for your Browser to open with any available helper Application.

Creating DVD Hybrid Discs in DVD Studio Pro

In DVD Studio Pro, files may be added to the DVD disc in the DVD-ROM zone. These files may be accessed by the *DVD@ccess* function, or just opened by any Computer that reads the DVD disc. Adding files to the DVD does not mean you *must* use DVD@ccess, but you can.

The *DVD-ROM* property in the Disc Inspector General Tab will allow you to select the DVD-ROM folder. Once selected, DVD Studio Pro will calculate the folder size and include this information in the Total Disc size property at the bottom of the Format dialog. Files included in this folder will be accessible to the *DVD@ccess* function.

You may also select a DVD-ROM folder to be added to an existing build in the *DVD-ROM Data* property in the *Format Dialog General Tab*.

Remember that to create your DVD@ccess Hybrid Disc:

- Collect the ROM files into a single folder while authoring;
- Enable the DVD-ROM property in the *Disc Inspector*, or
- Enable the DVD-ROM property in the *Format Dialog*;
- Select your collection folder as the DVD-ROM source;
- Enable any *DVD@ccess* links you need for the project, then

- Build and burn the project as you would normally.

> **NOTE!**
> Remember that some DVD players cannot handle a large number of files or folders in the root directory. Less is better! (See Figs. 17-3 and 17-4.)

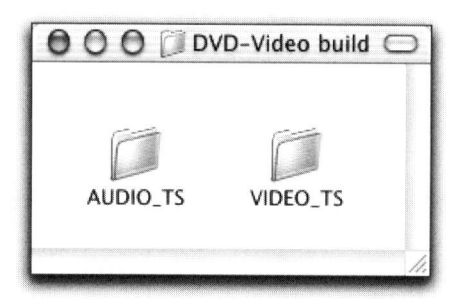

Figure 17-3 — A standard DVD-Video Build folder.

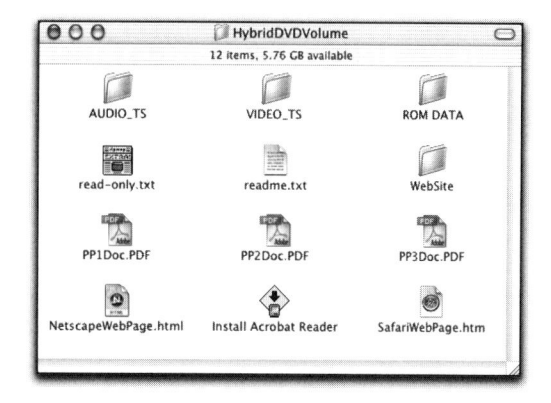

Figure 17-4 — A Hybrid DVD Volume (Video + DVD-ROM data).

Closed Captioning

Very different from Subtitles, Closed Captioning is the specially prepared coding that provides dialog, stage direction, and sound and music notes in visible form overlaid on the video program that has been Captioned. This form of captioning is called Closed because it requires a special decoder to reveal the

457

DVD Studio Pro 4

captions that are buried within the video signal itself, generally on Line 21 of the Video Master.

Closed Captioning is not something that most beginning DVD authors choose to take on, but if you have the proper software, there would be nothing preventing you from starting to learn about it.

There is a program out called *Mac Caption* that runs in OS X and provides a great number of Closed Captioning features.

To properly use Closed Captioning in DVD, you will need a Caption file in one of the standard formats typically provided for DVD authoring.

DVD Studio Pro 2 can accept these files in these formats:

- .SCC format
- .CC format

To add a Closed Caption file to a Track:

- Control-Click in the Media Clip that you wish to add the CC file to:
- Select "Import Line 21 file…"
- From the Open "Choose Line 21 file" dialog box, select your CC file.

That's it—except for the fact that you will want to burn a DVD disc in order to check the Closed Caption data on a TV set equipped with the appropriate Closed Caption decoder. Apple DVD Player now has Closed Caption decoding, but under some circumstances, projects that are built to the Hard Drive may not properly display the Closed Captions, while discs burned from these projects will display them. Go figure!

User Operations (UOPs)

As should be expected of a professional DVD authoring application, DVD Studio Pro offers extensive control over the User's Remote Control functions, called User Operations (usually just called "UOPs").

The ability to disable certain functions during the playback of a specific Element is the purpose of these UOPs. There are UOPs properties available on each Track, Marker, Menu, and Story Inspector.

You are probably already familiar with how UOPs work because the FBI warnings on most commercial Movie DVDs are usually locked to prevent skipping past it, or scanning it at high speed, or even jumping out using the Title or Menu Remote Control Buttons.

The UOPs that you select (check) will be disabled for that specific Element. There is a *Disable All* function, which will select (and *disable*) all UOPs for a particular element. Once disabled, certain UOPs may easily be re-enabled. There is also an *Enable All* function, which will deselect (and *enable*) all UOPs for a particular element.

Markers have a "Same as Track" default function, which allows the UOPs to follow the Track UOPs, while allowing specific markers to have unique UOP settings.

To Set User Operations:

- Select the Track, Marker, Story, or Menu you wish to set;
- Select the *User Operations* Tab;
- Select the desired UOPs you wish to *disable*, or
- Select *Disable All* to disable ALL UOPs;
- If you wish, you may uncheck certain UOPs to re-enable them, or

Advanced Authoring

- Select *Enable All* to re-enable all UOPs (they will be unchecked).

About User Operations Properties

User Operations are shown in the Inspector for the specific Element. UOPs are organized into Groups, and we'll describe each UOP within the Group it belongs to (see Fig. 17-5):

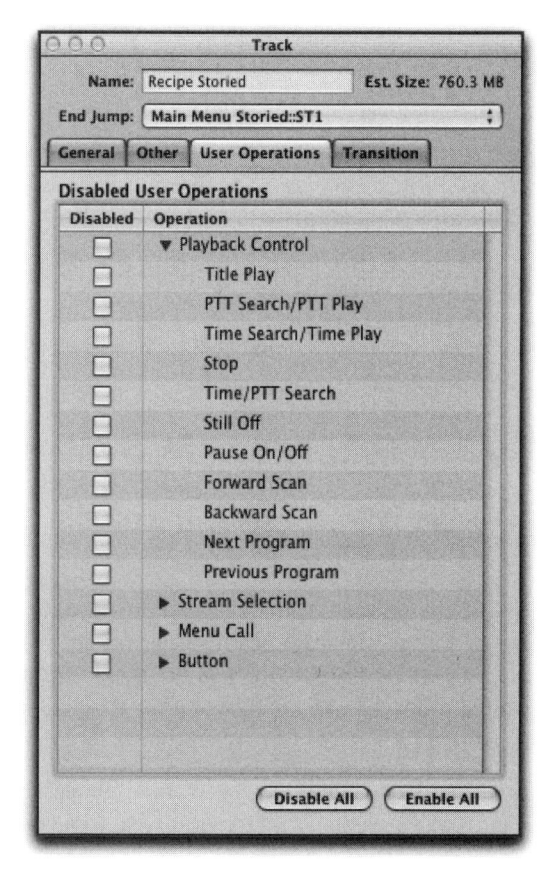

Figure 17-5 — The User Operations Tab in Track Inspector.

> **NOTE!**
> Because UOPs may be set for *Tracks*, *Markers*, *Menus*, and *Stories*, the selected UOP functions will be valid only for the duration of the specific *Track*, *Marker segment*, *Menu*, or *Story* onto which they are set.

Playback Control

In general, *Playback Control* disables certain transport-related functions, including searches and straight playback.

- *Title Play*: Disables entering and jumping to a new title number by using the DVD player remote control.

- *PPT Search/PTT Play*: Disables entering and jumping to or playing from a new chapter marker by using a remote control's numeric keypad. PTT stands for *part of title* and means "Chapters."

- *Time Search/Time Play*: Disables entering and jumping to or playing from a new time by using a remote control's numeric keypad.

- *Stop*: Disables the "stop" function while a selected menu, track, marker, or story is playing.

- *Resume*: Disables the "resume playback" function while a selected menu is displayed or while a track, marker, or story is playing.

- *Time/PTT Search:* Disables entering a specific time in order to locate to a specific point in a video stream.

- *Still Off*: Checking this UOP allows a still image to be paused.

- *Pause On/Off:* Disables the pause/unpause function while a Track, Marker segment, or story is playing. Also functions in Motion Menus.

459

DVD Studio Pro 4

- *Forward Scan*: Disables fast-forwarding through the element.

- *Backward Scan*: Disables fast-rewinding through the element.

- *Next Program*: Disables skipping to the next Chapter.

- *Previous Program*: Disables skipping to the previous Chapter.

- *Go Up*: Disables the Return button function on the remote control.

Stream Selection

- *Audio Stream Change*: Disables changing to a different Audio stream while a Track, Marker segment, or Story is playing.

- *Subpicture Stream Change*: Disables changing to a different subtitle stream while a Track, Marker segment, or Story is playing.

- *Angle Change*: Disables changing to a different Video Angle stream while a Track, Marker segment, or Story is playing.

- *Video Presentation Mode Change*: Disables any change of aspect ratios (4x3 and 16x9). Disables switching between pan-scan and letterbox modes, as well.

Menu Call:

- *Root Menu (DVD Menu)*: Disables the Remote Control Menu button to access a DVD menu that has been set in the Connections tab.

- *Title Menu*: Prevents using the Title button on a remote control to access a DVD menu that you have previously set up in the Connections tab.

> **NOTE!**
> You will want to use the following functions carefully, and test them to see if they function on the DVD players you are deploying your project on.

- *Subpicture Menu:** Prevents accessing the subtitle menu during movie playback.

- *Audio Menu:** Prevents accessing the audio menu during movie playback.

- *Angle Menu:** Prevents accessing the angle menu during movie playback.

- *PTT Menu:** Prevents accessing the PTT menu (chapter menu) during movie playback.

> ***NOTE!**
> Many DVD Remote Controls do NOT have the above functions available.

Button:

- *Selection/Activation*: Disables Menu button selections.

Obviously, this is NOT advisable on Menus!

Display Conditions

A Display Condition may be attached to a Track, Story, or Menu, and can control whether or not that Element will be displayed, or if an alternate Element will be displayed instead, based on the Condition. (See Fig. 17-6.)

A Display Condition may be used, for example, to control which version of a Story is played, based on the value of a GPRM, or which version of a Track is played, based on the Parental Rating value. Additionally, Stream selections may be made based on Language Preferences set in the DVD player's SPRMs.

Advanced Authoring

Setting a Display Condition

Display Conditions are in many ways similar to the Conditional Commands created in DVD Studio Pro Scripting (see Chapter 13). This similarity exists in that both *Conditional Commands* and *Display Conditions* make navigation "decisions" based on the status of some specified element, or Parameter, as ascertained by testing it.

Figure 17-6 — Display Condition dialog.

There are four parts to setting up a Display Condition:

- The *Element*—The Element determines which feature, status, or parameter is being used as the source of the Display Condition test.

- The *Relationship*—The Relationship defines the test that will be performed between the *Element* and the *State* to determine if the Display Condition should execute or not.

- The *State*—Depending on the Element chosen to test, the State may be a Stream number, a value in a GPRM or SPRM, or the name of a menu language (these are just a few possibilities). This argument defines the value the test is seeking.

- The *Alternate Target*—Defines which Element will be played if the Display Condition test is not true. This may be any element in the DVD.

Be sure to set all four of these when establishing a Display Condition.

To Set up a Display Condition:

- Choose the Element for the Display Condition: Menu, Track, or Story.

- Activate that element's Inspector, if not already visible.

- Select the proper Tab for the Display Condition property:

- Menu Tab for Menu Elements,

- General Tab for Story or Track elements.

- Activate the Display Condition checkbox, then set up the condition:

  – First, Select the *Element* to be checked—see the list of *Conditional Elements and their Values* further on in this section for a summary. A SPRM or GPRM is generally selected as the conditional *Element*.

  – Second, Select the *Relationship* to be used to test the conditional Element against the State to be defined in the following step—there are a variety of different Relationship tests, outlined in the *Conditional Relationships* list appearing later in this section.

  – Third, Select the State of the Conditional *Element* you wish to test for. You may again consult the list of *Conditional Elements and their Values*, as a guide.

  – Fourth, and last, select an *Alternate Target Element*. This is the Element that will be played if the defined display condition is not true.

461

If you wish to turn off the Display Condition function, uncheck the *Display Condition* checkbox.

> **NOTE!**
> Stories may have their own unique Display Condition defined, or they may use the Display Conditions set for a Track. Select *Apply to Stories* in the Track Inspector's General Tab to have the Story use the Track's Display Condition.

An Example Display Condition: Setting a Language-Based Element

Let's suppose you are authoring a DVD to be released in Canada, which has bilingual programming, both English and French. Although the English Language FBI warning is the First Play, you'd like the *French* FBI/Interpol warning Menu screen to be selected automatically instead, if the DVD player has been set to "French" language. If not, the English should play normally. Since SPRM 16 holds the audio language choice selected for the DVD Player, it's a simple thing to read this SPRM to see which language code has been set by the DVD user. That's the basis of our Display Condition example.

- Select the English Language FBI/Interpol warning Menu element.
- In the Menu Tab, locate the Display Condition property.
- Activate the Display Condition checkbox.
- Set the *Element* (*to test*) as *Audio Language* (this will test SPRM16).
- Set the *Relationship* to "=" (equal to).
- Set the *State* to "*English*" from the pop-up menu. (Since we're testing an Audio Language, the state will be one of the Language Names).
- Finally, set the *Alternate Target* to the French FBI menu.

Figure 17-7 shows what this completed Display Condition would look like.

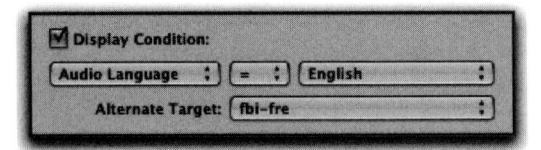

Figure 17-7 — Our completed Display Condition example.

If we wanted to express this Display Condition in English, it would be, "Display this Element IF the Audio Language is currently set to English; otherwise display the FBI-FRE Element (the French FBI warning)."

> **BONUS!**
> You will find a practice project for this in the DVD accompanying this book. Look for the "Display Condition" folder.

> **NOTE!**
> The Display Condition setting does NOT show up in the Connections tab, regardless of the settings.

> **TIP!**
> To make an Audio Stream choice on a Track, do not use Display Condition, but consider a Pre-Script that will test SPRM16 and set the Audio Stream value based on the Language Name in SPRM16. Same basic principle, but utilized in a different method.

Conditional Elements and Their States

The conditional *Elements* are the various System Parameters (SPRMs) or any one of the 8 General Parameters (GPRMs 0–7).

The conditional *States* are determined by the element. For example, the menu language element has all supported languages as the possible states. If the state is a numeric entry, you need to enter a decimal-based number (binary and hex numbers are not supported).

Advanced Authoring

- *Menu Language*: Uses the value in SPRM 0—the DVD player's menu language setting. Valid values include all supported languages.

- *Audio Language*: Uses the value in SPRM 16—the DVD player's audio language setting. Valid values include all supported languages.

- *Subtitle Language*: Uses the value in SPRM 18—the DVD player's subtitle language setting. Valid values include all supported languages.

- *Audio Stream Number*: Uses the value of SPRM 1—the currently selected audio stream. Valid values include audio stream numbers 1 through 8. "Not set" appears next to any streams that currently have no assets assigned.

- *Subtitle Stream Number*: Uses the value of SPRM 2—the currently selected subtitle stream. Valid values include subtitle stream numbers 1 through 32. "Not set" appears next to any streams that currently have no subtitles assigned.

- Player Aspect Ratio: Uses the value in bits 8 and 9 of SPRM 14—the DVD player's aspect ratio setting (4:3 or 16:9). The states include 4:3, 16:9 Pan-Scan, 16:9 Letterbox, and 16:9 Pan-Scan and Letterbox.

- *Last Track Played*: Uses the value of SPRM 4—the most recently played track. The states include all track numbers in the project.

- *Last Chapter Played*: Uses the value of SPRM 7—the most recently played chapter. The states include all chapter marker numbers in the track (1–99).

- *Last Button Selected*: Uses the value of SPRM 8—the most recently selected button. The states include numbers 1 through 36.

- *Player Region Code:* Uses the value of SPRM 20—the DVD player's region setting. The states include all eight regions, although region 7 really isn't used.

- *Parental Level*: Uses the value of SPRM 13—the DVD player's parental level setting. The states include all eight parental levels plus a Not Rated setting.

- *Video Player Config*: Uses the value of SPRM 14—the DVD player's aspect ratio setting (4:3 or 16:9) and display mode (letterbox or pan-scan). The state requires a decimal number entry.

- *Audio Player Config:* Uses the value of SPRM 15—the DVD player's audio configuration—PCM, AC-3, and MPEG. The state requires a decimal number entry.

- *GPRM 0–7*: Uses the value stored in one of the eight GPRM locations. The values must first be placed into the GPRM by a separate script that you have run. The state requires a decimal number entry. Valid values are 0–65535.

Conditional Relationships

You can set one of these possible relationships when testing a *Conditional Element versus a State*:

- = *(equal)*: The element and state are equal.

- != *(not equal)*: The element and state are not exactly equal.

- >= *(greater than or equal)*: The element is either greater than the state, or they are equal.

463

DVD Studio Pro 4

- *> (greater than):* The element is greater than the state.

- *<= (less than or equal):* The element is either less than the state, or they are equal.

- *< (less than):* The element is less than the state.

- *& (and):* Provides a binary bit-wise "and" function.

(For more on these functions, see Chapter 13.)

Display Condition versus Pre-Scripts

While Display Conditions are quite versatile, there may be times when a Pre-Script can better do what you want. That said, remember:

- Pre-Scripts execute only when the root element they are attached to is the target of navigation (e.g., the Menu, Track, or Slideshow).

- Navigating inside of the Element, to a Marker, Menu Button, or Slide directly will not trigger the Pre-Script.

- Display conditions will be executed any time any part of the attached Element is the target of navigation.

- When an Element has both a Pre-Script and a Display Condition, the Pre-Script will execute first, then the Display Condition.

About Next and Previous Remote Control Buttons

When playing Tracks and Slideshows, the Next and Previous buttons may exhibit unpredictable behavior if one attempts to navigate "Previous" from the first Marker in a Track, or first Slide in a Slideshow, or navigate "Next" after the Last Marker in the Track has been played, or while on the last Slide in a Slideshow. This is due to aberrant playback standards in some DVD players.

To avoid these strange remote control behaviors:

- Disable the *Previous* Remote function on the Start Marker in a Track.

- Disable the *Next* Remote function on the last Marker in a Track.

- Disable the *Previous* Remote function on the first slide in a Slideshow.

- Disable the *Next* Remote function on the last slide in a Slideshow.

These remote functions may be disabled in the Connections Tab.

Chapter 18

Future Developments

So Where Do We Go from Here?

If you've been following the development of the next generation of DVDS, you can see a lot of great things on the horizon, and yet some uncertainties still exist. The biggest problem is that we don't yet have a single unified standard that we can all rally behind, as we did with SD DVD. So much promise, and yet so much remains to be codified.

The powers that be in DVD are working feverishly on great new developments for us DVD authors, and creating great new features to use in our new enhanced DVD players, but the lack of a locked-down standard causes some difficulty for we who try to provide accurate information to you who seek it.

Developments in HD DVD, Blu-ray Disc, EVD, and Holographic discs are all potentially very interesting, and yet at the time this book is being finalized, are still so up in the air that it is impossible to accurately report on these new technologies.

To that end, the **www.Recipe4DVD.com** website will be undergoing a makeover quite soon, and will emerge shortly thereafter, reborn withnew content, and fresh weekly updates on the newest developments in next-generation DVD authoring. We invite you to drop in and participate in our activities, and partake of our news.

We will also be looking to build an even **bigger** community of DVD enthusiasts than we already have at our recipe4dvd Yahoo group.

Purchasers of this book will receive a complimentary first year of membership into the new DVD Pro Club at www.Recipe4DVD.com, where we will look forward to providing up-to-the-minute information and insight into the latest developments in next-generation DVD, as well as exclusive forum postings and hints and tips on how to use DVD Studio Pro to its fullest extent.

I really think you will like the new version of the website.

Login at: **http://www.recipe4dvd.com**
Username: bookbuyer
Password: dvd2006

...and by the way, we'll be bringing our recipe4DVD onlne store into the website during the upgrade, so you will be able to find all kinds of DVD goodies there, like projects, scripts, tutorials and the like, as well as project consulting for DVD Studio Pro.

Appendix A

Command Reference

DVD Studio Pro 4

If you're looking here for the many pages that used to describe all of the DVD Studio Pro commands, you're in for a surprise—they have been moved to our new *Online Reference Section*, in the Recipe4DVD.com website.

We did this for a very good reason—space!

In order to put valuable information into the book and keep its page count (and cost to you) reasonable, we opted to devote the printed pages of the book to information you needed from me that you could not get anywhere else. Of course the guide is still part of the book, it's just online—and you can download it as a PDF once you get there!

The DVD Studio Pro Command Reference will be waiting for you when you enter the new, Private Book Area of recipe4dvd.com.

Just point your web browser to: **http://www.recipe4dvd.com**. Once there, you will see the link.

Appendix B

Alternative Encoders

DVD Studio Pro 4

Besides the Hardware encoders we discussed in the book, there are several Mac-based software encoders up for consideration, even though recent advances in Apple's Compressor encoder has rendered some of these either less effective than in previous years, or just plain undesirable. On the other hand, there are some new entries to this list, and more and more folks are jumping on the Mac encoding bandwagon!

While you may never need to access any encoder other than Compressor, it doesn't hurt to know what's out there.

In Alphabetical order, but not order of preference, here's the cast:

- Innobits' *BitVice*
- *Cleaner 6* from discreet
- Digigami's *MegaPEG-X* and *MegaPEG-X-QuickTime plugin*
- Heuris' *MPEG Power Professional*
- *Main Concept Encoder* from Main Concept AG
- Popwire's *CompressionMaster*

BitVice from Innobits

BitVice has been on the scene for some time now, and reports are that it does very good MPEG encoding (see Fig. B-1). A free downloadable demo is available from its website, and we've also included some demo software and BitVice information on the DVD accompanying this book. Look in the folder called "Encoding" for "BitVice."

A free tryout version is downloadable from the BitVice website: **http://www.bitvice.com**.

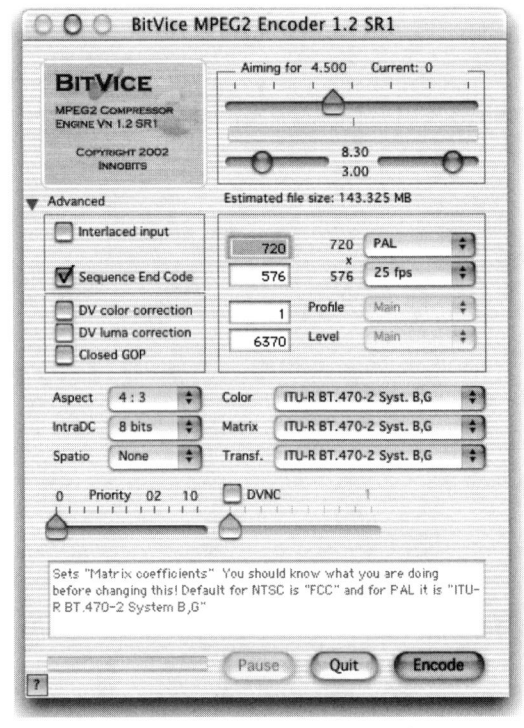

Figure B-1 — BitVice settings window.

System Requirements for BitVice

For BitVice PPC G4 and BitVice PPC G4 Lite:

- A Macintosh with Power PC G4 or G5 processor
- Mac OS 9.1 or later, Mac OS X (Jaguar or Panther)
- QuickTime 5.0 or later
- 180 MB RAM

For BitVice PPC G3:

- A Macintosh with Power PC G3 processor
- Mac OS 9.1 or later, Mac OS X (Jaguar or Panther)

Alternative Encoders

- QuickTime 5.0 or later
- 180 MB RAM

Summary of BitVice MPEG2 Encoder Features

- Scene change detection
- True Variable Bit Rate
- Adaptive GOP structure
- IBP GOP sequences
- Special still picture encoding
- Adaptive Quantization and Bit Rate Control
- Interleaved rate control
- Extensive block matching
- DV input color and gamma correction
- Good control of the generated file size

Heuris MPEG Power Professional

A well-respected piece of software, Heuris MPEG Power Professional comes in a few different versions, depending on the depth of MPEG capability you need. The correct one for DVD is *MPEG Power Professional DVD*. It is available now in an OS X version, as well as the original OS 9 version.

About MPEG Power Professional—DVD

MPEG Power Professional software gives comprehensive control of many aspects of MPEG encoding, and can create compliant MPEG streams for use with DVD Studio Pro. Check out the HEURIS website for more info: **http://www.heuris.com**.

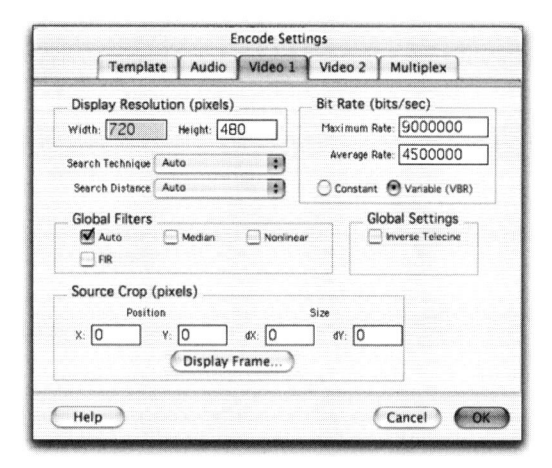

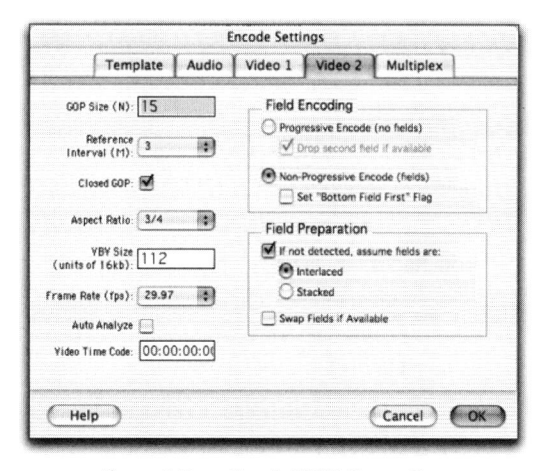

Figure B-2 — Heuris MPEG Power Pro.

- MPEG-1 constant bit rate encoding
- MPEG-2 constant & variable bit rate
- Support for popular DVD authoring systems
- VOB file creation feature
- Support for multi-angle DVD playback
- Converts QuickTime™ files to MPEG-1, MPEG-2

DVD Studio Pro 4

- Auto Analysis feature can be used to detect scene changes, place, I-frames, apply filters and set search ranges.

- Encoding Control Lists (ECLs) can be created with Auto Analysis or manually to record and reuse encoding events.

- Inverse Telecine feature can be used for 3:2 pulldown. The user can also control telecine phase changes.

- User has frame-specific control of filters, reference frames, frame rates, frame sizes, and more.

- Crop Source feature can be used to remove overscan lines.

- Automatic audio frequency scaling.

- Easy-to-use built-in PAL/NTSC-specific templates allow user to set parameters for specific applications with just one click.

- Includes online and context-sensitive help, interactive multimedia help screens, an MPEG tutorial CD-ROM, a demo DVD, and a comprehensive user's manual.

- Support for popular CD burning packages such as Roxio Toast™ and Easy CD Creator™.

- Source File Concatenation abilities to allow user to circumvent source file size restrictions imposed by some operating systems.

- Full D1 and half D1 MPEG-2 encoding.

- Special built-in templates for MPEG-2 decoders and laptops.

Cleaner 6 from Autodesk

Recently purchased by Autodesk (makers of discreet*), Cleaner 6 is the latest version of the longtime bedrock of Web Compression tools, and Cleaner 6 includes MPEG-1 and MPEG-2 capabilities that may be of interest for DVD authors (Fig. B-3).

A free tryout version is downloadable from the Autodesk website: **http://www.autodesk.com**.

Cleaner 6 Features:

- **Support for QuickTime® 6 (with MPEG-4 and AAC)**

- **Direct Support for Native Mac Architecture**
  - Up to 150% performance increase in filtering and encoding speeds over cleaner 5 on dual G4 processors.
  - Up to 100% performance increase in filtering and encoding on single G4 processor.

- **MPEG-1 & 2 with 2-pass Variable Bit Rate Encoding**
  - Creates high quality video for DVDs and Video-CDs.
  - Tweak your video with new controls to get the highest quality.

- **Watch Folders**
  - Encode directly from the editing station for a streamlined workflow.
  - Drag and drop files into Watch Folders for automated encoding.

Alternative Encoders

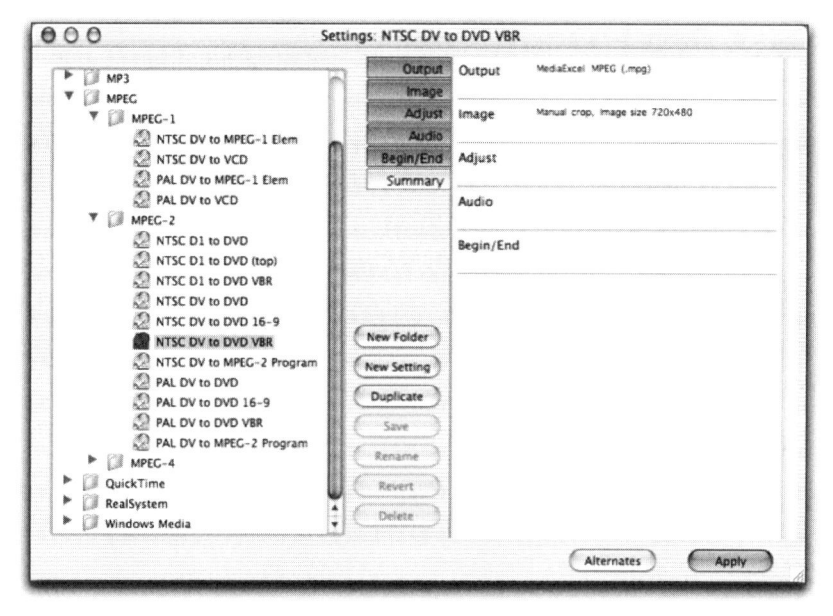

Figure B-3 — The Cleaner 6 Settings Window. Note support for both MPEG-1 and MPEG-2 Elementary streams in both PAL and NTSC video formats.

- **Filtering Engine**
  - Output content faster.
  - Preserve high fidelity between YUV source and its encoded output.

- **Improved Settings Management**

  –Organize and manage all your filtering and encoding settings easily.

  –Create custom folders and settings for consistency and re-use over time.

- **Filtering & Encoding Presets**
- **Aqua® User Interface**

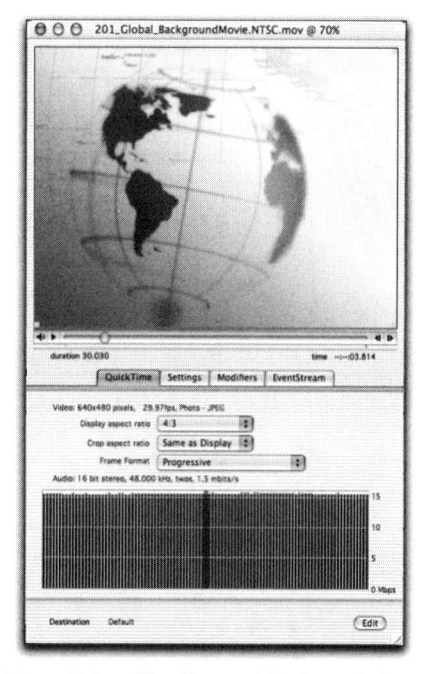

Figure B-4 — The Cleaner 6 Project window.

473

MegaPEG.X Pro HD and MegaPEG.X Pro HD-QT from Digigami

Digigami have been quietly plugging away at really cool tools for compressionists, and their Encoder line featuring, MegaPEG.X Pro SD and HD is worth a look. See Figure B-5. A free tryout version is downloadable from the Digigami website: **http://www.digigami.com/download/**.

MegaPEG.X Pro SD and HD Features:

New and Improved in Digigami MegaPEG.X Version 3.0:

- Up to four times faster than version 2.0 (which was already one of the fastest Mac encoders around).

- Unbelievable low bitrate for S-VCD encoding from feature films. 1.0 Mbs avg (typical) for drama, 1.5 Mbs avg (typical) for action.

- New for HD: Create VBR High-Definition MPEG-1 files for playback on Windows and Macintosh computers—no extra downloads required for HD movies.

- New for HD: Convert HDTV broadcasts, HDV Movies, and DVD discs into DVD, S-VCD, and VideoCD.

- New for HD: De-multiplex DTV transport streams back into .m2v and audio elementary streams.

- New optical-flow and motion-detection for faster encodes and better quality at low bitrates.

- New built-in "MPEG Reference Decoder" reads MPEG-1 (.m1v) and MPEG-2 (.m2v)

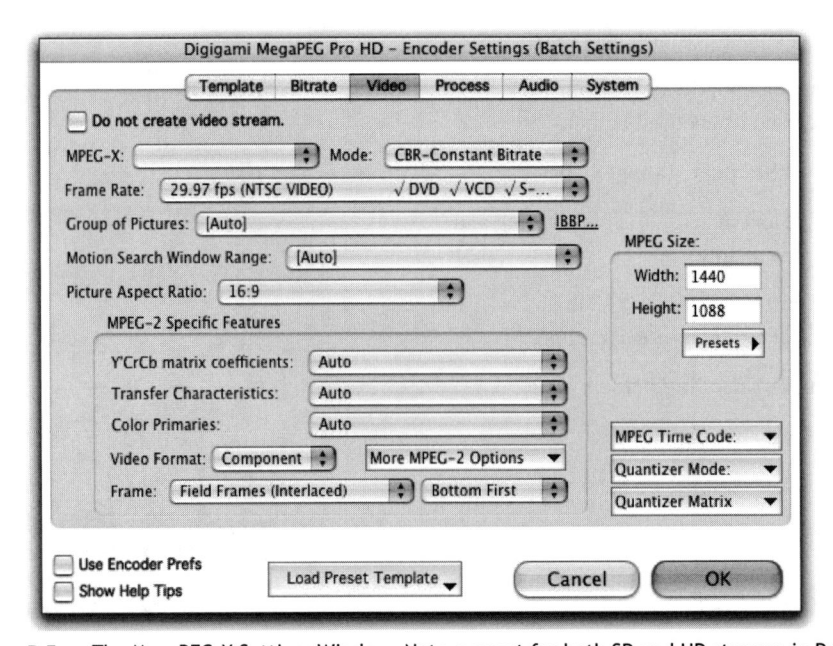

Figure B-5 — The MegaPEG-X Settings Window. Note support for both SD and HD streams in Pro HD

Alternative Encoders

for transcoding HDV/DVD/DTV material to DVD, S-VCD, and VideoCD.

- New built-in "Movie Player Reader" provides transcoding from Macromedia Flash (.swf) and other QuickTime-compatible formats.

- New built-in "Movie Codec Reader" provides high-quality transcoding from QuickTime-native codecs.

- New Digigami MPEG-1 and MPEG-2 built-in "videophile" decoder for ultra high-quality transcoding.

- New "Mastering" quality takes five times longer, but with unmatched 32-bit floating point picture quality.

- Improved frame size and noise-reduction for best quality and low bitrate.

- Improved "Best" quality features both improved speed and picture quality.

- Improved MPEG movie analysis features multi-threaded, multi-document user experience.

- Improved MPEG movie analysis (Pause & Review) serves as "reference playback" for quality assessment.

- Improved redesign of Preset Encoding Templates—almost no tweaking required for picture-perfect results.

- Improved Quantization Matrix presets outperform other encoders, especially for "progressive" encoding.

Features in All versions of MegaPEG.X (Introduced in 2.0):

- MPEG-2 movies compatibile with major DVD authoring software including DVD Studio Pro.

- MPEG-1 movies compatibile with both Windows (MediaPlayer), Macintosh (QuickTime), and Linux (VideoLAN).

- Every conceivable combination of encoding algorithm and precision, wrapped in an easy-to-use interface.

- Pause & Review feature allows you to pause the encoder at any time and review the results "in great detail."

- Analyze compression and bitrate profiles of Hollywood DVD discs with built-in analysis tool (bitrate, stuffing, and quant).

- Over 40 preset encoding templates, which are easily customized with sliders. Pick something close and refine to suit your needs.

- Includes a compliant DVD template at each 1 Mbits/s bitrate notch from 1 Mbits/s to 7 Mbits/s—all completely reconfigurable.

- Supports WhiteBook VideoCD, Super-VideoCD, DVD, and general MPEG-1 and MPEG-2 movie compression with MPEG-1 Layer 2 audio.

- Saves Apple DVD Studio Pro (DVDSP) compatible .AIFF audio track with the in and out points for the video selection.

- Exports both MPEG-1 and MPEG-2 video files to DVD Studio Pro.

- Reads most kinds of movie files readable by QuickTime, including Final Cut Pro "reference" movies.

- De-multiplex iDVD VOB files back into .m2v and .aiff files for use with DVD Studio Pro.

- De-multiplex VOB files back into .m2v and .ac3 files for use with DVD Studio Pro.

DVD Studio Pro 4

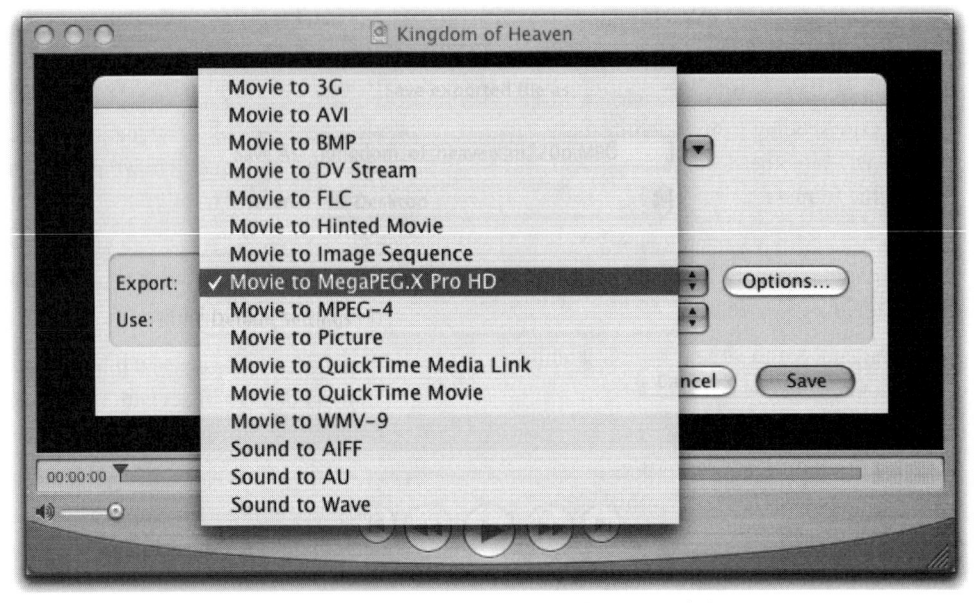

Figure B-6 — The MegaPEG.X Pro HD QT export dialog.

- Creates MPEG-1 files playable on both QuickTime and WindowsMedia—no extra downloads required for www movies.

Also available as a QuickTime export component, if you are feeling the lack of the QT component (see Fig. B-6).

One other product that really bears mention is Digigami's incredible stream analysis tool called Mpressionist.

This program will tell you things about your compressed streams that you never knew existed, and it does it so easily you will wonder why it took this long for someone to create this! (See Fig. B-7.)

Main Concept

The Main Concept encoder is somewhat new to the Mac. It is a stand-alone application with the standard complement of expected features for video encoding (see Figs. B-8 and B-9). It has been a stalwart on the PC side for a while—in fact, you will find the Main Concept encoder inside of Adobe Encore DVD, performing the MPEG encoding function for that application. My sources also tell me that it is the encoding engine that drives the Sorenson Squeeze Suite's MPEG capabilities.

Features of the Main Concept Encoder:

- MPEG-1 and MPEG-2 export
- Ultra-high quality
- Ultra-fast encoding; faster than real-time in many cases (depending on system specs)
- Two-pass encoding
- Enhanced SVCD quality
- Smart rendering

476

Alternative Encoders

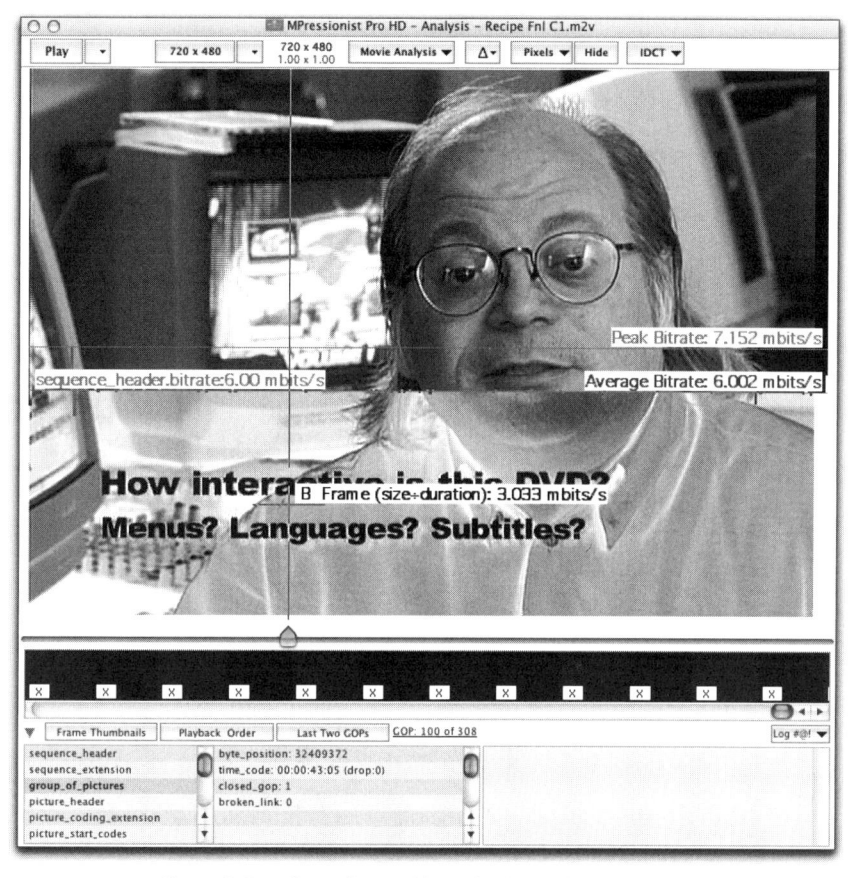

Figure B-7 — One of many Mpressionist analysis displays.

- Powerful batch encoding and file splitting options
- Enhanced scaling and deinterlacing for improved quality
- Multi-processor support
- PowerPC optimizations
- 4:2:2 support
- Versatile QuickTime import module
- Cropping, scaling, and image flip

- Enhanced preview—watch the video file before export
- Easy, multi-layered interface
- Presets for DVD, Video CD, and Super Video CD
- Support for custom presets
- Convenient hotkeys for many functions
- Drag-and-drop—just drag video or audio files into the main window

477

DVD Studio Pro 4

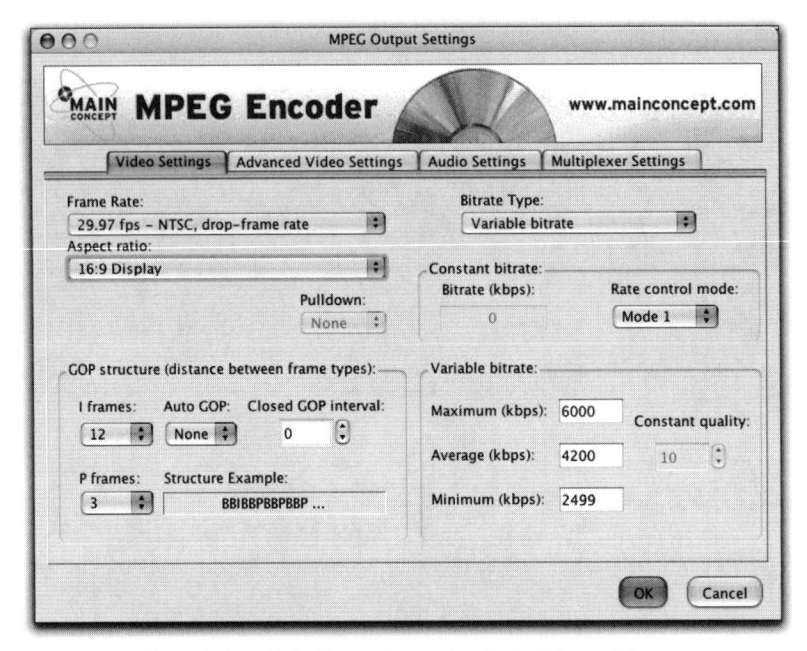

Figure B-8 — Main Concept encoder Basic Video settings.

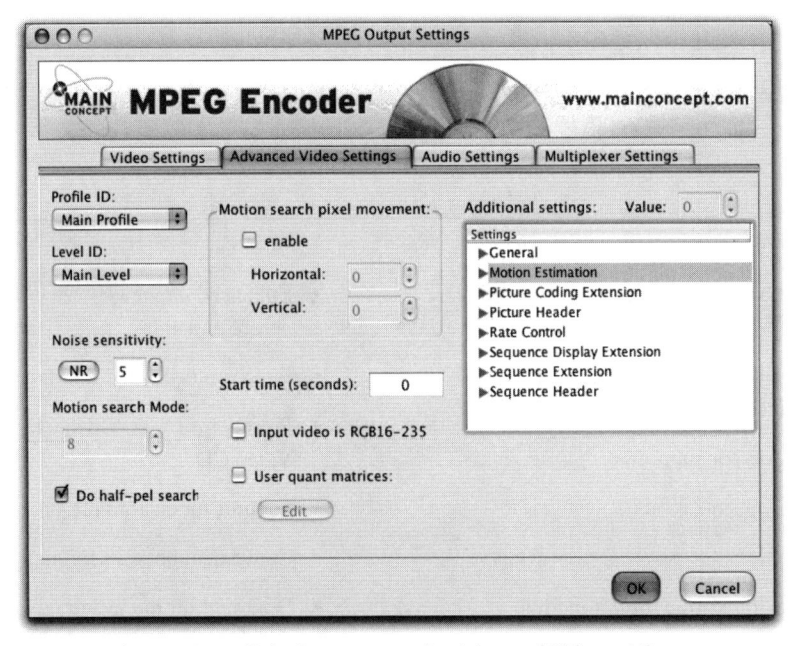

Figure B-9 — Main Concept encoder Advanced Video settings.

Alternative Encoders

Compression Master from Popwire

From Popwire in Sweden comes *Compression Master* for Macintosh OS X (see Fig B-10). Now at version 3.2, it features an impressive list of capabilities:

Features:

- Batch: processes in the background; save multiple batches; prioritize encoding and add files and settings to the batch while running; job editing; create MBR jobs; set CPU priority and memory usage; create bookmarks to source folders; copy and paste settings onto multiple sources.
- HD support
- Multiprocessor support
- Native YCbCr (YUV) and RGB processing
- Settings manager
- Metadata support
- More than 80 professional setting templates
- Automatic and configurable naming convention of output files
- Support for pass-through of tracks (copy)
- Automatic optimizing of color conversion
- Hinting (streaming) according to IETF, ISMA, and 3GPP
- Drag-and-drop support
- Automatic input format detection: video codec, audio codec, color sampling, frame rate, pixel aspect, audio sampling 2-pass encoding
- Support for Time Code
- Preview: A/B comparison, auto-update

Filter:

- Frame rate conversion: 3:2 pulldown, 2:3 pulldown, intelecine, NTSC to/from PAL
- Deinterlace: blending, adaptive, edge detection interpolation, chroma
- Resize: scale fields independently, crop, save crop area, pixel aspect ratio detection
- In and out points
- Time Code: read TC track, burn TC in picture, create TC track
- Noise reduction: median, average, temporal, support for multiple runs
- Black and White Restoration
- Contrast
- Smoothing
- Gamma correction
- HSV: Hue, Saturation, Brightness (value)
- RGB filter
- Audio: balance, channels, high/low pass, fade in and out
- Audio volume: normalize, manual adjust
- Audio equalizer
- Sharpening
- Watermark: PiP (Picture in Picture), apply a movie as watermark, GIF,
- TIFF, BMP, JPEG, alpha channel, opacity, positioning, scaling and cropping of watermark source
- Fade: video in/out, audio in/out

479

DVD Studio Pro 4

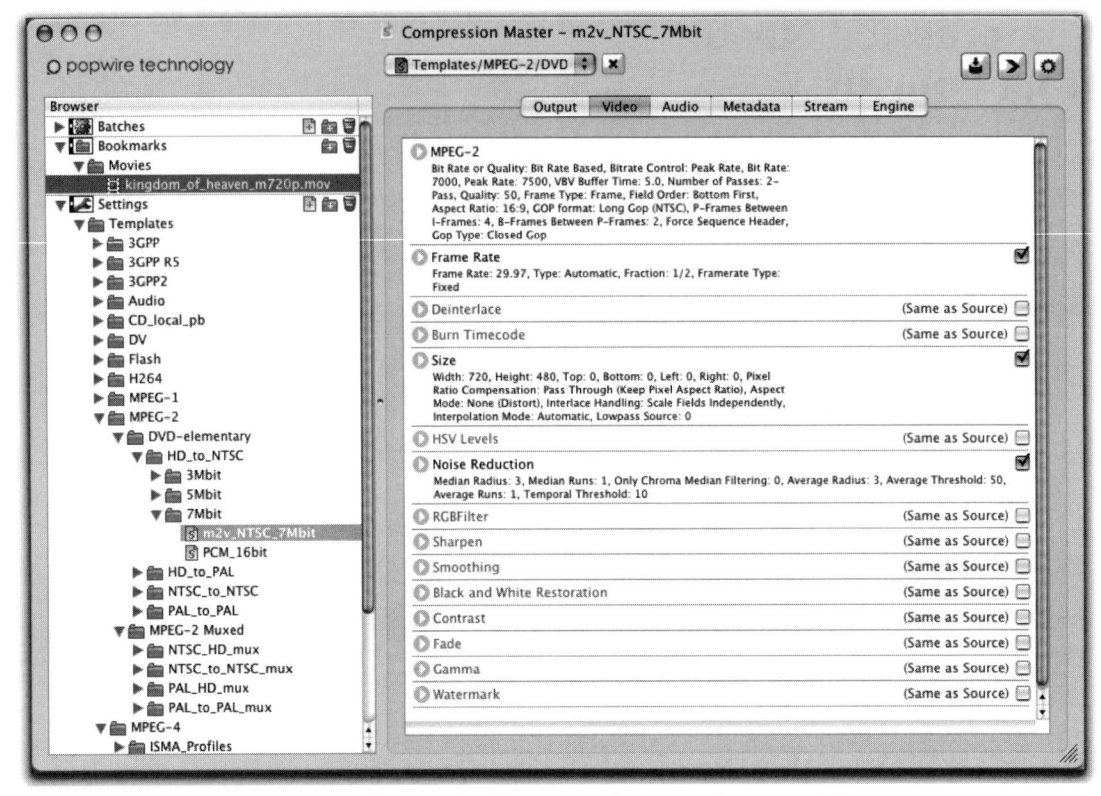

Figure B-10 — Compression Master encoder presets/settings window.

Look on the DVD for More Information

We've included information for as many of these encoder products as possible, including tryouts or demos when they are available. Enjoy!

Appendix C

DVD Media Reference Guide

DVD Studio Pro 4

As you probably already know, there is a bewildering array of DVD media floating around out there, making it difficult if not downright impossible to know WHICH disc is appropriate for you to choose, and what drive will burn them! Hopefully, this appendix will help sort this out for you.

DVD-R (Authoring Media)—Write-Once, not eraseable:

- 3.95 GB or 4.7 GB capacity—The original DVD-Recordable format
- Only burnable on the Pioneer DVR-S101 or S201
- Media are still around, but getting harder to find
- 3.95 GB format gives the best all-around DVD-R compatibility, if you can afford the discs and burner

DVD-R (General Media)—Write-Once, not eraseable:

- 4.7 GB capacity—The consumer DVD-Recordable format
- Burnable on any General Media DVD-R burner—There are dozens!
- Can also be burned on dual-format (±R) drives like Pioneer DVR-106, 107, 108, 109
- Media are still around, but slower speed media (1x, 2x) are getting harder to find as drive speeds climb to 16x
- In general, compatibility is good, depending on the media
- Interestingly, Mini-DVD-R formats exist: 3" DVD-1 disc
- Caution—Media smaller than the standard 5" discs (especially credit card sized discs) are known to cause problems in slot loading DVD players. In short, they go in, but many don't come out.

DVD-RW (re-writeable media)—Eraseable, re-recordable:

- 4.7 GB capacity—The original DVD-Re-Recordable format
- Usable on many DVD-R General Media drives, dual-mode drives
- Media are still around, not hard to find, often more expensive than DVD-R media, and often not quite as compatible as –R, but can be read in many common DVD drives.

DVD-R DL (Double Layer)—Write-Once, not eraseable:

- 8.54 GB capacity—New format like DVD+R DL media described below.
- Not as mature as the DVD+R DL format, but gaining popularity
- Allows almost twice the capacity of standard DVD-R discs
- Great as an initial test disc for DVD-9 projects
- Burnable on DVD recorders that support DVD-R DL, or ±R DL
- DVD Studio Pro 4 supports this media, but only on Mac systems that have double-layer support enabled for the supplied DVD Burner ("SuperDrive"). Check the System Profiler.
- Compatibility is good, depending on which media you use.

DVD Media Reference Guide

DVD+R (Recordable media)—Write-Once, not eraseable:

- Originally available only in 4.7 GB capacity
- Is the alternative DVD-Recordable format
- Burnable only on a DVD+R burner or dual-format (±R) drive
- Drive speeds have climbed quickly from 2x, now at 16x
- Compatibility is good, depending on which media are used and the length of the recording.
- Not recommended for use where the overall run-time of the video is less than 5 minutes.

DVD+RW (re-writeable media)—Eraseable, re-recordable:

- 4.7 GB capacity—The alternative DVD-Re-Recordable format;
- Usable on many DVD+R drives, and dual-format (±RW) drives
- Media are available, not hard to find, about same price as -RW
- Compatibility not quite as good as +R, but can be read in many common DVD drives.

DVD+R DL (Double Layer)—Write-Once, not eraseable:

- 8.54 GB capacity—Two layers one side—the hot new format
- Allows almost twice the capacity of standard DVD+R discs
- Great as an initial test disc for DVD-9 projects
- Burnable only on DVD recorders that support DVD+R DL
- DVD Studio Pro 4 supports this media, but only on Mac systems that have double-layer support enabled for the supplied DVD Burner ("SuperDrive"). Check the System Profiler.
- Other programs ("Toast") require software driver updates.
- In general, compatibility is good, depending on the media.

DVD-RAM:

- Exists in both Mini (3"–8cm) and fullsize (5"–12 cm) formats
- DVD-RAM is recordable, eraseable, rewriteable
- Capacities of 1–9.4 GB, depending on # of sides
- DVD-RAM is also featured in some DVD settop recorders, and some DVD-RAM based video cameras.
- DVD-RAM cartridges may have a protective carrier/cartridge
- DVD-RAM Cartridges are available in three formats, four sizes:
  - 2.6 GB, Single-Sided, 5.2 GB Double-Sided
  - .7 GB, Single-Sided, 9.4 GB Double-Sided
  - Type I Cartridges—media not removable; (5.2 DS)
  - Type II Cartridges removable single-sided media; (2.6, 4.7 GB)

- Type IV Cartridges do provide removable media (9.4 GB)

DLT (Digital Linear Tape):

Available in sizes from 10–20 GB native capacity, DLT tapes are a streaming tape format used almost exclusively for delivery of DDP-formatted DVD Master tapes for replication. Most replicators and many of the leading distributors insist on delivery of disc images on DLT.

DLT Type III tape has a 10 GB Native capacity

DLT Type III XT tape has a 15 GB Native capacity

DLT Type IV tape has a 20 GB Native capacity

(Note—Make sure you have compression turned OFF if delivering a DVD Master.)

1. Type III tape is required to deliver a DVD-5 Master
2. Type III tapes are required to deliver a DVD-9 Master, 1 per layer

(Note—You CANNOT deliver 2 layers on a single high capacity tape.)

For more information about DLT drives, visit: **http://www.recipe4dvd.com/DLT.html**.

Appendix D

DVD Glossary

4x3 (see also Aspect Ratio) The aspect ratio of Standard Definition Television Signals in the NTSC format 4 units wide, by 3 units high. This may also be expressed as 1.33:1.

16x9 (see also Aspect Ratio) The aspect ratio of Widescreen (High Definition) Television Signals: a picture that is 16 units wide, by 9 units high. This may also be expressed as 1.78:1.

5.1 Channel Surround (see also Dolby digital or DTS) A six-channel theatrical sound format using one channel each for left, center, right, left surround, right surround, and subwoofer (Low Frequency Effects).

525/60 An abbreviation for NTSC video derived from the number of scan lines (525) and the reference scanning frequency (60 Hz).

625/50 An abbreviation for PAL video derived from the number of scan lines (625) and the reference scanning frequency (50 Hz).

A

AAC (Advanced Audio Codec) An audio-encoding standard for MPEG-2 that is not backward-compatible with MPEG-1 audio.

AC-3 (See also Dolby Digital) Properly, the original name for a Dolby digital audio encoding algorithm used for many different channel formats from Mono to 5.1. Commonly misreferenced as Dolby 5.1, this format also includes a six-channel theatrical sound format (using one channel each for front left, center, and right, left surround, right surround, and a subwoofer), AC-3 was subsequently adapted for home use and has now become one of the most common sound formats for DVD. It now goes by its proper name of Dolby Digital. AC-3 is the successor to Dolby's AC-1 and AC-2 audio-coding techniques.

Amaray Case The "standard" packaging for a DVD, as designated by the VSDA (Video Software Dealers Association).

Amplifier A device that increases the sound level of an audio signal so it can be played through speakers.

Analog Digital video or audio information is stored in numbers; analog uses a continuously variable signal of any kind.

Analog Controls Controls that move in a smooth progression, as opposed to the discrete steps that characterize digital. Unlike digital controls, analog controls must be reset every time.

Anamorphic In DVD, used to indicate widescreen video content (typically 16 × 9) that has been squeezed into a 4 × 3 visual space for encoding. In film, an anamorphic movie is a widescreen movie in which the image is squeezed to fit on a standard 35 mm film negative.

Anamorphic Mastering Anamorphically mastered titles can play widescreen (16 × 9) or standard TV (4 × 3) formats with black letterbox bars on the top and bottom of the screen. The instructions about which version to use with which television set are imbedded in the disc's MPEG data stream. For DVD, the image can be digitally unsqueezed.

Angle A specific view of a scene, sometimes recorded from a certain camera position, sometimes just a different edited cut. DVD players allow viewers to choose from as many as nine different camera angles on titles offering multiple angles, but few titles really take advantage of this standard DVD player feature.

Angle Menu An optional DVD Menu used to select the video angle number.

Artifact A flaw in the MPEG encoded video image that mars the picture quality, and was not present in the original material.

Asset list A complete list of the various audio, video, graphic, text, and subpicture elements included in a DVD title.

DVD Glossary

Assets   The audio, video, graphic, text, and subpicture elements included in a DVD title.

Aspect Ratio   The ratio between the width and height of a screen. The aspect ratio for standard TV is 4x3; widescreen TV is typically 16 × 9. This is sometimes also show, using the colon, as in 4:3, 16:9.

Authoring   The generally-used name for preparing a DVD-Video disc for replication, or for DVD-R burning. The physical process of assembling assets and navigation into a playable DVD.

Audio Menu   An optional DVD Menu used to select the audio stream.

Audio Stream Number   The consecutive numbers assigned to the audio streams for a title in a VTS. These numbers range from 0 to 7 in the order described in the VTS manager area. For menus, the number of audio streams is limited to 0 or 1.

Audio_TS   The UDF file name used for a DVD-Audio directory on disc volume. On SD DVD-Video-only discs, this directory is empty.

B

B-Frame (or B-Picture)   One of three picture types used in MPEG video. An MPEG "Bi-directionally Predictive" frame. Compressed both spatially and temporally, and with some regard for motion estimation, a B frame uses information from its neighboring I- and P-frames to properly reconstruct the video frame at that instant. An MPEG Chapter may not be placed on this frame.

Bandwidth   The amount of information that can be carried by a circuit or a signal. The larger the bandwidth, the sharper and more complex the image can be. In digital technologies, bandwidth is referred to as "data rate."

Bit Rate   The rate at which digital information is presented or encoded. Since there is room for only so many bits on a DVD, the bit rate can vary. The rate is kept lower for simpler sequences so that extra space will be available for more complex ones.

Bitmap   A format for storing still graphic images. In DVDs, bitmaps are used for menus and subpicture overlays.

Blocking   A visible artifact of digital compression in which blocky rectangular areas appear in the picture due to an encode rate insufficient to properly capture the picture details.

Burst Cutting Area   A circular section near the center of a DVD disc where ID codes and manufacturing information can be inscribed in bar code format.

Button   This is a rectangular area in the subpicture display area highlighted by the Highlight Information (HLI) that is used to define the active area on a menu associated with a specific action.

Button Number   The consecutive numbers assigned to every button on a menu. These numbers range from 1 to 36.

Buss   A pathway over which data travels.

Button Highlight Marker   A Cell Marker that is used to turn on or turn off the Highlights for Buttons over Video.

C

Caption   Text that appears on screen during a program, such as foreign language subtitles, commentary, program notes, etc.

CBR   Constant Bit Rate encoding, which digitizes and compresses a video source using a single, unvarying encode rate, as opposed to VBR (see Variable Bit rate).

DVD Studio Pro 4

Cell A contiguous length of video within a DVD presentation object (i.e. a "Track") having a specific beginning marker and a specific ending marker. The video in between the markers is defined as the "cell". A Track may have one cell, or many cells.

Cell Command A DVD command inserted into the beginning marker of a DVD video cell, which defines the navigation action taken when that cell has completed its playback. This command does not execute until the playback has reached the cell's end marker. In DVD Studio Pro, these are the commands programmed into Track Markers, and cause interruptions in the track's playback at the next marker.

Cell Marker A Marker that is not defined as a Chapter Marker or a Program Marker.

CGMS (Copy Guard Management System) A method of preventing copies or controlling the number of sequential copies allowed. CGMS/A is added to an analog signal, such as line 21 of NTSC. CGMS/D is added to a digital signal, such as IEEE 1394.

Channel One section of an audio track, usually carrying the sound for one speaker.

Chapter Properly known as a "Part of Title," (PTT) a point on a DVD that the viewer can instantly jump to, like a track on a CD. In DVD Studio Pro, this is a Marker within a Track (and also the division point between slides in a Slideshow).

Chapter Marker Chapter Markers define Chapter locations, which may be navigated to from a Menu, or may be skipped to by the DVD viewer while that Track is playing. In DVD Studio Pro, these are the highest priority of all of the various Markers.

Chroma The color part of a video signal, as well as a description of how intense the color is in a given frame.

Closed Caption A caption that is only visible when specifically decoded (and requiring a decoder of some kind), as opposed to "open captions," which are a permanent part of the picture image. DVD Studio Pro can process .CC files to add Closed Captions to your project.

CMF (Cutting Master Format) A protocol, similar to DDP, that describes data that will be recorded onto an optical disc. Developed by Pioneer, CMF can now be recorded directly to DVD-R for Authoring (4.7 GB) media using the Pioneer DVR-S201 recorder, but is not supported by the DVD-R General Media Format (i.e., the Superdrive). This allows a single DVD-R (A) disc to serve as both a check disc for testing playback, and as the replication master for saving time and ensuring accuracy of the data. DVD-R Cutting Master Format is written to the lead-in area, which has been specially reserved for mastering applications and is defined in version 2.0 of the DVD-R for Authoring specification. Special software and some handling at a press facility are required to use the DVD-R Cutting Master. CMF may also be used to prepare a DLT master tape, but DDP is more commonly used, and more generally acceptable at replicators.

Coaxial Cable A two-conductor cable with a hot center wire and a neutral shield wire running along the same axis. Used to link a television to a DVD player, route digital audio signals, or to provide cable TV access.

CODEC Short for Compression/Decompression, a codec is a set of rules that govern the encoding (or compression) and decoding (or decompression) of media.

Color Balance The balance of different colors on a video monitor. Improperly balanced, the image will appear weighted toward one hue or another. The right balance can be achieved by using color reference bars.

Colorist The individual responsible for the look of a title when it is mastered to video, including adjusting not only the color and contrast, but also occa-

488

DVD Glossary

sionally the framing of the image (sometimes called a telecine colorist).

Combo Drive A DVD-ROM drive capable of reading and writing CD-R and CD-RW media. May also refer to a DVD-R, DVD-RW, or DVD+RW drive with the same capability.

Component Video A video system in which the video signal is divided into three parts, and transferred through three individual wires, so that the color signals don't interfere with one another. Many television sets are now equipped to receive a component video signal, which offers higher quality pictures than composite video.

Composite Video A video system in which all three colors (R, G, & B) and brightness ("Luminance") are transferred over one wire. Single-cable outputs from both NTSC (the United States television standard) and PAL (the European television standard) DVD players and VCRs are typical examples of composite video, which offers a lower quality picture than component video. This is not a good format to use for encoding MPEG (unless you have nothing else). Composite video cables are typically color-coded Yellow.

Compression Any method of reducing the amount of space needed to record or transmit information. In DVDs, video is compressed using a process called variable bit rate encoding, which allots a changing number of bits to enhance resolution in a given scene. Scenes with lots of light or little action require less hits than dark scenes or those with lots of action.

CPRM (Content Protection for Recordable Media) Copy protection for writable DVD formats.

CSS "Content Scramble System" One of the two types of encryption available in DVD, CSS scrambles the DVD data in such a manner as to require a decrypting key for proper playback. The decrypting keys are inserted during replication, and is typically not copyable if the DVD contents are digitally extracted without special software. CSS-encrypted masters are typically delivered on DLT, not DVD-R.

D

DDP (Disc Description Protocol) Acronym for the protocol that describes collections of data that are recorded onto a CD or DVD. DDP allows for automated transfer of data from data publishers to disc manufacturers. DDP is the de-facto standard in the DVD industry for delivering disc image data to the replication plant for manufacturing. Originally developed by Doug Carson & Associates, Inc. (DCA) for CD manufacturing, DDP provides descriptive information about the disc to be manufactured, such as the physical size, number of sides, etc. DVD Studio Pro utilizes both version 2.0 and 2.1 of this specification, information about which can be requested via the DCA website at www.dcainc.com.

Decode To take compressed, encoded sources and turn them into a data stream that can be played back by another device.

Decoder Hardware or software that can render an encoded file or bitstream playable. DVD Playback always requires an MPEG decoder of some kind.

Digital Compression Any scheme, method, or algorithm that reduces the storage space or data bandwidth required to record or transmit information. MPEG and JPEG are both digital compression schemes, as are Dolby Digital AC-3, MP3, AAC, MPEG-4, and other compression schemes.

DLT (Digital Linear Tape) A digital archive standard using half-inch tapes. Commonly used for submitting a premastered DVD disc image to a replication service. Drives are almost universally manufactured by Quantum, but marketed under a variety of trade names.

489

DTS (Digital Theatre Systems) Another well-established format in the 5.1 channel surround sound market besides Dolby Digital. DTS is a Stereo or 5.1 channel surround format available for DVD or Audio CD. This format is similar to Dolby Digital, except that it uses less compression, and by consequence, a higher encoding rate. These features theoretically provide higher quality sound. Many DVD players do not provide integral DTS audio decoding, hence they require an external DTS decoder. DTS audio streams are supported in DVD Studio Pro 3 and 4, but not 2.

DSP Digital Signal Processor (or Processing) DSP is usually found on surround sound receivers, and in computerized DAWs (Digital Audio Workstations). DSP uses digital algorithms to alter the way sound is heard. For example, setting the DSP to "Hall" would create an effect that sounded like you were in a Concert Hall. DSP can also be used to modify the level, frequency response, and many other characteristics of a sound file.

Direct Satellite System (DSS) DSS is a method used in broadcasting television. A small 18" dish can receive hundreds of channels with high quality digital picture and sound.

Direct View Television Unlike Projection systems, a TV in which the picture is directly scanned on a CRT tube. This is the most typical television found in American homes. These TVs can vary in size from less than 12 to 40 inches or more.

Discrete Surround Sound Sound that has an individual channel for each speaker, with each channel handled separately.

Dolby Digital (a.k.a. Dolby 5.1) (see also AC-3) A six-channel digital surround sound system, using three speakers in the front (left, right, and center), split surrounds (left and right surround) and a subwoofer for bass frequencies. The six-channel format was originally created by Dolby Labs for theatrical soundtracks and subsequently adapted for home use.

Originally known as AC-3, Dolby Digital is the most prevalent 5.1 channel surround sound format available on DVD, and (with PCM) is one of the two worldwide DVD sound standards. Dolby Digital can encode a 5.1 channel surround sound program into a single bitstream, and then decompress this stream back into five discrete channels of sound, and a ".1" subwoofer (LFE) channel. This recreates the original 5.1 soundtrack.

Dolby Pro Logic An audio coding system that allows four channels of sound to be encoded into two channels (stereo) and then decoded back into four channels (left, center, right, and surround). The center channel enables Dolby Pro Logic to reproduce theatrical soundtracks, especially Dialog, more fully than traditional Left/Right Stereo, while the Mono surround gives the "you are there" feeling. Dolby Pro Logic does not provide as rich a listening experience as Dolby Digital in the 5.1 mode.

Dolby Surround A encode/decode process created by Dolby that encodes four channels of audio (left, center, right, and surround) onto two audio channels, which are then decoded into three channels (left, right, and surround). Not as elaborate a sound system as either Dolby Pro Logic or Dolby Digital.

DTS (Digital Theater Systems) An audio format like Dolby. DTS has its own six-channel (5.1) encode/decode system for movie theaters and also sells a line of DVD videos and DVD-Audio titles that use its compression technique. Audio track DTS, however, requires almost four times as much space on a DVD disc than a Dolby track. DTS will only play in 5.1-channel sound on a system that has a DTS decoder.

Dual-Layer Breakpoint Marker A Marker used to designate the Chapter Marker at which a dual-layer DVD may shift playback from Layer 0 to Layer 1.

DVD Digital Video DiscDigital Versatile Disc DVD is the size of a standard CD, but it can hold

DVD Glossary

more information (4.7 gigabytes compared to 650 megabytes). DVD's primary features include: Dolby Digital soundtracks, multiple aspect ratios, multiple spoken languages with multiple language subtitles, multiple camera angles, parental lockout, and additional features that are at the discretion of the studio producing the disc. In addition, DVD is the highest resolution, consumer format available, with 500 lines of resolution. In comparison, VHS has 240, Super VHS has 400, and LD has 425 lines. For more information, check out the DVD FAQ by Jim Taylor.

DVD (see also Mini-DVD) Originally *Digital Video Disc, then Digital Versatile Disc*, DVD is the acronym for the 8 or 12 cm digital disc format providing from 1 to about 17 billion bytes of data storage—commonly misused as a catch-all acronym for Digital Home Video, DVD is also a general format for data storage and delivery that includes several different subformats: see also *DVD-ROM, DVD-Audio, DVD-Video, DVD-R, DVD-RW, DVD-RAM*.

DVD-1, 2, 3, 4 (also Mini-DVD) There is a standard specified for the creation of Mini-DVD (8 cm) discs, with data structures identical to its big brother, but reduced capacities (1.46 GB SS/SL, 2.92 GB SS/DL). As of late 2000, the first Mini-DVD discs are finally making their way into some replication plants as a viable format.

DVD-5 Single-Sided, Single-Layer DVD disc, which can hold 4.7 billion bytes of data.

DVD-9 Single-Sided, Dual-Layer DVD disc, which can hold 8.5 billion bytes of data.

DVD-10 Dual-Sided, Single-Layer DVD disc, which can hold 9.4 billion bytes of data.

DVD-14 Dual-Sided DVD with one Single-Layer side, and one Dual-Layer side, which can hold around 14 billion bytes of data.

DVD-18 Dual-Sided, Dual-Layer DVD disc, which can hold around 17 billion bytes of data.

DVD-Audio Again using the DVD-ROM format for delivery, DVD-Audio is a specialized form of DVD that provides better-than-CD audio quality for music releases, while adding graphics and even some video capability to the disc. Previous forms of this (CD-I, CD-G) were built to enhance 44.1 kHz PCM Audio, and used the Compact Disc format. DVD-Audio allows sampling rates of up to 192 kHz (!) and data sample widths of up to 24 bits—the increased sampling rates provide better frequency response (fidelity) while the larger bit widths provide better dynamic range.

DVD-ROM In its most basic incarnation, DVD-ROM in the underlying technology that allows 4.7 Billion Bytes of data to be stored and retrieved from a 12 cm data disc. DVD-ROM can be used to distribute computer-readable data, which are cross-platform compatible (Mac/PC). No, this doesn't mean your Mac will suddenly be able to run a Windows program, but it DOES mean that both platforms (and all others with DVD-ROM drives) will be able to read compatible data from the same DVD disc. The cross-platform magic is accomplished by the Micro UDF file format.

DVD-Video Built on the DVD-ROM disc (above) as a delivery vehicle, DVD-Video is a specialized form of DVD that can provide multiple selectable broadcast-quality video streams, digital audio streams, and subpicture (graphics) streams. Originally designed for Home Entertainment (Movies on DVD), the DVD Video format is being increasingly used for Corporate and Industrial Communications, Kiosks, point-of-interest displays, and, refreshingly, education and training!

DVD-R (recordable) A form of DVD that allows the creation of a DVD-5 equivalent by a laser-equipped writing device ("DVD burner"). Typically used in conjunction with a DVD formatting program (Adaptec TOAST on Macs, Prassi DVD-Rep (or many similar) on Windows), and utilizing a specific

491

DVD Studio Pro 4

type of hardware device (Pioneer DVRS-201, for example). This combination of hardware and software will allow a user to output a unique custom-built DVD-ROM, or DVD-Video disc. This is a "write-once" medium, and the discs cannot be erased or altered.

DVD-RAM (random-access read/write/erase) A form of DVD that allows the creation of a DVD-5 or -10 equivalent by a laser-equipped writing device. Using a different technology than DVD-RW, DVD-RAM nonetheless provides a rewritable (reusable) format where the media (DVD-RAM discs) can be reused. First Generation DVD-RAM only allowed the creation of a 2.6 GB disc, 5.2 GB on two sides.

DVD-RW (rewritable) A form of DVD that allows the creation of a DVD-5 or -10 equivalent by a laser-equipped writing device (a "DVD-RW burner")—either single-sided or double-sided, the DVD-RW is analogous to CD-RW in that both formats allow re-recording of the medium once it has been recorded. This re-recordable disc format is one of two being used for Digital Home Video recording on Disc (the new DVD-VR format).

DVD Laptop A portable PC or Mac equipped with a DVD drive of some kind, typically DVD-ROM for DVD Video playback. A marvelous portable entertainment unit that is also capable of balancing your checkbook, or reading your email if you get bored on a long plane flight.

DVD Portable player ("Handheld DVD") A miniature DVD player, which typically contains both a DVD player and an LCD, display screen of diminutive size. Designed for battery operation while traveling or away from AC power.

DVD Settop Player ("Settop box") A stand-alone DVD player designed to use a Television set or Video Monitor as its output device.

DVD Remote Control (a.k.a. Remote) The somewhat standardized independent control unit that allows a DVD user to control navigation, interactivity, and the choice of video, audio, and subtitle presentation options while a DVD title is playing. Note that on a computer equipped with DVD playback, the mouse and keyboard will typically emulate the remote control menu selection functions using "rollovers."

DVD Remote Control Angle Select key The key on the DVD remote that allows the DVD player to change the video stream currently being displayed. Takes its name from the misnomer "Video Angles," which characterize the individual DVD MPEG streams that are available for playback in a DVD project. Multi-Angle programming is becoming more available, as DVD authors and producers utilize this standard DVD function more often. Adult DVDs especially have popularized this feature.

DVD Remote Control Audio Select key The key on the DVD remote that allows the DVD player to change the audio stream that is being decoded/played. Frequently used to change between multiple spoken language tracks, alternate audio tracks are more frequently being provided in contemporary DVDs. Some examples of this are: commentary tracks, music-only tracks, Descriptive Video tracks, and even Klingon language tracks!

DVD Remote Control Return key The RETURN is an author-programmable navigation aid, allowing an additional method of interactive navigation to be programmed into a particular DVD title. Normally, the RETURN function is not active on the DVD remote unless the DVD Author has specifically programmed it.

DVD Remote Control Subtitle Select key The key on the DVD remote that allows the DVD player to change the subpicture stream currently being displayed. While widely used to display its namesake, subtitles, DVD's inherent graphic overlay capability is also being exploited for more exotic uses each

DVD Glossary

month. See Columbia/Tri-Star's Ghostbusters, Special Edition DVD for several such innovative uses.

DVD Remote Control Menu Navigation Arrows
The up, down, left right arrow keys used to navigate around an interactive DVD menu. The arrows follow the pre-programmed navigation paths in the menu to allow access to all menu selections (see Easter Eggs).

DVD Remote Control Enter key The ENTER key is used to activate a DVD menu selection, after navigating to it.

DVD Remote Control Transport Controls

Play—initiate playback

Pause—temporarily stop playback

Stop—stop playback; clear GPRMs

Skip Chapter Forward (Next)

Skip Chapter Backwards (Previous)

Fast Fwd

Fast Rewind

DVDA The DVD Association—a worldwide group of interested DVD Authors, Producers, and enthusiasts. **http://www.dvda.org**

DVD@cess Marker A marker used to trigger a DVD@ccess event.

E

Element Element is the generic term for what used to be known as an Item in DVD Studio Pro 1. Elements in DVD Studio Pro are the containers used to contain Assets—Tracks, Menus, Slideshows, and Scripts.

Encode To convert a signal from one form to another, such as analog to digital. In DVD, Encode is most used to refer to the conversion of video into MPEG.

Encoder Any device or program that compresses a signal for audio or video and outputs a digital signal. Encoders can be either software- or hardware-based. Encoders may compress either analog video signals from videotape, or captured digital video files.

F

Field Effectively, one-half of a Fame. *Interlaced video* displays each frame that appears on screen in two passes, or "fields." The first pass lays down half of the scan lines that make up the on-screen picture—the second pass paints in the remaining ones. These two fields are interlaced to compose one frame. Which set of lines are scanned first is known as "Field Dominance."

File System How data are organized in digital playback systems. In DVD, the UDF file system (Universal Digital Format) forms the foundation of the DVD disc.

FireWire A digital video interface standard developed by Apple, formally known as IEEE-1394. FireWire allows a digital video (DV) camera to feed its signal directly in and out of a computer with virtually no loss of picture quality.

Frame The basic unit of video (or film). One single still image that, when played in rapid succession with other frames, creates the illusion of motion.

Frame Doubling A method of doubling the vertical lines present in a given frame, which allows for better resolution.

Frame Rate The number of frames that appear on screen per second. In the world of feature film (universally), the frame rate is 24 frames per second; in U.S. video (see NTSC), the frame rate is 30 frames per second (but more accurately, 29.97 fps); in European film and video (see PAL), the frame rate is 25 frames per second.

Frequency Number of cycles per second in an electronic signal, measured in Hertz (Hz), KiloHertz (KHz), MegaHertz (MHz), GigaHertz (Ghz). Named for Heinrich Hertz, an early German electronics pioneer.

G

General Parameters (GPRMs) The second of the two sets of DVD parameters kept within the player, General Parameters are a set of 16 variables that may be used by the DVD author to maintain values specific to the operation of a particular disc, not the player. Unlike System parameters, GPRMs have no relation to the operation of the DVD player; DVD Studio Pro gives you access to 8 of the 16 GPRMs.

Gbps (gigabits per second) In DVD parlance, decimal billions of bits per second.

Gigabyte (GB) A unit of data storage, equal to 1024 megabytes.

GOP (Group of Pictures) The number of frames contained in a particular MPEG encode sequence, as determined by the compressionist or encoder at the time of compression. Typically no less than 4, nor more than 15 frames, the GOP LENGTH is an important determining factor in the effectiveness of MPEG-2 compression.

GOP Length The number of frames contained in an MPEG Group of Pictures (GOP).

GOP Header (see I-Frame header)

H

HDTV (High Definition Television) HDTV is the highest quality version of digital television currently available. There are several different formats of video accepted as HDTV, including 720 lines Progressive (720P), and 1,080 lines Interlaced (1080i).

I

I-Frame Header The first Intra-frame in an MPEG Group of Pictures (GOP). The frame on which a DVD Chapter point must be anchored or set.

I-Frames An MPEG "Intra" frame—a frame that contains a complete frame of video information, but which is compressed only spatially, and not temporally (see Spatial Compression, Temporal Compression)—Typically, the I-frame is the first frame in the MPEG GOP, and is frequently called the "header" or "Key Frame." Each DVD CHAPTER (Part of Title) must begin at an I-Frame HEADER.

In-line Script A Script attached to another Element, which executes in the normal course of navigation.

Interlace To display a frame from top to bottom using two passes. They are then interlaced into a single frame. The method used in nearly all television monitors now on the market.

Interleave Arrangement of digital data in alternating packets of information.

Interpolate Creating new pixels, lines, or frames by averaging information from those on either side of a given pixel, line, or frame. Photoshop's de-Interlacing filter offers this as one option.

Intraframe Information about a frame of video that does not rely on information from other frames.

ITU-R BT.601 The International Standard for digital video, as designated by the ITU (International Telecommunication Union). Previously known as CCIR-601, this is the standard signal format for digital transmission of broadcast video along a coaxial cable. Many good MPEG encoders use 601 digital video as the preferred source. This is the Serial Digital V/A format (SDI) used in a Digital Betacam.

DVD Glossary

J

Jewel Box   A plastic storage case for a CD—in some cases, this form of packaging is also used for DVDs. (see also Amaray Case)

K

Kbps (Kilobits per second)   A means of expressing data rate in thousands of bits per second.

Key Frame   A video picture that contains all the information about the image (intraframe information), rather than just the difference between it and another image.

KiloHertz (kHz)   A measure of frequency, equal to 1000 cycles per second.

Kilobyte (KB)   A measure of data storage, equal to 1024 bytes.

L

Laserdisc (LD)   A 12-inch optical disc, that resembles a metallic phonograph record. This disc holds audio and visual information, but each side is limited to only 60 minutes. A precursor to DVD, Laserdisc uses an analog video format stored on a 12-inch disc, with either analog or digital sound. A dying (if not dead) breed of device, LD originally offered some of the same features DVD now offers, but not to the same extent.

Low Frequency Effects (LFE)   The ".1" channel on a 5.1 channel surround sound mix. This channel is typically fed to a subwoofer and produces frequencies below 150 Hz. The LFE subwoofer send is also extracted from the front and surround channels in many home systems.

LCD (liquid crystal display)   The technology behind flat panel displays, laptop computer screens, and even digital watches. The on-screen image is formed when an electric current passes through a liquid crystal solution, causing the crystals to align so that light cannot pass through them. Can use either passive or active matrix technology; active matrix produces sharper color images, but is more expensive.

Letterbox   Format in which black stripes at the top and bottom of a television screen make up the difference in size between the aspect ratio of the program being shown and that of the video screen. (Movies shown on 4 × 3 televisions are sometimes letterboxed to preserve the original widescreen aspect ratio.)

Line Doubler   A processor that doubles the number of lines on screen, making the scan lines smaller and, therefore, less visible and resulting in better video resolution.

M

Macrovision™   A proprietary anti-copy system (as opposed to an encryption scheme) that works by degrading the video program in a specific manner, and rendering it unpleasant or impossible to watch if it is recorded and played back. In DVD, Macrovision consists of a "bit" set in the finished DVD, which turns on the Macrovision circuit built into every DVD player.

Marker   A location notated within a clip in a DVDSP Track ("Title"), which defines a specific location where one of several possible functions may be enabled. Which function is enabled depends on the type (or types) of Marker established at that location. This location MUST be defined at an I-frame Header location within the MPEG encode.

Matrix Surround System   A surround sound system, such as Dolby Pro Logic or Dolby Surround, in which two sound channels hold more than two encoded channels of sound and are then decoded upon presentation.

Mbps (Megabits per second) Millions of bits per second.

Megabyte (MB) A measure of data storage, equal to 1024 kilobytes.

Meridian Lossless Packing An audio coding system by Meridian Associates that compresses sound information so that it can be relayed more efficiently yet still recovered exactly as it was before compression with no loss of fidelity. This is known as "lossless" compression, as opposed to "lossy" compression in which some information is irretrievably lost during the encoding/decoding process.

Moiré A ring-like visual artifact in composite analog video.

Menu An Element used to create and display Interactive choices allowing the DVD user to control navigation.

Monophonic Mix (a.k.a. "Mono") A sound scheme in which all of the sound is combined into a single channel. Mainly found in older movies and older television sets.

MPEG *The Moving Picture Experts Group*—NOT the Motion Picture Editors Guild! The industry experts group who created the rules for compression and decompression that bears their name ("the MPEG codecs").

MPEG-1 A *codec* for the compression and decompression of Video at a maximum bit rate of 1.856 Mbps. The pixel size ("resolution") for MPEG-1 is 352 × 240 for NTSC video, and 352 × 288 for PAL video.

MPEG-2 A codec for the compression and decompression of Video for DVD at a bit rate of a maximum 1.856 Mbps. The pixel size ("resolution") for MPEG-1 is 352 × 240 for NTSC video, and 352 × 288 for PAL video. MPEG-2 is the form of compression generally used for DVD. MPEG stands for *Moving Picture Experts Group*, an international committee that defines the standards for the data compression of moving pictures. The MPEG moniker is used to refer to compression methods for moving pictures.

Multiplexing The act of combining multiple data streams into one, as when MPEG video, digital audio and Subtitle events are joined into a single data stream during the "Build" phase of DVD authoring.

N

Non-interlaced See Progressive Scan.

P

P-Frame An MPEG "Predictive" frame. Compressed both spatially and temporally, and with some regard for motion estimation, a Predictive frame uses information from its neighboring I- and P-frames to properly reconstruct the video frame at that instant. An MPEG Chapter may not be placed on this frame.

Pan & Scan When a widescreen movie is shown on a 4x3 screen without letterboxing, the sides of the original image are lost. If important action occurs in those areas, the viewable image area is shifted to show the action, a process known as pan & scan.

Parental Control Similar to the scheme used in Region Coding, this is a method of coding a DVD with a particular Parental Guidance Rating (i.e., PG, PG-13, R) so as to allow the DVD player to enable or disable playback based on the Player's programmed Parental Control level.

PCI (Presentation Control Index) DVD feature that specifies how the program will be presented, with categories such as aspect ratio, multiangle, etc.

PCM (Pulse Code Modulation) A method of digitizing analog audio without compression. PCM is

DVD Glossary

the audio format used in Audio CDs, laserdiscs, and some DVDs.

Pixel Short for picture element, pixels are the points of light that make up a video image. Each pixel has four components (red, green, blue, and a transparent alpha channel) that combine to make one point.

Pixel Format The shape of the pixel in a particular monitor. In computer monitors, they're square, while in video monitors they're rectangular.

Pixel Frequency How many pixels appear on screen per second (the width of the screen in pixels, times the height in lines, times the number of frames per second). The greater the pixel frequency, the better the video resolution.

Port Any kind of input/output jack.

Post-Command DVD navigation commands programmed by the DVD author, that [may] activate after (post) the playback of the media contained in the presentation object. In DVDSP, these can be created using Scripts.

Pre-Command DVD navigation commands programmed by the DVD author, that [may] activate before (pre) the playback of the media contained in the presentation object. In DVDSP, these can be accomplished using Pre-Scripts.

Pre-Script A Script attached to another Element, which executes prior to the attached element is processed.

Presentation (also PGC) One individual "container" of Video cells (possibly with Audio(s) and Subtitle[s]), along with the (inherent) rules for selection and playback. Typically, a presentation also may include pre-commands and post-commands to control or alter the navigation flow prior to, or after the playback of the media contained in that presentation.

Pro Logic See Dolby Pro Logic.

Progressive Scan Scanning a frame from the top to bottom of the frame, taking each line in turn. The standard technology for computer monitors.

Program Chain—also PGC (abbr.) A chain of video cells (programs) contained within one DVD presentation object, along with the rules for playback. May also contain pre- and post-commands.

Program (Program Marker) The starting frame of one of many video cells within a DVD presentation object. A Program may define a DVD chapter point, and frequently is used for that purpose. It may also define a cell marking point that is not a chapter.

R

Raster The pattern of horizontal lines (made up of pixels) that form the image on a video monitor screen.

Raster Scan The constant scanning of the horizontal lines on a video monitor that, when seen as a whole, creates the on-screen images.

Rasterizing Converting a digital bit stream to black, white, or 16 shades of gray that make up a picture.

Region Coding To prevent DVD releases from encroaching on worldwide theatrical releases, DVD Titles can be "Region Coded" so they will play only in certain regions of the world. The six international regions are:

1. The U.S., U.S. territories, and Canada
2. Europe, Japan, the Middle East, Egypt, South Africa, Greenland
3. Taiwan, Korea, the Philippines, Indonesia, Hong Kong
4. Mexico, South America, Central America, Australia, New Zealand, Pacific Islands, the Caribbean

5. Russia, Eastern Europe, India, most of Africa, North Korea, Mongolia

6. China

Resolution   The number of pixels (digital formats) or horizontal lines (analog formats) that can be displayed on screen in a particular format. The higher the resolution, the sharper the picture.

Root Menu   Specifically, the primary menu at the top of a particular Video Title Set. Generally, the location a DVD's navigation will seek when the "MENU" button is pressed on the DVD Remote Control, although variations on this behavior can be programmed into each DVD player by its manufacturer. Variations on this may also be specifically programmed into a DVD by its Author.

S

Scanning   The process through which a video image is displayed on a monitor. The pixels that make up the video image are arranged on screen in horizontal scan lines; an electronic signal travels left to right along each scan line, from the top line to the bottom, eventually creating the on-screen image.

Script   A container for DVD commands. The Script Element can be linked to practically every other element including other Scripts.

SDDS®   The acronym for "Sony Dynamic Digital Sound®." While a somewhat popular theatrical sound system format, this 7.1 digital audio format has yet to make any inroads into the world DVD. http://www.sdds.com.

SDI   Serial Digital Interface—a method of communicating digitized video (and possibly audio) information along a coaxial cable (typically using BNC connectors). This interface can be found on many professional video decks, like Digital Betacam, DVCAM, etc.

Slideshow   A container for a series of up to 99 still images which may or may not have individual audio streams, or which may contain one overall audio stream.

Subwoofer   A low frequency speaker possibly containing a built-in powered amplifier. This speaker is a standard feature of 5.1 channel sound systems. The subwoofer reproduces the low frequency information of the original soundtrack, typically only the frequencies below a specific frequency (like 100 Hz).

Super Video (a.k.a. S-VHS, Super-VHS, S-Video)   A video input/output designed to significantly increase the quality of the video signal over standard composite video cables. In this format, Brightness (Luminance) and Color (Chrominance) are separated, but carried within the same cable. The S-Video connection uses one cable with a 4-pin circular mini-DIN connector, and is included with most DVD Players, and higher quality TVs.

System Parameters (SPRMs)   One of two sets of DVD parameters kept within the DVD player—System Parameters keep track of values important to the operation of the player (which Region it is coded for, which Title and Chapter it's currently playing, etc.) DVD authoring can access certain of these parameters, to dynamically control the operation of the DVD player. DVD Studio Pro allows you to read all of the currently used SPRMs, but you can only write to four of them.

T

Telecine   The transfer of film to video (also, the machine used in the process). As a film image goes through telecine, its color, contrast, and sometimes aspect ratio are adjusted by the telecine colorist.

THX™   A program run by Lucasfilm, Ltd. that ensures certain minimum standards are being met in the presentation of moving pictures, regarding sound

DVD Glossary

and picture. THX establishes certain criteria, then licenses or specifies devices (crossovers, amps, speaker systems and physical dimensions and designs) that meet those guidelines.

Title Used variously within DVD Authoring—a DVD "title" may be:

- The DVD project (disc) itself ("We delivered six new Titles last week")

- One of many logical groupings of DVD objects inside of a DVD project ("All these presentations belong to Title 1")

- An individual DVD presentation object ("Put the Movie in Title 1 by itself")

Title Set A collection of multiple DVD Title objects. There can be only 99 combined Title Sets and Titles in any given DVD project. In DVDSP, this means you may only use a total of 99 Tracks, Stories, and Slideshows.

Title Menu The "top menu" of a particular DVD project or disc. Generally, the location a DVD's navigation will seek when the "TITLE" or "TOP MENU" button is pressed on the DVD Remote Control.

Track An Element container for up to 9 Video Angles, 8 Audio Streams, and up to 32 Subtitle Streams.

V

VBR Encoding To improve the overall compressed image quality, Variable Bit Rate encoding works by allocating more bits to the more complex sections of video and less bits to the less complex sections, while maintaining an overall Average Bit Rate appropriate to the running time of the DVD disc size.

Video Compression Any method of reducing the amount of space needed to record or transmit video information. In DVDs, the video is compressed using MPEG-2 encoding, There are two types of bit-rate encoding: variable bit rate (see VBR) and constant bit rate (see CBR).

W

Weave Up-and-down or side-to-side motion of an image, originating from a lack of stability in the original film. It can be tracked digitally and eliminated, so that the image appears stable.

Widescreen Displays To better mimic the theatrical feature film experience, digital television sets today have wider displays screens than the original, nearly square television monitors that have been standard since the beginning of television. Instead of the width-to-height ratio being 4x3, as before, widescreen monitors have an aspect ratio of 16x9, allowing them to display films in a form much closer to that seen in theaters.

Appendix E

DVD Studio Pro Install/Upgrade

DVD Studio Pro 4

Installing DVD Studio Pro 4 is pretty simple—after all, it's a Macintosh application—but there are a few things you will want to check before you just "jump into the deep end":

Before You Install— Check Your Current System Configuration

There are certain minimum requirements your Mac must have in order to run DVD Studio Pro 4 effectively. These will be shown in one of the Installer screens, and are listed in the printed guide *"Installing your Software"* included in the DVD Studio Pro 4 package, but I've reproduced them here for convenience, along with some comments and clarifications:

Hardware Requirements

- A Macintosh computer with a PowerPC G4 or G5 processor with an AGP (Accelerated Graphics Pro) graphics card

- 733 megahertz (MHz) or faster processor*

  (* **Note**—You will find that CPU speed is really an advantage in running DVD Studio Pro 4—even a 733 MHz system seems slow.)

- 8 megabytes (MB) of video memory (32 MB recommended)

- 512 MB of random-access memory (RAM) (1GB required for HD)

- For application installation: DVD drive, 20 gigabytes (GB) of disk space*

  (* **Note**—Not all of this is for the application—much of this is recommended

free space to contain assets [up to 8.5 GB] and a Build [up to 8.5 GB] for a typical DVD-9 project.)

- For writing finished projects to disc: Apple SuperDrive or other DVD-R recorder*

  (* **Note**—DVDSP 4 can write to both DVD-R/RW and DVD+R/RW media. Some recent G5 machines may also write to DVD+R DL (Double Layer) media.)

- For preparing projects for replication (especially DVD-9 projects): DLT drive

  (* **Note**—There are many different kinds of DLT drives but all have one thing in common—they are SCSI devices, so you will need to install a SCSI card before you can use this peripheral)

- For transporting projects on disc that contain copy-protection or dual-layer features to a replicator: Authoring DVD drive (Pioneer DVR-S201)*

  (* **Note**—The Pioneer DVR-S201 is also an SCSI device, so you will need to install an SCSI card before you can use this peripheral.)

Software Requirements

- Mac OS X v10.3.9 or later
- QuickTime 7.0 (or higher)

To Upgrade Your Mac System Software:

> **NOTE!**
> This will NOT upgrade you from 10.3.X to 10.4.X—you will need to purchase Mac OS Tiger from Apple to do so.

DVD Studio Pro Install/Upgrade

- In System Preferences, select *Software Upgrade* (Fig. E-1):

Figure E-1 — OS X System Preferences "Software Update" pane.

Use the "*Check Now*" button (Fig. E-2) to initiate a check of Software status on your System files, QuickTime installation, and other components that may be needed by DVD Studio Pro 4.

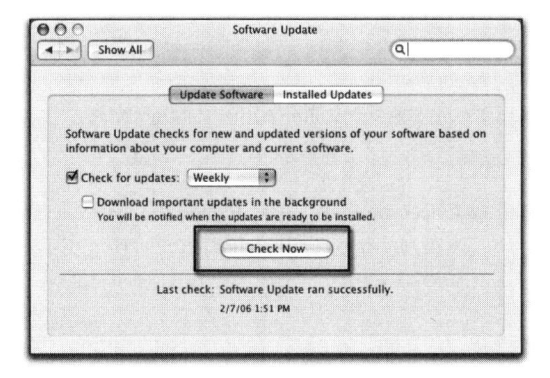

Figure E-2 — OS X Software Update "Check Now" button.

Once your software has been brought current (you may need to restart), you can continue with the DVD Studio Pro 4 installation:

- Locate the DSP2 Installer DVD, and insert it into your DVD drive

NOTE!
You will need to have a DVD-ROM drive, at least, in order to use DVD Studio Pro 4 properly, and in order to be able to preview your DVD project.

- Open the DSP Installer DVD; locate the Installer Package alias.

Before you attempt to install DVD Studio Pro 4, Apple recommends accomplishing the following steps, as listed in the "Before You Install DVD Studio Pro 4.rtf" document included in the Installer DVD (Fig. E-3).

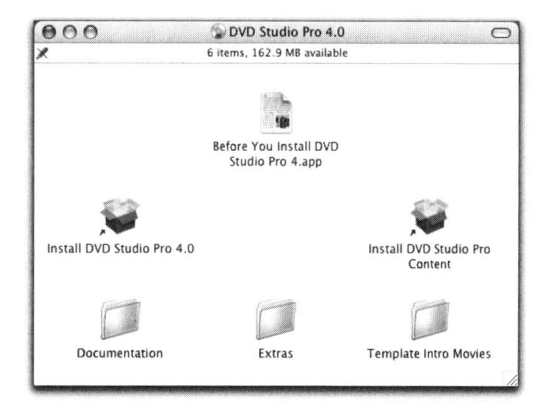

Figure E-3 — The DVD Studio Pro 4 Installer DVD contents.

Make sure you do the following before you install the software:

- Turn off any security software that you may have installed on your computer.

- Turn off any virus-protection software that may be installed on your computer.

- Upgrade Mac OS X and QuickTime, if necessary. (It is no longer necessary to unlock QuickTime Pro manually, as that function is built into the DVD Studio Pro 4 installer.)

503

DVD Studio Pro 4

- Double-Click the "Install DVD Studio Pro 4" alias to begin the install (Fig. E-4a).

Figure E-4a — The Installer Icon.

- You may need to *Verify the installation packages* (Figure E-4b); click CONTINUE to proceed.

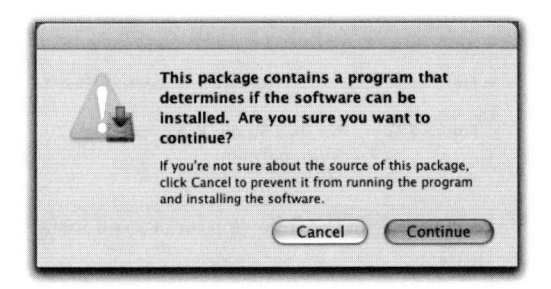

Figure E-4b — Verifying Installer Packages.

- You may need to *Authenticate* that you have permission to perform the installation (Fig. E-5); Enter a valid Administrator Name and password, then click OK.

Figure E-5 — Installer Authentication Dialog.

- Once you have authenticated the install, you will begin to see Installer screens (Fig. E-6 and forward).

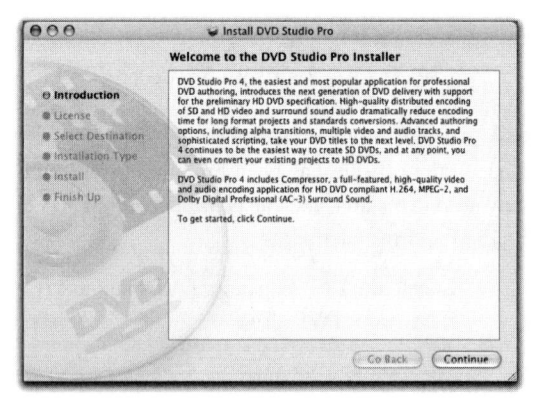

Figure E-6 — Installer "Welcome" Dialog.

DVD Studio Pro Install/Upgrade

- Click *Continue* to display the Software License Agreement screen (Fig. E-7).

- Click *Continue* to bring up the "Agreement" dialog; you must agree to the terms in order to continue the installation (Fig. E-8).

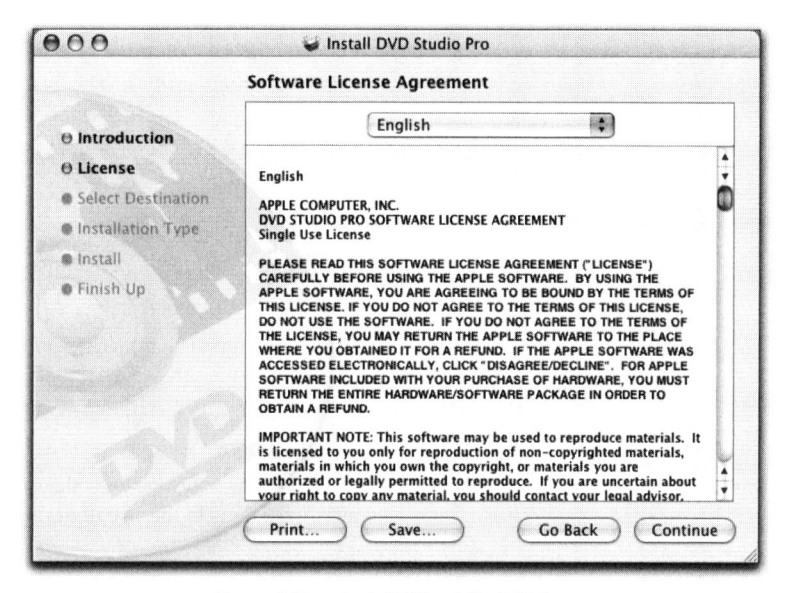

Figure E-7 — Install "Read Me" Dialog.

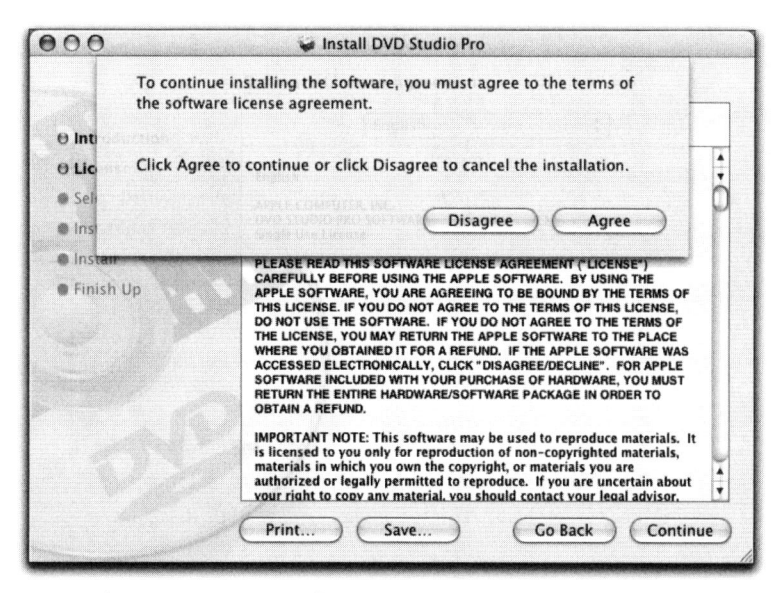

Figure E-8 — Agree to the License Agreement text to continue.

DVD Studio Pro 4

- Once you have agreed, the Installer will present the Licensing Dialog, if it cannot find a valid License Key already installed (Fig. E-9).

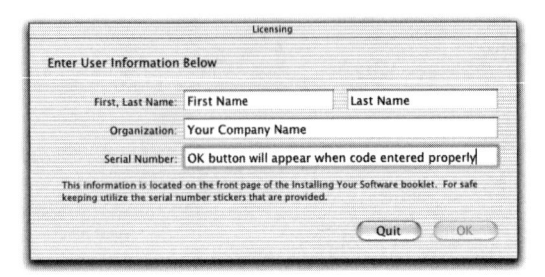

Figure E-9 — Using the License Key info to Unlock the Installer.

> **NOTE!**
> You will find a valid License Key for DVD Studio Pro 4 in the Installation package provided; the numbers are usually affixed to the front of the *Installing Your Software* pamphlet.

- Once successfully Licensed, the Installer will scan your system to find available hard disk volumes, then display them (Fig. E-10).

While all attached hard disk drives will be shown, only your Boot Volume is eligible for this Install (note the green arrow). The installer dialog will reflect the actual space required, which could be more than 2.0 GB, depending on the options you select next:

- Select the Installation Options you prefer—the default is an "Easy Install," but Custom installation is possible (Fig. E-11).

The DVD Studio Pro 4 Installer DVD includes many installer packages which comprise the complete DVD Studio Pro 4 install. You may not wish to install everything (for example, you may have no wish to install iDVD Theme Elements). You can deselect any undesired items in the Install Screen.

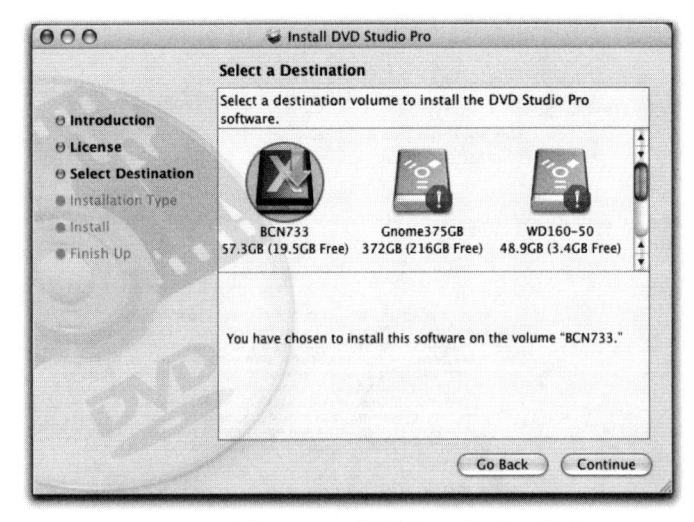

Figure E-10 — Selecting a valid Volume for installation.

DVD Studio Pro Install/Upgrade

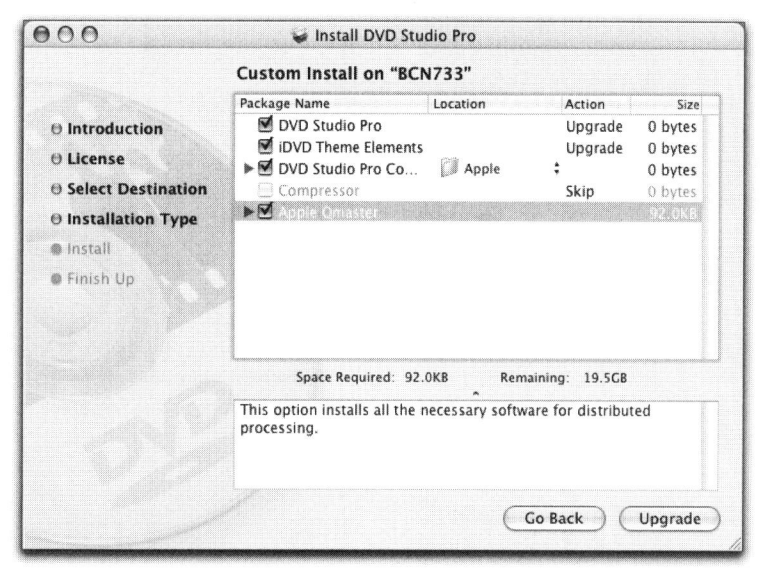

Figure E-11 — The Install screen showing "Custom Install" upgrade.

- To select from the available packages, click to select (check) the packages you would like installed—each package will indicate the disk space it requires; a total is also shown (Figure E-11). (The installation shown in Fig E-11 was a re-installation, so it knew the media data was already still installed, hence the "zero bytes" messages).

- Once the packages have been selected, click the *"Install"* button to begin the installation or upgrade. (If you already have an installation of DVD Studio Pro 4, the *"Install"* button will read *"Upgrade,"* as in Figure E-13.

- You will see a dialog asking you to authenticate (Figure E-14).

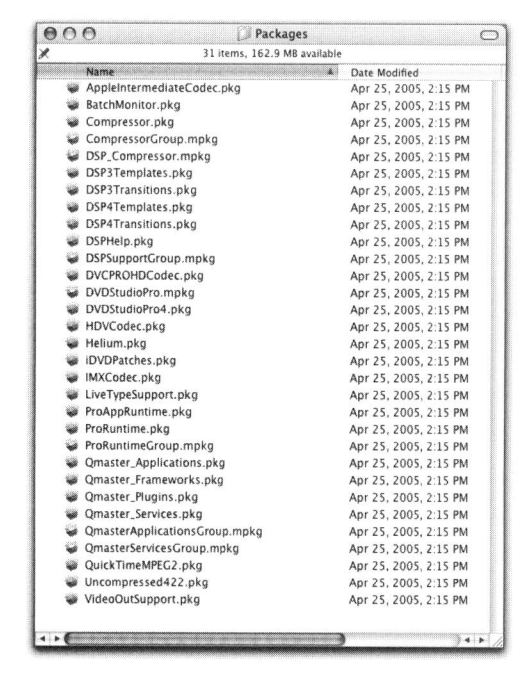

Figure E-12 — Your Installation will select from these packages.

507

DVD Studio Pro 4

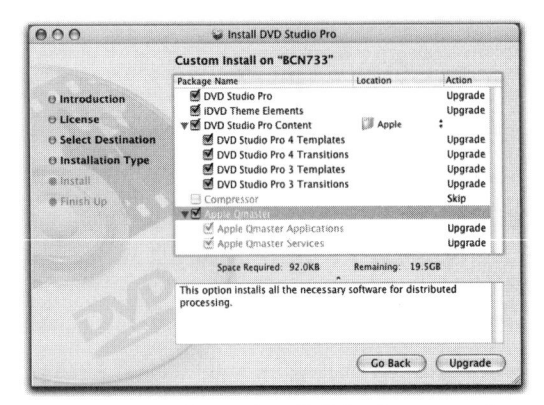

Figure E-13 — Custom Install: Selecting among the Packages to install.

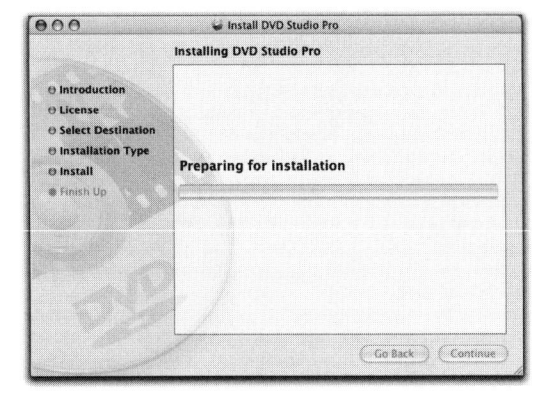

Figure E-15 — Preparing the "Packages" for installation.

Figure E-14 — Authentication—enter user name, password.

- Selecting "*OK*" will begin the installation.
- A series of "Preparing…" screens will appear, as in Figure E- 15.
- Once each package has been prepared, the Installer will begin installing the prepared files—it will take a few minutes, but the installer will display its estimate of the time required (Figure E-16).

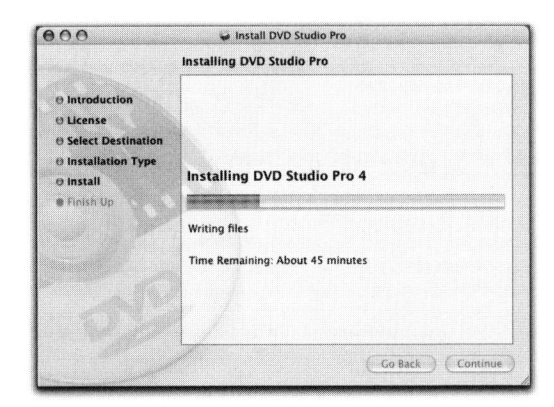

Figure E-16 — Installing the software packages.

- Once the installation has finished, the Installer will tidy up after itself (Figure E-17).
- Upon completion, *Success* will be reported (Fig. E-18).

Once the installation cleanup has concluded, you will be able to launch DVD Studio Pro 4 and begin checking the configuration to be sure the Preferences are set to correspond with your needs.

508

DVD Studio Pro Install/Upgrade

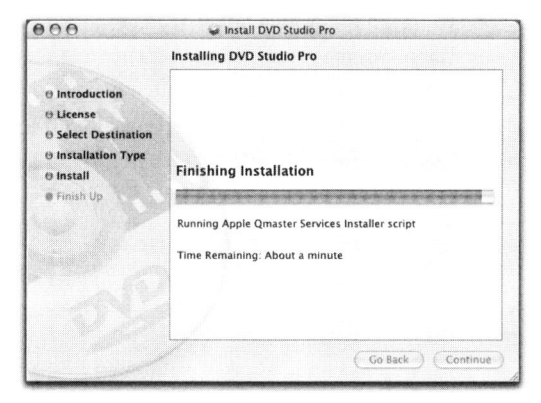

Figure E-17 — Finishing the installation.

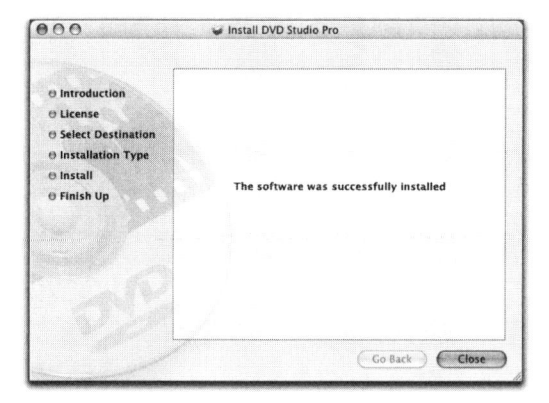

Figure E-18 — Success! Time to restart and make DVDs!

> **NOTE!**
> The default DVD Studio Pro 4 installation does NOT automatically install any Tutorial media. You can copy the Recipe4DVD source content (NTSC) from the book DVD if you want some content to use for practice. You may also wish to copy the Template Intro movies in your choice of video format: NTSC or PAL—both are provided

Configuring DVD Studio Pro Software

Before you can begin using DVD Studio Pro effectively, you should specify your work environment's video and audio settings and your preferred settings for many things.

To Configure DVD Studio Pro preferences:

1. Double-click the DVD Studio Pro icon on the Macintosh desktop to launch DVD Studio Pro. The DVD Studio Pro Splash Screen appears (Fig. E-19).

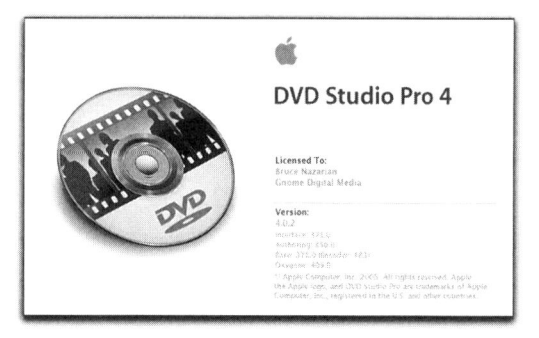

Figure E-19 — DVD Studio Pro 4 splash screen.

2. From the DVD Studio Pro Menu, select **Preferences...** (Fig. E-20).

509

DVD Studio Pro 4

Figure E-20 — DVD Studio Pro 4 "Preferences" command.

3. Select the **Preference**... dialogs appropriate to the settings you need to adjust or personalize. You will find a complete rundown of the Preferences screens in Chapter 3. Below are a few that you may wish to verify.

- To Install the Apple-supplied content for Templates and Transitions, you must run the Content Installer—you may already have some of this installed, but DVD Studio Pro 4 includes new HD templates you will not have installed with DVD Studio Pro 3. (Figure E-21).

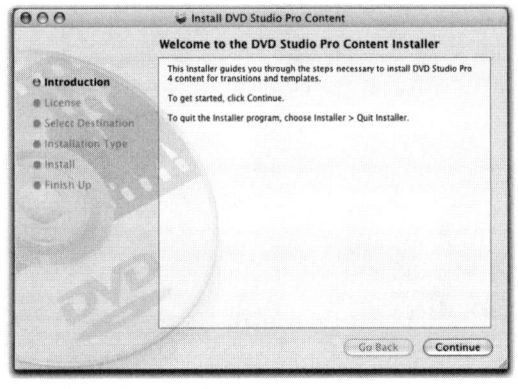

Figure E-21 — The Content Installer screen.

General:		
	Project:	DEFAULT LANGUAGE
	Slides:	DEFAULT SLIDE LENGTH
	Subtitle:	DEFAULT FADE IN, LENGTH, FADE OUT
Menu:		
	Motion Duration:	DEFAULT 30 secs
	Drop Palette Delay:	Adjust pop-up time
Track:		
	Marker Prefix (root):	Adjust as needed
	Default Language:	English(?)
Alignment:		
	Rulers:	Show
	Units:	Pixels

510

DVD Studio Pro Install/Upgrade

Simulator:
    Default Language Settings:
        Audio: **Adjust as needed**
        Subtitle: **Adjust as needed**
        DVD Menu: **Adjust as needed**
    Default Language Settings:
        Features: **Enable DVD@ccess**

This checkbox enables the DVD@ccess URL functions during preview, allowing DVDSP to launch your Web browser to preview DVD@ccess events in your DVD.

Encoding:
    Video Standard: NTSC or PAL

Appendix F

QuickTime™ Installation Guide

DVD Studio Pro 4

An integral part of DVD Studio Pro 4 is QuickTime™—Apple's innovative multimedia architecture that handles the movies, audio and MPEG files used in DVD Studio Pro. In order for the application to function properly, you need to have the proper version of QuickTime™ installed, and the QuickTime™ Pro Player unlocked. DVD Studio Pro 4 will take care of the rest.

If you are going to install Final Cut Studio, or DVD Studio Pro, do so before attempting to perform any QuickTime upgrade.

Checking the Current QuickTime™ Version in OS X

To check the status of your QuickTime player in OS X:

- Launch the System Preferences (Fig. F-1).

Figure F-1 — System Preferences icon.

- Locate and select the QuickTime™ Icon within the *Internet & Network* Preferences Pane (Fig. F-2).

Figure F-2 — Select the QuickTime™ icon.

- Click on the QuickTime Icon to open the QuickTime prefs panel (Fig. F-3).

Figure F-3 — Click on the "Blue Q."

- Once the QuickTime™ prefs pane has opened, click the "*Register*" button to reveal the QuickTime registration information (Fig. F-4).

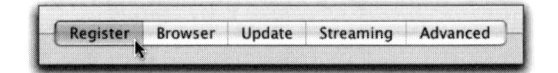

Figure F-4 — Click Registration to display the Registration pane.

- The QuickTime registry pane will drop down into view, revealing the current status of your QuickTime install (Fig. F-5a). Look for the words "Pro Player" in the QuickTime™ edition that is indicated. If it isn't Pro, you need to upgrade QuickTime™.

Figure F-5a — The QuickTime™ registration pane showing version.

514

QuickTime™ Installation Guide

Figure F-5b — Click "About QuickTime..." to verify QuickTime version.

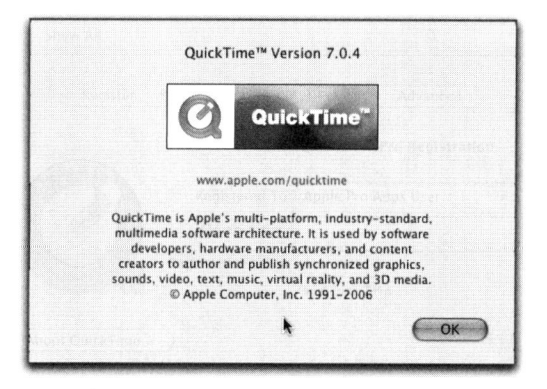

Figure F-5c — QuickTime version display.

- If you have already installed Final Cut Studio, Final Cut Pro, or DVD studio Pro, you will have automatically been upgraded to QuickTime™ Pro, and no upgrade key code is required.

- If the registration is incorrect, for any reason, click in the *Registration Code* field and enter a valid QuickTime™ Pro upgrade key.

If you are going to install Final Cut Studio, or DVD Studio Pro, do so before attempting to perform any QuickTime upgrade.

- Once you have located the key code, enter it carefully (pay attention to the **CaSe** of letters) and click the *OK* button. Your QuickTime™ should now be upgraded, and say "Pro Player" in the edition name.

> **NOTE!**
> You DO NOT need to buy the QuickTime MPEG decoding license to use DVD Studio Pro—it's included in the package.

IMPORTANT: DO NOT REGISTER ONLINE or PURCHASE an upgrade code if you are installing DVD Studio Pro 4 or upgrading a previous installation. A QuickTime™ upgrade will be performed automatically as part of the installation or upgrade.

Yes, It's That Simple!

That's all there is to it—with a properly upgraded and current QuickTime™ installation, you'll be all set to install DVD Studio Pro 4 and start making DVDs.

515

Index

Note: Boldface numbers indicate illustrations; *t* indicates a table.

3dB attenuation setting
   A.Pack, 158, **158**
   Compressor 2, 179, **179**

A.Pack, 14, 26, 67, 112, 147–167, 181
   3dB attenuation setting in, 158, **158**
   AIFF to Dolby Digital using, 148–149
   audio coding mode setting in, 151, **151**
   Audio Production Information settings in, 154, **154**
   Audio Tab settings in, 149–152, **151**
   Batch List in, 148, 158–161, **159–161**
   bit stream mode parameter setting in, 152, **152**
   Bitstream Tab settings in, 150, 153–155, **153**
   Center Downmix level setting in, 153, **153**
   Compression Factor setting in, 155, **155**
   Copyright settings in, 154, **154**
   data rate setting in, 151, **152**
   DC filter setting for, 157, **157**
   De-emphasis parameter setting in, 156, **156**
   dialog normalization setting in, 152, **152**
   Dolby Digital (AC3) files and, 147–167
   Dolby Surround mode setting in, 153–154, **154**
   File Grid (Input Channels) in, 150, **150**
   Instant Encoder in, 147–148, 149–150
   log window in, 162, **162**
   Low-Pass filter setting for, 156, **156**, 157, **157**
   monitoring Dolby Digital encodes in, 161–162, **162**
   Original Bitstream settings in, 154, **154**
   Overmodulation Protection setting in, 155, **156**
   Peak Mixing Level setting in, 154
   preparing to use, 149
   Preprocessing Tab settings in, 150, 155–158
   Room Type setting in, 155
   stereo files in, quick encode guide for, 148–149
   Surround Downmix level setting in, 153, **153**
   Surround Phase Shift setting in, 157–158, **158**
   Target System setting in, 150–151, **151**

A.Pack *(continued)*
   uses for, 148
   WAV to Dolby Digital using, 148–149
AC3 files, 67, 94. *See* Dolby Digital (AC3) files
AC3 integrated encoding, 26
acronym of DVD, 2
Action Safe area/zone, 245
   graphics for DVD and, 419, **419**
   slideshows and, 297
Activated state of buttons, 241, 272–273, **272, 273**
Add Language tool in, 74
Add Layered Menu tool, 74
Add Menu tool, 74
Add Script tool, 74
Add Slideshow tool, 74
Add Story tool, 74
Add Submenu, Slideshow, Track tools, 76
Add Track tool, 74
Adobe After Effects
   menus and, 239
   motion menus and, 427
Adobe Photoshop, 68, 95. *See* Photoshop
Advanced Audio Code (AAC), 144
Advanced Configuration, 51–52, **52**
Advanced Video Codec (AVC). *See* H.264 (AVC)
Advanced workspace, 29
After Effects. *See* Adobe After Effects, 427
AIFF files, 94, 140–141, **141**, 144–147, **145–147**, 181
   Dolby Digital conversion from, using A.Pack, 148–149
aliasing, in graphics for DVD, 422, **422**
alignment preferences, 38–39, **38**
alternative encoders, 135–136, 469–480
AltiVec, 19
Angle Button, Remote Control, **5**
angle playback in Viewer, 216, **216**
angles, 2
Apple DVD Player
   DVDaccess and, testing links to, 456
   emulation, 391–392, **392**
Apple Qmaster, 25, 28, 46

Index

Apple Superdrive, 2, 8, 15, 20, 26
   building and formatting and, 399
Arrange Controls tool, 75
aspect ratio, 44
   slideshows and, 297–298
Asset Inspector, trimming streams using, 205–206, **206**
Asset Tab, 88
asset-based markers and timelines, 204, 227
assets and Asset Tab, 51, 78, 83, **83**, 89, 91–106, **97**. *See also* audio; slideshows; stills; video
   adding, to Assets Tab, 98–99, **98, 99**
   Assets Tab for, 96–98, **97**
   Assets Tab for, creating now folders within, 98
   audio, 94
   definition of, 92
   details about, 101, **101**
   displaying more data about, 101–102, **101, 102**
   drag and drop to add, 99–100, **100**, 104–105
   DVD Studio Pro organization of, 100, **100**
   folder creation for, 98
   hybrid DVD-ROM deliverable files as, 95–96
   importing, 96, 97–98, **97**, 99, **99**, 100, 106
   Inspector for, 103, **103, 104**
   kinds of, 92–96
   locating, in Finder, 100–101, **101**
   missing, 101–104, **101**
   organization of, using Palette, 68
   organizing, 96
   Palette for, 96
   properties of asset files in, 103, **103, 104**
   relinking missing folders in, 103, **103**
   relinking one missing asset file in, 103, **103**
   removing file or folder from, 104
   required for DVD projects, 14
   Slideshow Editor and, dragging assets into, 105
   still, 94–95
   subtitle, 95
   Track Editor and, dragging assets into, 105
   usability of, checking, 100, **100**
   video, 92–94
ATI Rage, 421
audio, 3, 89, 94, 139–182
   A.Pack and, 147–167, 181. *See also* A.Pack
   adding by hand, 193
   Advanced Audio Code (AAC), 144
   AIFF files in, 140–141, **141**, 144–147, **145–147**
   BD and, 144
   Compressor 2 encoding of, 163–167, **164**
   Digital Theatre System (DTS) and, 140, 143–144, **144**, 179–181
   Dolby Digital (AC3) in, 140, 141–142, **142**, 144–145, 147–167, 181

audio *(continued)*
   Dolby Surround and, 140, 142–143, **143**, 167–171, **168**
   external audio monitor simulation for, 27
   formats for, 140, 144–145
   high definition, 144
   Inspector for, 166, **166**
   Meridian Lossless Packing (MLP), 144
   MPEG (MPA) and, 140, 143, **143**, 181
   pulse code modulation (PCM) and, 140–144
   slideshows and, 304–307, **305–307**
   stereo file encoding for, using Compressor 2, 164–167
   stereo files in, A.Pack quick encode guide for, 148–149
   stillframe menu with, 242–243
   tracks and, 189
   video automatically matched to, 192, **192**
   WAV files in, 141, **141**, 144–145
Audio Button, Remote Control, **5**
Audio Coding Mode setting, in Compressor 2, 172, **172**
audio files matching, Tracks, 38
Audio Production Information setting
   A.Pack, 154, **154**
   Compressor 2, 176, **176**
Audio, Palette, 64–65, **65**, 67, **67**
Aurora Igniter, 421
authoring, 2, 3–4, 15
   advanced techniques in, 451–464. *See also* DVDaccess
   HD-DVD, 19–20
   menus. *See* menus
   tracks, 183–218. *See also* tracks
Autodesk. *See* Cleaner 6 (Autodesk), 472
AVC, 93, 94
average bit rate, 17

background
   in menus, 35–36, 240, 242, 254, 255, 261–262, 269, **270**
   in Slideshows, 33
background encoding, 45
backup files. *See* BUP files
bandwidth. *See* data rates
Basic Configuration, 29, 50, **50**, 68–69, 68
Batch List
   A.Pack, 148, 158–161, **159–161**
   Compressor 2, 163, **164**
BD audio, 144
bit budgeting, 14, 17–18, **17**, 20
bit rates, 17, 18, 18*t*, 20, 44, 120, 121
bit stream mode parameter setting
   A.Pack, 152, **152**
   Compressor 2, 174, **174**
BitVice (Innobits), 93, 137, 470–471, **470**
   motion menus and, 430

518

Index

black levels, in graphics for DVD, 420–421, **420, 421**
Blu-Ray Disc, 2, 9, 18–19, 466
blue laser burners, 26
BMP files, 68, 297, 415
bonding, in duplication and replication, 446, **446**
breakpoint, dual-layer discs, 27, 221, 225–226, 395
Brisbin, John, 117
Browse clip function, 212
Build tool, 75
Build/Format tool, 75
building (multiplexing, muxing) and formatting, 15, 383–412. *See also* simulation and Simulator
   Advanced Tab of Disc Inspector for, 397–398, **397**
   Apple DVD player emulation in, 391–392, **392**
   Apple Superdrive formatting in, 399
   break point setting in, 395
   build and format in one operation, 406, **406**
   building in, 384–385, 398, 401–403, **402**
   burning in, 384, 400–401, **401**
   checking project before, 384
   configuring hard drive for, 403
   copy protection settings in, 396–397, **396**, 406, **406**
   digital linear tape (DLT) in, 399
   Disc Information setting, 395–396, **395**, 405, **405**
   Disc Property Inspector for, 393–398, **393**
   disc property setting before, 393–398
   Disc/Volume Tab of Disc Inspector for, 395–396, **395**, 405, **405**
   DVD-ROM setting for, 394–395
   DVDaccess and, 397, 410
   DVDaccess emulation in, 392–393, **393**
   embedded text in, 397
   emulation in, 384, 391–393
   fast disk drives for, 403
   formatting process for, 385, 399–400, 403–405, 404, **405**
   General Tab of Disc Inspector for, 394
   GPRM partitions and, 398
   hard disk space needed for builds in, 402
   hybrid (video + ROM data) DVDs in, 399, **400**, 408–410, **408–410**
   Joliet files in, 410
   language selection in, 398
   messages pertaining to, 403, **403**
   optimizing disk space for, 403
   outputting a finished DVD project in, 398–407
   preview in, 384
   region code settings, 396, **396**, 405–406
   Region/Copyright Tab of Disc Inspector for, 396, 405–406
   Remote Control setting for, 394
   reusing build folders in, 406–407, **407**

building (multiplexing, muxing) and formatting *(continued)*
   simulating in, 384, 385–391, 385
   Simulator for, 385–391
   size limits, DVD-R, 411
   speed of, maximizing, 402
   Streams setting for, 394
   testing DVD before, 385
   Toast Titanium in, 409, **409**, 411, **411**
   video-only volume in, 398–399, **398, 399**
   Viewer for, 385
   Volume information in, 405, **405**
bundle, project, 43
BUP files, 3
Burn tool, 75
burners, 384
burning DVDs, 2, 15, 384, 400–401, **401**
   blue vs. red laser, 26
   dual-layer, 26
button defaults, 40
Button Outlet tool, 76
Button Shapes, 95
Button State Selection tool, 76
Buttons, 63, 94
buttons and Button Inspector, 76, 274–275, **274**, 460. *See also* menus
   Activated state of, 241, 272–273, **272, 273**
   adding to menu, 262
   Button Inspector for, 274–275, **274**
   button styles dragged to, 286, **286**
   connections made through, 319, **319**
   connections to, 274–275
   creating, 262, **262**
   drawing, 264, **264**
   duplicating, 264, **264**
   layered menus, dragging assets to, 290–292, **290–292**
   markers and, 225
   menus and, 240, 242, 246, 254–255, 262–265
   menus dragged to, 284, **284**
   moving and aligning, 262–263, **263**
   naming, 265
   Normal state of, 241, 271–272, **271**
   renaming, 265
   resizing, 263–264, **263**
   Resume function and, 319
   scripts dragged to, 284, **284**
   Selected state of, 241, 272, **272**
   slideshows dragged to, 283–284, **283**
   state of, 241
   stories dragged to, 283, **283**
   templates dragged to, 285, **285**
   tracks dragged to, 283, **283**

519

Index

Buttons over Video
  markers and, 220, 229
  subtitles and, 339

capturing image from screen, 431–432, **431**, 431
cells, 51
  markers and, 226
Center Downmix level setting
  A.Pack, 153, **153**
  Compressor 2, 174, **174**
Central Stock, motion clips, 429
chapters and chapter markers, 6, 51, 123–124, **123**, 186, **187**, 188–189, 220, 225
Cinematize, 136
Cleaner 6 (Autodesk), 137, 472–473, **473**
Clip Inspector, 210–212
closed captions, 208, 457–458. *See also* subtitles
  subtitles and vs., 340–341, **340**
color, 40–41, **41**
  in graphics for DVD, 416, 417, 419, **419**, 424–425
command reference, 467–468
Compare test, scripts and, 370
component parts of DVDs, 2–3
components of DVD Studio Pro 4, 24–25
Compression Factor setting
  A.Pack, 155, **155**
  Compressor 2, 176–177, **177**
Compression markers, 123, **123**, 123
Compression Master (Popwire), 93, 479–480, **480**
Compressor 2, 14, 25, 28, **28**, 46, **108**, 109, 111–117, **111**, 137
  3dB attenuation setting in, 179, **179**
  A.Pack and, 112
  AC3 integrated encoding for, 26
  Audio Coding Mode setting in, 172, **172**
  audio files and, 94, 163–167, **164**
  Audio Production Information setting in, 176, **176**
  Audio Tab settings in, 169–174, **171**
  Batch List in, 163, **164**
  batch monitor in, 112, **112**
  Bit Stream Mode parameter setting in, 174, **174**
  Bitstream Tab settings in, 170–171, 174–176
  Center Downmix level setting in, 174, **174**
  Compression Factor setting in, 176–177, **177**
  Copyright settings in, 175–176, **175**
  Data Rate settings in, 173, **173**
  DC filter setting in, 178, **178**
  De-Emphasis Parameter setting in, 177–178, **177**
  Destination and Output filename settings in, 165–166, **166**
  Dialog Normalization setting in, 173–174, **173**
  Dolby Digital AC3 encoding and, 112–113, 163–167, **164**

Compressor 2 *(continued)*
  Dolby Surround encoding using, 167–171, **168**
  Dolby Surround Mode setting in, 175, **175**
  Encode Format Preset for, 164, **165**, 166–167
  encoding settings for, 112, **113**
  exporting into MPEG-2 from Final Cut Pro using, 124–125, **124, 125, 126**
  File Grid in, 171, **171**
  Final Cut Studio for, 112
  H.264 (AVC) and, 112
  Inspector for, 166, **166**
  launching, 163
  loading Surround Sound group in, 168–171, **168–170**
  Low-Pass filter setting in, 178, **178**
  manuals available for, 126
  motion menus and, 430
  MPEG-2 Encoder and, 112, 115
  NTSC/PAL conversion and, 113, **114**
  Original Bitstream setting in, 176
  Overmodulation protection setting in, 177, **177**
  Preprocessing Tab settings in, 171, 176–179
  Qmaster distributed encoding and, 113, **114**
  QuickTime and, 115–116, **116**
  Sample Rate settings in, 173
  stereo file encoding using, 164–167
  submitting job for encoding in, 167
  Surround Downmix level setting in, 174–175, **175**
  Surround Phase Shift setting in, 179, **179**
  target system setting in, 171, **172**
  uses for, 163
  video and, 93
conditional elements and states, 462–464
Conditional Jump command, 380, **380**
conditional vs. unconditional commands in scripts, 367–370, **368–370**, 462–464
configuration settings, 29, 49
connections and Connections Tab, 51, 78, 81–82, **82**, 89, 317–336
  Advanced view for, 323
  Basic view for, 323
  breaking, 327–328, **327**
  buttons for making, 319, **319**
  checking, 328–329
  Connect button for, 325, **325**
  Connections Tab for, 320, **320**, 323
  Connections Tab Layout for, 321, **322**
  Connections Tab tools for, 321, **321**
  Control-Click for, 325, **326**
  Delete Key for, 327
  Disconnect Button for, 327, **327**
  drag-and-drop for, 325, **326**
  Filter pop-up in, All, Connected, Unconnected in, 324

520

Index

connections and Connections Tab *(continued)*
   First Play setting for, 328, **328**
   Inspectors for making, 319, **319**
   jumps in, Next, Previous, 324
   keyboard for, 326–327
   making, 324–327
   modifying, 327–328
   Recipe4DVD sample projects for, 324
   remote control key problems with, 329
   Resume function and, 319
   sources for, 329, 330, 330*t*, 331*t*, 332–333*t*
   Standard view for, 323
   targets for, 334–335
connectivity, 111
constant bit rate (CBR) encoding, 17, 110
Control-Click, moving Tabs with, 60
copy protection, 396–397, **396**, 406, **406**
   A.Pack, 154, **154**
   Compressor 2, 175–176, **175**
Customize Toolbar, 71, **71**, 72

data rate and file size, 117–118, 212–213
data, making DVD discs using, 8–9
DC filter setting
   A.Pack, 157, **157**
   Compressor 2, 178, **178**
De-emphasis parameter setting
   A.Pack, 156, **156**
   Compressor 2, 177–178, **177**
de-interlacing, in graphics for DVD, 433–434, **434**
default configuration setting, 29, **29**
Destination and Output filename settings, Compressor 2, 165–166, **166**
Destinations preferences, 42–43, **43**
dialog normalization setting
   A.Pack, 152, **152**
   Compressor 2, 173–174, **173**
Digigami MegaPEG Pro QT, 93, 116, **116**, 474–476, **474**, **476**
digital audio workstations (DAWs), 181
Digital Cinema Desktop, 26, 89
digital linear tape (DLT), 20, 484
   building and formatting and, 399
   mastering, 2
   video and, 110
   writing, 15
Digital Theatre System (DTS), 89, 94, 140, 143–144, **144**, 179–181
   DTS 6.1 audio format, 27
   encoding packed file into DTS bitstream in, 180–181, **180**, **181**
   packing in, 180

Digital Theatre System (DTS) *(continued)*
   preparing files for packing in, 180, **180**
Digital Versatile Disc. *See* DVDs
Digital Video Disc. *See* DVDs
Digital Voodoo D1 Desktop, 421
direct-to-customer distribution of DVDs, 447
Disc Meter tool, 74
Disc Property Inspector, 95
disc running times in seconds, 17, 18*t*
disc size, 31, **32**. *See also* size issues
disc space, 20
display conditions, 460–464, **461**
   pre-scripts vs., 463
Display Mode, 33, 34
distribution routes for DVD, 447–448
DL+, 20
DLT-4000, **385**
DLT-8000, **385**
documentation files, 28
Dolby Digital (AC3) files, 20, 26–28, 46, 67, 89, 94, 112–113, 120, 140, 141–142, **142**, 144–145, 147–167, 181
   A.Pack and, 147–167. *See also* A.Pack, 147
   AC3 integrated encoding, 26
   adding to track, 193
   AIFF to, using A.Pack, 148–149
   Batch List for (A.Pack), 148
   Compressor 2 encoding of, 163–167, **164**
   Instant Encoder for (A.Pack), 147–150
   monitoring Dolby Digital encodes in A.Pack, 161–162, **162**
   stereo files in, A.Pack quick encode guide for, 148–149
   uses for, 148
   WAV to, using A.Pack, 148–149
Dolby Surround, 140, 142–143, **143**, 167–171, **168**
   Compressor 2, 167–171, **168**, 167
   mode setting, in A.Pack, 153–154, **154**, 153
   mode setting, in Compressor 2, 175, **175**, 175
double-sided DVD, size limitations of, 8, **8**. *See also* dual-layer DVD
drag-and-drop, 99–100, **100**, 104–105, 195–196, **195**, 275–279
Drop Palette, 76, **77**
   drop palette delay, 35
   menus and, drag-and-drop editing in, 249–256
drop shadow text, 255, 423
drop zone, 63
   menus and, dragging to drop zone in, 287
   menus and, dragging to empty area in, 286, **286**
dual-layer DVDs, 46
   break points in, 27, 221, 225–226, 395
   burning support, 26

521

Index

dual-layer DVDs *(continued)*
   duplication and replication, 447, **447**
duplication and replication, 13, 15–16, 437–449
   bonding in, 446, **446**
   compatibility of process with commercial DVD players, 441
   complexity of each process in, 441
   distribution routes for, 447–448
   duplicator machines for, 441–443
   glass master preparation in, 444, **444, 445**
   labeling in, 446–447
   metallizing glass master in, 445, **445**
   molding in, 445, **446**
   numbers of copies required vs. time/cost in deciding, 439–440
   packaging in, 447
   physical formatting of DVD for, 444
   plating and stamping in, 445, **445**
   replication process in, steps for, 443–447
   single- vs. dual-layer discs in, 447, **447**
   sputtering in, 446, **446**
DVCPRO, 19
DVCPRO HD, 25
DVD Pro Club, 466
DVD recorder as encoder, 135–136
DVD standard, 31, **31**, 32–33, **33**
DVD Volume, 15
DVD+R, 9–11, 10*t*, 483
DVD+R DL, 9–11, 10*t*, 483
DVD+R/RW, 20
DVD+RW, 9–11, 10*t*, 483
DVD-1 (SS/SL), 8
DVD-2 (SS/DL), 8
DVD-3 (DS/SL), 8
DVD-4 (DS/DL), 8
DVD-9 recording, 8
DVD-Audio, 9, 20
DVD-Hybrid discs. *See* hybrid
DVD-R, 9–11, 482
   size of, 10–11
DVD-R DL, 482
DVD-R/RW, 20
DVD-RAM, 9–11, 10*t*, 20, 483
DVD-ROM (hybrid) deliverable files, 8–9, 20, 95–96, 106. *See also* hybrid
DVD-RW, 9–11, 10*t*, 482
DVD-Video, 2, 9, 20
DVDaccess, 41, 452–457
   building and formatting and, 397, 410
   configuring computers for, 452
   demo files and test projects for, 456–457
   emulation using, 392–393, **393**

DVDaccess *(continued)*
   HTML files and, 455
   hybrid (video + ROM data) DVDs, 457, **457**
   img file mounted on Mac desktop and, 456
   installer for, 453
   interactivity enabled through, 452
   link activation in, URLs and, 454–455
   local file access and, limitations to, 455–456
   Macintosh configuration for, 452
   markers and for, 221
   markers and, 228
   menus and, 260, 454, **455**
   PC configuration for, 452
   property settings for, 453–454, **453**
   slides in, 454, **455**
   syntax for, 454
   testing links for, 456
   track markers in, 453–454, **455**
   uses for, 452
DVDs, 1–21
   acronym of, explained, 2
   Blu-Ray, 2
   component parts of, 2–3
   double-sided, size limitations of, 8, **8**
   DVDS
   formats for, 9–12
   functioning of, 3–4
   HD type, 2
   mini-, size limitations of, 8
   mouse navigation of, 6–7
   play back of, 4
   Remote Control navigation of, 6–7
   single-sided, size limitations of, 7–8, **7**
   size limitations of, 4, 7–8, 12–13
   video type, 2
   workflow of, 13–16, **16**

earlier versions of Studio Pro, using projects from, 24
Echo Fire, 421
elementary streams, 14, 15, 110, 189
Elements, 89
embedded text in DVD, 397
emulation
   Apple DVD player emulation in, 391–392, **392**
   building and formatting and, 384, 391–393
encode bitrate, 18, 18*t*, 20, 120, 121
Encode Format Preset, Compressor 2, 164, **165**, 166–167
encode on build, 45
encode status indicators, 128, **128**, 128
encoding and encoders. *See also* Compressor 2; Final Cut Pro
   alternative encoders for, 135–136, 469–480
   BitVice (Innobits), 470–471, **470**

522

Index

encoding and encoders *(continued)*
   Cleaner 6 (Autodesk), 472–473, **473**
   Compression Master (Popwire), 479–480, **480**
   constant bit rate (CBR), 110
   content vs. encoding quality in, 120–121
   Digigami MegaPEG Pro QT, 116, **116**, 474–476, **474, 476**
   Dolby Digital AC3, 112–113, 120
   DVD recorder as encoder in, 135–136
   encode bitrate and, 120, 121. *See also* bitrate
   encode status indicators in, 128, **128**
   encoder selection for, 120
   expectations for, 120
   Final Cut Pro and, 115, **115**, 121–122
   format for, 119
   hardware, 110–111, 116–117
   Heuris MPEG Power Professional, 471–472, **471**
   image enhancement before or during, 119
   Main Concept encoders, 476–477, **477, 478**
   masters for, 119
   motion menus and, 430
   MPEG Append utility for, 117
   MPEG Encoder for, 115, 127–128, **127**
   Optibase Inc. Master Encoders for, 116–117
   preferences for, setting, 44–45, **44**
   Qmaster distributed, 113, **114**
   QuickTime and, 115–116, **116**
   software, 111–117. *See also* Compressor 2
   Sonic Solutions Inc. SD-xxx encoders for, 117
   technical standards for, 119–120
   variable bit rate (VBR), 110
   video, 14–15, 109–111
   Wired Inc. Mediapress encoders for, 117
End Action, 80
End Jump, 186, 188, 226, 229, 230, 232, 233, 234
Enter key, Remote Control, 5, **5**
Entry Points, 229
EVD, 466
Exit command, 375
Exit Pre-Script command, 375
exporting from Final Cut Pro using QuickTime, 128–131
Extended Configuration, 29, 50–51, **51**
external audio monitors, 27
external video monitor simulation, 26–27

fade in/out, subtitles, 34–35
field order, 44
file organization, 3
file sizes by data rate, 117–118
Filter pop-up, connections and, All, Connected, Unconnected in, 324
Final Cut Express, 70

Final Cut Pro, 19, 51, 70, 93, 111, 112, 115, **115**, 121–122, 137
   AIFF file creation in, 145–146, **145, 146**
   batch encoding in, 133
   Chapter markers in, 123–124, **123**
   Compression markers in, 123, **123**
   exporting from, using QuickTime, 128–131
   exporting into MPEG-2 from, using Compressor 2, 124–125, **124–126**
   in and out time setting and, 122
   menus and, 239
   motion menus and, 428–430, **429**
   rendering and, 122
   resolution and, 122
   setting DVD markers in, 122–124, **122**
Final Cut Studio, 28, 111, 112
Finder, for assets, 100–101, **101**
FireWire, 15, 27, 93
First Play
   connection, setting, 328, **328**
   simulation and Simulator, 385–386
folders, 3
font effects, 28, 63, 423, **423**
   in graphics for DVD, 422, **422**
Format tool, 75
formats for DVD, 9–12
   audio, 140
   Blu-ray Disc, 9
   HD-DVD, 9
   logical formats for, 9
   physical formats for, 9–11, 10*t*
   recordable, 9, 10
   rewritable, 10
   video and, 119
formatting, in duplication and replication, 444
formatting a DVD. *See* building and formatting
FPS collection, motion clips, 429
future developments in DVD, 465–466

G4, 19, 20
G5, 19, 20, 26
general parameters (GPRMs)
   partitions for, 375–377, **376**, 398
   scripts and, 357, 367–370, 372–376, **374**, 460, 461, 462–464, 462
getting started with DVD Studio Pro 4, 27–28
   documentation files in, 28
   extras installation in, 28
   installing applications for, 28
   templates, shapes, patches, fonts and, 28
gigabyte (GB) vs. billions of bytes (BB), 7, 12–13, 13*t*
Giles, Darren, 112

Index

glass master, 16, 444, **444, 445**
glossary, 485–499
GOP. *See* group of pictures
Goto command, 373, **373**
GPRM partitioning, 27. *See also* general parameters (GPRMs)
Graphical View window, 49, 76–79, **77, 78**, 89
graphics for DVD, 95, 413–435
    aliasing and, 422, **422**
    basic concepts in, 414
    black/white level control in, 420–421, **420, 421**
    color and, 416, 417, 419, **419**, 424–425
    composition limitations and Safe Zones in, 418–419, **419**
    de-interlacing in, 433–434, **434**
    flaws and limitations in, revealing, 421–422, **421**
    font effects in, 423, **423**
    font size and styles in, 422, **422**
    formats for, 68
    image size constraints for, 415–418, **417, 418**, 422, 426–427
    layer styles (layer effects) in, 422–423
    motion menus and, 425–434. *See also* motion menus
    NTSC and PAL standards for, 414–415, 417, **417**, 421, 424–425, **425**, 433–434, **434**
    Photoshop and, 414, **414**, 425
    PICT, TIFF, BMP, PSD, TARGA files in, 415
    pixel shapes and, 415–416, **415, 416**
    rendering layer styles for layered menus using, 423–424, **423, 424**
    resolution for, 417
    rules for, 416–417
    screen capture of image, 431–432, **431**
    trimming screen capture images for, 432–433, **432, 433**
Great Video, 136
group of pictures (GOP), 37, 133
guides, 39, 76
    menus and, 246–247
Guides button tool, 76

H.264 (AVC), 25, 46, 93, 96, 109, 112
hardware encoders, 110–111, 116–117
hardware requirements for DVD Studio Pro, 502
HD-DVD. *See* high-definition
Heuris, 93, 137
Heuris Extractor, 136
Heuris MPEG Power Professional, 471–472, **471**
    motion menus and, 430
high definition (HD), 2, 13, 18–19, 20, 31, 32–33, **33**, 44, 46
    audio, 144
    authoring in, 19–20, 25

high definition (HD) *(continued)*
    integrated encoding for, MPEG2, 26
    NTSC frame rates/aspect ratios for, 19*t*
    PAL frame rates/aspect ratios for, 19*t*
    slideshows and, 297
    templates for, 26
    video and, 109, 112
Highlight Maps, 95
highlights in menus, 255, 256–257, **257**, 258–259, 265–269, **265**, 270, **271**, 273, **273**
holographic discs, 466
hot zones, 7
HTML files, DVDaccess and, 455
hybrid (video + ROM data) DVDs, 8–9, 399, **400**, 408–410, **408–410**
    deliverable files, 95–96
    DVDaccess and, 410, 457, **457**
    Joliet files in, 410
    viewing on set-top boxes, 411

IDesc files, 364
iDVD, 70, 137
IFO files, 3
image enhancement before or during encoding, 119
img file mounted onMac desktop, for DVDaccess, 456
iMovie, 70
    AIFF file creation in, 146, **146, 147**
Import Asset tool, 74
importing assets, 100, 106
    destination of, 97–98, **97**
    drag and drop method for, 99–100, **100**
    organization of, 96
    removing file or folder from, 104
    usability of, checking, 100, **100**
Info Pane, 89
information file. *See* IFO files
inline scripts, 360–361, **361**
Innobits, 93, 470. *See also* BitVice
Inspectors, 49, 50–53, **53**, 74, 86–88, 89
    assets, 103, **103, 104**
    Compressor 2, 166, **166**
    connections made through, 319, **319**
    selecting project asset for, 88
    selecting project element for, 87–88, **87**
    show/hide, 52–53
install/upgrade for DVD Studio Pro, 501–512
installing applications in DVD Studio Pro 4, 28
Instant Encoder, A.Pack, 147–150
interface configurations, 49

Joliet files, 410
JPEG files, 68, 297

Index

jump action, in menus, 252
JUMP command, 371, **371, 372**
Jump Indirect command, 375
jumps, Next/Previous, 324

keyboard shortcuts to select/display Tabs/Windows, 61
Keynote, 70
Keyspan Media Remote, 7

labeling, in duplication and replication, 446–447
language selection, 37, 41, 88
    building and formatting and, 398
    display conditions and, 462, **462**
    menus and, 255
    simulation and Simulator, 387
    subtitles and, 338
    tracks and, 201
launching DVD Studio Pro 4, 29–30
    default configuration setting, 29, **29**
    options and option setting in, 29–30
Layer Break, 51
layout styles, menus and, dragging to Menu Editor in, 287
Layouts, 63
Log Tab, 51, 78, 84, **84**, 89
log window, 51, 162, **162**
logic diagram example, in scripts, 378–380, **379**
logical DVD formats, 9, 20
logo, DVD Studio Pro 4, **115**
loop points, 27
Low-Pass filter setting
    A.Pack, 156, **156**, 157, **157**
    Compressor 2, 178, **178**

Mac system software upgrade for DVD Studio Pro, 502–509, **503–509**
Macrovision, 208, 228, 396–397. *See also* copyright protection
Main Coder, 93
Main Concept encoders, 476–477, **477, 478**
markers and Marker Inspector, 51, 220–229
    adding, 222
    adding, at end of clip, 223
    asset-based, 227
    button highlight, 225
    Buttons over Video and, 220, 229
    cells and, 226
    Chapter markers in, 123–124, **123**
    chapters and, 220, 225
    colors to denote types of, 225
    Compression markers in, 123, **123**
    creating, in tracks, 221–224
    definition of, 220

markers and Marker Inspector *(continued)*
    deleting, 223–224, **224**
    displaying information on, 222
    do's and don'ts for, 228–229
    dual-layer breakpoint, 221, 225–226, 395
    DVDaccess and, 221, 228, 453–454, **455**
    embedded, importing clips with, 224
    End Jump, 226
    Final Cut Pro and, setting, 122–124, **122**
    functions for, 226
    General Tab of Marker Inspector for, 226–228, **227**
    importing, 224
    invalid, fix, 38
    locations of, 221–222
    Macrovision and, 228
    Marker Inspector for, 226–228
    moving existing, 223
    multiple clips and, defining, 223, **223**
    name generation, 36–37, 226
    Operations Tab of Marker Inspector for, 228, **228**
    playback options and, 227–228
    Playhead positioning on, 223
    positioning of, automatic, 222, **222**
    prefix (root) name, 36
    previewing, 228, **228**
    snap to, 37
    still as, 227
    stories and, 221, 229, 231–232
    subtitles and, 225
    text files, importing from, 224
    thumbnails for, 226
    top area of Marker Inspector in, 226, **226**
    Track Editor for, 221–222, **222**
    tracks and, 193–194, **194**, 217
    types of, 225
    zero-based, 227
mastering, 16, 20. *See also* glass master
masters, video, 119
media reference guide, 481–484
Mediapress encoders, 117
MegaPEG Pro, 93
Menu button, 5–6, **6**
menus and Menu Editor, 3, 14, 28, 46, 50, 69–70, **69**, 74, 94, 237–293
    Action Safe area in, 245
    Activated state of buttons in, 241, 272–273, **272, 273**
    Add Submenu, Slideshow, Track tools in, 76
    adding assets to, 249
    adding button to, 262
    Adobe After Effects and, 239
    advanced color (chroma) settings in, 258–259, **259**
    advanced color (grayscale) setting in, 257–258, **258**

Index

menus and Menu Editor *(continued)*
  Advanced Tab of Menu Inspector for, 259–260, **259**
  Arrange Controls tool in, 75
  audio assets in, dragging to empty area of, 278, **278**
  auto assign options in, 35, 245
  background for, 240, 242, 254, 255, 261–262
  backgrounds for, dragging in, 269, **270**
  button defaults for, 40
  Button Outline tool in, 76
  Button State Selection tool in, 76
  button styles in, dragging to button in, 286, **286**
  button styles in, dragging to empty area in, 285–286, **285**
  buttons in, 240, 246, 254–255, 262–265
  buttons in, dragging assets to, 279–281, **279, 280, 281**
  buttons in, state of, 241
  calling, 460
  capturing image from screen for, 431–432, **431**
  color set 2 and 3, in highlights, 267
  color settings for, 243, 255–256
  connections made through, 319, **319**
  connections to buttons in, 274–275
  creating, 260–261
  creative functions in, 239
  display condition settings for, 260
  Display Mode for, 33, 34, 255
  dragging assets and elements into, 275–279
  drawing buttons in, 264, **264**
  Drop Palette in, 76, **77**
    delay in, 35
    drag-and-drop editing in, 249–256
  drop shadows in, 255
  drop zone in, dragging assets to, 280–281, **280**
  drop zone in, dragging to drop zone in, 287
  drop zone style in, dragging to empty area in, 286, **286**
  duplicating buttons in, 264, **264**
  DVDaccess and, 260, 454, **455**
  editing, 248
  extra tools for, 75–76, **76**
  Final Cut Pro and, 239
  final rendering setting in, 35
  General Tab of Menu Inspector for, 250–251, **251**
  graphics for DVD and, rendering layer styles for, 423–424, **423, 424**
  guides in, 76, 246–247
  highlights in, 94, 255–257, **257**, 258–259, 265–269, **265**, 270, **271**, 273, **273**
    trouble seeing, 266, **266**
  hot zones in, 7
  interactivity between user and, 238
  jump action in, 252
  language selection in, 255

menus and Menu Editor *(continued)*
  layered type, 241, 244, 247, 252–254, **253**, 261, 271–273
    dragging assets to buttons in, 290–292, **290–292**
    dragging assets to, 287–290
  layout style in, dragging to Menu Editor, 287
  Loop points, 27
  Menu Editor for, 239, 244–247, **244, 246**
  Menu Editor Settings pop-up for, 245, **245**
  Menu Inspector for, 250, **250**
  Menu Language pop-up selector in, 70
  Menu Tab of Menu Inspector in, 51, 78, 81, **81**, 254–255, **254**
  menus in
    dragging to button in, 284, **284**
    dragging to empty area in, 282, **282**
    dragging to layered button in, 292, **292**
    dragging to layered empty area in, 290, **290**
  Motion button tool in, 76
  motion duration setting in, 35
  motion menus, 239, 243, 251–252. *See also* motion menus
    graphics for, 425–434. *See also* motion menus
  moving and aligning buttons in, 262–263, **263**
  naming, 248
  naming buttons in, 265
  navigating, using Remote Control, 5–6
  Navigation Arrows, 5, **5**
  navigation using, 240
  Normal state of buttons in, 241, 271–272, **271**
  number pad for, 255
  offset for buttons in, 255
  outside editors for, round-trip links to, 239
  overlay colors for highlights in, 256, 265–266, **265**
  overlay layer in, 254
  Photoshop layers in, 271–273
  pixel shape and, 245
  playback option setting for, 259–260
  preferences setting for, 35–36
  project elements in
    dragging to buttons, 283–284, **283, 284**
    dragging to empty area in, 281–283
    dragging to layered button in, 291–292, **292**
  PSD layered still images in, dragging into, 277–278, **278**
  renaming, 248
  renaming buttons in, 265
  resizing buttons in, 263–264, **263**
  resolution for, 34, 255
  scripts in
    dragging to button in, 284, **284**
    dragging to empty area in, 282, **283**

526

Index

menus and Menu Editor *(continued)*
    dragging to layered button in, 292, **292**
    dragging to layered empty area in, 290, **290**
  Selected state of buttons in, 241, 272, **272**
  Settings pop-up selector in, 70
  shapes in
    dragging to button or drop zone in, 284–285, **285**
    dragging to empty area in, 284, **284**
  slideshows in
    dragging to button in, 283–284, **283**
    dragging to empty area in, 282, **282**
    dragging to layered button in, 292, **292**
    dragging to layered empty area in, 289–290, **289**
  standard type, 241–247, 260–261
  still/stillframe images in, 95, 242–243, 273–274, **274**
    dragging into, 277, **277**
    dragging to button, 279, **279**
    dragging to drop zone, 281, **281**
    dragging to layered button in, 291, **291**
    dragging to layered empty area in, 288, **288**, 289, **289**
    multiple, dragging to button, 280, **280**
    with audio, 242–243
  stories in
    dragging to button in, 283, **283**
    dragging to empty area in, 281, **281**, 282, **282**
    dragging to layered button in, 292, **292**
    dragging to layered empty area in, 289, **289**
  submenus in, 246, 248, 275
  subpicture overlay for, 240, 242, 243, 267–268, **268**
    dragging in, 269–270, **269**
  templates for, 275
    from Palette, 250
    dragging to button in, 285, **285**
    dragging to empty area in, 285, **285**
    dragging to, 284–290
    shapes, patches, fonts and, 28
  text setting for, 40
  text style in
    dragging to empty area in, 286, **286**
    dragging to text object in, 286, **286**
  timeouts in, 252
  Title Safe area in, 245
  Toolbar in, 70–75, **70, 71**
  tracks in, 197, **198**
    dragging to button, 283, **283**
    dragging to empty area in, 281, **281**
    dragging to layered button in, 291–292, **292**
    dragging to layered empty area in, 289, **289**
  Transition Tab in Menu Editor for, 255, **256**
  uses for, 238

menus and Menu Editor *(continued)*
  video assets in
    dragging into, 276–277, **276**
    dragging to button, 279, **279**
    dragging to drop zone, 280, **280**
    dragging to layered button in, 290, **290**
    dragging to layered empty area in, 287, **287**
  video asset with audio in
    dragging to button in, 280, **280**
    dragging to empty area of, 278–279, **278**
    dragging to layered button in, 290–291, **291**
    dragging to layered empty area in, 288, **288**
  video background color in, 35–36
  View pop-up selector in, 70
Meridian Lossless Packing (MLP) audio, 144
metallizing glass master, 445, **445**
Micro Orbit Duplicator, **441**
Microboards QD-DVD Duplicator, **442**
mini-DVDs, size limitations of, 8
mixed-angle tracks, 213, 214
mmcoder, 93
mode, 44
molding, in duplication and replication, 445, **446**
Motion, 28, 239
Motion button, 76
motion clips, 429
motion estimation, 45
motion menus, 243, 251–252, 425–434. *See also* menus
  Adobe After Effects for, 427
  assembling motion sequence for, 429–430, **429**
  capturing menu image from screen for, 431–432, **431**
  customizing sequence in, 430, **430**
  de-interlacing in, 433–434, **434**
  Final Cut Pro for, 428–430, **429**
  graphics formats for, 425–426
  highlights in, 434, **434**
  image size constraints for, 426–427
  motion clips for, 429
  MPEG encoding for, 430
  nonlinear edit tools for, 428–429
  overlay pictures for, 431
  transitions in, 428, **428**
  trimming screen capture images for, 432–433, **432, 433**
mouse, DVD navigation using, 6–7
movie files, 67, 89, 93
  tracks and, 197
MPEG, 14–15, 19, 20, 25–28, 42–46, 69, 92, 93, 94, 137
  audio (MPA), 94, 140, 143, **143**, 181
  encoding and, selecting version of, 109, 112
  exporting from Final Cut Pro using Compressor 2, 124–125, **124, 125, 126**

527

Index

MPEG *(continued)*
   motion menus and, 430
   MPEG Append utility for, 117
   parsing, 42
   QuickTime and, 129, **130**
   video and, 117–121, 133, **133**
   video encode guidelines for, 131
MPEG-2 Encoder, 115, 127–128, **127**
MPEG Append utility, 117
MPEG Power Professional, 93
MPEG Stream, 14
multi-angle tracks, 213, **213**
multiplexing. *See* building and formatting
music files, 67

name of disc, 30, **30**
navigation, 20
Next button, 463
nonlinear edit (NLE), 181
   graphics for DVD and, 428–429
NOP command, 371
normal state of buttons, 241, 271–271, **271**
NTSC, 30–31, 33, 69, 93, 119–120, 137
   audio and, 94
   de-interlacing in, 433–434, **434**
   frame rates/aspect ratios for HD-DVD, 19*t*
   graphics for DVD and, 414–415, 417, **417**, 421, 424–425, **425**, 433–434, **434**
   PAL conversion to/from, 113, **114**
   QuickTime and, 131
   slideshows and, 297
number pad, menus and, 255

offset, menu buttons, 255
on-demand distribution of DVDs, 448
one-pass mode, 44
operations (logical and math), in scripts, 372–373
Optibase Inc., 94, 116
   Master Encoders, 116–117
options and option setting, 29–30
   disc name in, 30, **30**
   disc size in, 31, **32**
   DVD standard in, 31, **31**, 32–33, **33**
   preferences setting in, 32–38
   project preferences setting in, 32–33
   video standard in, 30–31, **30**, 33
Orbit II Duplicator (Microboards), **442**
Original Bitstream setting
   A.Pack, 154, **154**
   Compressor 2, 176
OS 10.4 Tiger, 20
OS 9 and QuickTime encoder, 134–135, **135**

OS X, 70
Outline Tab, 51, 87–88, **87**
Outline VST Editor Tab, 78, 79, **79**, 89
overflow, in scripts, 357
overlay picture file, 254, 431. *See also* menus
Overmodulation protection setting
   A.Pack, 155, **156**
   Compressor 2, 177, **177**

packaging, in duplication and replication, 447
packing, Digital Theatre System (DTS) and, 180
PAL, 30–31, **30**, 33, 69, 93, 119–120, 137
   audio and, 94
   de-interlacing in, 433–434, **434**
   frame rates/aspect ratios for HD-DVD, 19*t*, 19
   graphics for DVD and, 414–415, 417, **417**, 433–434, **434**
   NTSC conversion to/from, 113, **114**
   QuickTime and, 131
   slideshows and, 297
Palette, 49, 50–54, **53**, 63–68, 89
   adding/removing elements/folders to, 65–66
   asset organization using, 68, 96
   Audio in, 64–65, **65**, 67, **67**
   color setting for, 41
   default entries in, tied to current user, 65
   Drop Palette in, 76, **77**
   movie files in, 67
   music and sound files in, 67
   quitting the application, contents of, 66
   show/hide, 53–54
   Stills in, 64–65, **65**, 68, **68**
   Tabs in, 63, **63**
   templates, styles, and shapes in, 63–64, **63**, **64**
   thumbnail size in, 66
   Video in, 64–65, **65**, 66–67, **66**
Panasonic, 15
partitions, in scripts, 27, 375–377, **376**, 398
patches, 28, 64
Peak Mixing Level setting, A.Pack, 154
Photoshop, 68, 95
   graphics for DVD and, 414, **414**, 425
   layered menus from, 271–273
   trimming screen capture images in, 432–433, **432**, **433**
physical DVD formats, 9–11, 10*t*, 20
physical formatting of DVD for replication, 444
PICT, 68, 297, 415
Pioneer, 15
Pioneer A03/04/05/06, **384**
Pioneer DVR-S201, **384**
Pioneer DVR-S201/S101 burners, 11
pixel shape, 245
   graphics for DVD and, 415–416, **415**, **416**

528

Index

planning the DVD project, 14, 17–18, 20
plating and stamping, 445, **445**
Play All Script, 377–380, **378**
play/pause, space bar toggles between, 37
playback, 4
Playback Control, 459–460
Playhead, markers and, positioning on, 223
Playhead in Viewer, 216–217
Popwire Technologies, 93, 479. *See also* Compression Master
PowerMac, 2, 20
pre-scripts, 361, **362**
 display conditions vs., 463
 Exit Pre-Script command in, 375
preferences setting, 32–38
 alignment, 38–39, **38**
 colors, 40–41, **41**
 Destinations, 42–43, **43**
 encoding, 44–45, **44**
 general, 33–38, **34**
 Menu, 35–36, **35**
 project, 32–33
 Simulator, 41–42, **42**
 text, 39–40, **40**
 Tracks, 36–38, **36**
Presentation Options, Remote Control, 4, **5**
preview, in building and formatting, 384
Previous button, 463
Primera Bravo DVD Duplicator, **438**
Primera Bravo DVD Production System, **443**
program streams, 14, 110
project basics, 30–45
project bundle, 43
project preferences setting, 32–33
PSD files, 68, 297, 415
pulse code modulation (PCM), 20, 94, 140–144

Qadministrator, 113, **114**
Qmaster distributed encoding, 113, **114**
Qmaster. *See* Apple Qmaster
QTI, 297
quadrants, 49–52, 54, 59
 dragging linked, 58–59
 dragging Tabs between, 59, **59**
 dragging unlinked boundaries of, 56, **56**
 hide/reveal, 56
 hiding, 56–57, **57**
 resize all, 55, **55**
 revealing, 57–58, **57**
 unlink, 55–56, **56**, 58, **58**
Quantum DLT-4000, **384**

QuickTime, 14–15, 19, 44, 68, 93, 109, 115–116, **116**, 137
 audio and, 94
 current version of, checking, 514–515, **514**
 Digigami MegaPEG Pro QT and, 116, **116**
 Encode Progress window in, 132, **132**
 encoder performance benchmarks for, 133–135
 exporting from Final Cut Pro using, 128–131
 Group of Pictures (GOPs) in, 133
 installation guide, 514–515
 MPEG encode guidelines for, 131
 MPEG encoding in, 25, 28, 129–133, **130**, **132**
 NTSC/PAL conversion and, 131
 OS 9 and, 134–135, **135**
 QuickTime Pro vs. QuickTime Player in, 134–135
 RAM allocation for, in OS 9, 134
QuickTime Pro Player, 115–116, **116**

Read DLT tool, 75
Recipe4DVD Basic Tutorial, 70, 96, **96**
recordable DVDs, 9, 10
recording, 8–9
 data for, 8–9
 DVD-9, 8
 DVD-Hybrid discs and, 8–9
 DVD-ROM discs and, 8–9
region code, 41, 396, **396**, 405–406
 simulation and Simulator, 387
remote control, 3, 4–6, **7**
 Angle Button on, **5**
 Audio Button on, **5**
 connections and, avoiding problems with, 329
 DVD navigation using, 6–7
 Enter key on, 5, **5**
 Menu Navigation Arrows on, 5, **5**
 menu navigation using, 5–6
 mouse use vs., 6–7
 Next button on, 463
 Presentation Options for, 4, **5**
 Previous button on, 463
 Subtitle Button on, **5**
 Transport controls for, 4, **4**
rendering
 hardware- vs. software-based, 35
 video and, 122
replication master, 15
replication vs. recordable, 9
replication. *See* duplication and replication
resolution
 graphics for DVD and, 417
 HD menus, tracks, slideshows, 34
 menus and, 255
 video and, 122

529

Index

Resume command, 374
Resume function, 319
Return button, 5–6, **6**
rewritable DVDs, 10
RGB color, in graphics for DVD, 419–420, **419**
Richo, 15
Rimage ProtÇgÇ II DVD Production System, **443**
ROM zone, ROM assets, 95–96, 106
Room Type setting, A.Pack, 155
Roxio Toast Titanium, 8
    building and formatting and, 409, **409**, 411, **411**
rulers, 38–39
running time, 20

Safe Zones, graphics for DVD and, 418–419, **419**
SCR format, 95, 341
screen capture of image, 431–432, **431**
screen resolution, 26
script description (IDesc) file, 364
Script Editor, 85–86, **85**
Script Tab, 78, 85–86, **85**, 89
scripts, 14, 46, 51, 355–381
    command line reordering in, 365
    command options in, 367
    command selection in, 367
    command syntax in, 365
    commands for, 370–377
    comments fields in, 367
    Compare test in, 370
    comparing commands in, 367
    Conditional Jump command in, 380, **380**
    conditional vs. unconditional commands in, 367–370, **368, 369, 370**, 462–464
    duplicating, 363
    editing, 367
    enhanced interactivity through use of, 356
    Exit command in, 375
    Exit Pre-Script command in, 375
    general parameters (GPRMs) in, 357, 367–376, **374**, 460–464
    Goto command in, 373, **373**
    GPRM mode in, 374–375, **374**
    GPRM partitioning for, 27
    how it works, 356–357
    inline, 360–361, **361**
    JUMP command in, 371, **371, 372**
    Jump Indirect command in, 375
    loading, 364
    logic diagram example for, 378–380, **379**
    making, 362–363, **363**
    menus and
        dragging to button in, 284, **284**

scripts *(continued)*
    dragging to empty area in, 282, **283**
    dragging to layered button in, 292, **292**
    NOP command in, 371
    operations (math and logical) used in, 372–373
    overflow in, 357
    partitions in, 375–377, **376**
    Play All Script using, 377–380, **378**
    pre-scripts in, 361, **362**
    renaming, 363
    Resume command in, 374
    saving, 364
    script description (IDesc) file in, 364
    Script Editor for, 364–365
    Script Inspector for, 365–366, **366**
    Set GPRM command in, 372–373
    Set Stream command in, 374, **374**
    Set System command in, 374, **374**
    system parameters (SPRMs) in, 357–359, 367–370, 461, 462–464
    tool for, 375
    underflow in, 357
    uses for, 359–360, 362
    uses of, 377
    volatility of GPRMs in, 357
SCSI Adapters, 11
Selected state of buttons, 241, 272, **272**
Set GPRM command, 372–373
Set Stream command, 374, **374**
Set System command, 374, **374**
shadowed text, 423
shapes, 28, 89
    menus and, dragging to button or drop zone in, 284–285, **285**
    menus and, dragging to empty area in, 284, **284**
    Palette, 63, 64, **64**
Show and Hide option, 38
simulating and Simulator, 26–27, 82, 89, 385–391
    building and formatting and, 384
    Digital Cinema Desktop and, 26
    Display Mode setting in, 390–391, **390**
    Display Window in, 387
    DVDaccess and, testing links to, 456
    entire project, First Play and, 385–386
    external video monitor simulation through, 26–27
    features of, 387
    from specific starting point, 386
    Information Drawer in, 391, **391**
    language selection in, 41, 387
    Live or Active state of, 391
    Menu Control in, 380–390, **389**
    playback options in, 41–42, 387

530

Index

simulating and Simulator *(continued)*
    preference setting in, 41–42, **42**, 386–387, **386**
    region code for, 41, 387
    Remote Control Panel in, 387–389
    Resolution setting in, 390–391, **390**
    Simulator Window in, 387, **388**
    stopping, 386
    Stream Selection and Display in, 390, **390**
    Transport Control in, 389, **389**
single-layer discs, in duplication and replication, 447, **447**
single-sided DVD, size limitations of, 7–8, **7**
size issues, 4, 7–8, 12–13, 31, **32**
    bit budgeting in, 14, 17–18, **17**
    disc running times in seconds for, 17, 18*t*
    DVD-R, 10–11, 411
    file sizes by data rate, 117–118
    gigabyte (GB) vs. billions of bytes (BB) in, 7, 12–13, 13*t*
    graphics for DVD and, 414–418, **417, 418**, 422, 426–427
    slideshows and, 297–298
sizing windows, 49–50
Skip Back, 6
Skip Forward, 6
Slideshow Tab, 78, 84–85, **85**, 89
slideshows and Slideshow Editor, 3, 14, 46, 51, 74, 94, 295–315, **296**
    action safe zones and, 297
    adding slides to, 301–302, **301–303**
    Advanced Tab of Slideshow Inspector for, 309, **309**
    aspect ratio of slides in, 297–298
    assets dragged into, 105
    assets for, selecting, 300–301
    background color, 33
    creating, 298–304, **298**
    creative issues of, 296–298
    default slide length in, 33
    deleting slides from, 301
    Display Mode for, 33, 34
    duration of images in, 296, 302–303, **304**
    DVDaccess and, 454, **455**
    empty, creating, 298–299
    Fit to Audio option for, 307
    Fit to Slides option for, 307
    formats usable for, 297
    General Tab of Slideshow Inspector for, 308–309, **308, 311**
    high definition and, 297
    keyboard shortcuts for Slideshow Editor in, 314
    Manual Advance setting for, 304, **304**, 304
    menus and
        dragging to button in, 283–284, **283**

slideshows and Slideshow Editor *(continued)*
        dragging to empty area in, 282, **282**
        dragging to layered button in, 292, **292**
        dragging to layered empty area in, 289–290, **289**
    NTSC standards and, 297
    number of images in, 296
    PAL standards and, 297
    resolution setting, in HD, 34
    size limitations of, 4
    size of slides in, 297–298
    Slide Inspector for, 310
    Slideshow Editor controls for, 299–300, **300**
    Slideshow Inspector for, 308–310, **308–310**
    sound/audio added to, 304–307, **305–307**, 310–311, **311**
    sound/audio deleted from, 305, **305**
    title safe zones in, 297
    tracks from, conversion of, 311–312, **312**
    Transition Tab of Slideshow Inspector for, 309–310, **310, 311**
    transitions between, 307–308, **308**
    uses for, 297
    viewing and simulating, 312–313, **313**
snap to marker in, 37
Software Architects, 136
software encoders, 111–117. *See also* Compressor 2
software requirements for DVD Studio Pro, 502
SON format, 95, 341
Sonic Solutions, 94, 116
    SD-xxx encoders, 117
Sony, 15
sound files, 67
space bar toggle between play/pause and, 37
splash screen for DVD Studio Pro 2, **510**
SPU files imported for subtitles, 342, **342**
sputtering, in duplication and replication, 446, **446**
stamping, in duplication and replication, 16, 445, **445**
standard definition (SD), 20, 31, 32–33, **33**, 46. *See also* high definition
stereo file encoding using Compressor 2, 164–167
stills/stillframe images, 89, 94–95
    markers and, 227
    menus using, 242–243, 273–274, **274**
        dragging to layered button in, 291, **291**
        dragging to layered empty area, 288, **288**
        dragging to layered empty area, 289, **289**
    Palette, 64–65, **65**, 68, **68**
STL format, 95, 341
stories and Story Editor, 46, 51, 74, 187–189, 229–236
    adding marker for, 231–232
    adding, 81, 230
    creating, 230, **230**

531

Index

stories and Story Editor *(continued)*
    defining segments of, 231, **231**
    deleting markers for, 232
    End Jump and, 229, 230, 232, 233
    Entry Points and, 229
    General Tab settings for, 233–234, **234**
    how they work, 229–230
    marker property setting in, 234–235, **235**
    markers and, 221, 229
    menus and
        dragging to button in, 283, **283**
        dragging to empty area in, 281, **281**, 282, **282**
        dragging to layered button in, 292, **292**
        dragging to layered empty area in, 289, **289**
    moving markers for, 232
    order of, 229, 233
    playback options for, 233–234
    reassigning track markers for, 232
    remote control settings for, 235
    renaming, 233
    size limitations of, 4
    Story Editor Tab for, 231–233, **231**
    Story Inspector for, 233–235, **233**
    Stream options for, 233
    top area of Story Inspector for, 233
    tracks and, 217
    User Operations (UOPs) and, **234**
Story Inspector, 233–235, **233**
Story Tab, 78, 80–81, **80**, 89
streams, 2, 3
    areas for tracks, 199–203, **200**
    selection of, 460
styles, 89
    menus and, dragging to button in, 286, **286**
    menus and, dragging to empty area in, 285–286, **285**
    Palette, 63, **64**
submenus, 246, 248, 275
subpicture overlay, 94, 95, 240–243, 267–270, **268**, **269**.
    See also menus; overlay files
Subtitle Button, Remote Control, **5**
subtitles, 2, 3, 34–35, 51, 95, 106, 337–354, **340**
    advanced actions using, 353
    Button Tab of Subtitle Inspector for, 347–348, **347**
    Buttons over Video using, 339
    checking, 343
    closed captions vs., **340**
    color selection for, 344, **344**, 348–349, 352–353
    Colors Tab of Subtitle Inspector for, 348–349, **349**
    creating, 350–351
    demonstration videos using, 339
    duration of, 350
    entering text for, 351

subtitles *(continued)*
    fade in/out for, 34–35
    font selection for, 351–352, **352**
    formats for, 95, 341
    formatting, 346–347
    General Tab of Subtitle Inspector for, 345–347, **346**
    global settings for, 345, **345**
    graphic files for, 347, 353
    how they work, 341
    importing text files for, 341
    individual entry of, 341
    language selection in, 37, 338
    length of, 34
    markers and, 225
    naming conventions for, 339
    options for, 341
    position of, 350, 353
    preferences setting for, 344, **344**
    previewing, switching streams for, 343
    repositioning, 351
    selecting, 351
    SPU files imported for, 342, **342**
    stream selection for, 347
    Subtitle Inspector for, 345–349, **345**
    text selection for, 344, **344**
    text setting for, 39–40
    timing attributes for, 344, **344**
    tools for, 342–343
    tracks and, 202
    uses for, 339
    video commentaries using, 339
    watermarks as, 342, **342**
SuperDrive, 20, 26
SuperDrive. *See* Apple SuperDrive, 2
Surround Downmix level setting
    A.Pack, 153, **153**
    Compressor 2, 174–175, **175**
Surround Phase Shift setting
    A.Pack, 157–158, **158**
    Compressor 2, 179, **179**
Surround Sound. *See* Dolby Surround
Synthetic-Aperture Inc., 421
system parameters (SPRMs), in scripts and, 461–464, 357–359, 367–370
system streams, 110

Tabs, 49, 50, 54, 59–60, 78–86
    Asset, 78, 83, **83**, 88, 89, 96–98, **97**
    Connections, 78, 81–82, **82**, 89
    Control-Click to move, 60
    dragging between quadrants, 59, **59**
    freestanding windows from, 60

Index

Tabs *(continued)*
   Graphical View, 78, 89
   keyboard shortcuts to select/display, 61
   Log, 78, 84, **84**, 89
   Menu, 78, 81, **81**, 89
   Outline VST Editor, 78, 79, **79**, 89
   Outline, 87–88, **87**
   restoring torn-off, 60
   Script, 78, 85–86, **85**, 89
   Slideshow, 78, 84–85, **85**, 89
   Story, 78, 80–81, **80**, 89
   tearing off, 60
   Track, 78, 84, **84**, 89
   Viewer, 78, 82–83, **82**, 89
Targa files, 68, 415
target system setting, Compressor 2, 171, **172**
targets for connections, 334–335
Taylor, Jim, 136
technical standards, video, 119–120. *See also* NTSC; PAL
templates, 28, 89
   high definition (HD), 26
   menus and, 275
      dragging to button in, 285, **285**
      dragging to empty area in, 285, **285**
      dragging to, 284–290
   Palette, 63, **63**
text preferences, 39–40, **40**
text style
   menus and, dragging to empty area in, 286, **286**
   menus and, dragging to text object in, 286, **286**
TGA files, 68, 297
thumbnails
   markers and, 226
   offset, 37
   palette for, 34
   size of, 34, 66
   Slideshows and, 34
TIFF files, 68, 297, 415
Tiger. *See* OS 10.4 Tiger
timeline, in Track Editor, 198
   markers and, 227
   zero- vs. asset-based, 204, **204**
timeouts, menus and, 252
timestamps, 208
Title button, 5–6, **6**
Title Safe area/zone, 245
   graphics for DVD and, 419, **419**
   slideshows and, 297
Toast Titanium. *See* Roxio Toast Titanium
Toolbar, 70–75, **70, 71**
   Add Language tool in, 74
   Add Layered Menu tool in, 74

Toolbar *(continued)*
   Add Menu tool in, 74
   Add Script tool in, 74
   Add Slideshow tool in, 74
   Add Story tool in, 74
   Add Track tool in, 74
   adding/removing tools in, 72–73, **72, 73**
   available tools for, 74–75
   Build tool in, 75
   Build/Format tool in, 75
   Burn tool in, 75
   Customize selections in, 71, **71**, 72
   Disc Meter tool in, 74
   Format tool in, 75
   Import Asset tool in, 74
   Inspector in, 74
   locking, 73, **73**
   Menu Editor in, 74
   moving tools around, 73–74, **73, 74**
   Read DLT tool in, 75
   saving configuration of, 74
   Slideshow Editor in, 74
   Story Editor in, 74
   Track Editor in, 74
   Viewer tool in, 75
Top Menu, 5
Track Inspector, 189, 194
Track Tab, 78, 84, **84**, 89
tracks and Track Editor, 2, 3, 14, 46, 51, 74, 183–218, **184**
   additional clips added to, 196
   angle playback in, 216, **216**
   asset-based timelines and, 204
   assets dragged into, 105
   assets for, 189–190
   audio file matching in, 38
   audio in, adding, 193
   authoring, 183–218
   Browse clip function in, 212
   Chapter Markers and, 186, **187**, 188–189
   Clip Inspector for, 210–212
   closed captions and, 208
   configuring streams for preview in, 215, 509–510
   conflicts between clips in, 211
   creating, 194–198, **195, 196**
      in Track Editor, 189–194
   current system configuration check for, 502
   customizing stream area display of, 200–203, **201, 202, 203**
   data rates and, 212–213
   definition of, 184–185
   Display Mode for, 33, 34

533

Index

tracks and Track Editor *(continued)*
   Dolby Digital (AC3) audio in, adding, 193
   dragging assets in, 195–196, **195**
   duplicating clips in, 196, **197**
   elementary streams and, 189
   empty, adding video asset to, 190–191, **190–192**
   empty, creation of, 189–190
   End Jump in, 186, 188
   hardware requirements for, 502
   invalid marker fix, 38
   language selection in, 37, 201
   limits to, 185
   Mac system software upgrade in, 502–509, **503–509**
   Macrovision and, 208
   marker name generation for, 36–37
   marker prefix (root) name for, 36
   markers and, 193–194, **194**, 217, 221–224, **222**
   matching audio to video automatically in, 192, **192**
   Menu Editor to creation, 197, **198**
   menu-less, 198
   menus and
      dragging to button in, 283, **283**
      dragging to empty area in, 281, **281**
      dragging to layered button in, 291–292, **292**
      dragging to layered empty area in, 289, **289**
   mixed-angle, 213, 214
   movie files in, 197
   multi-angle, 213, **213**
   number of, on DVD, 185
   organizing stream display of, 200, **200**
   playback options setting for, 207–208
   Playhead of Viewer for, 216–217
   preferences setting for, 36–38, **36**, 510, **510**
   previewing angles of, 216
   previewing, in Viewer, 214–217
   Recipe4DVD practice assets for, 193, **193**
   resolution setting, in HD, 34
   separate audio and video for, 189
   size limitations of, 4
   slideshows and, conversion to, 311–312, **312**
   snap to marker in, 37
   software requirements for, 502
   space bar toggle between play/pause and, 37
   stories and, 187, 188–189, 217
   stream areas for, 199–203, **200**
   stream information and settings in, 212
   structure of DVD and, 186
   structure of, 185
   subtitles and, 202
   thumbnail offset for, 37
   timeline for, in Track Editor, 51, 198, 204, **204**
   timestamps for, 208

tracks and Track Editor *(continued)*
   Track Editor for, **184**, 189, 198–199
   Track Editor Tab controls in, 198–199, **199**
   Track Inspector for, 189, 194, 206–209, **207**
   Track Inspector General Tab for, 207, **207**
   Track Inspector Other tab for, 208
   Track Inspector User Operations Tab for, 208–209, **209**
   Transition Inspector for, 209–210
   transitions between, 188, 209–210
   trimming assets in timeline of, 204–205, **205**
   trimming streams using Asset Inspector of, 205–206, **206**
   unique marker names for, 36
   user operations (UOPs) and, 208–209, **209**
   video assets in, Track Inspector for, 194
   video clips as, 211–212
      playing individually and sequentially, 187–188, **187, 188**
      playing individually, 186, **186**
      playing sequentially, 186
   Viewer controls, keyboard commands for, 215
   Viewer for, 214–217
   viewing clip properties in, 210–212
   zero-based timelines and, 204
transcoding, 14
transitions and Transition Inspector, 209–210
   motion menus and, 428, **428**
   slideshows and, 307–308, **308**
Transport controls, Remote Control, 4, **4**
Transport Stream, 14, 110
trimming assets in timeline, 204–205, **205**
trimming streams using Asset Inspector, 205–206, **206**
24P compatibility, 25
two-pass mode, 44
TXT format, 95
Type Styles, 63

unconditional commands in scripts, 367–370, **368–370**, 462–464
underflow, in scripts, 357
units of measure, 38–39
update alert dialog box, 24, **24**
upgrade for DVD Studio Pro. *See* install/upgrade for DVD Studio Pro, 501
USB, 27
user operations (UOPs), 208–209, **209**, 458–459, **459**
   stories and, 234, **235**

variable bit rate (VBR) encoding, 110, 17
Velocity Engine, 19

Index

video, 107–137, 211–212. *See also* Compressor 2; Final Cut Pro; graphics in DVDs; NTSC and PAL standards
　A.Pack and, 112
　alternative encoders for, 135–136, 469–480
　audio automatically matched to, 192, **192**
　Browse clip function in, 212
　Chapter markers in, 123–124, **123**
　Clip Inspector for, 210–212
　color in, 419, **419**
　Compression markers in, 123, **123**
　Compressor 2 and, 109, 111–117, **111**
　constant bit rate (CBR) encoding for, 110
　content for, 109
　content vs. encoding quality in, 120–121
　digital linear tape (DLT) and, 110
　Dolby Digital AC3 encoding and, 112–113, 120
　DVD recorder as encoder in, 135–136
　elementary streams in, 110, 189
　encode bitrate and, 120, 121
　Encode Progress window in, 132, **132**
　encode status indicators in, 128, **128**
　encoder performance benchmarks for, 133–135
　encoder selection for, 120
　encoding for, 109, 110–111
　expectations for, 120
　exporting from Final Cut Pro using QuickTime in, 128–131
　exporting into MPEG-2 from Final Cut Pro using Compressor 2, 124–125, **124–126**
　file sizes by data rate, 117–118
　Final Cut Pro and, 111, 112, 115, **115**, 121–122
　format for, 119
　Group of Pictures (GOPs) in, 133
　H.264 (AVC), 109, 112
　hardware encoders for, 110–111, 116–117
　high definition, 109, 112
　hybrid (video + ROM data) DVDs in, 399, **400**, 408–410, **408–410**
　image enhancement before or during encoding of, 119
　in and out time setting and, 122
　markers in, Final Cut Pro, 122–124, **122**
　masters for, 119
　menus and, dragging to layered button in, 290–291, **290, 291**
　menus and, dragging to layered empty area in, 287, **287**, 288, **288**
　motion clips for, 429
　MPEG 1, MPEG 2, MPEG 4 for, 109–110, 112, 117–118
　　Append utility for, 117
　　encode guidelines for, 131
　　Encoder for, 127–128, **127**

video *(continued)*
　encoding hints for, 119–121
　encoding in QuickTime player and, 129, **130**
　encoding in, 132, **132**, 133, **133**
　MPEG-2 Encoder for, 115
　NTSC/PAL conversions for, 113, **114**, 119–120
　Optibase Inc. Master Encoders for, 116–117
　PAL and, 119–120
　playing individual clips of, 186, **186**, 186
　playing individual and sequential clips of, 187–188, **187, 188**
　playing sequential clips of, 186
　program streams in, 110
　Qmaster distributed encoding and, 113, **114**
　QuickTime and, 109, 115–116, **116**
　QuickTime MPEG encoder option setup for, 130–132, **130, 131**
　rendering and, 122
　resolution and, 122
　RGB graphics colors and, 419, **419**
　software encoders for, 111–117
　Sonic Solutions Inc. SD-xxx encoders for, 117
　system streams in, 110
　technical standards for, 119–120
　tracks and, 189
　transport streams in, 110
　types of assets in, 92–93
　using, as assets in DVD, 93–94
　variable bit rate (VBR) encoding for, 110
　video-only DVD volume using, 398–399, **398, 399**
　Wired Inc. Mediapress encoders for, 117
Video Object. *See* VOB files
video standard, 30–31, **30**, 33. *See also* NTSC and PAL standards
video streams, 3
Video Title Sets (VTS), 3
Video, Palette, 64–65, **65**, 66–67, **66**
video-only encode rate, 17
Viewer, 214–217, **215, 216**
　building and formatting and, 385
　Viewer Tab, 51, 78, 82–83, **82**, 89
　Viewer tool, 75
VOB files, 3
volatility of GPRMs in scripts, 357
VTS Editor, 26, 49

Waggoner, Ben, 112
watermarks, 4, 94, 342, **342**
WAV files, 94, 141, **141**, 144–145, 181
　Dolby Digital conversion from, using A.Pack, 148–149
white levels, in graphics for DVD, 420–421, **420, 421**
Window Buttons, 61–62

535

Index

windows, 54
  buttons for, 61–62
  close, 61
  dragging linked quadrants in, 58–59
  dragging unlinked quadrant boundary in, 56, **56**
  hide/reveal quadrants, 56
  hiding visible quadrants, 56–57, **57**
  keyboard shortcuts to select/display, 61
  maximize, 61
  minimize, 61
  quadrants in, 54
  re-linking unlinked boundaries in, 59
  reduce/enlarge size of, 54, **54**
  resize all quadrants simultaneously, 55, **55**
  resizing halves of, 54, **55**
  revealing hidden quadrants, 57–58, **57**
  sizing of, 49–50, 54
  unlink quadrants in, 55–56, **56**
  unlink quadrants in, 58, **58**
Wired Inc., 94, 116
  Mediapress encoders, 117
workflow of DVDs, 13–16, **16**, 46
  authoring in, 15
  bit budgeting in, 14, 17–18, **17**
  building (multiplexing, muxing) in, 15
  burning the DVD or writing the DLT in, 15
  disc running times in seconds for, 17, 18*t*
  duplication in, 15–16
  encoding video in, 14–15
  planning the project and acquiring assets in, 14, 17–18
  replication in, 16
workspace, 49
  adding, deleting, renaming configurations for, 62
  Advanced Configuration in, 29, 51–52, **52**
  apply configuration for, 62

workspace *(continued)*
  Assets Tab in, 78, 83, **83**, 88, 89
  Basic Configuration of, 29, 50, **50**, 68–69
  configuration management for, 62, 68–70
  Connections Tab in, 78, 81–82, **82**, 89
  Extended Configuration of, 29, 50–51, **51**
  Graphical View window in, 76–77, **77, 78**, 78, 89
  Inspector in, 52–53, **53**, 86–88
  keyboard shortcuts to select/display Tabs/Windows in, 61
  Log Tab in, 78, 84, **84**, 89
  Menu Editor in, 69–70, **69**
  Menu Tab in, 78, 81, **81**, 89
  Outline Tab in, 87–88, **87**
  Outline VST Editor Tab in, 78, 79, **79**, 89
  Palette in, 53–54, **53**, 63–68
  Quadrants in, 54
  recalling configurations for, 62
  Script Tab in, 78, 85–86, **85**, 89
  sizing windows in, 49–50
  Slideshow Tab in, 78, 84–85, **85**, 89
  Story Tab in, 78, 80–81, **80**, 89
  Tabs in, 54, 59–60, 78–86
  Toolbar in, 70–75, **70**
  Track Tab in, 78, 84, **84**, 89
  Viewer Tab in, 78, 82–83, **82**, 89
  Window Buttons in, 61–62
  Windows in, 54
writing DLTs, 15
WYSIWYG environment, 69

Yahoo groups for DVD production, 466

zero-based timelines and markers, 204, 227

ABOUT THE AUTHOR

Bruce Nazarian is President and CEO of Digital Media Consulting Group, Inc, through which he produces, consults, and instructs on DVD. A world-renowned expert on DVD authoring and a Master Trainer, he is also a member of the DVD Association National Board of Directors, and the busy webmaster of www.recipe4dvd.com and www.dvda.org.

DISK WARRANTY

This software is protected by both United States copyright law and international copyright treaty provision. You must treat this software just like a book, except that you may copy it into a computer in order to be used and you may make archival copies of the software for the sole purpose of backing up our software and protecting your investment from loss.

By saying "just like a book," McGraw-Hill means, for example, that this software may be used by any number of people and may be freely moved from one computer location to another, so long as there is no possibility of its being used at one location or on one computer while it also is being used at another. Just as a book cannot be read by two different people in two different places at the same time, neither can the software be used by two different people in two different places at the same time (unless, of course, McGraw-Hill's copyright is being violated).

LIMITED WARRANTY

McGraw-Hill takes great care to provide you with top-quality software, thoroughly checked to prevent virus infections. McGraw-Hill warrants the physical diskette(s) contained herein to be free of defects in materials and workmanship for a period of sixty days from the purchase date. If McGraw-Hill receives written notification within the warranty period of defects in materials or workmanship, and such notification is determined by McGraw-Hill to be correct, McGraw-Hill will replace the defective diskette(s). Send requests to:

McGraw-Hill, Inc.
Customer Services
P.O. Box 545
Blacklick, OH 43004-0545

The entire and exclusive liability and remedy for breach of this Limited Warranty shall be limited to replacement of defective diskette(s) and shall not include or extend to any claim for or right to cover any other damages, including but not limited to, loss of profit, data, or use of the software, or special, incidental, or consequential damages or other similar claims, even if McGraw-Hill has been specifically advised of the possibility of such damages. In no event will McGraw-Hill's liability for any damages to you or any other person ever exceed the lower of suggested list price or actual price paid for the license to use the software, regardless of any form of the claim.

McGRAW-HILL, INC. SPECIFICALLY DISCLAIMS ALL OTHER WARRANTIES, EXPRESS OR IMPLIED, INCLUDING, BUT NOT LIMITED TO, ANY IMPLIED WARRANTY OF MERCHANTABILITY OR FITNESS FOR A PARTICULAR PURPOSE.

Specifically, McGraw-Hill makes no representation or warranty that the software is fit for any particular purpose and any implied warranty of merchantability is limited to the sixty-day duration of the Limited Warranty covering the physical diskette(s) only (and not the software) and is otherwise expressly and specifically disclaimed.

This limited warranty gives you specific legal rights; you may have others which may vary from state to state. Some states do not allow the exclusion of incidental or consequential damages, or the limitation on how long an implied warranty lasts, so some of the above may not apply to you.